Biostatistics for the Health Sciences

R. Clifford Blair

University of South Florida

Richard A. Taylor

PEARSON
Prentice
Hall

Upper Saddle River, New Jersey 07458

Library of Congress Cataloging-in-Publication Data
Blair, R. Cliff
Biostatistics for the health sciences.
1st ed./R. Cliff Blair, Richard A. Taylor
    p. cm.
Includes references and index.
ISBN 0-13-117660-9
1. Numerical Analysis

CIP data available

Executive Acquisitions Editor: *Petra Recter*
Editor-in-Chief: *Chris Hoag*
Project Manager: *Michael Bell*
Production Editor: *Raegan Keida Heerema*
Assistant Managing Editor: *Bayani Mendoza de Leon*
Senior Managing Editor: *Linda Mihatov Behrens*
Executive Managing Editor: *Kathleen Schiaparelli*
Manufacturing Buyer: *Maura Zaldivar*
Manufacturing Manager: *Alexis Heydt-Long*
Director of Marketing: *Patrice Jones*
Marketing Manager: *Wayne Parkins*
Marketing Assistant: *Jennifer de Leeuwerk*
Editorial Assistant: *Joanne Wendelken*
Art Director/Cover Designer: *Jayne Conte*
Creative Director: *Juan R. López*
Director of Creative Services: *Paul Belfanti*
Manager, Cover Visual Research & Permissions: *Karen Sanatar*
Cover Image: *Istockphoto.com*

© 2008 Pearson Education, Inc.
Pearson Prentice Hall
Pearson Education, Inc.
Upper Saddle River, New Jersey 07458

Printed in the United States of America
10 9 8 7 6 5 4 3 2

ISBN 0-13-117660-9

Pearson Education LTD., *London*
Pearson Education Australia PTY, Limited, *Sydney*
Pearson Education Singapore, Pte. Ltd
Pearson Education North Asia Ltd, *Hong Kong*
Pearson Education Canada, Ltd., *Toronto*
Pearson Educación de Mexico, S.A. de C.V.
Pearson Education - Japan, *Tokyo*
Pearson Education Malaysia, Pte. Ltd

To Pal

Contents

Preface

As the title implies, this book provides an introduction to biostatistics directed toward students and workers in the health sciences. It will typically be used in advanced undergraduate and graduate level courses designed for students who are majoring in health-related disciplines other than biostatistics. That is, this book does not provide the theory or mathematical rigor typically found in a first course for biostatistics majors. It is appropriate for virtually any other health-related discipline that requires a fundamental knowledge of biostatistics. Mathematical prerequisites are minimal with a working level of algebra being sufficient for most students.

What led the authors to produce yet another introductory biostatistics book? Our motivation was threefold. The first was to produce a text that would function as effectively in, what to this point has been, non-traditional learning environments as in the traditional classroom setting. By this we mean settings in which student-instructor contact is limited by physical proximity and/or method of instructional delivery. Chief among these are distance learning courses based on Internet or satellite delivery systems or directed/self-study courses. Our desire then, was to produce a book so clearly done as to allow student acquisition of statistical knowledge under the direction of an instructor whose face-to-face interaction with students may or may not be, limited.

For this purpose, explanations are elaborated upon to a much greater degree than would be found in most other texts. The result is a book that some might characterize as "wordy" but that we see as preoccupied with clarity of exposition. Additionally, we have included step-by-step solutions to exercises rather than simply supplying answers. Such solutions incorporate references to pages in the text as well as specific equations. As a result, students unable to obtain answers to exercises can turn to the solutions portion of the book thereby acquiring a step-by-step discourse leading to the problem resolution.

A second motivating factor was to introduce students to equivalence testing. Equivalence tests are now commonly used in clinical trials and other contexts so that students may gain some familiarity with their use. Additionally, as noted by Hoenig and Heisey [23] in an article in the *American Statistician*, introducing students to equivalence testing at an early stage may help cement their understanding of hypothesis testing by driving home the point that failure to reject a null hypothesis does not constitute evidence of its validity. Students learn that if one wishes to establish the null condition, special (equivalence) tests are required.

A third motive was to present nonparametric methods in a more modern light than the current fare and to put certain myths related to such techniques to rest. With the advent of powerful computers and fast algorithms, permutation-based methods have seen a dramatic rise in level of use in recent years. For this reason, we forego the traditional (and in our opinion) outdated approach and present nonparametric techniques in the broader context of permutation-based methods. This approach has several advantages. (1) Even the mathematically unsophisticated student can clearly discern the logic underlying the construction of sampling distributions related to such methods. (2) The student can see that certain traditional nonparametric methods (e.g., the Wilcoxon-Mann-Whitney test) are simply rank-based analogs of tests conducted on original scores. (3) By presenting nonparametric methods as rank transform [8] versions of parametric statistics with which they are already familiar, much of the mystery of such tests is removed. In other words, the student is not presented with new, unfamiliar methods referencing odd appearing and at times counter-intuitive tables of critical values, but rather is shown that distribution-free tests can often be conducted by applying parametric methods which they learned in earlier chapters to ranks. This also allows tables of critical values to take forms with which the student is already familiar. In short, nonparametric methods are demystified.

This book is conceptually divided into three component parts. The first four chapters lay the foundation for all that follows. Chapter 1 provides the conceptual framework that encompasses the remainder of the book. Chapter 2 takes a rather traditional approach to descriptive statistics but includes some innovative views of the median, percentiles, and percentile ranks. Chapter 3 provides a non-theoretical view of probability and lays the foundation for the probability-based models that underpend the inferential methods that follow. Chapter 4 covers the foundations of inference and is the keystone of the book. This chapter also introduces the logic and method of equivalence testing.

Chapters 5 through 9 present specific techniques used with continuous and binary data. In addition to the traditional paired samples, two group, k group, correlation, and regression methods, equivalence methods associated with paired samples and two group methods are presented and elaborated upon.

Chapter 10 gives an overview of permutation-based methods and addresses specific tests related thereto. This is a large chapter which, with some supplementation, could be used as the basis for a short course in nonparametric methods.

There is more material in this book than can be covered in a single semester. However, the instructor should not hesitate to assign portions of the book for which class/lecture time cannot be allocated. As previously indicated, although text assignments ideally will be supported with lectures or other means, the elaborate levels of exposition allow material to be covered that might otherwise be omitted because of limited class time. In short, the intent of this book is to teach rather than to merely instruct.

Supplemental materials for this text can be found at:

$$\texttt{http://www.biostats-hs.com.}$$

These materials include short chapters dealing with Kaplan-Meier estimates and logrank tests, logistic regression, and factorial ANOVA. There are also a number of downloadable software manuals that can be used as the basis of a computer lab to accompany the course. Many other helpful materials can be found there as well. Readers with questions/comments are encouraged to contact the authors via an email link found at this site.

Additionally, let us say that this text has benefited from extensive field testing. It has been used in draft form in both classroom and distance learning settings taught by a variety of instructors. Much effort was made to collect student and instructor input which led to modifications and to a general style of presentation that differs from most other texts.

Finally, this book was reviewed and commented upon by the following experts who represent a variety of health-related disciplines: Sara Vesely, University of Oklahoma Health Sciences Center; Jessica L. Thomson, Louisiana State University Health Sciences Center, School of Public Health; Lynn E. Eberly, University of Minnesota, School of Public Health; Lisa M. Sullivan, Boston University; Hua Yun Chen, University of Illinois at Chicago;

Stephen C. Alder, University of Utah; Kenneth R. Hess, Rice University; Heather A. Young, George Washington University; Bonnie Davis, University of New England; Margaret Louis, University of Nevada, Las Vegas; Reg Arthur Williams, University of Michigan; Sudipto Banjeree, University of Minnesota. Their suggestions, corrections, and thoughtful insights are reflected throughout this text though they cannot be held responsible for the final form. We express our heartfelt appreciation to all those who contributed to this endeavor. Of special note in this regard are Drs. James Mortimer and Lakshminarayan Rajaram of the College of Public Health at the University of South Florida who led the field testing effort.

R. Clifford Blair, PhD
Richard A. Taylor, PhD

Foundations of Biostatistics

1.1 INTRODUCTION

Researchers in health-related disciplines use a wide variety of tools in order to gain an understanding of the phenomena they study. Perhaps the most important of these is biostatistics. Biostatistics plays a fundamental role in the analysis of data collected in the context of clinical trials as well as from studies in other areas such as epidemiology, health policy, community and family health, and environmental and occupational health.

What then is biostatistics? First it must be said that biostatistics is a branch of the broader field of statistics. **Statistics** is a discipline concerned with (1) the organization and summarization of data, and (2) the drawing of inferences about the characteristics of some collection of persons or things when only a portion of these characteristics are available for study. **Biostatistics**, then, is that branch of statistics that deals primarily with the biological sciences and medical/health-related disciplines. Thus, this book is concerned with the study of statistics with an emphasis on its application to the health sciences.

When approaching the study of any organized body of knowledge, especially one as diverse and complex as biostatistics, it is important that some framework be identified from which the component material can be viewed. Without such an organizing structure the concepts to be learned will likely appear to the learner as unrelated topics whose purposes are only vaguely perceived. To some degree, this situation must simply be tolerated. Many of the important elements of biostatistics cannot be fully appreciated until they are juxtaposed with other such elements. Thus, their role and utility in the grand scheme of the discipline becomes clear only when they are viewed as parts of the whole.

Fortunately, biostatistics has a rather natural framework that helps to alleviate this problem to some degree. Most of the material in an introductory biostatistics course can be structured around the concepts of populations and samples. These concepts provide the foundation upon which this book is organized.

1.2 POPULATIONS AND SAMPLES

You might think that, because of the fundamental role of populations in statistics/biostatistics, there would be consensus on its definition. Unfortunately, this is not the case. Compare the following statements regarding populations taken from two different statistics texts.

A population is a set of persons (or objects) having a common observable characteristic [29].

Note that the word *population* refers to data and not to people [36].

These two statements are clearly at odds and reflect the impreciseness with which the term is often used. Much of the confusion regarding populations stems from the fact that statisticians use the term in two different senses. The first refers to what we will call **popular populations** and the second to **statistical populations**. Popular populations are made up of persons or things. Thus, in common usage we may refer to the population of persons living in Florida who test positive for hepatitis C, or to the population of deer in a particular county in Michigan that carry the tick responsible for Lyme disease. These populations are clearly made up of persons or things.

By contrast, statistical populations are made up of the *characteristics* of persons or things. To see the distinction, consider the following. A popular population might be composed of the students at some university. A statistical population might then consist of the blood pressures of these same students. Likewise, the statistical population might be made up of an indicator as to whether each student had experienced some form of sexual abuse in their lifetime or their opinion regarding the quality of the education they are receiving.

It appears, then, that the first author quoted above was attempting to define a popular population while the second statement was in regards to a statistical population. There is a further problem in regards to the second definition that must be clarified. Statistical populations consist of characteristics of persons or things regardless of whether they have been measured or not. The word **data** refers to the recordings of measurements made on characteristics. Thus, if the blood pressures of some or all of the students are measured and recorded in some fashion the result is data. The distinction to be made here is that statistical populations are made up of the characteristics themselves rather than the recordings of those characteristics.

When such characteristics take on different values they are referred to as **variables**. While it is possible for a population to be made up of a characteristic that does not vary (i.e., a constant), this would be of little interest in a statistical context and so will not be dealt with in this book. For present purposes the terms "characteristic" and "variable" will be used interchangeably.

Obviously, the sizes of populations can vary. In the discipline of statistics it is useful to distinguish between finite and infinite populations because the methods used to deal with each differs somewhat. **Infinite populations** can be thought of as large populations while **finite populations** are those that are smaller. Admittedly, the distinction is arbitrary. The methods described in this book are generally appropriate for use with infinite populations.

A **sample** is some subset of a population. For example, the blood pressures of the students in some particular class at the aforementioned university would constitute a sample (though not a random sample).

The concept of a population is often much more abstract than the above discussion implies. For example, in a clinical trial the population may be conceived of as the blood pressures of all males over the age of 65 who will ever take some newly developed antihypertensive medication. In this situation it would be impossible to enumerate the population because no one knows exactly who will and who will not take the new drug. By contrast, the sample is almost always better defined. In a study of the drug's efficacy, the medication might be administered to 50 men over the age of 65 who meet the study protocol. In this case, the sample is readily defined because it is easy to identify persons who are or are not in the sample.

In a typical research setting the researcher will have measured or observed the characteristics that make up the sample and will have recorded them in the form of data. But the same will not be true of the population. In the case of the students, it may be impractical to measure the blood pressures of the entire student body in a large university but quite feasible to take measurements on a sample of 50 such blood pressures.

In this section a clear distinction has been made between the word population when used in a popular sense and when used in a statistical sense. This distinction is commonly blurred in statistical texts. It is not uncommon to read, "A sample of 50 subjects was used in the study." Clearly this sample was taken from a population of persons, which implies that the term is being used in the popular sense. You will also encounter statements such as, "The sample mean is 121." Here, the sample referred to is from a statistical population. You will encounter both uses of the word in this book as in most other statistics texts. In most instances the context will make the meaning clear. Understanding the difference between the statistical and popular meanings of the word "population" avoids a potential source of confusion for beginning statistics students.

1.3   PARAMETERS AND STATISTICS

Closely related to populations and samples are the concepts of parameters and statistics. A **parameter** is defined as any summarization of the elements of a population while a like summarization of the elements of a sample is referred to as a **statistic**. (Do not confuse "statistics" when used in this sense with the word "statistics" when used to refer to the discipline of study. Again, context will generally make the meaning clear.) By these definitions, the average blood pressure of all the students at the aforementioned university would be a parameter while the average blood pressure of the students in one particular class at that university would be a statistic. Likewise, the median blood pressure of all men over age 65 who will ever take the antihypertensive drug would be a parameter while the median blood pressure of the 50 men who participated in the study would be a statistic. Note that in order to obtain the value of a parameter or statistic, one would have to measure or observe the elements of the corresponding population or sample, record these measurements or observations in the form of data, and then carry out the summarization on these data.

An important point that follows directly from the above is that the values of parameters are usually not available to the researcher while the values of statistics are readily available. Much of the material that follows in this book relates to this fact.

The distinction between parameters and statistics is so fundamental to statistical thinking that two different conventions are commonly employed for their representation. In the first, parameters are represented by Greek letters while statistics are depicted by the Roman alphabet or some form thereof. For example, the average (or mean) of a population is often represented by the Greek letter μ (pronounced "mew") while the same summarization of the data from a sample is represented by $\bar{x}$ (pronounced "x bar"). A second convention represents parameters with capital letters from the Roman alphabet and places a character, called a "hat," over the same letter(s) to represent statistics. An example of this convention is the use of RR to represent the parameter form of the risk ratio (to be discussed in Chapters 3, 5, and 6 and $\widehat{RR}$ (pronounced "RR hat") to represent the statistic. Sometimes these two conventions are combined so that the parameter is represented by a Greek letter and the statistic by a Roman letter with a hat.

Both conventions will be used in this book in order to conform with common practice. We will point out which convention is being used if there is danger of confusion.

1.4   DESCRIPTIVE AND INFERENTIAL STATISTICS

With this background it now becomes conceptually convenient to portray the discipline of statistics/biostatistics as consisting of two component parts. The first component is referred to as descriptive statistics while the second is termed inferential statistics. **Descriptive statistics** is made up of various techniques used to summarize the information contained in a set of data. Consider the following problem.

Suppose a study is conducted to assess the serum lead levels of 150 children living in older houses in a particular urban neighborhood. If one were to inquire as to the findings of this study the answer could be a listing of the individual test results. Thus it would be reported that the first result showed a level of 20 ug/dl while the second showed a level of 25 ug/dl and so forth. After listening to the 150 test results the inquirer would likely have little understanding of the information provided. This unsummarized information would overwhelm the listener's ability to arrive at a meaningful conclusion. A more cogent response might have been that "The *average* serum lead level found in the children included in this study was 30 ug/dl." Other summarizations might include the highest and lowest values and various graphical representations of the data. Thus, descriptive statistics deals with exactly what the term implies, the description of data. To reiterate, summarizations of data related to the elements of a sample (statistics) will usually be easily obtained by the researcher while those related to the population (parameters) will not.

In contrast to descriptive statistics, **inferential statistics** is made up of various techniques used to provide information about parameter values based on observations made on the values of statistics. Opinion polls are a common example of this form of inference. In an opinion poll a sample of opinions obtained from a relatively small group of persons is used to draw conclusions concerning opinions held in some population. For example, 1,000 persons might be asked whether or not they favor some federally administered health plan. If 65 percent of the polled opinions were favorable then the pollster would attempt to use this information to draw conclusions about the proportion of favorable opinions in the country as a whole. Notice that in this case the value of a statistic (proportion of favorable opinions in the sample) is being used to gain insight into an unavailable parameter value (proportion of favorable opinions in the country). Returning to a previous example, researchers might compute the average blood pressure of the 50 patients given the antihypertensive medication

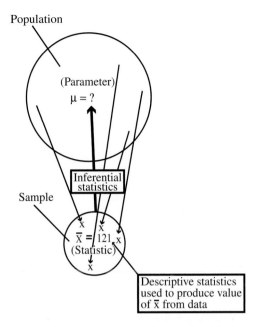

FIGURE 1.1: Schematic showing relationship between populations and samples, parameters and statistics, and inferential and descriptive statistics.

(value of a statistic) in order to estimate the average blood pressure of all men over age 65 who will ever take the drug (a parameter value).

The relationship between populations and samples, parameters and statistics, and inferential and descriptive statistics is depicted in Figure 1.1. Note that the proportion of a population represented by a sample is usually quite small and not nearly as large as this figure might imply. Notice also that the elements that make up the sample (represented by x in the figure) are randomly sampled from the population.

1.5   WHY POPULATIONS AND SAMPLES?

Earlier we asserted that statistics is an important tool used by researchers to gain insight into the subject of their inquiries. We also stated that the foundation of statistics is based on the concepts of populations and samples.[1] By now you must be wondering how this foundation can be used to assist in answering the various questions posed by researchers. There is the benefit to pollsters mentioned above, but exactly how does inference from a sample to a population help determine whether a newly developed drug is effective or whether exposure to a potential risk factor is related to the manifestation of some disease?

Unfortunately, this process is difficult to discern until more pieces of a rather elaborate mosaic have been put together. For this reason, you will be asked to master certain introductory concepts whose utility will not be fully appreciated until a more complete picture

---

[1]An exception is nonparametric statistics to be dealt with later in this book.

is formed. For the moment, simply understand that data collected in research studies can be quite complex and can involve numerous elements of chance. It will be shown that the concepts of populations and samples can be used to help separate these chance elements from the underlying reality.

1.6 WHAT HAPPENS NOW?

We have long maintained that many statistics textbooks are guilty of rushing through the foundational concepts in order to get to "the good stuff." As a result, students taking second or even third courses in statistics are often unclear as to the rationale behind the methods they are studying. Consequently, in this book a careful, methodical attempt will be made to lay a proper foundation from which the remaining content will be logically developed. This process has already begun.

In Chapter 2, after some preliminaries, you will take up the study of descriptive statistics. Chapter 3 will discuss some essentials of probability, the mechanism that underlies inference. Chapter 4 will introduce the logic of inference as well as some simple methods of hypothesis testing and confidence interval construction. The chapters that follow delve into various statistical methods that are commonly employed in health science research.

KEY WORDS AND PHRASES

After reading this chapter you should be able to demonstrate familiarity with the following words and phrases.

biostatistics 1	characteristic 2
data 2	descriptive statistics 4
finite population 2	inferential statistics 4
infinite population 2	parameter 3
popular population 2	sample 2
statistic 3	statistical population 2
statistics 1	variable 2

EXERCISES

1.1 Upon what tasks is the discipline of statistics primarily focused?

1.2 Differentiate between the following:

(a) samples and populations,

(b) statistics and parameters,

(c) popular populations and statistical populations,

(d) descriptive and inferential statistics, and

(e) infinite and finite populations.

1.3 What is meant by the term data?

1.4 Explain why populations are not made up of data as asserted by some authors?

A. The following questions refer to Case Study A (page 469)

1.5 Identify the sample in this study. Would you say that this sample is best characterized as being from a statistical or popular population?

1.6 Is the population in this study well identified? Explain your answer.

1.7 Describe the population to the best of your ability.

1.8 Are any descriptive statistics reported in this study? If so, give examples.

1.9 Are any parameters reported in this study? If yes, give examples, if no, explain why not.

1.10 Are any data to be found in this study? If so, give examples.

B. The following questions refer to Case Study B (page 470)

1.11 Identify the sample in this study. Would you say that this sample is best characterized as being from a statistical or popular population?

1.12 Is the population in this study well identified? Explain your answer.

1.13 Describe the population to the best of your ability.

1.14 Are any descriptive statistics reported in this study? If so, give examples.

1.15 Are any parameters reported in this study? If yes, give examples, if no, explain why not.

1.16 Are any data to be found in this study? If so, give examples.

F. The following question refers to Case Study F (page 473)

1.17 Do you believe that the results of this study are applicable to cases of tuberculosis in the United States? Explain the basis for your answer.

Descriptive Methods

2.1 INTRODUCTION

In Chapter 1 we pointed out that attempts to gain information from large, diverse data sets are likely to end in frustration unless some form of summarization is used to reveal salient aspects of the data. In this chapter we introduce some of the more commonly used of these methods. The topics to be addressed here can be divided into distributional, graphical, and numerical methods. While they apply equally well to data derived from populations and samples, they are almost always applied to sample data.

Before addressing these topics it will be useful to first gain an understanding of two related subjects often referred to as the scales of measurement and summation notation. After acquiring insight into these prerequisites we will return to the three topics listed above.

2.2 SCALES OF MEASUREMENT

Earlier we said that populations and samples are made up of variables that in turn are measurable or observable characteristics of persons or things that take on different values. We also said that once measurements were carried out and recorded the result was called data. But what is meant by the word *measure*? Simply put, it means that we assign numbers, letters, words or some other symbol to persons or things in order to convey information about the characteristic being measured. Thus, we might assign the number 220 to a person in order to represent their level of serum cholesterol or an "F" or "M" to represent their gender.

Often not recognized is the fact that measurements taken on variables can yield different amounts of information depending upon the scale employed in the measurement process. This means measurements that produce the numbers 1, 2, and 3 on one scale may convey a very different amount of information about the variable than would the same numbers obtained from use of a different scale. This in turn has implications for the statistical treatment of such data. All of this will become clear from what follows.

The scales discussed in this section were described by S.S. Stevens [44]. According to Stevens the measurement process can be conceived of as existing on four different levels which he referred to as the nominal, ordinal, interval (or equal interval), and ratio scales. Each of these will be discussed in turn.

2.2.1 The Nominal Scale

The nominal scale is the least sophisticated of the four and possesses two primary characteristics. First, it produces classifications of persons or things based on a qualitative assessment of the characteristic being considered and second, no information regarding quantity or amount is imparted by its use. Consider the following example regarding classification by blood type.

In this case the nominal scale is used to assign the designations of blood type A, B, AB, or O to persons based on certain hematological criteria. Notice that these designations merely classify persons into one of four blood type categories. Thus, all persons with the same blood type are given the same designation while those with other blood types are given other designations. Notice also that there is no concept of "greater than" or "less than" implied by these classifications. This means that nominal level measurements[1] do not permit comparisons of persons or things on the basis of *more* or *less* but rather on the basis of *similar* or *dissimilar*.

Designations produced by nominal scales may appear numeric in nature but should not be treated as such. When conducting a telephone or mail survey, households may be categorized by area or zip code for sampling purposes. In this case area codes of 813 or 272 would simply be indicators of geographic location of households. Certainly it would make no sense to maintain that residences with area code 813 have more "area code" than do those with area code 272. Likewise, arithmetic operations on such "numbers," e.g., calculating an average area code, would not yield a meaningful result. One could, however, count the number of households falling into each category.

2.2.2 The Ordinal Scale

Like the nominal scale, the ordinal scale classifies persons or things on the basis of the characteristic being assessed. Unlike the nominal scale, however, the classifications produced by this scale incorporate the very important attributes of "greater than" and "less than."

For example, suppose that in the course of a study dealing with pain management, patients are asked to rate their perceived pain as "none," "mild," "moderate," or "severe." This scheme categorizes patients into one of four categories that are ordered in terms of pain intensity. It is readily seen that the category "severe" represents *more* perceived pain than does the category "moderate" and so on. In this sense the ordinal scale can be said to provide more information about the characteristic being measured than does the nominal scale. Other examples include the categorization of some pathology as being stage 1, 2, 3, or 4 or the ranking of patients in a triage situation.

Notice that while this system can order categories in terms of more or less of the characteristic being measured, it does not allow for an indication as to *how much* more or less. Severe pain represents more pain than does moderate pain but how much more? A patient

---

[1]Many psychometricians object to use of the word "measure" in relation to nominal scales since many definitions of the term imply quantity.

triaged into one category may be in more need of care than a patient in another category but how much more?

Ordinal data are common in health-related research but have traditionally caused certain analytic difficulties. A common solution is to statistically treat these data as though they were at a nominal level. While in a sense correct, this practice is usually wasteful of information and therefore not entirely satisfactory. We will return to this problem in later chapters.

2.2.3 The Interval (or Equal Interval) Scale

Just as the ordinal scale adds the attributes of greater than and less than to those of the nominal scale, so the interval (also called equal interval) scale adds the attributes of *how much more* and *how much less* to those of the ordinal scale. While there are numerous examples of interval scales, their discussion is fairly complex and would necessitate a closer look at the field of psychometrics[2] than is warranted in a statistics book. For this reason the example most often given is that of the Fahrenheit thermometer which is quite straightforward.

Temperature when obtained from a Fahrenheit thermometer is measured in equal units which permits the quantification of differences. A reading of 70 represents five more degrees of temperature than does a reading of 65. The same is true of readings of 100 and 95. So this scale not only permits comparisons of the more than and less than variety but also indicates how much more or how much less one reading is than another.

A shortcoming of the interval scale is its lack of a true zero point. Said differently, the zero point on this scale is an arbitrary designation which means that it does not represent an absence of the characteristic being measured. Thus, it is possible to have a temperature of zero on one particular day and a reading of minus ten the next. The zero reading did not mean that there was no temperature but rather was merely another point on the scale. It follows that this scale does not permit the formation of meaningful ratios. It cannot be validly claimed that an 80 degree reading represents twice as much temperature as does a 40 degree reading.

2.2.4 The Ratio Scale

The ratio scale is similar to the interval scale with the exception that it possesses a true zero point. Physical measures such as height and weight are common examples. When something has zero weight the zero indicates that there is no weight present.

2.2.5 Continuous and Discrete Data

A simpler view of data holds that they are either continuous or discrete. A **continuous** *variable* is one that, theoretically at least, can take on any value in some specified range. For example, one person may weigh 160 pounds while another weighs 161 pounds. But it is possible to find a weight between these two, for example, 160.5 pounds. We can also find a weight between 160 and 160.5 pounds one example of which is 160.25 pounds. Theoretically this process could go on forever though we would eventually find that we did not have a scale sensitive enough to make the necessary distinctions. Weight then, is a continuous variable.

---

[2]The psychological theory or technique of mental measurement.

A **discrete** variable is one that is not continuous. For example, the number of persons living in households in some geographic area may be 1, 2, 3, 4, and so on, but cannot be 2.1367. Simply, discrete variables exist in discrete units rather than on a continuum.

Discrete variables that can take on only one of two values, e.g., male or female, dead or alive, positive or negative are referred to as **dichotomous** variables. Some statistical methods are specifically designed for use with dichotomous data.

It might be argued that all *data* are discrete because all measurement methods are limited by their level of precision thereby producing data in discrete units rather than on a continuum. Be that as it may, data obtained from continuous variables are usually considered and treated as continuous while data from discrete variables are treated as discrete. Sometimes researchers may measure a continuous variable but purposely record their findings as discrete data. This would happen, for example, if blood pressures were recorded as being in the normal range or not in the normal range. The classification of data (as opposed to variables) as discrete or continuous has an admittedly arbitrary component.

2.2.6   Further Comments on Scales

The conceptualization of the four scales presented above was first formulated in the context of psychometric rather than statistical theory. Their inclusion in, and potential contribution to, the statistics literature has not been without controversy [33]. At a minimum, they have provided a useful framework from which to view various analytic strategies. For example, some analytic methods are clearly appropriate for use with nominal data while others are more profitably employed with interval or ratio data. Ordinal data present another set of analytic questions. Opinions on some of these topics vary.

2.3   SUMMATION NOTATION

The statistical treatment of data often requires that they be summed in some fashion. A common example is the calculation of the average (or mean) of a data set. In this case the data are summed and then divided by the number of observations in the data set. But not all summation is so simple. Sometimes only part of the data are to be summed or the data are to be squared before summing or the data are to be summed with the sum then being squared. **Summation notation** is notation that is used to convey exactly how summation is to be carried out. By understanding a few simple rules of summation, you will gain significant insight into the formulas to be presented later.

2.3.1   Basic Notation

Suppose five numbers are written down in arbitrary order. Call the first number x_1, the second x_2 and so forth. If we wanted to indicate that these numbers are to be summed we could write the instruction

$$x_1 + x_2 + x_3 + x_4 + x_5.$$

A shorter form of this notation can be written as

$$\sum_{i=1}^{5} x_i.$$

The notation $\sum x$ indicates that the x values are to be summed while the i subscript on the x acts as a place holder for the numbers 1 through 5. The notation $i = 1$ shows that the summation is to begin with x_1 while the 5 indicates that the summation is to end with x_5. In other words, all the numbers in the set are to be summed.

Suppose now that the summation

$$\sum_{i=2}^{4} x_i$$

is to be carried out on the numbers 3, 0, 5, 9, 2, and 7. In this case the sum is to begin with x_2 and end with x_4 yielding $0 + 5 + 9 = 14$ as the result. If you wish to indicate that the sum is to include the last number in the data set but it is not known how many numbers will be involved, an n is used in place of the ending numeral. Consider the following

$$\sum_{i=2}^{n} x_i^2$$

This indicates that the squared values are to be summed with the sum beginning with x_2 and continuing to the last number. Using the example data this would be

$$0^2 + 5^2 + 9^2 + 2^2 + 7^2 = 159$$

Note also that

$$\left(\sum_{i=1}^{n} x_i\right)^2 = (3 + 0 + 5 + 9 + 2 + 7)^2 \neq \sum_{i=1}^{n} x_i^2 = 3^2 + 0^2 + 5^2 + 9^2 + 2^2 + 7^2$$

Many statistical calculations require that data be ordered with some partial sum then being computed. In this book sums will almost always include all values thereby allowing us to dispense with the additional notation. It will be specifically noted when this is not the case.

2.3.2   Some Rules of Summation

The four rules given below will assist in understanding the formulas to be presented later.

1. $\displaystyle\sum_{i=1}^{n} c = nc$

2. $\displaystyle\sum_{i=1}^{n} cx_i = c \sum_{i=1}^{n} x_i$

3. $\displaystyle\sum_{i=1}^{n} (x_i + y_i) = \sum_{i=1}^{n} x_i + \sum_{i=1}^{n} y_i$

4. $\displaystyle\sum_{i=1}^{n} (x_i - y_i) = \sum_{i=1}^{n} x_i - \sum_{i=1}^{n} y_i$

The first rule states that the sum of a constant (c) is equal to n (the number of constants) times the constant. That is

$$\sum_{i=1}^{n} c = \overbrace{(c + c + \cdots + c)}^{n \text{ values}} = nc$$

Suppose the constant to be summed has the value three and that there are four of them. Then we have

$$\sum_{i=1}^{4} 3 = \overbrace{(3 + 3 + 3 + 3)}^{4 \text{ values}} = 4 \cdot 3 = 12$$

The second rule states that the sum of the product of a constant and a variable (x) is equal to the product of the constant and the sum of the variable. That is

$$\sum_{i=1}^{n} c x_i = (cx_1 + cx_2 + \cdots + cx_n)$$

$$= c (x_1 + x_2 + \cdots + x_n)$$

$$= c \sum_{i=1}^{n} x_i$$

Again letting the constant take a value of 3 and the variable take the values 3, 0, 5, 9, 2, and 7 we have

$$\sum_{i=1}^{6} 3x_i = (3 \cdot 3 + 3 \cdot 0 + 3 \cdot 5 + 3 \cdot 9 + 3 \cdot 2 + 3 \cdot 7)$$

$$= 3 (3 + 0 + 5 + 9 + 2 + 7) = 3 \cdot 26 = 78$$

The third rule states that the sum of the sum of two variables (x, y) can be expressed as the sum of the first plus the sum of the second. If we let y represent the second variable then we have

$$\sum_{i=1}^{n} (x_i + y_i) = (x_1 + y_1) + (x_2 + y_2) + \cdots + (x_n + y_n)$$

$$= (x_1 + x_2 + \cdots + x_n) + (y_1 + y_2 + \cdots + y_n)$$

$$= \sum_{i=1}^{n} x_i + \sum_{i=1}^{n} y_i$$

Note that this result follows even if x and/or y are constants. Now suppose the variable sums $(x_i + y_i) = (3 + 4) + (0 + 2)$ and $(5 + 1)$ are to be summed. This can be done as

$$\sum_{i=1}^{3} (x_i + y_i) = (3 + 4) + (0 + 2) + (5 + 1)$$

$$= (3 + 0 + 5) + (4 + 2 + 1)$$

$$= 8 + 7 = 15$$

Rule four follows directly from rule three.

Example Application

These rules can be used to find a simple, yet important result. Suppose the mean of some data set is to be subtracted from each element of the set and the result summed. What will the result be?

$$\sum_{i=1}^{n}(x_i - \bar{x}) = \overbrace{\sum_{i=1}^{n}x_i - \sum_{i=1}^{n}\bar{x}}^{\text{rule 4}}$$

$$= \sum_{i=1}^{n}x_i - \overbrace{n\bar{x}}^{\text{rule 1}}$$

$$= \sum_{i=1}^{n}x_i - \sum_{i=1}^{n}x_i = 0$$

The last result follows from the definition of $\bar{x}$. (See Equation 2.1 on page 25.)

With the preliminaries completed we now turn to the primary focus of this chapter, the description of data. To this end we will examine in turn some distributional, graphical, and numerical techniques that are commonly used for this purpose.

2.4 DISTRIBUTIONS

Table 2.1 shows the (fictitious) responses of 60 post-operative patients who were asked to rate their perceived pain on a four-point ordinal scale as part of a pain management study. As may be seen, these unorganized data are largely uninformative insofar as patterns of response are concerned. Did some levels of pain dominate? Was severe pain common? What proportion of the patients had no pain? What proportion suffered mild pain or less?

With this small amount of data you could spend a few minutes scanning the table in order to formulate approximate answers to these questions. However, this strategy would not be effective with a larger data set. Even with this limited number of responses it would be convenient to have the data rearranged so as to make these answers more easily discernable.

2.4.1 Frequency Distributions

Table 2.2 shows these data arranged into frequency, relative frequency, cumulative frequency, and cumulative relative frequency distributions. The first column lists the scale categories from low to high. The second shows the frequency of response for each category which is obtained by counting the number of times each response occurs in the data set. The **frequency**, then, is the *number* of responses of each type.

2.4.2 Relative Frequency Distributions

The third column of Table 2.2 shows the relative frequency of response which is obtained by dividing each frequency by the total number of responses (60 in this case). The **relative frequency**, then, is the *proportion* of responses of each type.

TABLE 2.1: Measures of perceived pain collected from 60 patients.

Pt. Num.	Pain Level	Pt. Num.	Pain Level	Pt. Num.	Pain Level	Pt. Num.	Pain Level
1	moderate	16	mild	31	none	46	severe
2	none	17	mild	32	moderate	47	none
3	mild	18	moderate	33	none	48	none
4	none	19	none	34	none	49	mild
5	severe	20	none	35	mild	50	mild
6	none	21	mild	36	none	51	mild
7	moderate	22	none	37	moderate	52	none
8	none	23	none	38	mild	53	mild
9	none	24	mild	39	none	54	severe
10	mild	25	moderate	40	none	55	moderate
11	mild	26	moderate	41	none	56	none
12	none	27	none	42	none	57	none
13	mild	28	none	43	none	58	none
14	mild	29	mild	44	none	59	mild
15	none	30	severe	45	none	60	none

TABLE 2.2: Distributions of perceived pain measures.

Pain Category	Frequency	Relative Frequency	Cumulative Frequency	Cumulative Relative Frequency
Severe	4	.07	60	1.00
Moderate	8	.13	56	.93
Mild	17	.28	48	.80
None	31	.52	31	.52

You can quickly discern from the first two columns that the largest number of patients (31) indicated that they had no pain. This number represented .52 (or 52%) of the total sample. Severe pain was less common with only 4 persons (.07 or the sample) choosing this category. In general, the numbers of responses in categories declined as categories represented higher levels of pain.

2.4.3   Cumulative Frequency Distributions

The **cumulative frequency** column shows the *numbers* of patients who indicated that their pain was *less than or equal to* the level represented. For example, 48 patients (31 + 17) rated their pain as being mild or less than mild while 56 patients (31 + 17 + 8) perceived their pain as moderate or less then moderate. The cumulative frequency is obtained by adding the frequency in a given category to those of the categories showing a lesser level of the measured variable.

2.4.4 Cumulative Relative Frequency Distributions

The cumulative relative frequency is calculated by dividing each cumulative frequency by the total number of respondents. It can be seen that .80 of the patients believed that their pain was mild or less in intensity while .93 felt their pain was moderate or less. The **cumulative relative frequency** column, then, shows the *proportion* of patients who indicated that their pain was *less than or equal to* the level represented

The frequency, relative frequency, cumulative frequency, and cumulative relative frequency distributions shown in Table 2.2 were calculated for an ordinal level variable. The first two distributions could also be used for a nominal level variable. Obviously the cumulative distributions would not be appropriate in this case since there is no quantitative ordering for a nominal level variable. We will now look at some distributional issues related to continuous variables.

2.4.5 Grouped Distributions

Table 2.3 provides a frequency distribution for the (fictitious) systolic blood pressures of 144 moderately obese teenagers. In this table the frequencies relate to blood pressure values rather than discrete categories as was the case in Table 2.2. As a result, there are a large number of values for which frequencies are provided. This can cause interpretative difficulties, especially when individual frequencies are small and even include zero. In such instances it is sometimes helpful to reduce the number of values by forming them into groups. Frequency, relative frequency, cumulative frequency and cumulative relative frequency distributions can then be provided for these groups of values rather than for individual values.

Table 2.4 provides grouped distributions for the blood pressure data. As may be seen, blood pressure values have been formed into intervals that are technically referred to as **class intervals**. The various distributions have then been based on these intervals. By reducing the data in this manner patterns of response are more readily discerned. But the price paid for this interpretative convenience has been loss of information. For example, while it is easily seen that about 21.5 percent of values fall in the 135–139 interval, there is no information about the individual values in this interval.

In constructing tables of this sort two related questions must be answered. Into how many intervals should the values be grouped and how long should the intervals be? Too few intervals will result in the loss of too much information while too many intervals will defeat the purpose of the summarization. The size of the intervals will depend upon the number of intervals used and visa versa. There are no hard and fast rules in this regard. Basically you will want to present the data in the most meaningful way possible. There are, however, some rules of thumb that can be used for guidance. A common suggestion is that there be no fewer than six and no more than 15 intervals. Another helpful rule is that, when plausible, class interval widths of 5 units, 10 units or some multiple of 10 should be used in order to make the summarization more comprehensible.

In the case of Table 2.4 this latter suggestion seemed feasible. In order to determine whether or not class intervals of size five would produce a reasonable number of intervals,

TABLE 2.3: Frequency distribution of blood pressures of 144 moderately obese teenagers.

BP	Freq.	BP	Freq.	BP	Freq.	BP	Freq.
143	2	128	3	113	0	98	2
142	0	127	3	112	0	97	2
141	0	126	7	111	3	96	2
140	4	125	4	110	3	95	3
139	6	124	4	109	1	94	0
138	3	123	2	108	0	93	1
137	11	122	3	107	2	92	2
136	3	121	1	106	1	91	0
135	8	120	3	105	2	90	1
134	5	119	2	104	0	89	0
133	8	118	2	103	1	88	0
132	4	117	1	102	1	87	0
131	3	116	3	101	0	86	1
130	5	115	6	100	4		
129	3	114	2	99	1		

TABLE 2.4: Grouped distributions of systolic blood pressures using 12 intervals.

Interval	Frequency	Relative Frequency	Cumulative Frequency	Cumulative Relative Frequency
140-144	6	.042	144	1.000
135-139	31	.215	138	.958
130-134	25	.174	107	.743
125-129	20	.139	82	.569
120-124	13	.090	62	.431
115-119	14	.097	49	.340
110-114	8	.056	35	.243
105-109	6	.042	27	.188
100-104	6	.042	21	.146
95-99	10	.069	15	.104
90-94	4	.028	5	.035
85-89	1	.007	1	.007

TABLE 2.5: Grouped distributions of systolic blood pressures using eight intervals.

Interval	Frequency	Relative Frequency	Cumulative Frequency	Cumulative Relative Frequency
142-149	2	.014	144	1.000
134-141	40	.278	142	.986
126-133	36	.250	102	.708
118-125	21	.146	66	.458
110-117	18	.125	45	.313
102-109	8	.056	27	.188
94-101	14	.097	19	.132
86-93	5	.035	5	.035

the range[3] of the blood pressure values was divided by five. That is

$$\frac{(143 - 86)}{5} = 11.4$$

From this result we decided to use 12 intervals of width five.

Class intervals should also be equal in length whenever possible with the lower end of the first interval being less than or equal to the smallest measurement in the data set and the upper end of the last interval being greater than or equal to the largest measurement. It should also be noted that intervals must be contiguous and non overlapping. Table 2.4 follows these guidelines.

As was previously stated, the number and size of intervals is flexible. For comparative purposes, Table 2.5 shows the same data formed into eight intervals. The size of the intervals was estimated by dividing the range by eight which resulted in 7.125. This was rounded up to eight. Tables 2.4 and 2.5 will be compared in the next section.

It is not always necessary to form grouped distributions for continuous variables. When the number of values is not too large, distributions may be based on the ungrouped data.

2.5   GRAPHS

It is often more informative to present distributions as graphs rather than in the tabular form used in the previous section. Many graphical forms are available. In this section you will learn about the bar graph, histogram, polygon, and stem-and-leaf display.

2.5.1   Bar Graphs

Figure 2.1 shows the relative frequency distribution given in Table 2.2 in the form of a bar graph. As may be seen, response categories are depicted along the horizontal (x) axis while relative frequencies are noted along the vertical (y) axis. The relative frequency for

---

[3]The range which is defined as the highest minus the lowest value will be discussed later.

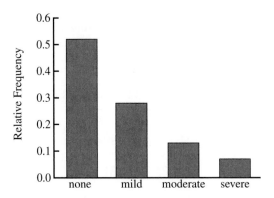

FIGURE 2.1: Relative frequency bar graph for pain scores.

each response category is read as the height, measured against the y axis, of a bar placed above the category. Frequency, cumulative frequency, and cumulative relative frequency bar graphs can be constructed in a like manner with the height of the bars indicating these quantities. It is easy to see from this figure why graphs are so commonly used to describe data. Even a cursory examination reveals a clear image of response patterns.

2.5.2 Histograms

Figure 2.2 is a relative frequency histogram for the blood pressure measures given in Table 2.4. Several differences between Figures 2.1 and 2.2 should be noted. First, the bars in Figure 2.2 are contiguous while those in Figure 2.1 are not. This is done to emphasize the fact that the data depicted in Figure 2.2 are continuous while those in Figure 2.1 are discrete. This then is the distinction between bar graphs and histograms. The former are used for discrete data and use noncontiguous bars while the latter are used for continuous data and use contiguous bars. The labeling of the x axis of Figure 2.2 also requires some explanation.

Theoretically at least, a measured value of a continuous variable is indicative of a range rather than a point on the measurement scale. For example, a blood pressure of 120 does not mean that the patient has a blood pressure of exactly 120 but rather that their blood pressure is somewhere between 119.5 and 120.5. The measurement device is not sufficiently precise to make distinctions between values of 119.7 and 120.1, for example, so these values are lumped together as 120. This conceptualization is not related to measurement error but rather to the level of precision produced by the measurement device. It follows that if a class interval is designated as being from 85 to 89, the blood pressures actually represented a range from 84.5 to 89.5. These values are termed the upper and lower **real limits** of the interval. A histogram of the eight interval relative frequency distribution of the blood pressure measures is shown in Figure 2.3 for comparison purposes.

2.5.3 Polygons

Another frequently used graphical form is the polygon. As with the histogram, polygons can be constructed for any of the distributions discussed here. The choice between the two

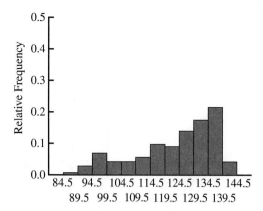

FIGURE 2.2: Relative frequency histogram of blood pressure measures in 12 categories.

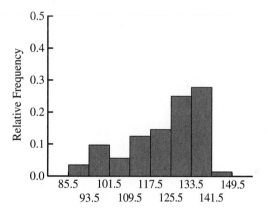

FIGURE 2.3: Relative frequency histogram of blood pressure measures in eight categories.

is largely a matter of taste and convenience. Figure 2.4 is a polygon of the same distribution shown in Figure 2.2. Polygons are constructed in a manner similar to histograms except that instead of placing a bar over each interval, a dot is placed at a height appropriate to the y axis. In the case of frequency and relative frequency polygons the dot is placed at the midpoint of the interval while for cumulative distributions the dot is placed at the upper real limit of the interval. These dots are then connected with straight lines that are joined to the x axis at the lower and, in the cases of frequency and relative frequency polygons, upper ends of the distribution. For frequency and relative frequency polygons the points at which the line is brought down to the x axis are at what would be the midpoints of an additional interval at each end of the distribution. This can be readily seen in Figure 2.5 on page 22 where Figure 2.4 is superimposed on Figure 2.2. Cumulative frequency and cumulative relative frequency polygons are not connected to the baseline at the upper end of the distribution and are connected at the lower end at the upper real limit of an additional interval added to

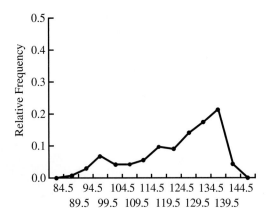

FIGURE 2.4: Relative frequency polygon of blood pressure measures in 12 categories.

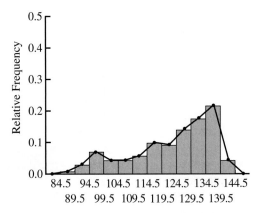

FIGURE 2.5: Relative frequency polygon imposed on histogram of blood pressure measures in 12 categories.

the lower end of the distribution. This may be seen in Figure 2.6. Polygons are particularly convenient when one wishes to compare two or more distributions as when male and female blood pressures are to be plotted in the same graph. (See also Figure 2.11 on page 33.)

2.5.4 Stem-and-leaf Displays

The stem-and-leaf plot is a graphical device that is related to, and sometimes used in place of, the frequency histogram. Its primary advantage over the histogram form is that it preserves the values of the displayed variable. Figure 2.7 on page 24 shows a stem-and-leaf display for the data in Table 2.3. The histogram-like appearance of this Figure is apparent. The following steps describe its construction.

1. Divide each observation into a "stem" and "leaf" component as described below.

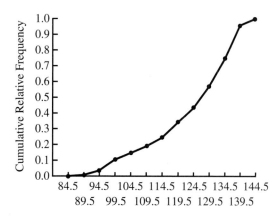

FIGURE 2.6: Cumulative relative frequency polygon of blood pressure measures in 12 categories.

2. List the stem components from lowest to highest values as would be done on the x axis of a histogram.

3. Place the leaf components associated with each stem over the stem in ascending order.

The **stem** of a number is defined as all of the digits in the number except for the right most. The **leaf** is then the right-most digit. Thus, the stem for the blood pressure value of 86 is 8 and the leaf is 6. Likewise the stem of 113 is 11 with a leaf of 3. The stems for the blood pressure measures are listed horizontally below the line in Figure 2.7. The leaves that are associated with each stem are placed above the stem in ascending order. The effect is similar to a frequency histogram but individual blood pressure values can still be recovered from the display.

A few points should be made before leaving this topic. First, stem-and-leaf displays are most effective with relatively small data sets. The 144 observations depicted in Figure 2.7 are probably near a maximum for this sort of graph. Second, displays of this type are usually not used for mass communications such as published research reports. Rather, they are used more informally by researchers trying to gain insight into their data. Third, stem-and-leaf displays can be more complex than the one shown here. For example, leaves can consist of two or more digits and stems can be grouped in a manner similar to grouped histograms. Last, it should be pointed out that traditionally stems are listed vertically with leaves being formed into rows. This convention was not followed here in order to emphasize the histogram-like nature of the display.

2.6   NUMERICAL METHODS

Distributional and graphical methods are excellent devices for conveying a general description of a data set. It is often desirable, however, to describe some particular characteristic of a data set numerically. Perhaps the most familiar measure of this sort is what is commonly referred to as the "average" or more precisely, the *arithmetic mean* of a set of data. In this section we will examine four distinct categories of such measures. They are, measures

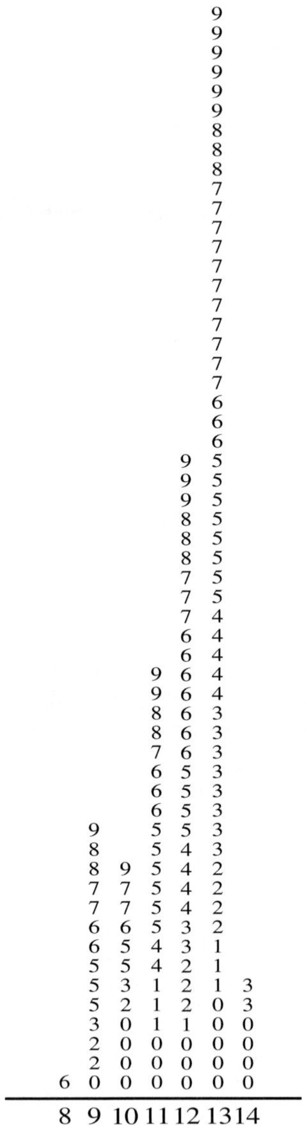

```
Stem | Leaves
-----+--------------------------------------------------------------
  8  | 9 8 8 7 7 6 6 5 5 5 3 2 2 6
  9  | 9 7 7 6 5 5 3 2 0 0 0 0
 10  | 9 9 8 8 7 6 6 6 5 5 5 5 5 5 4 4 1 1 1 0 0 0
 11  | 9 9 9 8 8 8 7 7 7 6 6 6 6 6 6 6 5 5 5 5 4 4 4 4 3 3 2 2 2 1 0 0 0
 12  | 9 9 9 9 9 8 8 8 7 7 7 7 7 7 7 7 7 7 7 6 6 6 5 5 5 5 5 5 5 5 5 5
       4 4 4 4 4 3 3 3 3 3 3 3 3 2 2 2 2 1 1 1 0 0 0 0 0
 13  | 0
 14  | 3 3 0 0 0 0
-----+--------------------------------------------------------------
        8   9  10  11  12  13  14
```

FIGURE 2.7: Stem-and-leaf display of blood pressure measures.

of central tendency, measures of variability, measures of relative standing, and measures associated with distribution shape.

2.6.1   Measures of Central Tendency

Measures of central tendency provide information about typical or average values of a data set. There are many such measures but we will consider only the mean, median, and mode as these are most commonly used.

Arithmetic Mean.   The arithmetic mean is the best known of the measures of central tendency and is what most people refer to as the "average." It is calculated by summing all the observations in the set of data and dividing this sum by the number of observations. The modifier "arithmetic" is used to distinguish it from other, less familiar, means. We will dispense with the modifier in this book since this is the only form of mean to be discussed here.

More formally the **sample mean** is defined as[4]

$$\bar{x} = \frac{\sum x}{n} \qquad\qquad (2.1)$$

while the population equivalent is given by

$$\mu = \frac{\sum x}{N} \qquad\qquad (2.2)$$

The notation for these expressions is as discussed in Section 2.3. The statistic and parameter forms differ only in the symbol used on the left hand side of the equation and the use of lower and upper case n which is a common convention to denote the number of observations in a sample and population respectively.

EXAMPLE 2.1

Find the mean of the numbers 3, 5, 4, 8, 7.

Solution

$$\bar{x} = \frac{3 + 5 + 4 + 8 + 7}{5} = 5.4$$

∎

Among the many properties of the mean are the following.

1. It is unambiguously defined in that its method of calculation is generally recognized.

2. It is unique in that a data set has one and only one mean.

3. Its value is influenced by all observations in the data set.

---

[4]As indicated earlier, subscripts will be omitted from formulas except in potentially ambiguous situations.

Median.   Unlike the arithmetic mean, there are a number of different ways to define and calculate the median. The most common definition maintains that the **median** is the value that divides a dataset into two equal parts so that the number of values that are greater than or equal to the median is equal to the number of values that are less than or equal to the median.

The most common way of computing the median when the number of values is odd is to order the observations in terms of magnitude and then choose the middle value as the median. A more formal way of expressing this is given by[5]

$$\boxed{\text{Median } (n \text{ odd}) = x_{\frac{n+1}{2}}}$$ (2.3)

where n is the number of observations and $\frac{n+1}{2}$ is the subscript of x.

When the number of observations in the data set is even, there is no middle value to choose as the median. In this case, the median is computed as the mean of the two middle values. A more formal way of expressing this is given by

$$\boxed{\text{Median } (n \text{ even}) = \frac{x_{\frac{n}{2}} + x_{\frac{n}{2}+1}}{2}}$$ (2.4)

where $\frac{n}{2}$ and $\frac{n}{2} + 1$ are the subscripts identifying the two middle values.

EXAMPLE 2.2

Find the median of the numbers 3, 5, 4, 8, 7, 0, 12.

Solution   The values are first arranged in order of magnitude as

$$
\overbrace{0 \quad 3 \quad 4}^{\text{3 values}} \quad \overbrace{5}^{\text{median}} \quad \overbrace{7 \quad 8 \quad 12}^{\text{3 values}}
$$ ∎

Because the number of observations is odd, the median will be the middle value which is five. Note that there are four numbers that are greater than or equal to five and four numbers that are less than or equal to five thereby satisfying the definition of the median given above.

Application of 2.3 with n equal to 7 yields $x_4 = 5$ which is the same value obtained above by inspection.

EXAMPLE 2.3

Find the median of the numbers 14, 8, 3, −1, 0, 12, 12, and 11.

Solution   Arranging the data and noting that the number of observations is even, we average the two middle values (i.e., x_4 and x_5) and obtain $\frac{8+11}{2} = 9.5$.

$$
-1 \quad 0 \quad 3 \quad \overbrace{8 \quad 11}^{\substack{\text{middle} \\ \text{values}}} \quad 12 \quad 12 \quad 14
$$ ∎

---

[5]We will make no distinction between the statistic and parameter forms of the median and mode since they will not be used in an inferential setting in this book. This policy will be followed for statistics in later sections as well.

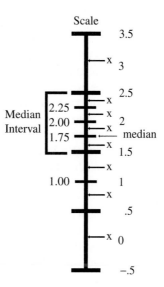

FIGURE 2.8: Conceptualization of the median of a continuous variable.

EXAMPLE 2.4

Find the median of the numbers 3, 2, 0, 2, 1, 1, 2, 2.

Solution   Again noting that the number of observations is even, we average the two middle values (i.e., x_4 and x_5) and obtain $\frac{2+2}{2} = 2$.

$$\underset{\substack{\text{middle}\\ \text{values}}}{0\ 1\ 1\ \overbrace{2\ 2}\ 2\ 2\ 3}$$

Notice that this value does not satisfy the definition of the median given above since there are seven values that are less than or equal to two and five that are greater than or equal to two. Clearly, two does not divide the data set into two equal parts. This difficulty often arises when there are tied observations near the middle of the distribution.   ∎

A less known but potentially more useful definition of the **median** states that it is a point on the scale of measurement located such that half the observations are above and below the point.[6] This definition does not provide a unique value for the median but when combined with the computational method described below does provide such a result.

The conceptualized scale shown in Figure 2.8 can be used to demonstrate this definition of the median as well as its calculation. You will recall from Section 2.5.2 that scores obtained from measurements of continuous variables are thought of as indicating values that lie within specified intervals of the underlying characteristic rather than at specific points. Thus, assuming intervals of length 1, a score of 0 indicates a value of the characteristic that

---

[6]See the discussion of percentiles on page 38 for a definition of the median as a percentile.

lies between $-.5$ and $.5$. These values are referred to, respectively, as the lower and upper real limits of the interval. A score of 1 indicates a value between the lower real limit of $.5$ and the upper real limit of 1.5 and so on. Thus, the upper real limit of one interval is also the lower real limit of the next greater value. Figure 2.8 shows these intervals for the data given in the last example. For present purposes, it is assumed that tied observations are uniformly distributed across the interval they represent.[7] Thus, since there are four 2s in the data set, it is assumed that the 1.5 to 2.5 interval can be divided into four equal parts with the four 2s being equally distributed among these subintervals as shown in the figure. Likewise, since there are two 1s the .5 to 1.5 interval is divided into two subintervals with one each of the two scores occupying each subinterval. The specific locations of the values, marked as "Xs" in the figure, are not known. Only the interval or subinterval within which the values lie is assumed known.

The problem is to find a point on this scale below which and above which half the observations fall. Since there is a total of 8 observations, half the observations would be $(.5)(8) = 4$. There is one observation that is less than .5, 3 observations that are less then 1.5 and 7 that are less than 2.5. The median, therefore, must lie in the interval 1.5 to 2.5. This interval is referred to as the **median interval**. Since there are three observations that are less than the lower real limit of the median interval, the median must lie at a point in the median interval that is greater than one of the four observations in the interval. As can be seen in the figure, this point would be at 1.75. Notice that given the assumption of equal dispersal across the interval, there are four observations below and above 1.75 thereby satisfying the definition given above. This method of calculating the median can be formalized by arranging the data into a cumulative frequency distribution and applying

$$\text{Median} = LRL + (w) \left[\frac{(.5)\,(n) - cf}{f} \right] \tag{2.5}$$

where LRL is the lower real limit of the median interval, w is the width of the median interval calculated as the difference between the upper and lower real limits of that interval, n is the total number of observations, cf is the cumulative frequency *up to* the median interval and f is the frequency of the median interval.

Applying equation 2.5 to the problem at hand yields

$$1.5 + (1.0)\, \frac{(.5)\,(8) - 3}{4} = 1.75$$

which is the result obtained by inspection.

Two exceptions to use of Formula 2.5 must be noted. The first of these is conceptualized in Figure 2.9. In this case there are eight observations, four of which are below the upper limit of 1.5 and four of which are above that point. There is then, no median interval since the point that (unambiguously) divides the data into two equal parts falls at the upper (or lower) limit of an interval. In situations of this sort the median is taken as the value at the real limit as shown in the figure.

The second exception is depicted in Figure 2.10. In this case there are four observations, two of which fall below and two of which fall above the upper real limit of .5. The problem

---

[7]This is sometimes referred to as the assumption of isomorphism.

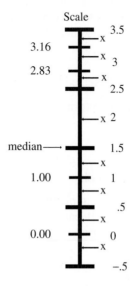

FIGURE 2.9: Conceptualization of the median of a continuous variable with no median interval.

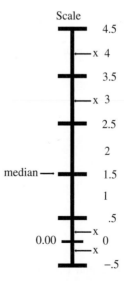

FIGURE 2.10: Conceptualization of the median of a continuous variable with zero frequencies in some key intervals.

is that this statement is true for any point between .5 and 2.5 on the scale. This results from the zero frequencies in intervals near the center of the distribution. In this case the midpoint of the interval(s) is taken as the median. This results in a median of 1.5.

EXAMPLE 2.5

Use the method presented here to find the medians of the values in examples 2.2 and 2.3.

Solution Applying formula 2.5 to the first example yields

$$4.5 + (1.0) \frac{(.5)(7) - 3}{1} = 5.0$$

which is the answer obtained previously. ∎

For the second example we note that there are four observations below 8.5 and 4 observations above 10.5. Because any point in the interval 8.5 to 10.5 satisfies the definition, the midpoint of the interval, which is 9.5, is taken as the median. This is the result previously obtained.

EXAMPLE 2.6

Find the median of the scores given in Table 2.3 on page 18.

Solution Arranging the data into a simple frequency distribution and then applying Equation 2.5 yields

$$125.5 + (1.0) \frac{(.5)(144) - 66}{7} = 126.36$$ ∎

EXAMPLE 2.7

Find the median of the data provided below.

Score	Frequency	Cumulative Frequency
2.4	3	145
2.3	40	142
2.2	36	102
2.1	21	66
2.0	18	45
1.9	8	27
1.8	14	19
1.7	5	5

Solution Noting that $(.5)(n) = 72.5$ and that the lower and upper real limits of the 2.2 interval have cumulative frequencies of 66 and 102 respectively, we identify the median interval as being from 2.15 to 2.25. Applying Formula 2.5 we obtain

$$2.15 + (.10) \frac{(.5)(145) - 66}{36} = 2.17$$ ∎

To summarize, when computing the median by the method described here, three different scenarios are possible. (1) When a median interval is identified, Formula 2.5 is applied. (2) When half the observations fall below and half above a real limit with the interval above the limit having nonzero frequency, the real limit is taken as the median. (3) When half the observations fall below and half above a real limit with the interval above the limit having zero frequency, the midpoint of the zero frequency interval(s) is taken as the median.

Among the many properties of the median are the following.

1. It may be defined and calculated in a number of different ways.

2. Given a specific definition and manner of calculation, it is unique in that a data set has one and only one median.

3. It is insensitive to extreme observations.

Mode.   The **mode** of a data set is the score or scores in the set that occur most frequently. If all scores in the set occur with equal frequency there is no mode. On the other hand, if two or more scores occur with equal frequency and that frequency is greater than that of the other scores in the set, then there will be more than one mode.

In the case of nominal or ordinal level data, the modal category can be found. The modal category is the one that has the greatest frequency. If two or more categories have equal frequencies and that frequency is greater than that of all other categories, then there will be more than one modal category.

EXAMPLE 2.8

Find the mode of the data depicted in Table 2.3 on page 18.

Solution   Reference to Table 2.3 shows that the blood pressure value of 137 occurs 11 times in the data set which is more than any other value. The mode is then 137.   ■

EXAMPLE 2.9

Find the mode(s) of the numbers 7, 8, 9, 7, 7, 4, 9, 5, 9, 3, 1, 9, 7, and 8.

Solution   The modes can be easily identified once the data have been formed into a frequency distribution as shown below. Because 9 and 7 both have frequency 4 which is greater than the frequency of any other value, 9 and 7 are the modes of the data.

Score	Frequency
1	1
2	0
3	1
4	1
5	1
6	0
7	4
8	2
9	4

■

EXAMPLE 2.10

Find the modal category of the data depicted in Table 2.2 on page 16.

Solution The modal category is "none" because it has frequency 31 which is greater than the frequency of any other category. ∎

A Comparison of the Properties of the Mean, Median, and Mode.

1. The mean and median are measures that locate the "middle" of a distribution in some sense. This is not necessarily true of the mode.

2. The mode can be unstable in small samples. For example, the mode of the numbers 1, 1, 1, 2, 3, 4, 5, and 5 is 1. But if one of the 1s is changed to 0 the set becomes bimodal. On the other hand if one of the 1s were changed to 5 the mode would be 5.

3. The median is *not* affected by the *size* of the scores at the upper and lower ends of the distribution for which it is computed. For example, the median of 1, 2, 3, 4, and 5 is 3. If the 5 is changed to 490 the median is still 3.

4. The mean *is* affected by the size of every value in the data set. For example, the mean of the original set of numbers given above was 3. After changing the 5 to 490 the mean was 100.

5. The mean plays an important role in inferential statistics while the median plays a lesser role and the mode virtually none at all.

2.6.2 Measures of Variability

Figure 2.11 shows the frequency distributions of two data sets each of which has mean, median, and mode equal to four. In spite of their common measures of central tendency, these data sets differ in an important respect. The scores in distribution A are less scattered or spread out than are those of distribution B. Said differently, the scores in A are more homogeneous than are those of B. As you will learn in later chapters, it is important to be able to quantify the degree of spread or scatter in a data set. Measures of this sort are referred to as measures of variability or dispersion.

As with measures of central tendency, there are many measures of variability. In this section you will learn about four of these, range, mean deviation, variance, and standard deviation.[8] While all of these are useful in given circumstances, the last two are by far the most important and will form the foundation of many of the methods you will study in later chapters.

Range. The range is a function of only the largest and smallest scores in a data set. Two forms of the range are often identified. The **exclusive range** is defined as the difference between the largest and smallest scores in the data set or more formally

$$\boxed{\text{Range (exclusive)} = x_L - x_S} \tag{2.6}$$

where x_L and x_S are the largest and smallest scores in the data set respectively.

---

[8]An additional measure of variability, the semi-interquartile range, is discussed in connection with percentiles in Section 2.6.3.

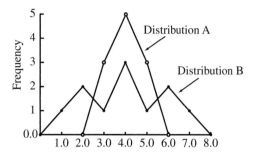

FIGURE 2.11: Two distributions with common mean, median, and mode.

EXAMPLE 2.11

Find the exclusive ranges of the data in Figure 2.11

Solution   The exclusive range for distribution A is $5 - 3 = 2$ while that for distribution B is $7 - 1 = 6$.   ■

The **inclusive range** takes into account the upper and lower real limits (see Section 2.5.2 on page 20 and the discussion beginning on page 27) of the highest and lowest scores and is expressed as

$$\text{Range (inclusive)} = URL_L - LRL_S \qquad (2.7)$$

where URL_L and LRL_S are the upper real limit of the largest and lower real limit of the smallest scores in the data set respectively.

EXAMPLE 2.12

Find the inclusive ranges of the data in Figure 2.11

Solution   The inclusive range for distribution A is $5.5 - 2.5 = 3$ while that for distribution B is $7.5 - .5 = 7$.   ■

The range is an unstable measure of variability due to the fact that it is based on only two values. A change in one of these scores can dramatically effect the range.

Mean Deviation.   Like the range, the mean deviation is a highly intuitive measure of variability. Unlike the range, however, mean deviation takes into account all the data for which variability is to be assessed thereby making it a more stable statistic.

A number of different measures of variability are based on the differences between the values in a distribution and some central point in the distribution. For example, suppose that the difference $x - \bar{x}$ is found for each score in a data set. This value, called a **deviation score** or simply a **deviation**,[9] gives the number of units between the score and the mean. When data are closely "clumped" around the mean, deviations tend to be small. For data that are more spread out, deviations will be larger. Plausibly, a reasonable representation of

---

[9]Deviations may also be taken from the median or other points on a distribution but we will use the term to refer to deviations from the mean.

TABLE 2.6: Calculation of mean deviation and variance of distribution A.

(1) x	(2) x^2	(3) $x - \bar{x}$	(4) $|x - \bar{x}|$	(5) $(x - \bar{x})^2$
3	9	-1	1	1
3	9	-1	1	1
3	9	-1	1	1
4	16	0	0	0
4	16	0	0	0
4	16	0	0	0
4	16	0	0	0
4	16	0	0	0
5	25	1	1	1
5	25	1	1	1
5	25	1	1	1
$\sum$ 44	182	0	6	6

variability could be based on the average of these deviations. When data are more spread out, the average of the deviations would be larger than for data with less spread. The difficulty with using deviations in this manner is that they always sum to zero. (See Section 2.3.2 on page 13 for details.) This problem can be overcome by taking the absolute values of the deviations. This then is the rationale for mean deviation. **Mean deviation** (MD) is the average of the absolute values of the deviations of a set of scores. The formal expression is

$$MD = \frac{\sum |x - \bar{x}|}{n}$$

(2.8)

EXAMPLE 2.13

Find the mean deviation of the data portrayed in Figure 2.11.

Solution Column (1) of Table 2.6 shows the scores that make up distribution A in Figure 2.11. Columns (3) and (4) of this table show, respectively, deviation scores and absolute values of deviation scores for the data in column (1). Using the sum of column (4) we obtain

$$MD = \frac{6}{11} = .55$$
■

Table 2.7 provides results for distribution B. Using the sum of column (4) from this table we obtain

$$MD = \frac{16}{11} = 1.45$$

Thus, the average deviation of the scores in distribution A was .55 units while that in distribution B was 1.45 units thereby confirming that distribution B has greater variability than does A.

TABLE 2.7: Calculation of mean deviation and variance of distribution B.

(1) x	(2) x^2	(3) $x - \bar{x}$	(4) $\lvert x - \bar{x}\rvert$	(5) $(x - \bar{x})^2$
1	1	−3	3	9
2	4	−2	2	4
2	4	−2	2	4
3	9	−1	1	1
4	16	0	0	0
4	16	0	0	0
4	16	0	0	0
5	25	1	1	1
6	36	2	2	4
6	36	2	2	4
7	49	3	3	9
$\sum$ 44	212	0	16	36

Variance.   Variance is a less intuitive but generally more useful measure of variability than is the range or mean deviation. As a descriptive statistic variance is less appealing than is mean deviation but is generally more useful because of its role in inference as will be seen in later chapters. Like mean deviation, variance uses deviations as its basis but squares them rather than using absolute values. The parameter form is given by

$$\sigma^2 = \frac{\sum (x - \mu)^2}{N} \tag{2.9}$$

Thus the parameter form of **variance** is the average of the *squared* deviations of the scores that make up the population. The statistic is

$$s^2 = \frac{\sum (x - \bar{x})^2}{n - 1} \tag{2.10}$$

You will notice that the devisor for the statistic is $n - 1$ while that for the parameter is N. This derives from the fact that s^2 is commonly used as an estimate of σ^2 in inferential settings as you will see later. It can be shown that if the devisor of s^2 were n, the resulting estimate would be biased. That is, on average the value of s^2 would be smaller than σ^2. By dividing by $n - 1$ this bias is removed making s^2 a better estimate of σ^2. This is all rather theoretical and beyond the scope of this book so you need not know more than is given here.

EXAMPLE 2.14

Find the variance of the data portrayed in Figure 2.11 on page 33.

Solution Column (5) of Table 2.6 shows the squared deviations of the data in distribution A. Using the sum of this column we obtain

$$s^2 = \frac{6}{10} = .60$$

Using the sum of the same column in Table 2.7 we obtain

$$s^2 = \frac{36}{10} = 3.60$$

for distribution B. ■

The equations given above for variance are called **conceptual equations** because they reflect the concept of variance. That is, they convey the idea that variance is based on squared deviations. Sometimes it is more convenient to employ computational equations. **Computational equations** are useful for computing variance but do not convey the concept. We can use the rules given in Section 2.3.2 to find a commonly used computational equation for variance. Using the numerator of 2.10 which is referred to as the **sum of squares** we note that

$$\sum (x - \bar{x})^2 = \sum \left(x^2 - 2x\bar{x} + \bar{x}^2 \right)$$

$$= \overbrace{\sum x^2 - \sum 2x\bar{x} + \sum \bar{x}^2}^{\text{rule 4}}$$

$$= \sum x^2 - 2\bar{x}\overbrace{\sum x}^{\text{rule 2}} + \overbrace{n\bar{x}^2}^{\text{rule 1}}$$

$$= \sum x^2 - 2\frac{(\sum x)^2}{n} + \frac{(\sum x)^2}{n}$$

$$= \sum x^2 - \frac{(\sum x)^2}{n}$$

This form of the sum of squares, as well as its population counterpart, can be divided by $n - 1$ or N to produce the following computational equations for variance.

$$\sigma^2 = \frac{\sum x^2 - \frac{(\sum x)^2}{N}}{N} \qquad (2.11)$$

$$s^2 = \frac{\sum x^2 - \frac{(\sum x)^2}{n}}{n - 1} \qquad (2.12)$$

EXAMPLE 2.15

Use Equation 2.12 to find the variance of the data portrayed in Figure 2.11.

Solution   Using the sums of columns (1) and (2) of Table 2.6 produces

$$s^2 = \frac{182 - \frac{(44)^2}{11}}{10} = .60$$

which is the same result obtained with the conceptual equation. Likewise, taking the same sums from Table 2.7 gives

$$s^2 = \frac{212 - \frac{(44)^2}{11}}{10} = 3.60$$

which is again the same value obtained by means of the conceptual equation.   ■

Variance may not possess much appeal as a purely descriptive statistic because it is expressed in squared units but as you are going to learn, it is one of the most important concepts in statistics.

Standard Deviation.   **Standard deviation** is defined as the square root of variance. It follows from the results presented above that the parameter is given by

$$\sigma = \sqrt{\frac{\sum (x - \mu)^2}{N}} \tag{2.13}$$

and

$$\sigma = \sqrt{\frac{\sum x^2 - \frac{(\sum x)^2}{N}}{N}} \tag{2.14}$$

The statistic is

$$s = \sqrt{\frac{\sum (x - \bar{x})^2}{n - 1}} \tag{2.15}$$

and

$$s = \sqrt{\frac{\sum x^2 - \frac{(\sum x)^2}{n}}{n - 1}} \tag{2.16}$$

EXAMPLE 2.16

Find the standard deviation of the data portrayed in Figure 2.11.

Solution   The standard deviation of distribution A is $\sqrt{.60} = .77$ and that of distribution B $\sqrt{3.60} = 1.90$.   ■

Variance is expressed in terms of the *square* of the original units of measurement. Thus, if measurements are in grams, variance is expressed as square grams. By contrast, standard deviation assesses variability in terms of the original units—grams in this case. We will have occasion to make much use of variance and standard deviation in the chapters to follow.

2.6.3   Measures of Relative Position

In this section you will learn about methods that locate the relative positions of observations in a distribution. For example, you may wish to find a point on the scale of measurement below which 25% of the observations in the distribution fall. Points characterized in this manner are called percentiles. On the other hand, you may wish to calculate the percentage of observations that fall below a specified point on the scale. These percentages are called percentile ranks. Finally, you may wish to determine the position of an observation relative to the mean of a distribution by means of what are called z scores. In this section we will in turn examine percentiles, percentile ranks, and z scores.

Percentiles.   When calculating the median you learned to find a point on the scale of measurement below which half (or 50%) of the observations fell. This point was termed the median. This concept can be broadened to find a point on the scale below which an arbitrary percentage of observations fall. For example, you may wish to find the point below which 25 or 75% of the observations fall. These points are called the 25th and 75th percentiles respectively. A **percentile** is a point on the scale of measurement below which a specified percentage of the observations are located. By this definition, the **median** can be defined as the fiftieth percentile.

Since the median is just one example of a percentile, you will not be surprised to learn that the method you learned for computing the median can be generalized to find any percentile. You would do well at this point to review the concepts related to the median presented on page 26 as they apply directly to the problem at hand. A slight modification to Equation 2.5 yields

$$P_p = LRL + (w) \left[\frac{(pr)\,(n) - cf}{f} \right] \tag{2.17}$$

where P_p represents the *pth* percentile, LRL is the lower real limit of the interval that contains the *pth* percentile, w is the width of the interval calculated as the difference between the upper and lower real limits of that interval, pr is p expressed as a proportion (i.e., $p/100$), n is the total number of observations, cf is the cumulative frequency *up to* the percentile interval and f is the frequency of that interval.

Notice that the only difference between Equations 2.5 and 2.17 is the substitution of pr for the constant .5. This permits the flexibility of finding any percentile rather than restricting calculation to the 50% percentile (median).

EXAMPLE 2.17

Find the 25th, 60th, and 75th percentiles of the data in Table 2.3 on page 18.

Solution   By forming a cumulative frequency distribution and noting that $(.25)\,(144) = 36$, the 25th percentile interval is identified as 114.5 to 115.5. This follows from the fact that the cumulative frequency up to the lower real limit of 114.5 is 35 and that for 115.5 is 41. The point on the scale below which 36 observations fall must be between these two limits. Using this information with Equation 2.17 gives

$$P_{25} = 114.5 + (1.0) \left[\frac{(.25)\,(144) - 35}{6} \right] = 114.67$$

Since $(.6)(144) = 86.4$, and the cumulative frequencies up to 129.5 and 130.5 are 82 and 87 respectively, P_{60} must lie in this interval. Again applying Equation 2.17 gives

$$P_{60} = 129.5 + (1.0) \left[\frac{(.6)(144) - 82}{5} \right] = 130.38$$

Using the same method, P_{75} is obtained by

$$P_{75} = 134.5 + (1.0) \left[\frac{(.75)(144) - 107}{8} \right] = 134.63 \qquad \blacksquare$$

EXAMPLE 2.18

Find P_{15}, P_{40}, and P_{69} for the data provided below.

Score	Frequency	Cumulative Frequency
.7	22	80
.6	26	58
.5	0	32
.4	0	32
.3	20	32
.2	7	12
.1	4	5
.0	1	1

Solution P_{15} is the point on the measurement scale below which 15% or $(.15)(80) = 12$ of the observations fall. Because the cumulative frequency at the upper real limit of .25 is 12, .25 is the 15th percentile. (Review the discussion of the median on page 26 if you don't follow this logic.)

Since $(.4)(80) = 32$ of the observations fall below any point between .35 and .55, the midpoint of .45 is taken as the 40th percentile. (Again, review the discussion of the median if the answer escapes you.)

Because the cumulative frequencies up to .55 and .65 are 32 and 58 respectively, the point below which $(.69)(80) = 55.2$ observations fall must be in the .55 to .65 interval. Applying Equation 2.17 gives

$$P_{69} = .55 + (.10) \left[\frac{(.69)(80) - 32}{26} \right] = .64 \qquad \blacksquare$$

Percentiles are often used to compare an individual's score to those of some larger group. For example, a pediatrician may be concerned that a baby's weight falls below the 5th percentile on a norm table of babies the same age because so few babies of that age have such low weights. Likewise, persons scoring above some specified percentile on an examination may be selected for special honors. Notice that in neither case are comparisons being made to some absolute standard but rather to other scores.

Percentiles are also useful for describing distributions. By using percentiles one can convey the fact that five percent of the observations fall below a given point (P_5) or that fifty percent of the observations fall between two points—P_{25} and P_{75}. Because percentiles are so commonly used for this purpose certain of them have specific designations. For example, P_{10}, P_{20}, P_{30} ... P_{90} are termed the first through ninth **deciles** because they divide the distribution into ten components with ten percent of the observations in each. Likewise, P_{20}, P_{40}, P_{60}, and P_{80} are termed the first through fourth **quintiles** (quint for five) and P_{25}, P_{50} and P_{75} are the first, second and third **quartiles**. (Notice that fifth decile, second quartile and median are different terms for the same point.) Collectively, percentiles, deciles, quintiles and quartiles are referred to as **quantiles**.

Finally, some descriptive statistics are based on percentiles. For example, a sometimes used measure of variability called the **semi-interquartile range** (or Q) is calculated by

$$Q = \left[\frac{P_{75} - P_{25}}{2} \right] \tag{2.18}$$

Percentile Ranks. As you now know, a percentile is a point on the scale of measurement below which a given percentage of observations fall. By contrast, a **percentile rank** is the percentage of observations that fall below a given point on the scale. Thus, percentiles are points and percentile ranks are percentages. Because of the close relationship between these two concepts they are often confused or used interchangeably. Given the same assumptions as was used for calculating percentiles, we can use the following equation to find percentile ranks.

$$PR_P = \frac{100 \left[\frac{f(P - LRL)}{w} + cf \right]}{n} \tag{2.19}$$

where P is the point on the scale for which the percentile rank is to be calculated, LRL is the lower real limit of the interval containing P, w is the width of the interval calculated as the difference between the upper and lower real limits of that interval, n is the total number of observations, cf is the cumulative frequency *up to* the interval containing P and f is the frequency of that interval.

EXAMPLE 2.19

Use the data in Table 2.3 on page 18 to find the percentile ranks of the scale points 114.67, 130.38, and 134.63.

Solution Noting that 114.67 falls in the 114.5 to 115.5 interval and using other information from Table 2.3 in Equation 2.19 gives

$$PR_{114.67} = \frac{100 \left[\frac{6(114.67 - 114.50)}{1.0} + 35.0 \right]}{144} = 25$$

Using the same method for the latter values yields

$$PR_{130.38} = \frac{100 \left[\frac{5(130.38 - 129.50)}{1.0} + 82.0 \right]}{144} = 60$$

and

$$PR_{134.63} = \frac{100 \left[\frac{8(134.63-134.50)}{1.0} + 107.0 \right]}{144} = 75 \qquad \blacksquare$$

None of these results should surprise you since these three points were previously identified as the 25th, 60th, and 75th percentiles.

Unfortunately, data analysts are usually not interested in finding the percentile rank of a *point* on the measurement scale, but rather want to know the percentile rank of a *score* or *observation*. This presents a difficulty since, as you now know, scores are not points on the scale but rather, represent intervals. How then do you find the percentile rank of a score? Basically, you must choose a point on the scale to represent the score. Three common choices are used.

1. The lower real limit of the score interval.

2. The midpoint of the score interval.

3. The upper real limit of the score interval.

When the lower real limit is used Equation 2.19 simplifies to

$$\boxed{PR_P = 100 \left[\frac{cf}{n} \right]} \qquad (2.20)$$

For the midpoint and upper real limit methods the simplifications are

$$\boxed{PR_P = 100 \left[\frac{(.5)\,(f) + cf}{n} \right]} \qquad (2.21)$$

and

$$\boxed{PR_P = 100 \left[\frac{f + cf}{n} \right]} \qquad (2.22)$$

where all terms are as previously defined.

EXAMPLE 2.20

Use the lower real limit, midpoint and upper real limit methods to find the percentile rank for a *score* of 134 in the distribution given in Table 2.3.

Solution   The lower real limit, midpoint, and upper real limit methods respectively produce

$$PR_{133.5} = 100 \left[\frac{102}{144} \right] = 71$$

$$PR_{134} = 100 \left[\frac{(.5)\,(5) + 102}{144} \right] = 73$$

and

$$PR_{134.5} = 100 \left[\frac{5 + 102}{144} \right] = 74$$

We have rounded the results given above because fractional parts of percentile ranks are usually not reported. ■

Before leaving this topic we want to emphasize our previous statement that there are a number of different definitions of, and computational methods for, percentiles and percentile ranks. Authors usually don't make this clear which sometimes results in confusion. This also explains why various software packages seem to produce conflicting results. Our attempt here has been not only to provide you with useful methods for dealing with these statistics but also to give you an intuitive basis for understanding the issues involved.

z **Scores.** Percentiles locate points relative to all the observations in a data set. For example, the 25th percentile is the point below which 25% of the data fall. By contrast, *z* scores locate points relative to the mean of the data. You have already seen one way of doing this. Deviation scores (see page 33) calculated as $x - \bar{x}$ show how far an observation is above or below the mean of the data. A deviation score of minus six would indicate that the score in question is located six units below the mean while a score of one would indicate that the score is one unit above the mean. A problem that arises with deviation scores is that they are scale dependent. Suppose that a patient's deviation scores for blood pressure and weight are 12 and 16 respectively when compared to other patients of the same age. It is difficult to compare these two scores because they are on different scales. Obviously the patient is above the mean of each measure but, relatively speaking, how far? It would be helpful if blood pressure and weight were on a common scale.

A **z score** indicates the distance and direction of a point from the mean in terms of standard units. More precisely, a sample z score is given by

$$z = \frac{x - \bar{x}}{s} \qquad (2.23)$$

The population equivalent is

$$Z = \frac{x - \mu}{\sigma} \qquad (2.24)$$

Notice that *z* scores are simply deviation scores divided by the standard deviation of the distribution. The effect of this division is to *standardize* the deviations or place them on a common scale. Instead of expressing the deviations in terms of their original units (blood pressure and weight in this example), they are now expressed in standard deviation units. When all of the scores in a distribution are converted to z scores, these z scores have a mean of zero and a standard deviation of one regardless of the original scale of the data.

Suppose now that the blood pressure deviation of 12 has a *z* score of 1.5 while that for weight is .6. This indicates that the patient is 1.5 standard deviations above the mean of the group in terms of blood pressure but is only .6 of a standard deviation above the mean in terms of weight. We will not say any more about z scores at this point because we will discuss them in greater detail when we begin our study of inference.

EXAMPLE 2.21

Convert the set of scores 1, 3, 3, and 9 to z scores. Then find the mean and standard deviation of the z scores.

Solution   Respectively, the mean and standard deviation of the original data are 4 and 3.46. Using these values in Equation 2.23 gives

$$z_1 = \frac{1 - 4}{3.46} = -.867$$

$$z_3 = \frac{3 - 4}{3.46} = -.289$$

$$z_3 = \frac{3 - 4}{3.46} = -.289$$

$$z_9 = \frac{9 - 4}{3.46} = 1.445$$

Because the sum of these scores is zero, the mean is also zero. With mean zero, Equation 2.15 simplifies to

$$s = \sqrt{\frac{\sum z^2}{n - 1}} = \sqrt{\frac{3}{3}} = 1$$

■

EXAMPLE 2.22

Convert the set of scores 11, 32, 13, 9, and 10 to z scores. Then find the mean and standard deviation of the z scores. Comparing the results of Example 2.21 to the results obtained here, which was the most extreme observation in terms of distance from the mean?

Solution $\bar{x} = \dfrac{75}{5} = 15,$ $s = \sqrt{\dfrac{1495 - \frac{75^2}{5}}{4}} = 9.62$

Using these results

$$z_{11} = \frac{11 - 15}{9.62} = -.416$$

$$z_{32} = \frac{32 - 15}{9.62} = 1.767$$

$$z_{13} = \frac{13 - 15}{9.62} = -.208$$

$$z_9 = \frac{9 - 15}{9.62} = -.624$$

$$z_{10} = \frac{10 - 15}{9.62} = -.520$$

Again, the mean is zero (with rounding) and the standard deviation (with rounding) is

$$\sqrt{\frac{4}{4}} = 1$$

The score of 32 is the most extreme in the two data sets as it is 1.767 standard deviations above the mean of its distribution. This is more than for any other value.   ■

While z scores are useful for describing data, their true importance is realized in inference.

2.6.4   Measures of Distribution Shape

Certain aspects of distribution shapes can be characterized numerically. The two most common of these, skew and kurtosis, will be discussed in this section. For the moment we will simply treat them as two more in the list of descriptive methods you have been learning. In subsequent chapters we will use them when we discuss violations of assumptions underlying certain inferential methods.

Skew.   Consider the shapes of the three frequency polygons depicted in Figure 2.12. Distribution A is symmetric in the sense that one half of the distribution is a mirror image of the other. Distribution B has most of its observations at the upper end of the scale but has a "tail" extending to the left. Distribution C is also non-symmetric but has its most frequent observations at the bottom of the scale with its tail extending to the right. Distributions of these three general forms are respectively characterized as **symmetric**, **negatively skewed**, and **positively skewed**.

Various methods have been developed to numerically describe the amount of skew (or lack thereof) that characterizes a distribution. **Skew** is generally defined as the degree of asymmetry in a distribution. The most common measure of skew is given by

$$\text{Skew}^{10} = \frac{\sum z^3}{n} \tag{2.25}$$

where z is the standardized deviation as described by Equation 2.23 and n is the sample size. Stated differently, this expression of skew is simply the average of the cubed z scores.

Consider distribution B in Figure 2.12. The mean of this distribution is approximately 4.83. Notice that deviations of scores that are greater than the mean will be positively signed while those below the mean will be negatively signed. When these deviations are standardized and cubed they will retain their sign and because of the extended tail below the mean, their sum will be negative. Thus, the Skew value will also be negative. The same logic indicates that Skew for distribution C will be positive while that for A will be zero. In general, the further from zero, the more severe the skew.

EXAMPLE 2.23

Compute Skew for each of the three distributions in Figure 2.12.

Solution   Table 2.8 shows values of the data that make up Figure 2.12 along with corresponding z scores and their cubes. As expected, the sum of the z values for each distribution is zero. The sum of the cubed z values is also zero for the symmetric distribution thereby giving a value for Skew of zero. This is always the case for symmetric distributions. Skew for distribution B is $\frac{-11.670}{18} = -.648$ while that for C is $\frac{11.670}{18} = .648$.   ■

---

[10] An unbiased estimate of the population skew is given by the expression $\frac{n \sum z^3}{(n-1)(n-2)}$.

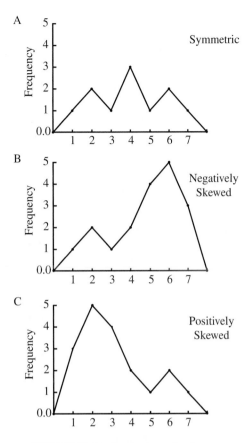

FIGURE 2.12: Variously shaped polygons.

Kurtosis.   Consider the two distributions depicted in Figure 2.13. Each has mean and median equal to four with variance and standard deviation of .73 and .85 respectively. Each is also symmetric. In spite of having these statistics in common, the two distributions have quite different shapes. Distribution A is peaked in the middle and has tails that extend further out than do those of distribution B. By contrast, distribution B appears flattened in the middle and has shorter tails.

The "peakedness" of a curve can be expressed by its kurtosis. More precisely, **kurtosis** refers to the peakedness of a distribution relative to the length and size of its tails. Distributions with sharp peaks such as distribution A are said to be **leptokurtic** while those that have flattened middles are said to be **platykurtic**.[11]

The formula for kurtosis is

$$\text{Kurtosis}^{12} = \frac{\sum z^4}{n}$$

(2.26)

---

[11] The prefix *lepto* means "slender" or "narrow" while *platy* means "flat" or "broad."

[12] An unbiased estimate of the population kurtosis is given by the expression $\frac{n \sum z^4}{(n-1)(n-2)}$.

TABLE 2.8: Data, z scores and cubed z scores from Figure 2.12.

Distribution A			Distribution B			Distribution C		
x	z	z^3	x	z	z^3	x	z	z^3
1	−1.581	−3.952	1	−2.103	−9.301	1	−1.188	−1.677
2	−1.054	−1.171	2	−1.554	−3.753	1	−1.188	−1.677
2	−1.054	−1.171	2	−1.554	−3.753	1	−1.188	−1.677
3	−.527	−.146	3	−1.006	−1.018	2	−.640	−.262
4	.000	.000	4	−.457	−.095	2	−.640	−.262
4	.000	.000	4	−.457	−.095	2	−.640	−.262
4	.000	.000	5	.091	.001	2	−.640	−.262
5	.527	.146	5	.091	.001	2	−.640	−.262
6	1.054	1.171	5	.091	.001	3	−.091	−.001
6	1.054	1.171	5	.091	.001	3	−.091	−.001
7	1.581	3.952	6	.640	.262	3	−.091	−.001
			6	.640	.262	3	−.091	−.001
			6	.640	.262	4	.457	.095
			6	.640	.262	4	.457	.095
			6	.640	.262	5	1.006	1.018
			7	1.188	1.677	6	1.554	3.753
			7	1.188	1.677	6	1.554	3.753
			7	1.188	1.677	7	2.103	9.301
$\sum$	0.000	0.000		.000[a]	−11.670		0.000	11.670

[a] This value is zero when calculations are carried out to a sufficient number of decimal places.

You will notice that while skewness was expressed as the average of the cubed z scores in a distribution, kurtosis is the average of z scores raised to the fourth power. In general, larger kurtosis values reflect sharper peaks than do smaller values. You should also note that kurtosis pertains to distributions with no more than one mode.

EXAMPLE 2.24

Find the kurtosis of the distributions depicted in Figure 2.13.

Solution Table 2.9 shows the data that make up the distributions in Figure 2.13 along with their z scores and the fourth powers of the z scores. Using the sums of the z scores raised to the fourth power the kurtosis of distributions A and B are respectively

$$\frac{60.5}{12} = 5.04 \quad \text{and} \quad \frac{15.125}{12} = 1.26.$$

These results confirm the earlier statement that, other things being equal, more peaked distributions have greater kurtosis values than do less peaked distributions. ∎

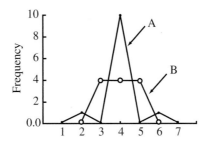

FIGURE 2.13: Two distributions with common mean and variance but different kurtosis.

TABLE 2.9: Data, z scores and their fourth power taken from Figure 2.13.

	Distribution A			Distribution B	
x	z	z^4	x	z	z^4
2	−2.345	30.250	3	−1.173	1.891
4	0.000	0.000	3	−1.173	1.891
4	0.000	0.000	3	−1.173	1.891
4	0.000	0.000	3	−1.173	1.891
4	0.000	0.000	4	0.000	0.000
4	0.000	0.000	4	0.000	0.000
4	0.000	0.000	4	0.000	0.000
4	0.000	0.000	4	0.000	0.000
4	0.000	0.000	5	1.173	1.891
4	0.000	0.000	5	1.173	1.891
4	0.000	0.000	5	1.173	1.891
6	2.345	30.250	5	1.173	1.891
$\sum$	0.000	60.500		0.000	15.125

2.7 A RE-ORIENTATION

You have just finished a long, and to some degree, tedious chapter that began with scales of measurement and summation notation and then proceeded through a host of descriptive statistics. With such intense study of "the trees" it is difficult to maintain proper perspective of "the forest." It is important for an understanding of biostatistics that you maintain the broad perspective while learning the details. To this end, you might want to skim Chapter 1 before continuing. You should pay special attention to Figure 1.1.

The focus of this book now moves from descriptive to inferential statistics. Before taking up inference, however, you must learn some fundamentals of probability. This will be the subject of Chapter 3.

KEY WORDS AND PHRASES

After reading this chapter you should be able to demonstrate familiarity with the following words and phrases.

bar graphs 19	central tendency 25
class intervals 17	continuous variables 11
cumulative frequency distributions 16	cumulative relative frequency distributions 17
dichotomous variables 12	discrete variables 12
frequency distributions 15	grouped distributions 17
histograms 20	kurtosis 45
lower real limit 20	mean 25
mean deviation 34	median 26
mode 31	percentile ranks 40
percentiles 38	polygons 20
range 32	relative frequency distributions 15
scales of measurement 9	semi-interquartile range 40
skew 44	standard deviation 37
stem-and-leaf displays 22	summation notation 12
upper real limit 20	variability 32
variance 35	z scores 42

EXERCISES

2.1 Table 2.10 on the facing page shows the number of annual sick days taken by nurses in a large urban hospital in 2003. Nurses are listed by seniority, i.e., nurse number 1 has least seniority, while nurse number 21 has most seniority.

(a) Which level of measurement is represented by the nurse number variable? Justify your answer.

(b) Is the sick days variable continuous or discrete? Justify your answer.

(c) Let x_i represent the number of annual sick days taken by the ith nurse where the index i is the nurse number. Find each of the following.

 i. x_3, x_9, x_{21}

 ii. $\displaystyle\sum_{i=1}^{10} x_i$

 iii. $\displaystyle\sum_{i=11}^{n} x_i$

 iv. $\displaystyle\sum_{i=1}^{n} x_i$

 v. $\displaystyle\sum_{i=1}^{n} x_i^2$

(d) Suppose that each nurse took exactly two more sick days than was reported in the table. Use summation notation to re-express the sum in 1(c)iv so as to reflect the additional two sick days taken by each nurse.

(e) Use the nurse annual sick days data to construct frequency, cumulative frequency, relative frequency, and cumulative relative frequency distributions.

(f) Use the nurse annual sick days data to construct a relative frequency histogram, a relative frequency polygon, and a cumulative frequency polygon.

(g) Use the nurse annual sick days data to calculate each of the following.

 i. mean, median, and mode

 ii. variance and standard deviation

 iii. exclusive and inclusive range

 iv. z scores for each of the x_i

 v. 15th, 50th, and 80th percentiles

 vi. percentile ranks of 2, 5, and 8 sick days

 vii. skew and kurtosis

TABLE 2.10: Table for Exercise 2.1.

Nurse Number	Sick Days	Nurse Number	Sick Days	Nurse Number	Sick Days
1	2	8	7	15	9
2	9	9	8	16	2
3	1	10	8	17	8
4	0	11	6	18	9
5	5	12	3	19	6
6	4	13	7	20	8
7	6	14	8	21	5

TABLE 2.11: Table for Exercise 2.2.

App. Numb.	Gender	Ws	Wp	App. Numb.	Gender	Ws	Wp
1	m	165	167	16	m	220	225
2	m	215	210	17	f	135	137
3	m	190	186	18	f	180	201
4	f	115	111	19	m	210	205
5	m	158	160	20	f	145	144
6	f	165	160	21	f	131	133
7	f	120	118	22	m	177	180
8	m	173	170	23	m	135	135
9	m	188	195	24	m	183	180
10	m	180	195	25	m	165	166
11	f	135	137	26	f	160	166
12	m	155	155	27	m	178	180
13	m	190	195	28	m	155	152
14	m	187	185	29	f	155	154
15	f	154	156	30	f	130	128

2.2 The data provided in Table 2.11 on the current page was obtained from applicants for employment at the Public Health Service. Recorded are the gender, self-reported weight (Ws) taken from the employment application and weight recorded at the time of the physical examination (Wp) of the applicants. Weights are recorded in pounds.

(a) Which level of measurement is represented by the gender variable? How else might this variable be characterized? Justify your answers.

(b) Construct a relative frequency bar graph for the gender variable. Would it be possible to construct a cu-mulative frequency polygon for the gender variable? Justify your answer.

(c) Using the applicant number as the index of summation (i), calculate the following.

i. $\sum_{i=1}^{15} \text{Ws}_i$ and $\sum_{i=1}^{15} \text{Wp}_i$

ii. $\sum_{i=1}^{15} \left(\text{Ws}_i - \text{Wp}_i \right)$

(Is this calculation really necessary to obtain the desired result? Why or why not?)

(d) Using class intervals of 110–119, 120–129, ..., 220–229, use the Ws variable to form a grouped: relative frequency histogram, relative frequency polygon, and cumulative relative frequency polygon.

(e) Use the Wp data to calculate each of the following.

    i. mean, median, and mode

    ii. variance and standard deviation

    iii. exclusive and inclusive range

    iv. z scores for each of the x_i

    v. first, second and third quartiles

    vi. percentile ranks of 111, 137, and 180

    vii. skew and kurtosis

A. The following questions refer to Case Study A (page 469)

2.3 Are any dichotomous variables mentioned in this study? If so, can you characterize them as belonging to one of the four measurement scales?

2.4 Are any continuous variables mentioned in this study? If so, can you characterize them as belonging to one of the four measurement scales?

2.5 Is the mean age of subjects reported in this study properly represented by $\bar{x}$ or μ? Why? What symbol should be used for the standard deviation of the age variable?

2.6 What is the modal yes/no category for the A.M. (morning) data reported in the table?

2.7 What would the age z score be for a subject who is 18 years of age? 75 years of age?

D. The following questions refer to Case Study D (page 471)

2.8 Construct a relative frequency distribution for the NPZ-8 scores.

2.9 Construct a relative frequency histogram and polygon for the NPZ-8 scores.

2.10 Find the mean, median and mode for the NPZ-8 data.

2.11 Find the mean, median and mode for the PBV data.

2.12 Find the variance and standard deviation for the NPZ-8 data.

2.13 Find the first, second and third quartiles of the NPZ-8 data.

2.14 Inspect the relative frequency histogram and polygon you constructed for the NPZ-8 data in 2.9. From your inspection, would you characterize this distribution as being positively skewed, negatively skewed or symmetric?

2.15 Calculate skew for the NPZ-8 data. Does this calculation confirm your observation made in 2.14?

2.16 Calculate skew for the NPZ-8 data using only data from the 15 infected subjects. Does including the five healthy subjects in the data set markedly increase skew in your opinion?

O. The following questions refer to Case Study O (page 477)

2.17 Which scales best describe the levels of measurement represented by the variables mentioned in point (a)?

Probability

3.1 INTRODUCTION

In Chapter 1 we pointed out that probability is the basis for inferential statistics. Said differently, it is the mechanism by which inference is carried out. This being the case, it is necessary that you gain a fundamental understanding of probability before beginning your study of inference. This chapter will provide that understanding.

Probability theory is a rather complex branch of mathematics whose roots are firmly embedded in games of chance. Indeed, much of the early work in this area was done in an attempt to gain some advantage in such games. Fortunately, your study of biostatistics will require only the most basic understanding of probability.

Most introductions to probability, at least those found in statistics texts, rely heavily on the antecedents of the discipline. Thus, many examples involving the rolling of dice and turning of cards are used to impart the foundational concepts. This is a reasonable approach to the study of probability. But our aim in this chapter is not to have you learn probability but to have you learn probability as it relates to inferential statistics. In our experience, students often have difficulty relating such examples to the ensuing study of inference. For this reason we will forgo tradition and instead present probability in specific contexts that are more closely related to the business of inference. Specifically, after defining a few terms, you will study probability as it relates to contingency tables and the normal curve. In the course of this study you will also be introduced to important descriptive methods not covered in Chapter 2 such as risk ratios, odds ratios, sensitivity, specificity, and positive and negative predictive values. We begin with a definition of probability.

3.2 A DEFINITION OF PROBABILITY

Definitions of probability sufficient for a rigorous study of the discipline can be complex and controversial. For purposes of this text, however, a simple intuitive definition will suffice. We begin with an example.

Suppose that four marbles, three of which are black and one of which is white, are placed in a bucket. We now vigorously shake the bucket then, with eyes closed, reach in and draw out a marble. (The shaking of the bucket and closing of eyes are done to ensure that the selection is *random*.) We now pose the question, "What is the probability that the marble drawn is black?" We suspect that your answer is something like "three out of four" or "three fourths" or "point seven five." We agree with these answers but what definition of probability did you use to arrive at them? A moments thought will convince you that you defined the probability of drawing a black marble as the number of black marbles in the bucket divided by the total number of marbles. What is the probability that the marble is white? Again, the probability is simply the number of white marbles divided by the total number of marbles or .25. Stated differently, the probability of a black marble was just the *proportion* of black marbles with the same being true for the probability of a white marble. Formally, we can define the **probability** of some occurrence,[1] let's call it *A* as

$$P(A) = \frac{N_A}{N} \tag{3.1}$$

The symbol $P()$ is read "The probability of ..." with *A* used to represent any event of interest, e.g., "drawing a black marble." The symbol $\overline{A}$ will be used to mean the compliment of *A* or in this case would mean "not drawing a black marble." Thus, in the present case $P(\overline{A})$ would be read, "The probability of not drawing a black marble." N_A is the number of events that meet the specified criterion (black marbles for example) and *N* is the total number of events.

Several important properties of probability can be deduced from the definition given in 3.1.

1. $P(A) \geq 0$. This follows from the fact that N_A is a count and cannot, therefore, be less than zero.

2. $P(A) \leq 1$. This follows from the fact that N_A can never exceed *N*.

3. $P(A) + P(\overline{A}) = 1$ or $P(\overline{A}) = 1 - P(A)$. This follows from the fact that $N - N_A$ outcomes fail to meet the stated criterion *A*. Thus $P(\overline{A}) = \frac{(N-N_A)}{N} = 1 - \frac{N_A}{N} = 1 - P(A)$.

With this definition in mind we now take up the study of probability in specific contexts. The first of these will relate to contingency tables.

3.3   CONTINGENCY TABLES

3.3.1   Sampling from the Population

Figure 3.1 depicts a conceptualized population made up of two characteristics associated with each of twenty persons.[2] Each of the 20 persons is characterized as being a smoker (*S*), not being a smoker $(\overline{S})$, having some particular disease (*D*), or not having the disease $(\overline{D})$.

We now pose the question, "What is the probability of selecting one person at random from this population and finding that they are a smoker?" This is just a variation of the

---

[1]This definition assumes equally likely events.

[2]In Chapter 1 we pointed out that the populations considered in this book are very large. The current depiction is small for pedagogical purposes.

Population

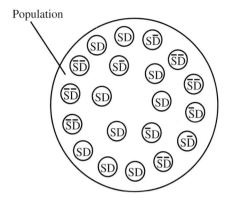

S = smoker, $\overline{S}$ = nonsmoker, D = Disease, $\overline{D}$ = no disease.

FIGURE 3.1: Population of two characteristics of 20 persons.

marbles in the bucket example given above and is solved in the same way. Counting the number of smokers and applying Equation 3.1 gives $P(S) = \frac{12}{20} = .60$. Likewise, the probability of selecting a person who is disease free is $P(\overline{D}) = \frac{9}{20} = .45$.

Probabilities involving both characteristics can also be calculated. For example, what is the probability of selecting a person who smokes *and* has the disease? Since nine people meet this criterion the probability is $P(SD) = \frac{9}{20} = .45$. Notice that $P(SD)$ is read "The probability of smoking *and* disease." Now, find $P(\overline{S}D)$. What does this represent? This is the probability of selecting a person who does not smoke and has the disease and is calculated as $P(\overline{S}D) = \frac{2}{20} = .10$

What is the probability of finding that the selected person smokes *or* has the disease? The criterion is met in this case by anyone who smokes *or* has the disease. Because there are fourteen people who either have an S or D (or both) associated with them, the probability is $P(S \cup D) = \frac{14}{20} = .70$. Notice that the notation $P(S \cup D)$ reads "The probability of smoker or disease."

In many statistical applications interest rests in only a specified portion of the population. For example, the question, "What is the probability of disease given that the selection is made *from among the smokers*?" implies that only smokers are to be considered. The symbol used for this type of probability is $P(D \mid S)$ which is read "The probability of disease given a smoker." This general form is referred to as **conditional probability**.

Figure 3.2 shows the redefined population, i.e., the original population with all non-smokers removed. The probability of selecting a person with disease from among the smokers is then $P(D \mid S) = \frac{9}{12} = .75$.

3.3.2   Frequency Tables

The sums used for calculations in Section 3.3.1 can be conveniently summarized in a contingency table as shown in Table 3.1.

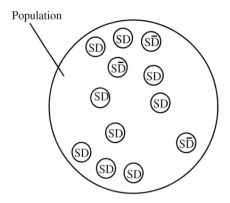

FIGURE 3.2: Population of disease status for smokers only.

TABLE 3.1: Contingency table showing population frequencies.

	D	$\overline{D}$	
S	9	3	12
$\overline{S}$	2	6	8
	11	9	

The numbers in the four cells are the counts of persons meeting both of the indicated criteria. That is, there were nine persons who smoked and had disease, three who smoked and had no disease, two who did not smoke and had disease and six who did not smoke and had no disease. The values at the table margins give the total count for the indicated characteristic. Thus, 12 persons smoked, eight did not smoke, 11 had disease, and nine were disease free.

EXAMPLE 3.1

Use the counts in the contingency table to find each of the following. Also state the meaning of each probability. Find $P\left(\overline{S}\right)$, $P\left(D\right)$, $P\left(S\overline{D}\right)$, $P\left(\overline{S}\,\overline{D}\right)$, $P\left(S \cup \overline{D}\right)$, $P\left(\overline{S} \cup D\right)$, $P\left(\overline{D} \mid S\right)$, and $P\left(\overline{S} \mid \overline{D}\right)$.

Solution $P\left(\overline{S}\right)$ is the probability of selecting a person who does not smoke. Using the appropriate count from the contingency table gives $\frac{8}{20} = .40$. Likewise, $P\left(D\right)$ is the probability of selecting someone with the disease and is $\frac{11}{20} = .55$.

$P\left(S\overline{D}\right)$ is the probability of selecting someone who smokes *and* does not have the disease while $P\left(\overline{S}\,\overline{D}\right)$ is the probability of selecting a non-smoker who does not have the disease. These are respectively, $\frac{3}{20} = .15$ and $\frac{6}{20} = .30$.

$P\left(S \cup \overline{D}\right)$ is the probability of selecting someone who smokes *or* is without disease. The number of persons meeting this requirement is $9 + 3 + 6 = 18$ which gives a probability of

TABLE 3.2: Contingency table showing population probabilities.

	D	$\overline{D}$	
S	.45	.15	.60
$\overline{S}$	.10	.30	.40
	.55	.45	

$\frac{18}{20} = .90$. Since $9 + 2 + 6 = 17$ persons are non-smokers *or* have the disease, $P\left(\overline{S} \cup D\right) = \frac{17}{20} = .85$.

The conditional probability $P\left(\overline{D} \mid S\right)$ is the probability of selecting someone who is without disease given that he/she is a smoker. Said differently, this is the probability of selecting someone who is disease free if the selection is deliberately made from among the smokers. Since there are 12 smokers, three of whom are disease free, the probability of selecting a disease free person from among the smokers is $\frac{3}{12} = .25$

Finally, $P\left(\overline{S} \mid \overline{D}\right)$ is the probability of a non-smoker given no disease. Again using the counts from the table, there are a total of nine persons without disease. Since six of these are non-smokers the probability of selecting a non-smoker from among the persons without disease is $\frac{6}{9} = .67$. ■

3.3.3 Probability Tables

Another common form of contingency table is obtained by dividing each count in a frequency table by N in order to obtain probabilities. The probability table represented by Table 3.2 was constructed from the frequency table shown in Table 3.1 on the facing page.

Notice that entries in the cells of this table represent $P\left(SD\right)$, $P\left(S\overline{D}\right)$, $P\left(\overline{S}D\right)$, and $P\left(\overline{S}\,\overline{D}\right)$ so that these values may be read directly from the table. The same is true for $P\left(S\right)$ and $P\left(\overline{S}\right)$ which are at the row margins and $P\left(D\right)$ and $P\left(\overline{D}\right)$ which are at the column margins. Probabilities of the form $P\left(S \cup \overline{D}\right)$ and $P\left(\overline{S} \cup D\right)$ are obtained by summing the appropriate cell entries. These values are respectively $.45 + .15 + .30 = .90$ and $.45 + .10 + .30 = .85$ which are the same answers obtained previously from the frequency table.

Conditional probabilities such as $P\left(\overline{D} \mid S\right)$ and $P\left(\overline{S} \mid \overline{D}\right)$ are calculated in the same manner as is done with frequency tables. Since $P\left(\overline{D} \mid S\right)$ is the probability of no disease given a smoker or equivalently, the proportion of smokers who do not have the disease, the fraction is $\frac{.15}{.60} = .25$ which is the result previously obtained. Likewise, the probability of selecting a non-smoker from among the persons without disease is simply the proportion of persons without disease who are non-smokers or $\frac{.30}{.45} = .67$ which again is the answer derived from the frequency table.

In general, if we let A, $\overline{A}$, B, and $\overline{B}$ represent arbitrary characteristics then the entries of the associated probability table are as follows

	B	$\overline{B}$	
A	$P(AB)$	$P(A\overline{B})$	$P(A)$
$\overline{A}$	$P(\overline{A}B)$	$P(\overline{A}\,\overline{B})$	$P(\overline{A})$
	$P(B)$	$P(\overline{B})$	

The entries within the cells of a probability table are referred to as **joint** probabilities while those at the margins are termed **marginal** probabilities. You should also note that the order in which characteristics are listed in expressions of joint probability are not important. That is, $P(AB)$ is equivalent to $P(BA)$. The same is true for expressions of the form $P(A \cup B)$ since this means the same thing as $P(B \cup A)$. It is important to understand that the same is *not* true for conditional probability. Thus, $P(A \mid B)$ is *not* the same as $P(B \mid A)$

Finally, we present two formulas useful in calculating and understanding probabilities related to contingency tables. They are

$$P(A \cup B) = P(A) + P(B) - P(AB) \tag{3.2}$$

and

$$P(A \mid B) = \frac{P(AB)}{P(B)} \tag{3.3}$$

3.3.4 Independence

Independence plays a fundamental role in many of the inferential methods you will study in later chapters as well as in the general methodology of health related research. Two events A and B are said to be **independent** if

$$P(A \mid B) = P(A) \tag{3.4}$$

If this equality does not hold, the events are said to be *non-independent* or *dependent*. The meaning of this statement can be understood by considering the probabilities in Table 3.2. The probability of disease in the general population (i.e., $P(D)$) is .55. But the probability of disease among smokers (i.e., $P(D \mid S)$) is $\frac{.45}{.60} = .75$. Since smokers have a higher probability of disease than does the general population it seems reasonable to conclude that some form of relationship exists between smoking and this disease.[3]

On the other hand, suppose the probability of disease among smokers had been the same as that in the general population. You would likely conclude that smokers were at no greater risk of having this disease than is the general population. In this circumstance smoking and disease would be independent.

The definition of independence given above is often presented in a slightly different form. If the right hand side of Equation 3.3 is substituted for the left hand side of Equation 3.4 and both sides multiplied by $P(B)$ the result is

$$P(AB) = P(A) P(B) \tag{3.5}$$

---

[3]This does not prove that smoking *causes* the disease, however.

In this form the definition of independence states that A and B are independent if the joint probability of A and B is equal to the product of their marginal probabilities. In reference to Table 3.2 we see that $.45 \neq (.60)(.55)$ so that we again see that smoking and disease are not independent.

EXAMPLE 3.2

Use the probability table given below to find the following, $P\left(\overline{B}\right)$, $P(A)$, $P\left(A\overline{B}\right)$, $P\left(\overline{A}\,\overline{B}\right)$, $P\left(\overline{A}\cup\overline{B}\right)$, $P\left(\overline{A}\cup B\right)$, $P\left(\overline{B}\mid A\right)$, and $P\left(A\mid\overline{B}\right)$. Are A and B independent? What is the justification for your answer?

	B	$\overline{B}$	
A	.28	.12	.40
$\overline{A}$	.42	.18	.60
	.70	.30	

Solution Reading from the table margins we see that $P\left(\overline{B}\right) = .30$ and $P(A) = .40$. From the table cells we get $P\left(A\overline{B}\right) = .12$ and $P\left(\overline{A}\,\overline{B}\right) = .18$. $P\left(\overline{A}\cup\overline{B}\right) = .12+.42+.18 = .72$ while $P\left(\overline{A}\cup B\right) = .28+.42+.18 = .88$. Using Equation 3.3 for finding the conditional probabilities produces $P\left(\overline{B}\mid A\right) = \frac{.12}{.40} = .30$ and $P\left(A\mid\overline{B}\right) = \frac{.12}{.30} = .40$.

Because $P(A\mid B) = \frac{.28}{.70} = .40$ and $P(A) = .40$, Equation 3.4 implies that A and B are independent. The same conclusion is reached by applying Equation 3.5 and noting that $P(AB) = .28$ which is equal to $P(A)P(B)$. ∎

3.3.5   Sensitivity, Specificity, and Related Concepts

Tests designed to establish the presence or absence of some condition are rarely perfect. For example, we would want a medical test for the presence or absence of some particular disease to be positive for those persons with disease and negative for those persons who do not have the disease. Unfortunately, it is usually the case that on occasion a person with the disease will receive a negative result or a person who does not have the disease will test positive.

How well or poorly a test performs in this regard can be assessed through computation of its sensitivity, specificity, positive predictive value and negative predictive value. We will use the probability table given below to explain each of these concepts. In this table a "+" indicates a positive test result and a "−" indicates a negative result. As with previous tables D is used to indicate disease and $\overline{D}$ to represent the absence of disease. By this table you can see that .015 of the population have the disease and test positive for the disease while .970 of the population do not have the disease and test negative.

	D	$\overline{D}$	
+	.015	.010	.025
−	.005	.970	.975
	.020	.980	

Sensitivity is the probability that a person with the disease will test positive or

$$\boxed{\text{Sensitivity} = P\left(+ \mid D\right)} \qquad (3.6)$$

For the above table sensitivity is $\frac{.015}{.020} = .75$. This might not be too comforting since only 75% of persons with the disease are correctly identified.

Specificity is the probability that a person who does not have the disease will test negative or

$$\boxed{\text{Specificity} = P\left(- \mid \overline{D}\right)} \qquad (3.7)$$

For the above table specificity is $\frac{.97}{.98} = .99$. This means that if you do not have the disease the test is almost (but not quite) certain to come back negative.

Positive predictive value is the probability that a person who tests positive has the disease or

$$\boxed{PPV = P\left(D \mid +\right)} \qquad (3.8)$$

The positive predictive value for the table is $\frac{.015}{.025} = .60$. This means that if a person tests positive for disease the probability is only .60 that she/he has the disease.

Negative predictive value is the probability that a person who tests negative does not have the disease or

$$\boxed{NPV = P\left(\overline{D} \mid -\right)} \qquad (3.9)$$

The negative predictive value for the table is $\frac{.970}{.975} = .99$. This means that if a person tests negative for disease the probability is .99 that she/he does not have the disease.

Finally, **prevalence** is simply the probability of disease or

$$\boxed{\text{Prevalence} = P\left(D\right)} \qquad (3.10)$$

In the present case, this value is .02.

EXAMPLE 3.3

Use the table given below to find sensitivity, specificity, positive predictive value, negative predictive value, and prevalence.

	D	$\overline{D}$	
$+$	.008	.011	.019
$-$	.001	.980	.981
	.009	.991	

Solution Applying Equations 3.6 through 3.10 yields the results Sensitivity $= \frac{.008}{.009} = .89$, Specificity $= \frac{.980}{.991} = .99$, $PPV = \frac{.008}{.019} = .42$, $NPV = \frac{.980}{.981} = .999$, and Prevalence $= .009$. ∎

EXAMPLE 3.4

Suppose you have been informed that your medical test has come back positive. Which *one* of the above test characteristics would probably be of most interest to you?

Solution   You would most likely want to know the probability of having the disease for a person who tests positive. Therefore, you would want to know the positive predictive value.   ■

3.3.6   Risk and Odds Ratios

Not all descriptive statistics of interest were covered in Chapter 2. Two of the most important of these, based on their common use in health-related research, are the risk ratio and the odds ratio. These will be explained in this section.

   The Risk Ratio.   Research questions commonly revolve around the issue of whether persons exposed to some potential risk factor are more or less likely to develop disease than are persons who have not experienced the exposure. Exposures such as smoking or working with asbestos may be suspected of increasing the probability of disease while exposure to some vaccine may be postulated to reduce the probability of disease. A common method to compare the probabilities of disease for exposed and unexposed persons is to form these into a ratio which is termed the risk ratio. Formally, the **risk ratio** is expressed as

$$RR = \frac{P(D \mid E)}{P(D \mid \overline{E})} \tag{3.11}$$

where D is as previously defined and E and $\overline{E}$ represent exposure and non-exposure respectively.[4] It follows from Equation 3.11 that a risk ratio of 2 would mean that the probability of disease for exposed persons is twice that of non-exposed persons. Likewise, a value of .5 would mean that the probability of disease for the exposed group is only half that of the non-exposed group. When RR is less than 1.0 the exposure is said to be **protective**. Note also that a risk ratio of 1.0 would mean that the probability of disease in the two groups is the same.

EXAMPLE 3.5

Use the probability table given below to calculate the risk ratio.

	E	$\overline{E}$	
D	.15	.10	.25
$\overline{D}$	.05	.70	.75
	.20	.80	

---

[4]The convention for the risk ratio is to have RR represent the parameter and $\widehat{RR}$ represent the statistic. See Section 1.3 on page 3 for an explanation.

Solution Since $P(D \mid E) = \frac{.15}{.20} = .75$ and $P(D \mid \overline{E}) = \frac{.10}{.80} = .125$, $RR = \frac{.75}{.125} = 6$. This means that the probability of disease for persons experiencing exposure is six times that of persons who are not exposed. ∎

The Odds Ratio. In certain research settings which you will learn about later, the risk ratio does not provide a meaningful comparison of exposed and non-exposed groups. In such instances the odds ratio is often used for comparison purposes. In order to understand the odds ratio it is first necessary to understand odds.

The odds of an event occurring is the ratio of the probability that the event will occur to the probability that the event won't occur. So the odds of disease in a certain group would be $P(D)/P(\overline{D})$. An odds of 2.0 would mean that the probability of disease is twice the probability of not having the disease. If the odds are computed for two groups and formed into a ratio, the result is, naturally enough, an odds ratio. If as above, you are comparing an exposed group to an non-exposed group, the **odds ratio** would be

$$OR = \frac{\frac{P(D|E)}{P(\overline{D}|E)}}{\frac{P(D|\overline{E})}{P(\overline{D}|\overline{E})}}$$

which simplifies to

$$OR = \frac{P(D \mid E) \, P(\overline{D} \mid \overline{E})}{P(\overline{D} \mid E) \, P(D \mid \overline{E})} \tag{3.12}$$

where E and $\overline{E}$ represent exposure and non-exposure respectively.

The odds ratio is not as intuitive as is the risk ratio but has the advantage of being applicable in a wider range of study designs. As with RR, a value of one for OR implies that the two groups are at the same level of risk. When the prevalence of disease is relatively small, the odds ratio provides a good estimate of the risk ratio. We will revisit this topic in a later chapter.[5]

EXAMPLE 3.6

Find the odds ratio for the table given in Example 3.5 on page 59.

Solution Since $P(D \mid E) = .15/.20 = .75$, $P(\overline{D} \mid \overline{E}) = .70/.80 = .875$, $P(\overline{D} \mid E) = .05/.20 = .25$ and $P(D \mid \overline{E}) = .10/.80 = .125$, $OR = (.75)(.875)/[(.25)(.125)] = 21$. This means that the odds of disease in the exposed group is 21 times that of the unexposed group. ∎

3.3.7 Bayes Rule

Bayes rule plays an important role in the general scheme of statistics but will not do so in this book. We present the rule here primarily for the sake of completeness of coverage and

---

[5]The convention for the odds ratio is to have OR represent the parameter and $\widehat{OR}$ represent the statistic.

because you will likely encounter it in your future studies of statistics. In its simplist form Bayes rule allows you to use $P(A \mid B)$ to find $P(B \mid A)$. The rule is expressed as

$$P(B \mid A) = \frac{P(A \mid B) P(B)}{P(A \mid B) P(B) + P\left(A \mid \overline{B}\right) P\left(\overline{B}\right)} \tag{3.13}$$

Notice that

$$P(A \mid B) P(B) = \frac{P(AB) P(B)}{P(B)} = P(AB)$$

and that

$$P\left(A \mid \overline{B}\right) P\left(\overline{B}\right) = \frac{P\left(A\overline{B}\right) P\left(\overline{B}\right)}{P\left(\overline{B}\right)} = P\left(A\overline{B}\right).$$

Also

$$P(AB) + P\left(A\overline{B}\right) = P(A).$$

Substituting these into Equation 3.13 gives

$$P(B \mid A) = \frac{P(AB)}{P(A)}$$

which satisfies the definition given in 3.3.

EXAMPLE 3.7

Given sensitivity of .83, specificity of .89 and prevalence of .05, find the positive predictive value.

Solution   Using the definition of positive predictive value given in Equation 3.8 and Bayes rule gives

$$P(D \mid +) = \frac{P(+ \mid D) P(D)}{P(+ \mid D) P(D) + P\left(+ \mid \overline{D}\right) P\left(\overline{D}\right)}$$

which can be rewritten as

$$PPV = \frac{(\text{Sensitivity}) (\text{Prevalence})}{(\text{Sensitivity}) (\text{Prevalence}) + (1 - \text{Specificity}) (1 - \text{Prevalence})}. \tag{3.14}$$

The terms in the numerator and left-hand side of the denominator of this expression follow directly from the definitions of sensitivity and prevalence. In order to understand the right-hand term in the denominator notice that one minus specificity equals

$$1 - \frac{P\left(-\overline{D}\right)}{P\left(\overline{D}\right)} = \frac{P\left(+\overline{D}\right) + P\left(-\overline{D}\right)}{P\left(\overline{D}\right)} - \frac{P\left(-\overline{D}\right)}{P\left(\overline{D}\right)} = \frac{P\left(+\overline{D}\right)}{P\left(\overline{D}\right)} = P\left(+ \mid \overline{D}\right)$$

and that

$$P\left(\overline{D}\right) = 1 - P(D) = 1 - \text{Prevalence}.$$

Substituting appropriate values into Equation 3.14 gives

$$PPV = \frac{(.83) (.05)}{(.83) (.05) + (1 - .89) (1 - .05)} = .28. \qquad \blacksquare$$

As you can now see, while sensitivity and specificity for this test appear adequate, positive predictive value is not very high.

EXAMPLE 3.8

Use the information in Example 3.7 to find negative predictive value.

Solution   By Bayes rule negative predictive value can be expressed as

$$P\left(\overline{D} \mid -\right) = \frac{P\left(- \mid \overline{D}\right) P\left(\overline{D}\right)}{P\left(- \mid \overline{D}\right) P\left(\overline{D}\right) + P\left(- \mid D\right) P\left(D\right)}$$

which can then be written as

$$NPV = \frac{(\text{Specificity})\,(1 - \text{Prevalence})}{(\text{Specificity})\,(1 - \text{Prevalence}) + (1 - \text{Sensitivity})\,(\text{Prevalence})}. \tag{3.15}$$

Substituting into this equation gives

$$NPV = \frac{(.89)\,(1 - .05)}{(.89)\,(1 - .05) + (1 - .83)\,(.05)} = .99$$

∎

3.4   THE NORMAL CURVE

3.4.1   Sampling from the Population

In Section 3.3 we posed and answered questions related to the probability of selecting a single observation from a population, observing the dichotomous characteristics associated with that observation and finding that those characteristics satisfied some specified criterion. The solutions to those questions were obtained by noting the proportion of characteristics in the population that met the criterion of interest.

The problems addressed in this section are similar but differ in two important respects. First, the characteristic to be observed is continuous[6] rather than dichotomous and second, we will assume that the proportions of observations in the population needed to calculate the desired probabilities are not available, thereby necessitating use of a mathematical model to estimate solutions. Figure 3.3 depicts a population of this sort made up of systolic blood pressures. The notation "···" indicates that the remainder of the population is made up of an unknown number of similar types of values.

The model we will use to estimate probabilities associated with this population is the familiar normal curve. Before demonstrating how the curve is used for this purpose, we will briefly describe some of its characteristics.

3.4.2   Some Characteristics of the Normal Curve

The **normal curve** is a mathematical function that is commonly used as a model of reality when that reality cannot be addressed directly. The use of the normal curve as a model will be discussed and demonstrated in the following sections. Here, we will simply describe some of its more notable characteristics.

---

[6]See Section 2.2.5 on page 11 for a discussion of continuous and dichotomous variables.

Population

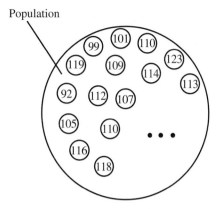

FIGURE 3.3: Population of an unknown number of systolic blood pressures.

The functional form of the normal curve is given by

$$f(x) = \frac{1}{\sigma\sqrt{2\pi}}e^{\frac{-(x-\mu)^2}{2\sigma^2}} \tag{3.16}$$

where e is a constant approximately equal to 2.718281828, μ and σ are constants that determine the mean and standard deviation of the distribution respectively and x is the variable whose function is to be determined.

The familiar "bell curve" is realized by setting μ and σ equal to desired constants and evaluating 3.16 for appropriate contiguous values of x. Plots of this type are shown in Figure 3.4 for selected values of μ and σ.

Notice that the upper panel of Figure 3.4 depicts two normal curves with identical values of σ but different values of μ. The effect of having different mean values is to locate the distributions at different points on the number line. For this reason, μ is sometimes referred to as a **location parameter**.

By contrast, the lower panel of Figure 3.4 shows two normal distributions with common values of μ but different values of σ. Because they share a common mean they are located at the same point on the number line but differ in degree of spread because one has larger σ than does the other. In this context, σ is referred to as a **scale parameter**.

Because normal curves can differ in location and scale, it is perhaps more appropriate to think of a *family* of normal curves rather than *the* normal curve. We will use this flexibility in a later section to obtain probability estimates. Other characteristics of normal curves are:

1. The mean, median, and mode are all located at the center of the distribution.

2. It is symmetric about its mean, median, and mode.

3. It is defined for all values of x between $-\infty$ and ∞. This means that depictions such as those in Figure 3.4 show only a segment of the curve since it stretches infinitely in either direction.

4. The area encompassed by the curve is equal to one regardless of the values of μ and σ.

3.4.3   Finding Areas Under the Normal Curve

Before using the normal curve to estimate probabilities, you must learn to find areas under the curve. In order to do this you must learn to use a normal curve table. You will learn both of these skills in this section.

To begin, suppose that, for some reason, you wish to discover the proportion of the total area of the normal curve that falls between some point on the x axis and the mean of the curve. This area is represented by the light gray shading in Figure 3.5.[7] By inspection you might guess that this area represents about 20 percent or .20 of the total area of the curve. On the other hand, you might wish to find the area represented by the darker shading. Again, you could estimate this area by visual inspection. Precise answers can be obtained by using the normal curve table in Appendix A. In this table, column one gives various points along the x axis, column two gives the areas between these points and the mean of the curve and column three shows the area in the tail as depicted by the darker shading. Note that the areas in columns two and three sum to .5 since together they represent half the curve. The following example will show how the table is used.

Given a normal curve with mean (μ) equal to 100 and standard deviation (σ) equal to 5, find the proportion of the curve that lies between 100 and 105. The area to be determined is represented by the gray shading in Figure 3.6. The normal curve table can be used to answer this question since it gives areas between points on the x axis, e.g., 105, and the mean of the curve which in this case is 100. We cannot look up the point 105 in column one of the table, however, because this is a scale dependent value. Obviously, the table cannot deal with units measured in inches, feet, pounds, IQ scores and all other values in which we may have an interest. Instead, column one expresses points along the x axis in terms of Z scores. You will recall from page 42[8] that Z scores indicate the number of standard deviations that a point on the x axis lies from the mean of the distribution.

Applying Equation 2.24 yields $Z = \frac{105-100}{5} = 1.00$. Thus, 105 has a Z score of 1.00 which means that 105 is one standard deviation above the mean of the distribution. Locating 1.00 in column one of Appendix A and reading the adjacent value in column two shows that the area between 105 and 100 constitutes .3413 of the total area of the curve. Note that .3413 of the normal curve lies between the mean and a point that is one standard deviation from the mean regardless of what values the mean and standard deviation take. Thus, if a normal curve is constructed with mean 10 and standard deviation 2, .3413 of the curve falls between 10 and 12. If the original problem had been to find the area *above* 105 we would have proceeded in the same fashion except that the answer, .1587, would have been found in column three.

EXAMPLE 3.9

Given a normal curve with mean 250 and standard deviation 25, what portion of the curve falls *below* 220?

Solution   It is usually helpful to sketch the problem as we have done in panel A of Figure 3.7. Notice that the Z score associated with 220 is $Z = \frac{220-250}{25} = -1.20$. This

---

[7]Because the total area under the curve is equal to one, you can think of the shaded portion as representing an area or as representing a proportion of the total area.

[8]You may wish to review this section before continuing.

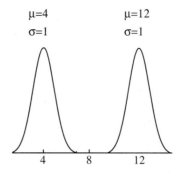

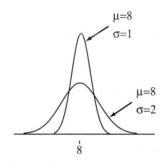

FIGURE 3.4: Normal distributions with selected values of μ and σ.

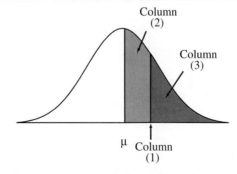

FIGURE 3.5: Areas of the normal curve given in Appendix A.

indicates that 220 is 1.2 standard deviations *below* the mean of the distribution. Though Appendix A does not provide negative Z scores, the symmetry of the normal curve can be used to find the area in question. That is, the areas associated with a Z score of -1.20 are the same as those associated with a Z score of 1.20.

Appendix A shows that the area in the tail of the curve associated with a Z score of 1.20 is .1151 which is the answer to the posed question. In order to familiarize you with the curve, we have labeled the other curve areas as well.   ∎

EXAMPLE 3.10

Given a normal curve with mean 80 and standard deviation 10, find the area between 65 and 85.

Solution   We have represented the problem in panel B of Figure 3.7. Notice that the shaded area cannot be read from the table directly because the table does not provide areas between any two arbitrary points but rather between a point and the mean or the tail area as previously described. Nevertheless, you can find the required area by finding each of the two component areas and summing.

The Z score for 65 is $Z = \frac{65-80}{10} = -1.50$. Column two shows the area between 65 and 80 to be .4332. Likewise, the Z score for 85 is $Z = \frac{85-80}{10} = .50$ which has an associated area of .1915. The area between 65 and 85 is then $.4332 + .1915 = .6247$.   ∎

EXAMPLE 3.11

Given a normal curve with mean 500 and standard deviation 50, find the area between 555 and 600.

Solution   As with the previous problem, the solution cannot be read directly from the table. Notice, however, that the Z score for 600 is $Z = \frac{600-500}{50} = 2.00$. Column two shows that the area between 600 and 500 is then .4772. But this is not the answer we seek because the area between 500 and 555 is included and is not part of the area to be found. This can be dealt with by finding the Z score for 555 which is 1.10 and noting that the (unwanted) area between 500 and 555 is .3643. The area between 600 and 555 is then $.4772 - .3643 = .1129$. The problem is outlined in panel C of Figure 3.7.   ∎

EXAMPLE 3.12

Given a normal curve with mean .05 and standard deviation .01, find the area below .0722.

Solution   The area to be located is depicted in panel D of Figure 3.7. As you can see from this sketch, the area consists of two component parts. The first is the area below .05. Since .05 is the mean of the distribution it has Z score 0.00 which has an associated tail area of .5. The second area lies between .0722 and .0500. The Z score for .0722 is $Z = \frac{.0722-.0500}{.01} = 2.22$. Column two shows that the area between .0722 and .0500 is .4868. The area below .0772 is then $.5000 + .4868 = .9868$.   ∎

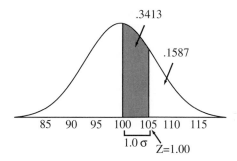

FIGURE 3.6: Finding an area under the normal curve.

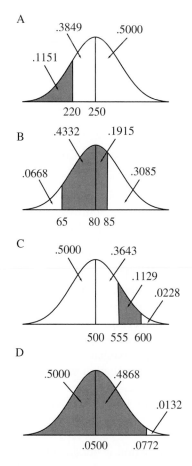

FIGURE 3.7: Finding areas under the normal curve.

TABLE 3.3: Population relative frequency distribution.

Blood Pressure	Relative Frequency	Blood Pressure	Relative Frequency
95	.004	110	.078
96	.000	111	.078
97	.003	112	.070
98	.004	113	.067
99	.008	114	.059
100	.009	115	.049
101	.013	116	.035
102	.025	117	.039
103	.031	118	.024
104	.033	119	.014
105	.050	120	.011
106	.057	121	.002
107	.066	122	.004
108	.074	123	.005
109	.085	124	.002
		125	.001

3.4.4 Using the Normal Curve to Approximate Probabilities

What is the probability of choosing an observation from the population represented by Figure 3.3 on page 63 and finding that the systolic blood pressure thus obtained is 111? In order to answer this question you would have to know the proportion of blood pressures in the population that are 111. Table 3.3 shows the relative frequency distribution of the blood pressures in this population. As this table shows, the proportion, and therefore the probability, associated with 111 is .078. What is the probability that the observation will be between 112 and 114 (i.e., 112, 113, or 114)? Again, using the information from Table 3.3 you can calculate the probability as .070 + .067 + .059 = .196.

Now suppose the proportions used in the above calculations are not available. In the circumstance that (1) the mean and standard deviation of the population are known[9] and (2) the population relative frequency distribution is *relatively* normal in shape, the normal curve can be used to approximate the desired probabilities. That is, the normal curve can be used as a model of the population relative frequency distribution.

In the present case the mean and standard deviation of the population are 110.023 and 4.970 respectively. Figure 3.8 shows a normal curve with this mean and standard deviation imposed on the population relative frequency distribution. As you can see, requirement (2) above is met.

The approximation is obtained by calculating the area under the curve that corresponds to the event of interest. For example, to estimate the probability that the selected blood

---

[9]The question of how these quantities might be known will be dealt with in Chapter 4.

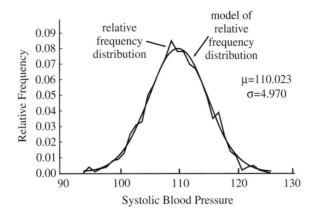

FIGURE 3.8: Comparison of a normal curve to a population relative frequency distribution.

pressure is 111 you calculate the area under a normal curve with mean 110.023 and standard deviation 4.970 that corresponds to 111. The area corresponding to 111 is the area between the lower real limit of 110.5 and the upper real limit of 111.5.[10] The Z score for 111.5 is (approximately) $Z = \frac{111.5 - 110.023}{4.970} = .30$ which has a corresponding area of .1179. The Z score for 110.5 is (approximately) $Z = \frac{110.5 - 110.023}{4.970} = .10$ with an associated area of .0398. The area between 111.5 and 110.5 is then $.1179 - .0398 = .0781$ which is quite close to the exact value of .078. The problem is sketched in Panel A of Figure 3.9.

An estimate of the probability that the randomly selected observation is between 100 and 105 (inclusive) is obtained by finding the area between 99.5 and 105.5. The Z score for 99.5 is (approximately) $Z = \frac{99.5 - 110.023}{4.970} = -2.12$ which has a corresponding area of .4830. The Z score and associated area for 105.5 are respectively $-.91$ and .3186. The estimated probability is then $.4830 - .3186 = .1644$ which compares favorably to the value of $.009 + .013 + .025 + .031 + .033 + .050 = .161$ obtained from Table 3.3. (See Panel B of Figure 3.9.)

EXAMPLE 3.13

Estimate the probability that the random observation discussed above will be greater than 103. Compare this estimate to the exact value computed from Table 3.3.

Solution   The estimate is obtained by finding the area under the previously used normal curve that lies above 103.5. (See Panel C of Figure 3.9.) In order to find this area, note that the Z score for 103.5 is (approximately) $Z = \frac{103.5 - 110.023}{4.970} = -1.31$ which has an associated area of .4049. Since this is the area between 103.5 and 110.023 and the area above 110.023 is .5000, the estimate is given by $.4049 + .5000 = .9049$. The exact result obtained from Table 3.3 is the sum of the probabilities associated with 104 and greater values which is .903. ∎

---

[10]You may wish to review Section 2.5.2 on page 20 if you've forgotten the rationale underlying the concepts of upper and lower real limits.

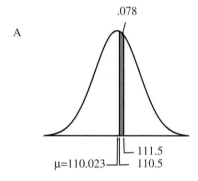

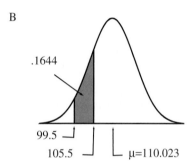

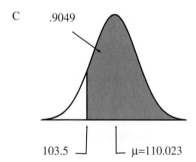

FIGURE 3.9: Using the normal curve to approximate probabilities.

You should not conclude that the normal curve always provides approximations as good as those presented here. If the population is skewed or differs substantially from the normal shape in some other fashion, the estimates obtained may not be very close to exact values. However, in Chapter 4 you will learn that the normal curve can still function to provide good probability estimates when populations are not normally distributed so long as the estimates involve certain statistics and not individual scores as was the case here.

KEY WORDS AND PHRASES

After reading this chapter you should be able to demonstrate familiarity with the following words and phrases.

Bayes rule 60
conditional probability 53
contingency table 53
frequency table 53
independence 56
joint probability 56
location parameter 63
marginal probability 56
negative predictive value 58
normal curve 62
normal curve area 64
odds ratio 60
positive predictive value 58
prevalence 58
probability 52
probability table 55
protective effect 59
risk ratio 59
scale parameter 63
sensitivity 58
specificity 58

EXERCISES

3.1 The probability table given here represents the results of a census conducted of all students in a large university. Each student was categorized by gender (M/F) and by whether or not they answered affirmatively to a question as to whether they had been inebriated in the past 30 days ($I/\bar{I}$).

	I	$\bar{I}$	
M	.22	.32	.54
F	.10	.36	.46
	.32	.68	

Use the probability notation of this chapter to characterize each of the following. Then use the table entries to find the indicated probability.

(a) The probability that a student, randomly selected from among the females, will not have been inebriated in the past 30 days.

(b) The probability that a randomly selected student will be female.

(c) The probability that a randomly selected student will be male or will have been inebriated in the past 30 days.

(d) The probability that a randomly selected student will be female and will have been inebriated in the past 30 days.

(e) The probability that a student, randomly selected from among those reporting having been inebriated in the past 30 days, will be female.

(f) The probability that a randomly selected student will be male or female.

3.2 Are gender and inebriation status as represented in the above table independent? What is the evidence for your conclusion?

3.3 Suppose that half the residents in a community are female and that 20% of the residents in that community support a tax increase aimed at providing funds for free inoculations against childhood diseases. If ten percent of the community residents are both female and supporters of the tax increase, can it be said that gender and support for the tax increase are independent events? What is the evidence to support your answer?

3.4 Using the information in Exercise 3.3, find $P\left(\overline{T} \mid M\right)$ where T indicates support for the tax increase and M represents a male resident.

3.5 Suppose that for events A, $\overline{A}$, B, $\overline{B}$, $P\left(B \mid A\right) = .22$, $P\left(\overline{A}\right) = .20$ and $P\left(B\right) = .72$. Use this information to construct a probability table showing all marginal and joint probabilities.

3.6 What do each of the following represent?

   (a) The proportion of persons who test positive for some disease who actually have the disease.

   (b) The proportion of persons with some disease that test positive for the disease.

   (c) The proportion of persons who do not have some disease who test negative for the disease.

   (d) The proportion of persons who test negative for some disease who do not have the disease.

3.7 Use the provided table to find each of the following.

	D	$\overline{D}$	
$+$	.065	.010	.075
$-$	.005	.920	.925
	.070	.930	

   (a) sensitivity

   (b) specificity

   (c) positive predictive value

   (d) negative predictive value

3.8 Use the provided table to calculate and interpret the risk and odds ratios.

	E	$\overline{E}$	
D	.18	.07	.25
$\overline{D}$	.02	.73	.75
	.20	.80	

3.9 Suppose that in a given community where 40% of the population is under age 40, it is found that the proportion of residents under age 40 who support mandatory inoculation of school age children against certain diseases is .72. The proportion of residents over age 40 who support the proposition is .52. Use this information to find the proportion of inoculation supporters who are under age 40.

3.10 Given sensitivity of .82, specificity of .93 and prevalence of .20, find positive predictive value.

3.11 Given a normally distributed variable with mean 80 and standard deviation 12, find the following curve areas.

   (a) The area between 80 and 98.

   (b) The area below 74.

   (c) The area below 82.

   (d) The area between 72 and 94.

   (e) The area between 56 and 60.

   (f) The area above 104.

   (g) The area below 54.

   (h) The area between 82 and 94.

3.12 Given an integer valued variable that is approximately normally distributed with mean 100 and standard deviation 8, use the normal curve to approximate the probability of randomly selecting a single observation from the population and finding that it meets each of the following criteria.

   (a) Is equal to 102.

   (b) Has a value between 92 and 108.

   (c) Has a value greater than 110.

   (d) Has a value below 105.

A. The following questions refer to Case Study A (page 469)

3.13 What is the probability of randomly selecting a participant in this study and finding that the selected participant is male?

3.14 Form the am (morning) data into a frequency and a probability contingency table.

3.15 Letting T represent "chose the treated lens," U "chose the untreated lens," Y "answered yes to the duration question," and N "answered no to the duration question," use the probability table formed in Exercise 3.14 to calculate each of the following.

(a) $P(T)$

(b) $P(Y)$

(c) $P(T \mid Y)$

(d) $P(T \mid N)$

(e) Express each of the above probabilities in the form of an English language statement.

(f) Are T and Y independent? Provide evidence for your answer.

Continue to use the data from the table formed in Exercise 3.14 for Exercises 3.16 through 3.20.

3.16 What is the risk of choosing the untreated lens for persons who answered yes to the duration question? For persons who answered no to the duration question?

3.17 For the a.m. group, form a risk ratio comparing the risk of choosing the untreated lens for subjects in the no group to the risk for subjects in the yes group. What implications does this ratio have for the study?

3.18 For the a.m. group, form a risk ratio comparing the risk of choosing the treated lens for subjects in the no group to the risk for subjects in the yes group. Are implications for this ratio the same as those for the ratio calculated in Exercise 3.17?

3.19 For the a.m. group, what are the odds of choosing the untreated lens for persons who answered yes to the duration question? For persons who answered no to the duration question?

3.20 For the a.m. group, form an odds ratio comparing the odds of choosing the untreated lens for subjects in the no group to the odds for subjects in the yes group. Why do you think this ratio might be of interest to the researchers who conducted this study?

E. The following questions refer to Case Study E (page 472)

3.21 Find the sensitivity, specificity, and positive and negative predictive values for the new test.

O. The following questions refer to Case Study O (page 477)

3.22 Comment on point (d).

P. The following questions refer to Case Study P (page 478)

3.23 You have learned how to characterize probabilities related to two dichotomous variables cast into a two by two contingency table. The following items are designed to further test your level of understanding of such probabilities by having you apply your knowledge to a situation that deals with *three* variables, one of which is not dichotomous.

Refer to Table J.10 for the following items and let M=male, F=female, I=type I diabetes, II=type II diabetes and (a-b)=age range a to b.

(a) Write a sentence to express $P(F\,I\,(25-44))$.

(b) Find $P(F\,I\,(25-44))$.

(c) Find $P(F \cup I \cup (25-44))$.

(d) Write a sentence to express $P(M \mid (15-24) \cup II)$.

(e) Find $P(M \mid (15-24) \cup II)$.

Introduction to Inference and One Sample Methods

4.1 INTRODUCTION

In Chapter 1 we briefly introduced the concept of inference and indicated its fundamental role in statistics. In this chapter you will learn to apply certain inferential methods as well as the rationale underlying their use. The methods introduced here are not commonly employed in research but are important pedagogically in that they are relatively simple and their mastery will open the way for understanding the more complex methods dealt with in following chapters.

The techniques you will learn may be divided into two broad categories, (1) tests of hypotheses and (2) confidence intervals. Before you can begin your study, however, you must understand the concept of sampling distributions which underlies both classes of inferential methods.

We must caution you that this is the most important chapter in this book. The material in this chapter is essential for all that follows. Time spent here will pay great dividends later. Now that we have your attention, we will take up the topic of sampling distributions.

4.2 SAMPLING DISTRIBUTIONS

4.2.1 Definition

Though somewhat abstract, the concept of sampling distributions is relatively simple. A **sampling distribution** is a distribution of sample statistics obtained from samples repeatedly drawn from one or more populations. For example, suppose you were to randomly choose a sample made up of five observations from the population represented by Figure 3.3. You then compute and record the sample mean. Now suppose you were to repeat this process many (actually, an infinite number of) times. If you were to then form these sample means into a relative frequency distribution you would thus have generated the sampling distribution of the mean or $\bar{x}$. You could, of course, use this same method to generate the sampling distribution of other statistics such as the sample standard deviation, median, fifth percentile or any of many other statistics.

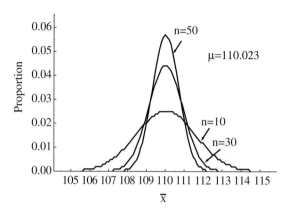

FIGURE 4.1: Sampling distributions of $\bar{x}$.

4.2.2 The Sampling Distribution of $\bar{x}$

Figure 4.1 shows three sampling distributions of $\bar{x}$ derived from the population whose relative frequency distribution is shown in Figure 3.8 on page 69. The method by which we generated these distributions will aid in your understanding of the general concept of sampling distributions.

In order to construct the sampling distributions shown in Figure 4.1 we created a computer program that (1) generated the aforementioned population, (2) randomly selected 10,000,000[1] samples of size ten from this population, computed the mean of each sample, (3) calculated the mean, variance, and standard deviation of the 10,000,000 means and finally, (4) generated the relative frequency distribution of the means. The process was then repeated using samples of size 30 and 50.

Several factors related to these sampling distributions should be noted. First, the mean of all three distributions is 110.023 which is the mean of the population as reported in section 3.4.4. The mean of a sampling distribution is termed the **expected value** of the statistic and is symbolized by $E[\ ]$ where $[\ ]$ contains an identifier of the statistic. In the present case the expected value of $\bar{x}$ is represented by $E[\bar{x}]$. In general $E[\bar{x}] = \mu$. You should not conclude from this that the expected value of all statistics is equal to its parameter counterpart. For example, $E[s] \neq \sigma$.

Second, you will notice that the standard deviation of the three sampling distributions decreases as the sample size increases. The standard deviation of sampling distributions is usually termed the **standard error** of the particular statistic. Thus, the standard deviation of the sampling distribution of $\bar{x}$ is termed the **standard error of the mean** which is symbolized $\sigma_{\bar{x}}$. The standard error of the mean can be expressed as

$$\sigma_{\bar{x}} = \frac{\sigma}{\sqrt{n}} \qquad (4.1)$$

[1]This was a bit of overkill but we wanted to emphasize the fact that sampling distributions are based on a large number of samples.

where σ is the population standard deviation and n is the sample size. For the case where n equals 10 in Figure 4.1, the standard error of the mean is $\frac{4.970}{\sqrt{10}} = 1.572$. You can see from Equation 4.1 that for a fixed population standard deviation the standard error of the mean decreases as sample size increases which is what is seen in the figure.

Since variance is the square of standard deviation, it follows that the variance of the sampling distribution of $\bar{x}$ is

$$\sigma_{\bar{x}}^2 = \frac{\sigma^2}{n} \tag{4.2}$$

Finally, you should observe that all three sampling distributions appear to have approximately normal shapes so that they would be well modeled with normal curves. While these models would have common means they would necessarily have different standard deviations. It is perhaps not surprising that the sampling distributions of $\bar{x}$ have approximately normal shapes given that the population from which they were derived is also approximately normal, but the question arises as to the shape of the sampling distribution when the parent population is not approximately normal.

The top panel of Figure 4.2 depicts a decidedly nonnormal population distribution while the bottom panel shows three sampling distributions of means taken from the same population. As may be seen, the sampling distributions are approximately normal in spite of the nonnormal shape of the parent population. This phenomenon is attributable to the **central limit theorem**.

Roughly speaking, the **central limit theorem** states that the sampling distributions of certain classes of statistics will approach normality as sample size (n) increases regardless of the shape of the sampled population. This means that for very small sample sizes the sampling distributions of these statistics may not be very normal but as sample size increases the shape of the sampling distributions will become more like the normal curve. This is an extremely important result because it implies that many sampling distributions may be modeled by the normal curve even when the parent population is not at all normal. The question arises as to how large n must be in order for the normal curve to be an apt model for a particular sampling distribution. That depends on several factors including the shape of the sampled population and the nature of the statistic whose sampling distribution is in question. As Figure 4.2 suggests, the central limit theorem usually produces approximately normal distributions for $\bar{x}$ even when sample sizes are modest.

EXAMPLE 4.1

Find the standard deviations of the three sampling distributions depicted in Figure 4.2.

Solution   Using Equation 4.1 and information from Figure 4.2, the standard errors of the mean for $n = 10$, 30, and 50 are respectively, $\frac{5.293}{\sqrt{10}} = 1.67$, $\frac{5.293}{\sqrt{30}} = .97$, and $\frac{5.293}{\sqrt{50}} = .75$.   ■

4.2.3   Using the Normal Curve to Approximate Probabilities Associated with $\bar{x}$

In Section 3.4.4 we posed the question "What is the probability of choosing an observation from the population represented by Figure 3.3 and finding that the systolic blood pressure thus obtained is 111?" You learned that this probability could be approximated by finding

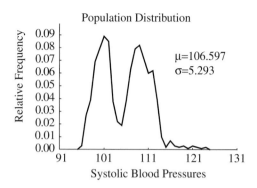

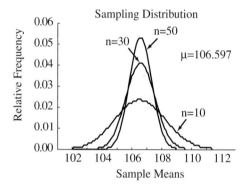

FIGURE 4.2: Nonnormal population with sampling distributions of $\bar{x}$.

the appropriate area under a normal curve model with such model having the same mean and standard deviation as the population.

In this section you will learn to use the normal curve model to answer questions such as "What is the probability of choosing a *sample* of size 15 from the population represented by Figure 3.3 and finding that the *sample mean* is greater than 111?" Note that if the sampling distribution of $\bar{x}$ were available for this population the exact value could be easily obtained. This follows from the fact that sampling distributions are relative frequency distributions which in turn represent probabilities.

Since sampling distributions are rarely if ever available, you can use the same technique you employed to find probabilities associated with single observations. That is, you can find the appropriate area under a normal curve. In this case the approximate probability would be the area above 111. Thus, you would simply find the Z score for 111 and locate the associated area in column three of the normal curve table. Note, however, that the Z score would take the form

$$Z = \frac{\bar{x} - \mu}{\frac{\sigma}{\sqrt{n}}}$$

(4.3)

rather than the form given in Equation 2.24. Because the denominator of this expression is the standard deviation of the sampling distribution, Z is the number of standard deviations between $\bar{x}$ and μ.

For the problem at hand,[2]

$$Z = \frac{111.0 - 110.023}{\frac{4.970}{\sqrt{15}}} = .76.$$

The associated area is .2236. So the probability of selecting a sample from this population and finding that the sample mean is greater than 111 is approximately .22. (We compute the actual probability to be .2157.) What do you think would happen to this probability if the sample size were increased—e.g., to 25?

EXAMPLE 4.2

What is the probability of randomly selecting 25 observations from the population depicted in Figure 4.2 on page 78 and finding that their mean value is less than 107.5?

Solution   The probability can be estimated by finding the appropriate area under a normal curve with mean 106.597 and standard deviation $5.293/\sqrt{25}$. The desired area would be that portion of the curve below 107.5. The Z score and associated area for the portion between 107.5 and 106.597 are respectively

$$Z = \frac{107.5 - 106.597}{\frac{5.293}{\sqrt{25}}} = .85$$

and .3023. The area below 106.597 is .50 so that the desired estimate is $.5000 + .3023 = .8023$. (We compute the actual probability to be .8034.)   ■

EXAMPLE 4.3

Suppose 100 observations are randomly selected from a population whose mean and standard deviation are respectively 100 and 20. What is the probability that the mean of these observations will be between 99 and 103?

Solution   The area of a normal curve with mean 100 and standard deviation $20/\sqrt{100}$ that lies between 99 and 103 is the sum of the areas between 99 and 100 and 100 and 103. The Z score and area between 99 and 100 are respectively,

$$Z = \frac{99.0 - 100.0}{\frac{20.0}{\sqrt{100}}} = -.50$$

and .1915. The same values for the area between 103 and 100 are

$$Z = \frac{103.0 - 100.0}{\frac{20.0}{\sqrt{100}}} = 1.50$$

and .4332. The probability estimate is then $.1915 + .4332 = .6247$.   ■

---

[2]Note that we do not employ upper and lower real limits because unlike individual scores, means are not restricted to discrete values.

4.2.4   The Sampling Distribution of $\hat{p}$

Consider now a population made up of some dichotomous characteristic such as lived—died, tumor remission—no tumor remission, pain—no pain, and so on. Traditionally, when speaking in a general sense, one of the two dichotomous outcomes is termed "success" and the other "failure." Now suppose that you randomly choose five observations from this population and record the proportion of successes in the sample. We will designate the proportion of successes in a sample as $\hat{p}$ and the proportion in the population as π. If you were to repeat this procedure many times and form the resulting proportions into a relative frequency distribution you would thus have generated the sampling distribution of $\hat{p}$.

If the members of the population with the characteristic "success" are designated with the number one and those with a "failure" characteristic with zero, then each sample will be made up of ones and zeros and $\hat{p}$ will simply be the mean of these values. For example, if a sample is made up of two 1s and three 0s, both the mean and proportion would be $2/5 = .4$. Likewise, π will be the mean of the 1s and 0s that make up the population. Because in this context $\hat{p} = \bar{x}$ and $\pi = \mu$, it is not surprising that $E[\hat{p}] = \pi$.

The population variance can be expressed as $\pi(1 - \pi)$. This can be seen by noting the consequence of using the computational formula for population variance (Equation 2.11 on page 36) with data whose values are restricted to 1s and 0s.

$$\frac{\sum x^2 - \frac{(\sum x)^2}{N}}{N} = \frac{N\pi - \frac{N^2\pi^2}{N}}{N}$$

$$= \frac{N\pi - N\pi^2}{N}$$

$$= \pi - \pi^2$$

$$= \pi(1 - \pi).$$

Squaring 1s and 0s does not change their values so that $\sum x^2 = \sum x$. Since $\pi = \frac{\sum x}{N}$ it follows that $\sum x^2 = N\pi$. The rest of the result follows from simple algebra.

By analogy with Equation 4.1 the **standard error of $\hat{p}$** is given by

$$\boxed{\sigma_{\hat{p}} = \sqrt{\frac{\pi(1 - \pi)}{n}}} \tag{4.4}$$

EXAMPLE 4.4

Panel A of Figure 4.3 shows the relative frequency distribution of a dichotomous population with $\pi = .10$ while panels B and C show sampling distributions of $\hat{p}$ derived from this population with sample sizes being 5 and 50 respectively. What is the standard error of $\hat{p}$ for the two sampling distributions?

Solution   By Equation 4.4 the standard error for the distribution in Panel B is

$$\sigma_{\hat{p}} = \sqrt{\frac{\pi(1 - \pi)}{n}} = \sqrt{\frac{(.10)(.90)}{5}} = .134$$

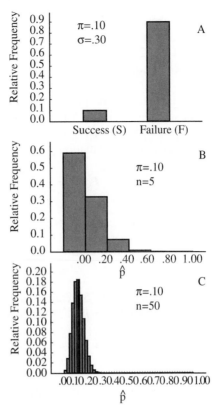

FIGURE 4.3: Dichotomous population with sampling distributions of $\hat{p}$.

and that for the distribution in Panel C

$$\sigma_{\hat{p}} = \sqrt{\frac{(.10)\,(.90)}{50}} = .042.$$ ∎

4.2.5   Using the Binomial Distribution to Approximate Probabilities Associated with $\hat{p}$

If the population is large and certain other conditions (to be discussed later) are met, the binomial distribution can be used to model the sampling distribution of $\hat{p}$. The **binomial distribution** is generated by the equation

$$P\,(y) = \frac{n!}{y!\,(n-y)!}\pi^{y}\,(1-\pi)^{n-y} \tag{4.5}$$

where n is the sample size, y is the number of successes, π is the proportion of successes in the population and $P(y)$ is the probability of y successes in a sample of size n taken from a population where the proportion of successes is π. The notation $n!$ is read "n factorial" and is calculated as $n(n-1)(n-2)\cdots(n-n+1)$. By this, $5!$ would be calculated as $5\cdot4\cdot3\cdot2\cdot1=120$ and $3!$ would be $3\cdot2\cdot1=6$. By definition $0!=1$. An example will clarify use of this equation.

Suppose we wish to find the sampling distribution of $\hat{p}$ for samples of size 5 drawn from a population in which the proportion of successes is .10. Notice that only six values of $\hat{p}$ are possible for samples of size 5. That is, there can be 0, 1, 2, 3, 4, or 5 successes which result in values of $\hat{p}$ of $0/5=.00$, $1/5=.20$, $2/5=.40$, $3/5=.60$, $4/5=.80$, and $5/5=1.00$. For $\hat{p}=.00$ the binomial equation yields

$$P(0) = \frac{5!}{0!(5-0)!}.10^0(1-.10)^{5-0}$$
$$= \frac{\cancel{5!}}{0!\cancel{5!}}.10^0.90^5$$
$$= .90^5$$
$$= .59049.$$

Notice that in line two the 5!s cancel, and that both $0!$ and $.10^0$ equal one so that the result in line three is obtained. For $\hat{p}=.20$ the equation yields

$$P(1) = \frac{5!}{1!(5-1)!}.10^1(1-.10)^{5-1}$$
$$= \frac{5\cdot\cancel{4!}}{1!\cancel{4!}}.10^1.90^4$$
$$= (5)(.10)(.6561)$$
$$= .32805.$$

Considerable effort can be saved by noticing that $5!=5\cdot4!$ so that $4!$ can be factored from the numerator and denominator of the expression. In the calculation for $\hat{p}=.40$ we save computational effort by noting that $5!=5\cdot4\cdot3!$ which allows for the cancelation of $3!$ from the numerator and denominator of the expression.

$$P(2) = \frac{5!}{2!(5-2)!}.10^2(1-.10)^{5-2}$$
$$= \frac{5\cdot4\cdot\cancel{3!}}{2!\cancel{3!}}.10^2.90^3$$
$$= (10)(.01)(.729)$$
$$= .0729$$

For $\hat{p}=.60$ and $.80$

$$P(3) = \frac{5!}{3!(5-3)!}.10^3(1-.10)^{5-3}$$
$$= \frac{5\cdot4\cdot3\cdot\cancel{2!}}{3!\cancel{2!}}.10^3.90^2$$

$$= (10) (.001) (.81)$$
$$= .0081$$
$$P (4) = \frac{5!}{4! \, (5 - 4)!} .10^4 \, (1 - .10)^{5-4}$$
$$= \frac{5 \cdot \cancel{4!}}{\cancel{4!} \, 1!} .10^4 .90^1$$
$$= (5) (.0001) (.90)$$
$$= .00045.$$

And finally for $\hat{p} = 1.0$

$$P (5) = \frac{5!}{5! \, (5 - 5)!} .10^5 \, (1 - .10)^{5-5}$$
$$= \frac{\cancel{5!}}{\cancel{5!} \, 0!} .10^5 .90^0$$
$$= .10^5$$
$$= .00001.$$

The above calculations are summarized in Table 4.1 and are graphically depicted in Panel B of Figure 4.3

TABLE 4.1: Sampling distributions of $\hat{p}$ for $n = 5$ and $\pi = .10$.

Proportion $\hat{p}$	Number of Successes y	Probability $P (y)$
.00	0	.59049
.20	1	.32805
.40	2	.07290
.60	3	.00810
.80	4	.00045
1.00	5	.00001

EXAMPLE 4.5

Given that 10% of the residents of the United States would test positive for a certain antibody, what is the probability of randomly selecting five residents of the United States and finding that all five test positive for the antibody? What is the probability that *at least* four (i.e., four or more) will test positive? What is the probability that at least one will be positive?

Solution   From Table 4.1, the probability that all five residents test positive is $P (5) = .00001$. The probability that at least four test positive is $P (4) + P (5) = .00045 + .00001 =$

.00046 and the probability that at least one tests positive is $P(1) + P(2) + P(3) + P(4) + P(5) = 1 - P(0) = 1 - .59049 = .40951.$ ■

EXAMPLE 4.6

Thirty-eight percent of all blood donors in the United States have type O positive blood. Suppose that, as the result of a theoretical genetics based argument, a researcher believes that this proportion is higher in the country of Iceland. In a preliminary study, the researcher randomly selects 10 blood donors in Iceland and notes their blood type. What is the probability that the number of Islandic donors in the sample with blood type O positive will be 9 or 10 if the researcher's theory is wrong (i.e., the proportion in Iceland *is* .38)? If the number of subjects with type O positive blood is in fact 9 or 10, what implications would this have for the researcher's theory?

Solution   Given a population proportion of .38, the probability that the sample will contain 9 or 10 donors with type O positive blood is $P(9) + P(10)$.

$$P(9) = \frac{10!}{9!\,(10-9)!} .38^9 \,(1 - .38)^{10-9}$$

$$= \frac{10 \cdot 9!}{9!\,1!} .38^9 .62^1$$

$$= (10)\,(.00017)\,(.62)$$

$$= .00105$$

$$P(10) = .38^{10} = .00006.$$

The probability that 9 or 10 donors in the sample will have type O positive blood is then $0.00105 + 0.00006 = 0.00111$. If the number of donors in the sample with type O positive blood is 9 or 10 then the researcher's theory is supported because the probability of achieving such a result from a population where the proportion is .38 is so small. It is likely, though not proven, that the proportion of type O positives in the Islandic blood donor population is greater than .38.   ■

4.2.6   Using the Normal Curve to Approximate Probabilities Associated with $\hat{p}$

The sampling distributions depicted in Panels B and C of Figure 4.3 on page 81 were generated from a common population (Panel A) but differ in that sample size for the distribution in Panel B is five while that for Panel C is 50. As you can see, the different sample sizes result in very differently shaped sampling distributions. Of particular importance is the fact that the distribution in Panel B is decidedly nonnormal while that in C appears approximately normal in shape. This is due to the central limit theorem which, as you will recall, guarantees that sampling distributions of certain statistics will approach normality as sample size increases.

As the two panels suggest, using a normal curve approximation for the distribution based on samples of size five would probably not produce satisfactory results but the same approach for the distribution based on samples of size 50 appears promising. The central limit theorem guarantees that the normal curve model will be appropriate for distributions based on *some* sample size, but what size might that be? There is no sure answer but an often used rule of thumb states that the normal curve model will be satisfactory so long as both $n\pi$ and $n(1-\pi)$ are greater than or equal to five.[3] For the distribution in Panel B this calculation is $(5)(.1) = .5$ and $(5)(.9) = 4.5$ so that the criterion is not met but for Panel C $(50)(.1) = 5$ and $(50)(.9) = 45$ so that the criterion is minimally met.

The normal curve model is applied to the distribution of $\hat{p}$ in the manner with which you are now familiar. That is, a Z score is calculated with the appropriate area then being located in the normal curve table. Substitution of results provided above into Equation 4.3 yields

$$Z = \frac{\hat{p} - \pi}{\sqrt{\frac{\pi(1-\pi)}{n}}} \tag{4.6}$$

As an example, suppose that a random sample of 50 observations is taken from the population shown in Panel A of Figure 4.3. What is the probability that the proportion of successes in this sample will be greater than .12?

The estimated probability will be the area under a normal curve with mean .10 and standard deviation

$$\sigma_{\hat{p}} = \sqrt{\frac{\pi(1-\pi)}{n}} = \sqrt{\frac{(.10)(.90)}{50}}$$

that lies above .13. Because the proportion of successes can only take values .00, .02, .04, ..., .12, .14, ..., 1.00, the upper real limit of the .12 interval (i.e., .13) is used rather than .12. The upper limit is employed because the problem is to find the probability that the proportion of successes is *greater* than .12. The lower limit would have been used if the problem required the probability of obtaining a proportion of *.12 or greater*. Upper and lower real limits of binomial proportions can be computed directly by adding and subtracting $.5/n$. For the present case the upper real limit is $.12 + .5/50 = .13$. Using upper and lower limits in this fashion when using a continuous distribution to approximate probabilities associated with a discrete variable is referred to as a **continuity correction**. The Z score is then

$$Z = \frac{\hat{p} - \pi}{\sqrt{\frac{\pi(1-\pi)}{n}}} = \frac{.13 - .10}{\sqrt{\frac{(.10)(.90)}{50}}} = \frac{.03}{.0424} = .71.$$

Reference to the normal curve table in Appendix A gives an associated area of .2389. The value as calculated by Equation 4.5 is $P(7) + P(8) + \cdots + P(50) = .2298$.

What is the estimated probability that the proportion will be .12? The estimate will be the area between the lower real limit of .11 and the upper real limit of .13. As calculated above, the Z score for .13 is .71 while that for .11 is

$$Z = \frac{.11 - .10}{\sqrt{\frac{(.10)(.90)}{50}}} = \frac{.01}{.0424} = .24.$$

Using these values in the normal curve table shows that the areas between .13 and .10 and .11 and .10 are .2611 and .0948 respectively. The area between .11 and .13 is then $.2611 - .0948 = .1663$. The probability as calculated by Equation 4.5 is $P(6) = .1541$.

[3]Some authors maintain that these values should be greater than or equal to 10.

EXAMPLE 4.7

Approximately 16 percent of men in the United States aged 60 to 64 who exhibit a particular risk profile will have a heart attack in the next 10 years [14]. If a random sample of 300 such men are observed over the next 10 years, what is the probability that less than 5% will experience a heart attack?

Solution Because the problem specifies that *less* than 5% will experience a heart attack, the lower real limit of the five percent interval will be used. This limit is $.05 - .5/300 = .048$. The Z score is then

$$Z = \frac{.048 - .16}{\sqrt{\frac{(.16)(.84)}{300}}} = \frac{-.112}{.021} = -5.33.$$

The normal curve table does not contain Z values of this magnitude but it can be safely concluded that the probability is less than .0002. (This is the tail area associated with $Z = 3.50$ which is the most extreme score in the table.) Using a continuity correction when samples are large usually has only minor influence on the result. In this case if .05 had been used rather than .048 the Z value would have been -5.24 which would make little difference in the result. ∎

EXAMPLE 4.8

Suppose it is believed that a large community is evenly divided in its opinion as to whether a cap should be placed on the amount that can be recovered in medical malpractice law suits. If this supposition is correct, what is the probability that a random poll of 200 community members will produce 55 percent or more favorable responses? Compute the probability with and without continuity correction.

Solution The continuity correction is $.5/200 = .0025$. Because the task is to find the probability that 55 percent *or more* will be favorable, the lower real limit of the 55 percent category or $.55 - .0025 = .5475$ will be used. Because the community is assumed evenly divided, the proportion favorable in the population is taken to be .50. The Z score is then

$$Z = \frac{.5475 - .50}{\sqrt{\frac{(.50)(.50)}{200}}} = \frac{.0475}{.035} = 1.36.$$

The area above 1.36 is .0869. Without continuity correction the Z score is $.05/.035 = 1.43$ which has an upper tail area of .0764. The probability as computed by Equation 4.5 is .0895. ∎

4.3 HYPOTHESIS TESTING

4.3.1 Introduction

In Chapter 1 we indicated that the discipline of statistics/biostatistics can be conceptually divided into two related areas of study—namely, descriptive and inferential statistics. With a few exceptions you have now completed your study of descriptive statistics and have mastered the prerequisite knowledge and skills for the study of inference. In this section you will begin your study of inference as it is expressed through hypothesis testing. Later you will study confidence intervals which is the other component of inference.

The tests of hypotheses you will learn to conduct in this chapter are not very important from one point of view but are critically important from another. They are not important in that they are not commonly employed in research practice. The reason for this is that they test simple hypotheses that do not reflect the sorts of questions that interest most health science researchers. But this simplicity is also their strength. Because they are relatively simple you can master their application and underlying logic without too much difficulty. The same methods of application and logic that underpend these tests also underlie the more sophisticated tests you will encounter in later chapters. Thus, by thoroughly mastering the simple tests presented here you will find the mastery of more complex methods rather painless. The tests you will learn to conduct in this chapter are the one mean Z test, the one mean t test, one proportion tests, and one mean equivalence test.

4.3.2   Rationale and Method

Hypothesis testing is essentially a method for decision making. The decision relates to the choice between two competing, mutually exclusive, statements regarding one or more population parameters.[4] The competing statements of fact are referred to respectively as the null and alternative hypotheses. The **null hypothesis**, symbolized H_0, is a precise statement[5] concerning the parameter(s) of interest while the **alternative hypothesis**, symbolized H_A, is a less precise competing statement. For example, the null hypothesis might state that the mean of the population (μ) is equal to 100[6] while the less concise alternative might assert that the mean is greater than 100. The term *null* is used because the null hypothesis is a statement of no difference. In the above example, the null hypothesis states that there is no difference between the population mean and the value of 100. On the other hand, the alternative hypothesis asserts that there *is* a difference between the population mean and the value of 100 and in this case goes on to specify that the mean is greater than 100.

The decision as to whether the assertion of the null hypothesis is to be abandoned and the alternative assertion established as the true condition is made by using a model of the sampling distribution of the involved statistic to establish a decision criterion. If the criterion is met the null hypothesis is rejected in favor of the alternative assertion. If the criterion is not met the null statement is retained. An example will clarify the matter.

Suppose you wish to test the null hypothesis that the population mean is equal to 100. This statement is symbolized as $H_0 : \mu = 100$. The alternative maintains that the mean is greater than 100 and is symbolized $H_A : \mu > 100$. The test is carried out by randomly selecting a sample of a specified size from the population and calculating the sample mean ($\bar{x}$). Because the sampling distribution of $\bar{x}$ tends to be approximately normal, the normal curve is chosen as the test model. Figure 4.4 shows this model.

---

[4]There are tests of hypotheses that do not relate to parameters but we do not deal with those here.

[5]Sometimes the null hypothesis is stated in a nonprecise manner but the test is conducted on a specific value.

[6]Some would prefer to state the null hypothesis as indicating that the mean is less than or equal to 100 but the test would be conducted on the precise value of 100.

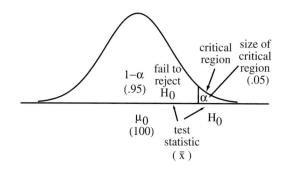

FIGURE 4.4: Example model for an hypothesis test.

Note first that the test is conducted under the assumption that the null hypothesis is true. Thus, the model used has the mean value specified by the null hypothesis which is symbolized μ_0 and is 100 in the present example. (A clear distinction must be made between μ, the population mean, and μ_0, the *hypothesized* population mean.) A small region in the tail of the curve, termed the **critical region**, is specified. The *size* of this region is symbolized by α (**alpha**). Alpha is traditionally set at .05 or .01 but may be set at other levels if so desired. The placement and size of the critical region are not arbitrary. The region is designed to fulfill two criteria.

First, the probability that the sample mean will take a value in the critical region *if the null hypothesis is true* is small. This is assured by choosing a sufficiently small value for α. If α is .05 then the probability that the test statistic will take a value in the critical region is only .05. Thus, it is unlikely that the test statistic will manifest a value in the critical region *if the null hypothesis is true*.

Second, the probability that the test statistic will take a value in the critical region must increase when the null hypothesis is false. More specifically, as the difference between μ_0 and μ increases, the probability of the test statistic being in the critical region must also increase. You can see that this criterion is met by placing the critical region in the tail of the distribution by considering the following. Suppose the null hypothesis is false because the mean of the population is 105. It follows that sample means from this population would tend to be larger (105 on average) than would be expected if the null hypothesis were true and would, therefore, have a higher probability of being large enough to fall in the critical region. If the mean of the population were even larger, say 200, the sample means would tend to be quite large and the probability of falling in the critical region would increase. This point will be elaborated upon later.

The decision as to whether the alternative hypothesis will be chosen as reflecting the true condition of the population mean depends upon whether or not the test statistic takes on a value in the critical region. If the test statistic takes a value in the critical region the null hypothesis is rejected in favor of the alternative. Why? Because it is unlikely that the test statistic will be in the critical region if the null hypothesis is true. In fact, the probability is only α. On the other hand, if the null hypothesis is false the probability is increased. Therefore, if the test statistic is in the critical region the logical explanation for this happenstance is that the null hypothesis is false.

But what if the test statistic is not in the critical region? You might think that the alternative hypothesis is rejected in favor of the null but that is not the case. The logic of hypothesis testing establishes the null as the condition that must be disproved. The required proof is in the form of a sufficiently small probability associated with the observed value of the test statistic under a true null hypothesis. Therefore, the conclusion reached when the test statistic is not in the critical region is that the null was not disproved. Using the formal language of hypothesis testing you either "reject the null hypothesis" or "fail to reject the null hypothesis." Another way of reporting the rejection of the null hypothesis is to state that the test was **statistically significant** while failure to reject is reported as was **not statistically significant**.

We suspect that by this time you are rather confused concerning hypothesis testing. Further, you are likely wondering "What's it good for?" The logic of hypothesis testing will become clear when we begin conducting tests and interpreting the results. We are now ready to begin this process. Before doing so, however, we suggest that you reread this section and carefully study Figure 4.4. A brief look at Figure 1.1 might also be helpful. As to the question of "What's it good for?", we will begin to address that issue in Chapter 5. For the moment, take heart—all will become clear as we continue.

4.3.3   The One Mean Z Test

The one mean Z test is used to test null hypotheses of the form $H_0 : \mu = \mu_0$ as we discussed above. The alternative hypothesis takes on one of two forms referred to as one- or two-tailed tests (also called one- or two-sided tests). The alternative hypothesis for one-tailed tests is *either* (but not both) $H_A : \mu > \mu_0$ or $H_A : \mu < \mu_0$. The two-tailed test has the alternative hypothesis $H_A : \mu \neq \mu_0$. We will illustrate and discuss each of these in turn.

One-Tailed Test: $H_A : \mu > \mu_0$.   Suppose that a census has been conducted in an African-American community located in a large metropolitan setting. The purpose is to assess the degree of satisfaction/dissatisfaction of community members with their access to medical care. The scale used in the census produces a score ranging from zero to 100 with lower scores indicative of less satisfaction. The census takers find that the average score in the community is 47.4 which is interpreted as a pronounced level of dissatisfaction. Analysis also shows that the standard deviation of these scores is 24.7.

Partially as a result of this census, a number of federal and non-federal grants are made available to improve medical access in the community. Some ten years after infusion of these monies, policy makers wish to determine whether community opinion concerning medical access has improved.

Obviously, this question could be answered by conducting a new census of the community to see if the mean score is higher now than it was ten years ago. But funding for such a large scale endeavor might not be available. In this situation, the evaluators might randomly select 100 community members to survey. The mean of this sample would then be used to perform a one mean Z test as follows.

The null hypothesis to be tested is that the mean value in the community (population) is 47.4 while the alternative states that the mean value is greater than 47.4. Stated more formally,

$$H_0 : \mu = 47.4$$
$$H_A : \mu > 47.4.$$

Notice that the null hypothesis asserts that the previously observed mean is still the true condition in the population while the alternative maintains that the mean in the community is now greater than was previously observed. The logic of the test requires that the null condition be either refuted or allowed to stand.

Because the sampling distribution of $\bar{x}$ tends to be normal in shape, the normal curve will be used as the model for the sampling distribution. Alpha is set at .05 and the critical region established as in Figure 4.4. The researchers know that the probability of the mean of the 100 sampled community members being large enough to fall in the critical region is only five in one hundred if the null hypothesis is true but will tend to be larger if the alternative is true. Therefore, a test of H_0 can be conducted by noting whether or not $\bar{x}$ is in the critical region. If it is in the critical region H_0 will be rejected, otherwise it will be retained. But how can we know if $\bar{x}$ is in the critical region? This is done by two different methods which we shall refer to as the p-value versus alpha and obtained Z versus critical Z methods. We will use each of these to conduct a one mean Z test for the problem outlined above.

p-value versus alpha method

Suppose the mean of the sample $(\bar{x})$ is found to be 52.8. Is this value in the critical region? The answer lies in Panel A of Figure 4.5. Notice that the mean of the model used to conduct the test is $\mu_0 = 47.4$ and that the critical region is in the right hand tail of the curve. The Z score for the sample mean is computed by

$$Z = \frac{\bar{x} - \mu_0}{\frac{\sigma}{\sqrt{n}}} \tag{4.7}$$

which is simply Equation 4.3 with μ_0 substituted for μ to indicate that a hypothesized mean value is being used rather than μ. By this equation

$$Z = \frac{52.8 - 47.4}{\frac{24.7}{\sqrt{100}}} = 2.19.$$

The p-value (or simply p) for a one-tailed test with alternative of the form $H_A : \mu > \mu_0$ is the probability of obtaining a test statistic as great as or greater than the value of the test statistic actually observed. In this case, the p-value is the probability of obtaining a sample mean whose value is 52.8 or greater. The area in column three of Appendix A for a Z score of 2.19 is .0143 which is the desired p-value (shown as the gray shading in the critical region of the curve in Panel A of Figure 4.5). The fact that the p-value is less than alpha (.05) implies that $\bar{x}$ is in the critical region which in turn means that H_0 is rejected. In general, the null hypothesis is rejected when $p \leq \alpha$ since this implies that the test statistic is in the critical region. It follows that H_0 is not rejected when $p > \alpha$.

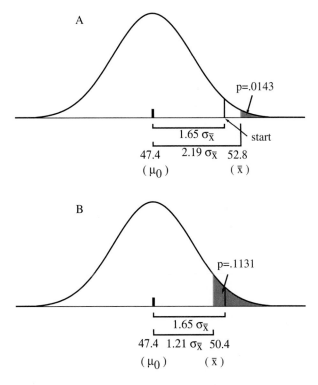

FIGURE 4.5: One mean Z tests of $H_0 : \mu = 47.4$ against the alternative $H_A : \mu > 47.4$ with $\alpha = .05$.

The logic of using the area of the curve above $\bar{x}$ (the p-value) to compare with α in order to determine if $\bar{x}$ is in the critical region can be seen by considering the following. Suppose that $\bar{x}$ were located at the starting edge of the critical region (labeled "start" in Panel A of Figure 4.5). In this case the area above $\bar{x}$ would be exactly .05. Any larger value of $\bar{x}$ would be in the critical region and would have a smaller area lying above it. By contrast, a smaller value of $\bar{x}$ would lie outside the critical region and would have a larger area above it.

At the risk of applying yet another stroke to this long-deceased equine, consider what the test result would have been had the sample mean been 50.4. The Z score in this case would be

$$Z = \frac{50.4 - 47.4}{\frac{24.7}{\sqrt{100}}} = 1.21.$$

Column three of Appendix A shows that the associated area is .1131 which is shown as the gray shading on the curve shown in Panel B of Figure 4.5. Because .1131 is greater than .05 it can be asserted that $\bar{x}$ is not in the critical region and that the null hypothesis is not rejected. Does this result indicate that the level of satisfaction/dissatisfaction is still 47.4? No, it means that we were unable to show that the level of satisfaction/dissatisfaction is greater than 47.4.

obtained Z versus critical Z method

The p-value versus alpha method of determining whether a test statistic is in the critical region is carried out by computing the Z score of the test statistic which is then used to obtain the area of the curve that lies above the statistic. Since alpha is also an area of the curve, the decision concerning the null hypothesis is made on the basis of a comparison of two areas. The obtained Z versus critical Z method begins in the same manner. That is, the Z score of the test statistic is computed. This value is termed **obtained Z**. You will recall that a Z score indicates the number of standard deviations (or standard errors in this case) that a score or statistic lies from the mean of the distribution. In the example depicted in Panel A of Figure 4.5 the computed Z score indicates that the sample mean of 52.8 is located 2.19 standard deviations above the mean of the distribution. **Critical Z** is the number of standard deviations between the mean of the distribution and the starting edge of the critical region. In other words, it is the Z score that lies at the leading edge of the critical region.

Critical Z can be obtained from Appendix A by looking up the appropriate area in order to find the associated Z value. Since we wish to find the Z value that has .05 of the curve above it, we look in column three of the table for .05. Unfortunately, .05 is not found. Instead, we find two areas that are equally close to .05 namely, .0495 and .0505. Generally, we would choose the area that is closest to the one desired but in this case the two areas are equally close to .05. A Z score of 1.64 cuts off .0505 in the tail while a Z of 1.65 cuts off .0495 in the tail. We could interpolate and use a Z of 1.645, but to simplify it is common practice to use the smaller area when two equally close areas are encountered. This means that we would take the Z value associated with .0495 which is 1.65.

So the distance from the mean of the distribution (47.4) to the leading edge of the critical region is (about) 1.65 standard deviations. The obtained Z indicates that $\bar{x}$ is 2.19 standard deviations from the mean of the distribution while critical Z indicates that the leading edge of the critical region is located 1.65 standard deviations above the distribution mean. Clearly, $\bar{x}$ is further from the mean of the distribution than is the leading edge of the critical region so that the test statistic is necessarily located in the critical region. (See Panel A of Figure 4.5.) Thus, while the p-value versus alpha method compares two areas, the obtained Z versus critical Z method compares two Z scores.

For the example depicted in Panel B, the test statistic is located 1.21 standard deviations from the distribution mean so that it is not as far as the leading edge of the critical region and so cannot be located therein. In general, for a one-tailed test with the alternative $H_A : \mu > \mu_0$ the critical region is located in the right hand tail of the curve so that the null hypothesis is rejected when obtained Z is greater than or equal to critical Z.

EXAMPLE 4.9

Use the information provided below to perform the indicated one mean Z test. Justify your decision regarding the null hypothesis on the basis of both the p-value versus alpha and obtained Z versus critical Z methods.

$$H_0 : \mu = 120 \qquad \sigma = 40 \qquad \bar{x} = 128.2$$
$$H_A : \mu > 120 \qquad n = 150 \qquad \alpha = .01$$

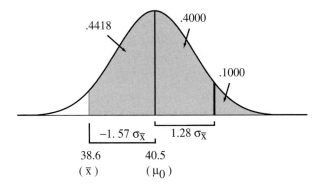

FIGURE 4.6: One mean Z test of $H_0 : \mu = 40.5$ against the alternative $H_A : \mu > 40.5$ with $\alpha = .10$.

Solution   The Z value associated with 128.2 is

$$Z = \frac{\bar{x} - \mu_0}{\frac{\sigma}{\sqrt{n}}} = \frac{128.2 - 120.0}{\frac{40.0}{\sqrt{150}}} = 2.51.$$

The p-value is the area above 128.2. Column three of Appendix A shows that the area above $Z = 2.51$ is .0060. Because this value is less than $\alpha = .01$, the null hypothesis is rejected. Column three of the normal curve table also shows that the closest value to .01 is .0099 which has an associated Z value of 2.33. Thus, the Z value at the leading edge of the critical region is 2.33. Since obtained Z of 2.51 is larger than critical Z of 2.33, the null hypothesis is rejected.   ■

EXAMPLE 4.10

Use the information provided below to perform the indicated one mean Z test. Justify your decision regarding the null hypothesis on the basis of both the p-value versus alpha and obtained Z versus critical Z methods.

$$H_0 : \mu = 40.5 \qquad \sigma = 11.5 \qquad \bar{x} = 38.6$$
$$H_A : \mu > 40.5 \qquad n = 90 \qquad \alpha = .10$$

Solution   The Z value associated with 38.6 is

$$Z = \frac{38.6 - 40.5}{\frac{11.5}{\sqrt{90}}} = -1.57.$$

The p-value is the area above 38.6 which in turn is composed of the area between 38.6 and the distribution mean of 40.5 and the area above 40.5. Using $Z = -1.57$, column two of Appendix A shows that the area between 38.6 and 40.5 is .4418. Since the area above 40.5 is .5000, the p-value is $.4418 + .5000 = .9418$ which is decidedly larger than $\alpha = .10$ so that the null hypothesis is not rejected. (See Figure 4.6.)

The closest value to .10 in column three of Appendix A is .1003 which has an associated Z value of 1.28. Thus, the leading edge of the critical region is located at a point that is 1.28

standard errors *above* the mean of the distribution. By contrast, the obtained Z of -1.57 indicates that $\bar{x}$ is located 1.57 standard errors *below* the mean of the distribution. Clearly, then, $\bar{x}$ is not in the critical region. (See Figure 4.6.). ∎

One-Tailed Test: $H_A : \mu < \mu_0$. A second form of the one-tailed test employs the alternative $H_A : \mu < \mu_0$ which implies a population mean that is less than the value specified by the null hypothesis. In order to increase the probability of attaining a significant test result when H_0 is false and H_A is true, the critical region must be located in the left rather than right tail of the distribution. Consider the following example.

The Maslach Burnout Inventory [35] is an instrument that measures three aspects of "burnout" as manifested by professional nurses. The norm table for the inventory reports the mean and standard deviation for US nurses on the Emotional Exhaustion subscale as 22.19 and 9.53 respectively. (Higher scores indicate a greater degree of emotional exhaustion.) Suppose that a nursing researcher believes that, after a long decline, working conditions for nurses in the United States have gradually improved since the norms were developed. The researcher further hypothesizes that because of this improvement, emotional exhaustion has declined. In order to test this hypothesis the researcher administers the inventory to 121 nurses in a local hospital.[7] The mean score of the 121 nurses on the emotional exhaustion subscale is found to be 20.58. The researcher conducts a one-tailed one mean Z test with $\alpha = .05$ and the following hypotheses

$$H_0 : \mu = 22.19$$
$$H_A : \mu < 22.19$$

Z is computed as

$$Z = \frac{\bar{x} - \mu_0}{\frac{\sigma}{\sqrt{n}}} = \frac{20.58 - 22.19}{\frac{9.53}{\sqrt{121}}} = -1.86.$$

p-value versus alpha method

The p-value for a one-tailed test with alternative of the form $H_A : \mu < \mu_0$ is the area *below* the test statistic. Column three of Appendix A reveals this value to be .0314. Because $p = .0314 < \alpha = .05$, the null hypothesis is rejected. (See Panel A of Figure 4.7.)

obtained Z versus critical Z method

Because the critical region for tests with this alternative are in the left tail of the curve, critical Z is negative. For the problem at hand critical Z is -1.65 which indicates that the leading edge of the critical region lies at a point 1.65 standard errors below the mean of the distribution. It follows that the null hypothesis is rejected when obtained Z is less than or equal to critical Z. Because $-1.86 < -1.65$, the null hypothesis is rejected. Panel A of Figure 4.7 depicts the elements of this analysis.

The researcher might conclude from this study that emotional exhaustion among nurses in the United States has declined. But this conclusion must be tentative since the 121 nurses in the study did not constitute a random sample from the US population of nurses. This lack of random selection of subjects for the study is the usual state of affairs so that conclusions gained in this type of study are rarely definitive.

---

[7]This is not a random sample but reflects the practical limitations of most research.

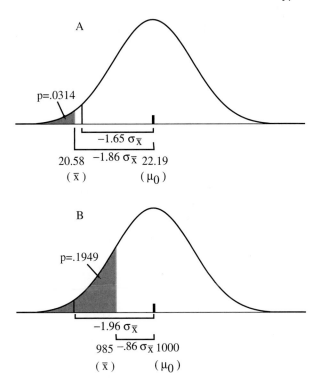

FIGURE 4.7: One mean Z tests with alternative $H_A : \mu < \mu_0$.

EXAMPLE 4.11

Use the information provided below to perform the indicated one mean Z test. Justify your decision regarding the null hypothesis on the basis of both the p-value versus alpha and obtained Z versus critical Z methods.

$$H_0 : \mu = 1000 \qquad \sigma = 235 \qquad \bar{x} = 985$$
$$H_A : \mu < 1000 \qquad n = 180 \qquad \alpha = .025$$

Solution   The Z value associated with 985 is

$$Z = \frac{\bar{x} - \mu_0}{\frac{\sigma}{\sqrt{n}}} = \frac{985 - 1000}{\frac{235}{\sqrt{180}}} = -.86.$$

The p-value is the area below 985. Using $Z = -.86$ Column three of Appendix A shows this area to be .1949. Since .1949 is greater than .025, the null hypothesis is not rejected.

Because a Z score of 1.96 cuts off .025 in the tail of the distribution and the critical region is below the mean of the distribution, critical Z is -1.96. Since $-.86$ is not less than -1.96, the null hypothesis is not rejected. The logic of this analysis is depicted in Panel B of Figure 4.7. ∎

Two-Tailed Test: $H_A : \mu \neq \mu_0$.   In many situations researchers are unwilling or unable to specify whether a false null hypothesis would imply a value of μ that is greater than or less than μ_0. In such instances a two-tailed test is conducted. In order to detect either eventuality, a critical region is placed in *both* tails of the distribution. The null hypothesis is rejected if the test statistic takes a value in either of the two regions. In order to maintain the probability of a test statistic falling in a critical region at α when H_0 is true, the size of each region is established at $\alpha/2$.[8]

Suppose that in the nursing example given on page 94, working conditions for nurses had been in flux for some time with some conditions improving while others appear to have deteriorated. The nursing researcher wishes to determine whether the level of Emotional Exhaustion on the part of US nurses has been impacted by these changes. Because some changes in the workplace have been positive and some negative, it is of interest to see if the overall impact has added to or ameliorated the average level of exhaustion. The researcher conducts a two-tailed one mean Z test with $\alpha = .05$ and the following hypotheses

$$H_0 : \mu = 22.19$$
$$H_A : \mu \neq 22.19.$$

As previously calculated, the Z score for the sample mean of 20.58 is -1.86.

p-value versus alpha method

The p-value for a two-tailed test is obtained by doubling the tail area defined by the test statistic. The tail area is the area above the test statistic if the test statistic is above the mean of the distribution or is the area below the test statistic if the test statistic is below the mean of the distribution. In the current example, the sample mean of 20.58 is below the distribution mean of 22.19 so that the tail area is the area below 20.58. Using $Z = -1.86$ in Appendix A shows this area to be .0314. The p-value is then $2 \times .0314 = .0628$. Since $.0628 > \alpha = .05$, the null hypothesis is not rejected. This p-value is shown as the combined shaded areas in Panel A of Figure 4.8. Notice that the area represented by the combined shaded areas (p-value) exceeds the combined areas of the two critical regions (α).

obtained *Z* versus critical *Z* method

Notice first that there are two critical Z values for a two-tailed test, one of which is positive and the other negative. This results from the fact that there is a critical region above and below the mean of the distribution. Because each of these regions encompass $\alpha/2$ of the curve, it is this value that is used in Appendix A to find critical Z. Since $\alpha = .05$ in the current example, we need to find the Z value that cuts off .025 in the tail of the curve. Column three of Appendix A shows this value to be 1.96. Thus, the critical *values* for this test are ± 1.96.

The null hypothesis is rejected if (a) obtained Z is less than or equal to the negative critical Z value *or* (b) obtained Z is greater than or equal to the positive critical Z value. This means that H_0 is rejected if the test statistic is in either critical region. Panel A of Figure 4.8 shows that the leading edge of the critical region in the left tail of the curve is 1.96 standard errors below the mean of the distribution while $\bar{x}$ is 1.86 standard errors below the mean of the distribution. Clearly, $\bar{x}$ is not far enough below the mean of the distribution to reach the leading edge of the critical region.

---

[8]Equal division of α between the two tails is not mandatory but is almost always done.

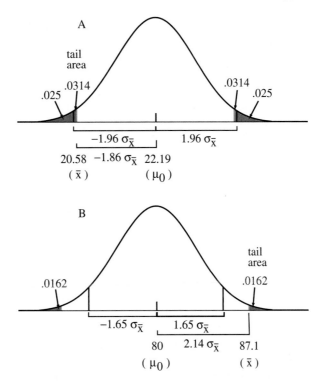

FIGURE 4.8: One mean Z tests with alternative $H_A : \mu \neq \mu_0$.

EXAMPLE 4.12

Use the information provided below to perform the indicated one mean Z test. Justify your decision regarding the null hypothesis on the basis of both the p-value versus alpha and obtained Z versus critical Z methods.

$$H_0 : \mu = 80 \qquad \sigma = 21 \qquad \bar{x} = 87.1$$
$$H_A : \mu \neq 80 \qquad n = 40 \qquad \alpha = .10$$

Solution   The Z value associated with 87.1 is

$$Z = \frac{\bar{x} - \mu_0}{\frac{\sigma}{\sqrt{n}}} = \frac{87.1 - 80.0}{\frac{21}{\sqrt{40}}} = 2.14.$$

Because 87.1 is above the distribution mean of 80.0, the tail area is the area above 87.1 which reference to Appendix A shows to be .0162. The p-value is then $p = 2 \times .0162 = .0324$. Because $p = .0324 < \alpha = .10$, the null hypothesis is rejected. The researcher can be confident (though not totally so) that the mean of the population is greater than 80. The details of this test are portrayed in Panel B of Figure 4.8.

Critical Z values for a two-tailed test at $\alpha = .10$ are the values that cut off .05 in each tail of the distribution. As noted above, these values are ± 1.65. Since the test statistic is 2.14

standard errors above the mean of the distribution and the right tail critical region begins at a point 1.65 standard errors above the mean of the distribution, the test statistic is clearly in the right tail critical region. Details of this logic are shown in Panel B of Figure 4.8. ■

EXAMPLE 4.13

Use the information provided below to perform the indicated one mean Z test. Justify your decision regarding the null hypothesis on the basis of both the p-value versus alpha and obtained Z versus critical Z methods.

$$H_0 : \mu = 150 \qquad \sigma = 50.4 \qquad \bar{x} = 142.4$$
$$H_A : \mu \neq 150 \qquad n = 105 \qquad \alpha = .01$$

Solution The Z value associated with 142.4 is

$$Z = \frac{142.4 - 150.0}{\frac{50.4}{\sqrt{105}}} = -1.55.$$

The tail region defined by the test statistic is the area below 142.4 which Appendix A shows to be .0606. The p-value is then $p = 2 \times .0606 = .1212$. Because $.1212 > .01$ the null hypothesis is not rejected.

Appendix A shows two tail areas that are equally close to .005. Namely, .0049 and .0051. As we noted previously, when two tail areas are equally close to the one being sought, we will use the smaller of the two areas. In this case, the result is a Z value of 2.58. Critical Z is then ± 2.58. Since the test statistic is 1.55 standard errors below the mean of the distribution, it does not reach the leading edge of the left tail critical region. Therefore, the null hypothesis is not rejected. ■

Assumptions Underlying the One Mean Z Test. As you are now aware, the rationale underlying hypothesis testing requires that one or more critical regions be established in such a manner as to make the probability of a test statistic taking on a value in the region(s) small when the null hypothesis is true. Thus, if a test statistic does take a value in the region(s), logic suggests that the null hypothesis is probably not true. But critical regions are constructed in conjunction with a *model* of the sampling distribution of the statistic rather than the actual sampling distribution. The question then arises as to the consequences if the model used for the hypothesis test is inappropriate so that estimated probabilities obtained from the model are inaccurate? In such circumstance, the probability of a statistic taking a value in the critical region(s) may be quite different from the estimated probability provided by the model. For example, suppose the researcher establishes α at .05. The researcher then believes that the probability of the test statistic taking a value in the critical region(s) is .05. But if the model does not accurately represent the sampling distribution of the statistic, the actual probability may be quite different, say .25. Obviously, the researcher cannot have confidence in the results of a hypothesis test in such a circumstance.

Under what circumstances can the researcher be sure that the model used for the hypothesis test will provide accurate probabilities in regards to the critical region(s)? These circumstances are referred to as the **underlying assumptions** of a test. The one mean Z test has two underlying assumptions, namely, the assumptions of (1) normality and (2) independence of observations. We will comment on each of these.

normality

The assumption of normality requires that the sample data used for the test be drawn from a normally distributed population. But as you know, the normal distribution is a mathematical model with very specific attributes one of which is that it extends infinitely in either direction. In reality, samples are never taken from *perfectly* normally distributed populations so in a sense, this assumption is always violated. It is more practical, if slightly less accurate, to say that the assumption of normality requires that data be drawn from an *approximately* normal population. We will say more on this topic later.

independence

The assumption of independence requires that each observation in the sample be unrelated to every other observation in the sample. Thus, for example, the blood pressure of one member of the sample neither influences nor is influenced by the blood pressure of another member of the sample. Likewise, the fact that one blood pressure has a specific value, say 120, does not change the probability that the next observed blood pressure will be some specific value, say 95. An example of how this assumption might be violated will help to clarify the concept.

Suppose 10 ophthalmology patients are sampled from a population of such patients. Visual acuity is measured for both eyes for all 10 patients. Thus, the sample consists of 20 visual acuity scores. But these scores are not all independent. Knowing that the visual acuity in the left eye of the first patient is 20/200 may (or may not) be indicative of the fact that the right eye measure also reflects reduced acuity. This stems from the fact that a person with reduced acuity in one eye often has reduced acuity in the other as might occur from diabetic retinopathy or cataracts.

As a second example, suppose that in a study of juvenile hypertension, a researcher randomly selects households for participation in the study. Once a household is selected, the blood pressures of all children in the household under the age of 18 are recorded. These data are likely to violate the assumption of independence because, due to genetic similarities, children in a given household are likely to exhibit similar tendencies in regard to expressed blood pressures. The assumption of independence can be violated in subtle ways so that researchers must always bear this assumption in mind when using statistical methods that require independence in order to assure accuracy of estimated probabilities.

Consequences of Assumption Violations.   It is not possible to specify the consequences of violating the assumptions underlying the one mean Z test because it is impossible to enumerate the forms such violations might take. For example, the consequences of violating the normality assumption depend upon, among other things, the shape of the nonnormal population. But there are an infinite number of such shapes. Having said this, however, it is possible to draw some general conclusions while bearing in mind such generalities may not hold for some specific circumstance.

violation of normality assumption

Perhaps the best way to understand the consequences of violating the normality assumption is to consider the consequence as it relates to the population depicted in Figure 4.2 on page 78. Before doing so, however, it will be helpful to distinguish between two types of α.

Nominal alpha (α_N) is the *intended* probability that a test statistic will take a value in the critical region(s) when the null hypothesis is true. It is the size of the critical region(s)

TABLE 4.2: α_E for the two-tailed one mean Z test when $\alpha_N = .05$ and sampling is from a nonnormal population.

n	Left Tail	Right Tail	Combined (α_E)
5	.020	.026	.046
10	.024	.028	.052
20	.023	.026	.049
30	.024	.027	.051
50	.024	.026	.050
100	.024	.026	.050

in the test model. We have previously referred to this value as α but will use α_N during the present discussion. **Empirical alpha** (α_E) is the *actual* probability that a test statistic will take a value in the critical region(s) when the null hypothesis is true. When both assumptions of the one mean Z test are met, $\alpha_E = \alpha_N$.

Table 4.2 shows values of α_E for the two-tailed one mean Z test when sampling is from the population shown in Figure 4.2. If there were no violations the probability of the test statistic taking a value in each of the two critical regions would be .025. Therefore, the difference between the observed probability and .025 in each of the tails is attributable to the nonnormality of the population. Likewise, in the absence of violations the combined probability would be .05.

As may be seen in this table, for samples of size five

$$\alpha_E - \alpha_N = .046 - .050 = -.004.$$

When $\alpha_E < \alpha_N$ a test is said to be **conservative** in the face of some violation. When $\alpha_E > \alpha_N$ the test is said to be **liberal** or **anti-conservative**. So, the one mean Z test is conservative when $n = 5$ but is liberal when $n = 10$. The important point to note is that the difference between α_E and α_N is never large, especially when $n \geq 10$. When the difference between α_E and α_N is small, a test is said to be **robust** to the particular violation. How small should this difference be in order for a test to be characterized as robust? While there are several rules of thumb[9] for making this determination, the issue is largely a matter of individual perception.

Notice also that the probabilities in the two tails are not equal. This is typical when sampling is from skewed populations. It often happens in such situations that the level in one tail is conservative while that in the other is liberal so that the combined probability is at or near the nominal level of the test. This implies that the one mean Z test is more robust for two-tailed than one-tailed tests.

In summary, the one mean Z test is reasonably robust to departures from normality and becomes increasingly so as sample size increases. For skewed populations, the two-tailed test is often more robust than is the one-tailed test. Empirical probabilities may not be satisfactory when samples are small and/or populations are highly skewed.

---

[9]One such rule declares a test robust if $.5\alpha_N \leq \alpha_E \leq 1.5\alpha_N$.

TABLE 4.3: α_E for the two-tailed one mean Z test when $\alpha_N = .05$, data are not all independent and sampling is from a normal population.

n	Small			Moderate		
	Left Tail	Right Tail	Com-bined	Left Tail	Right Tail	Com-bined
10	.037	.037	.074	.055	.055	.110
50	.037	.037	.074	.055	.055	.110
100	.037	.037	.074	.055	.055	.110

violation of independence assumption

Unlike the normality assumption, the one mean Z test is not generally robust against departures from the independence assumption. As an example, suppose that a naive researcher is conducting a study involving visual acuity. The researcher tests the visual acuity of both eyes of five patients and then conducts a two-tailed one mean Z test using the ten visual acuities. Table 4.3 shows the probability that $\bar{x}$ will take a value in the two critical regions when there are small and moderate dependencies between the visual acuities of the two eyes.[10]

As you can see from this table, when the test is conducted on the acuities of ten eyes and the dependency is small, the probability in each critical region is .037 producing an overall probability of .074. The probabilities increase to .055 in each tail and .110 overall when the degree of dependence is moderate. Notice that increasing the number of eyes to 50 and 100 does nothing to ameliorate the liberal test results. The lack of robustness of this test is reflected in the fact that moderate dependence more than doubles the intended overall probability of .05.

Violations of the assumption of independence can also cause conservative results though liberal results would seem more common. The important points to note are (1) the one mean Z test is generally not robust against violations of the independence assumption and (2) increasing sample size does not reduce the impact of the violation.

Summary.   The one mean Z test is used to test null hypotheses of the form $H_0 : \mu = \mu_0$ against one- and two-tailed alternatives. For the one-tailed alternative

$$H_A : \mu < \mu_0$$

the p-value is defined as the area below $\bar{x}$ while critical Z is the Z value that cuts off α in the lower tail. For the alternative

$$H_A : \mu > \mu_0$$

the p-value is the area above $\bar{x}$ and critical Z cuts off α in the upper tail. The two-tailed alternative is

$$H_A : \mu \neq \mu_0$$

---

[10]Small and moderate are operationally defined as correlations of .2 and .5 respectively. You will learn about correlation coefficients in Chapter 8.

The *p*-value for the two-tailed alternative is twice the tail area defined by $\bar{x}$. There are two critical Z values one of which cuts off $\alpha/2$ in the upper tail while the other cuts off the same area in the lower tail.

The test is generally robust to violations of the normality assumption and becomes increasingly so as sample size increases. Neither of these qualities are true insofar as the independence assumption is concerned.

4.3.4   The One Mean *t* Test

There is a difficulty associated with use of the one mean Z test that has both a logical and practical basis. The logical difficulty lies in the fact that the test is conducted in order to make a determination regarding the unknown parameter μ. But in order to conduct the test you must know σ (see Equation 4.7 on page 90) which is also a parameter. It seems likely that if σ is known, μ would also be known, thereby making use of an inferential test unnecessary. Although there are some situations where σ is known and μ unknown, they are not common. The practical difficulty, then, is how to conduct the hypothesis test without knowledge of σ.

The key to a solution was provided by William S. Gossett [45] (writing under the pseudonym "Student") in 1908 with enhancements/modifications later provided by R. A. Fisher [17].

The solution involves replacing the (usually) unknown parameter σ with the known statistic s in the expression for the standard error of $\bar{x}$. (You will recall that s is the sample standard deviation.) This results in the statistic

$$t = \frac{\bar{x} - \mu_0}{\frac{s}{\sqrt{n}}} \tag{4.8}$$

The most appropriate model for the sampling distribution of this statistic is not the normal curve but rather a model referred to as the **t distribution**. To be clear, the t distribution models the distribution that would be obtained if repeated samples were drawn from a population, t as represented by Equation 4.8, was computed on each sample, and the resulting statistics then formed into a relative frequency distribution. This distribution would have mean zero when the null hypothesis is true and mean $\mu - \mu_0$ when the null hypothesis is not true. While mathematically distinct, t curves are similar in appearance to the normal curve as can be seen in Figure 4.9.

Actually, there are an infinite number of distinct t distributions each of which is identified by what are termed its *degrees of freedom*. An in-depth discussion of degrees of freedom is beyond the scope of this book but, generally speaking, **degrees of freedom** refers to the amount of information available in s to estimate σ. The three and 20 degrees of freedom t curves are shown in Figure 4.9. Notice that the 20 degrees of freedom curve is more similar to the normal curve than is the three degrees of freedom curve. In general, as degrees of freedom increase, the associated t curve becomes more similar to the normal curve. Theoretically, at infinite degrees of freedom the associated t curve is identical to the normal curve.

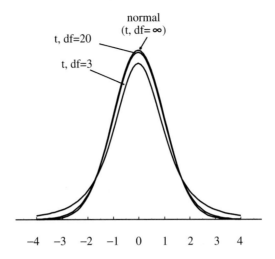

FIGURE 4.9: Normal curve with t distributions.

In order to find areas under any given t curve, you would need a table of areas (similar to the normal curve table in Appendix A) *for that particular curve*. This is impractical because we will be dealing with many different t curves. For efficiency, then, t tables do not provide areas as does the normal curve table, but rather gives commonly employed critical values for each curve.

Consider the t table in Appendix B. The first two rows of this table refer to confidence intervals which we will discuss in the latter part of this chapter. The third and fourth rows give values of α for one and two-tailed hypothesis tests respectively. The numbers in the main portion of the table are the critical values appropriate for each value of α. The table is used in the following manner.

Suppose you wish to find the critical value for a one-tailed test with alternative $H_A : \mu > \mu_0$ at $\alpha = .05$ for a t distribution with three degrees of freedom. Locating .05 in the third row and reading down to the three degrees of freedom row gives the appropriate critical value as 2.353. This means that a t value of 2.353 cuts off .05 in the upper tail of a t curve with three degrees of freedom. A lower tail version of the test would require a critical value of -2.353.

What would the critical value be for a two-tailed t test at $\alpha = .05$ for a t distribution with 10 degrees of freedom? Locating .05 in the fourth row and reading down to the 10 degrees of freedom row gives the value of 2.228. Note that 2.228 cuts off .025 in the tail of the curve. The critical values for the two-tailed test would then be ± 2.228.

Degrees of freedom are calculated differently for different tests. For the one mean t test the degrees of freedom are $n - 1$ where n is the number of observations in the sample.

Except for the manner in which obtained t is calculated, the mechanics of conducting a one mean t test are the same as those for a one mean Z test. Obtained t is calculated and compared with critical t. If obtained t is in a critical region the null hypothesis is rejected. Otherwise, the null hypothesis is retained. Because we do not have tables of the areas of the various t distributions, we will not be able to use the p-value versus alpha method and so will be restricted to testing via the obtained t versus critical t method. When conducting tests by means of computer software, the p-value will be supplied by the software so that the p-value versus alpha method will be the usual method employed.

EXAMPLE 4.14

Use the sample data provided here to conduct the indicated one mean t test.

$H_0 : \mu = 8.0$ Sample: 6.0 8.0 5.5 4.5 8.5 4.0 3.5
$H_A : \mu < 8.0$
$\alpha = .01$

Solution   The sample standard deviation can be obtained by application of Equation 2.16 to the summed scores and the sum of the squared scores as given below.

X	X^2
6.0	36.00
8.0	64.00
5.5	30.25
4.5	20.25
8.5	72.25
4.0	16.00
3.5	12.25
40.0	251.00

$$s = \sqrt{\frac{\sum x^2 - \frac{(\sum x)^2}{n}}{n-1}} = \sqrt{\frac{251 - \frac{(40)^2}{7}}{7-1}} = \sqrt{\frac{22.429}{6}} = 1.933$$

and

$$\bar{x} = \frac{40}{7} = 5.714.$$

By Equation 4.8 obtained t is then

$$t = \frac{\bar{x} - \mu_0}{\frac{s}{\sqrt{n}}} = \frac{5.714 - 8.00}{\frac{1.933}{\sqrt{7}}} = \frac{-2.286}{.731} = -3.127.$$

As can be seen from the table in Appendix B, a one-tailed test with degrees of freedom $7 - 1 = 6$ conducted at $\alpha = .01$ has an associated value of 3.143. As indicated by the alternative hypothesis, the critical region is in the lower tail so that the critical value is -3.143. Because the obtained value of -3.127 is greater than the critical value, the null hypothesis is not rejected. The relationship of the obtained value to the critical value is shown in Panel A of Figure 4.10.   ■

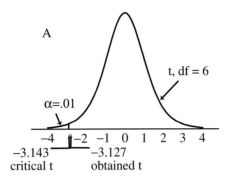

A

t, df = 6

α=.01

−4 −2 −1 0 1 2 3 4

−3.143   −3.127
critical t   obtained t

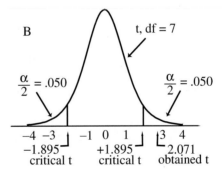

B

t, df = 7

$\frac{\alpha}{2}$ = .050 $\frac{\alpha}{2}$ = .050

−4 −3   −1 0 1   3 4

−1.895   +1.895   2.071
critical t   critical t   obtained t

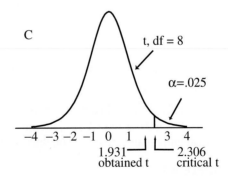

C

t, df = 8

α=.025

−4 −3 −2 −1 0 1   3 4

1.931   2.306
obtained t   critical t

FIGURE 4.10: Locations of obtained and critical
values for three one mean *t* tests.

EXAMPLE 4.15

Use the sample data provided here to conduct the indicated one mean t test.

$H_0 : \mu = 0.0$ Sample: -0.5 1.0 2.5 -1.0 3.5 -0.5 3.0 2.5
$H_A : \mu \neq 0.0$
$\alpha = .10$

Solution   The necessary sums for the calculation of s and $\bar{x}$ are

X	X^2
-0.5	0.25
1.0	1.00
2.5	6.25
-1.0	1.00
3.5	12.25
-0.5	0.25
3.0	9.00
2.5	6.25
10.5	36.25

So that

$$s = \sqrt{\frac{\sum x^2 - \frac{(\sum x)^2}{n}}{n-1}} = \sqrt{\frac{36.25 - \frac{(10.50)^2}{8}}{8-1}} = \sqrt{\frac{22.469}{7}} = 1.792$$

and

$$\bar{x} = \frac{10.5}{8} = 1.313.$$

Obtained t is then

$$t = \frac{\bar{x} - \mu_0}{\frac{s}{\sqrt{n}}} = \frac{1.313}{\frac{1.792}{\sqrt{8}}} = \frac{1.313}{.634} = 2.071.$$

Critical t for a two-tailed test at $\alpha = .10$ and seven degrees of freedom is ± 1.895. Because 2.071 is greater than 1.895, the null hypothesis is rejected. The relative positions of the critical and obtained t values on the test model are shown in Panel B of Figure 4.10. ∎

EXAMPLE 4.16

Use the sample data provided here to conduct the indicated one mean t test.

$H_0 : \mu = 20$ Sample: 22 19 17 26 21 20 29 27 22
$H_A : \mu > 20$
$\alpha = .025$

Solution   The necessary sums for the calculation of s are

X	X^2
22	484
19	361
17	289
26	676
21	441
20	400
29	841
27	729
22	484
203	4705

so that

$$s = \sqrt{\frac{4705 - \frac{(203)^2}{9}}{9-1}} = \sqrt{\frac{126.222}{8}} = 3.972$$

and

$$\bar{x} = \frac{203}{9} = 22.556.$$

Obtained t is then

$$t = \frac{22.556 - 20.000}{\frac{3.972}{\sqrt{9}}} = \frac{2.556}{1.324} = 1.931.$$

Critical t for a one-tailed test at $\alpha = .025$ and eight degrees of freedom is 2.306. Because obtained t of 1.931 is less than critical t of 2.306, the null hypothesis is not rejected. The relative positions of the critical and obtained t values on the test model are shown in Panel C of Figure 4.10.   ■

Assumptions Underlying the One Mean t Test.   The assumptions underlying the one mean t test are the same as those for the one mean Z test—namely, population normality and independence of observations.

Consequences of Assumption Violations.   Violations of the assumption of independence for the one mean t test produce the same consequences as for the one mean Z test and will not be reiterated here. Table 4.4 shows α_E for the one mean t test conducted on the same samples as were used in the construction of Table 4.2.

Comparison of Table 4.4 with Table 4.2 shows the same general patterns for the t test as were seen for the Z test. That is, (1) a lack of symmetry for the results in the two tails due to sampling from a skewed population, (2) results closer to α_N for the two sides combined than for the individual tails, and (3) increasing robustness with increasing sample size. Also notable is the fact that the t test is generally not as robust as is the Z test.

Summary.   The one mean t test differs from the one mean Z test in that the sample standard deviation (s) rather than the population standard deviation (σ) is used in the equation for the obtained value. As a result of this substitution, the appropriate model for the

TABLE 4.4: α_E for the two-tailed one mean t test when $\alpha_N = .05$ and sampling is from a nonnormal population.

n	Left Tail	Right Tail	Combined (α_E)
5	.032	.037	.069
10	.030	.023	.053
20	.029	.022	.051
30	.029	.022	.051
50	.028	.023	.051
100	.027	.023	.050

sampling distribution of the t statistic is the family of t distributions which are indexed by their degrees of freedom. The t test shares the Z test's lack of robustness to violations of the independence assumption and is somewhat less robust than the Z test to violations of the normality assumption.

4.3.5 One Sample Tests for a Proportion

In this section we will discuss two tests for a population proportion. The first, which we shall refer to as the *exact* test, is based on the binomial distribution (see sections 4.2.4 and 4.2.5 on pages 80 and 81 respectively) while the second, which we shall refer to as the *approximate* test, is based on the normal curve (see Section 4.2.6 on page 84). The reason for this terminology stems from the fact that, if certain assumptions are met, the binomial distribution provides exact probabilities while the normal curve method necessarily yields an approximate result.

Both methods can be used to test the null hypothesis

$$H_0 : \pi = \pi_0$$

where π is the proportion of observations in a population that meet some specified criterion and π_0 is the hypothesized proportion for the same population. One-tailed alternatives take the form

$$H_A : \pi < \pi_0$$

or

$$H_A : \pi > \pi_0$$

while the two-tailed alternative is expressed as

$$H_A : \pi \neq \pi_0.$$

The Exact Test.

one-tailed test

You will recall from Sections 4.2.4 and 4.2.5 that the binomial distribution can be used to model the sampling distribution of $\hat{p}$. It naturally follows that this same model can be used as the basis for hypothesis tests.

TABLE 4.5: Sampling distribution of $\hat{p}$ for $n = 10$ and $\pi = .38$.

Proportion $\hat{p}$	Number of Successes y	Probability $P(y)$
.00	0	.00839
.10	1	.05144
.20	2	.14188
.30	3	.23189
.40	4	.24872
.50	5	.18293
.60	6	.09343
.70	7	.03272
.80	8	.00752
.90	9	.00102
1.00	10	.00006

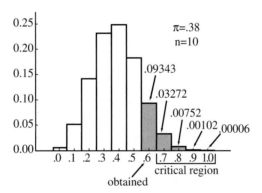

FIGURE 4.11: Binomial test with alternative H_A : $\pi > .38$.

As an example, suppose you randomly sample 10 observations from a dichotomous population and wish to conduct the following test.

$$H_0 : \pi = .38 \qquad \hat{p} = .60 \qquad n = 10$$
$$H_A : \pi > .38 \qquad \alpha = .05$$

You will recognize this scenario as the study of blood types in the Iceland blood donor population described in Example 4.6 on page 84. Equation 4.5 was used to construct the sampling distribution of $\hat{p}$ shown in Table 4.5. As with all hypothesis tests, this distribution was constructed under the assumption of a true null hypothesis (i.e., that $\pi = .38$). This distribution is shown graphically in Figure 4.11.

Discrete distributions such as the binomial present certain difficulties that must be recognized. We have indicated the critical region for the one-tailed test presently being considered in Figure 4.11. Notice that the probability of $\hat{p}$ taking a value in the critical region when the null hypothesis is true is not .05 as was intended but rather is $.03272 + .00752 + .00102 + .00006 = .04132$. If the critical region were expanded to include $\hat{p} = .60$, α would be $.04132 + .09343 = .13475$ which is much greater than the intended .05. When dealing with discrete sampling distributions the critical region is constructed so as to embrace the largest probability possible that does not exceed the intended α. That value in this case is .04132. Notice that for a test with alternative $H_A : \pi < .38$ the critical region would contain only $\hat{p} = .0$ which has an associated probability of .00839 (see Table 4.5). To expand the region to include $\hat{p} = .1$ would make the probability $.00839 + .05144 = .05983$ which is greater than the intended α.

The p-value for a one-tailed test with alternative of the form $H_A : \pi > \pi_0$ is the probability of obtaining a value of $\hat{p}$ that is greater than or equal to the value actually observed. In this case obtained $\hat{p}$ is .60 so that the p-value is $.09343 + .03272 + .00752 + .00102 + .00006 = .13475$. Because this value is greater than α, the null hypothesis is not rejected. The p-value is shown as the shaded region in Figure 4.11.

Once the critical region has been defined the obtained versus critical value approach may also be used for the test. The critical value is simply the value of $\hat{p}$ that defines the largest possible tail area that does not exceed α. As you can see from Figure 4.11 critical $\hat{p}$ is .70 in the current example. Because the obtained $\hat{p}$ of .60 is less than .70, the null hypothesis is not rejected.

Consider now the test

$$H_0 : \pi = .45 \qquad \hat{p} = .00000 \qquad n = 7$$
$$H_A : \pi < .45 \qquad \alpha = .20$$

The p-value for a test with alternative of the form $H_A : \pi < \pi_0$ is the probability of obtaining a value of $\hat{p}$ that is less than or equal to the value observed. In this case the p-value would be $P(0)$ which by Equation 4.5 is .01522. Because the p-value of .01522 is less than $\alpha = .20$ the null hypothesis is rejected.

The critical $\hat{p}$ value can be obtained by noting that $P(0) + P(1) = .01522 + .08719 = .10241 < \alpha$ and $P(0) + P(1) + P(2) = .01522 + .08719 + .21402 = .31643 > \alpha$. Therefore, critical $\hat{p}$ is the proportion associated with one success which is $1/7 = .14286$. Because the obtained proportion of .00000 is less than the critical proportion of .14286, the null hypothesis is rejected. The relationship between the p-value, α, critical and obtained $\hat{p}$ is shown in Figure 4.12. The p-value is shaded.

EXAMPLE 4.17

What is the smallest possible level of α that can be used for the following hypothesis test?

$$H_0 : \pi = .65 \qquad H_A : \pi* > .65 \qquad n = 9$$

What would the smallest possible level of α be if the alternative were $H_A : \pi < .65$?

Suppose a researcher decided to test the above null hypothesis against the alternative $H_A : \pi > .65$ using $\alpha = .15$. What would the *actual* level of α be? What would this value be for the alternative $H_A : \pi < .65$?

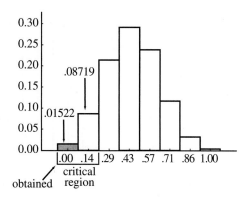

FIGURE 4.12: Binomial test with alternative H_A : $\pi < .45$.

Solution   The smallest possible level for the alternative $H_A : \pi > .65$ is by Equation 4.5

$$P(9) = \frac{9!}{9!\,(0)!}.65^9\,(1 - .65)^0 = .65^9 = .02071$$

while that for the lower tail alternative is

$$P(0) = \frac{9!}{0!\,(9)!}.65^0\,(1 - .65)^9 = .35^9 = .00008.$$

   The actual level of the upper tail test is defined by the largest possible upper tail area that does not exceed intended α. This value can be obtained by noticing that $P(7) + P(8) + P(9) = .21619 + .10037 + .02071 = .33727$ which is greater than intended α of .15. By contrast, $P(8) + P(9) = .10037 + .02071 = .12108$ which is less than intended α. The actual level for the upper tail test would thus be .12108.

   The actual level for the lower tail test would be .05359 because $P(0) + P(1) + P(2) + P(3) + P(4) = .00008 + .00132 + .00979 + .04241 + .11813 = .17173$ which is greater than $\alpha = .15$ while $P(0) + P(1) + P(2) + P(3) = 00008 + .00132 + .00979 + .04241 = .05359$ which is less than .15.   ■

EXAMPLE 4.18

Perform the following hypothesis test. Report results for both the p-value versus alpha and obtained $\hat{p}$ versus critical $\hat{p}$ methods. What would the result be if α were set at .01?

$$H_0 : \pi = .5 \qquad \hat{p} = 1.0 \qquad n = 5$$
$$H_A : \pi > .50 \qquad \alpha = .05$$

Solution   Because the alternative hypothesis specifies an upper tail test, the p-value is the probability of obtaining a value of $\hat{p}$ that is greater than or equal to the value actually observed. Because $\hat{p} = 1.0$, the p-value is $P(5) = .03125$ which is less than $\alpha = .05$ so that the null hypothesis is rejected.

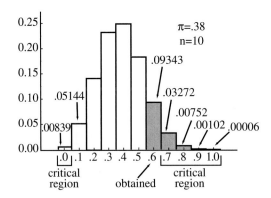

FIGURE 4.13: Binomial test with alternative H_A : $\pi \neq .38$, $\alpha = .10$.

Because $P(4) + P(5) = .15625 + .03125 = .18750 > \alpha = .05$ and $P(5) = .03125 < \alpha = .05$, critical $\hat{p} = 5/5 = 1.0$. Since obtained $\hat{p}$ equals critical $\hat{p}$, the null hypothesis is rejected. ■

two-tailed test

As with Z and t tests, two-tailed hypothesis tests based on the binomial distribution employ critical regions in both tails of the distribution. Unlike those tests, however, there is no one generally accepted method for constructing such regions or calculating two-tailed p-values. The method we present here is among the most commonly used.[11]

Because the binomial distribution is discrete, it will usually not be possible to construct critical regions with associated probabilities exactly equal to $\alpha/2$ or even to have equal probabilities in the two regions.[12] The strategy, then, is to construct regions whose associated probabilities are as close to $\alpha/2$ as possible without exceeding this value. For example, consider the test

$$H_0 : \pi = .38 \qquad \hat{p} = .60 \qquad n = 10$$
$$H_A : \pi \neq .38 \qquad \alpha = .10$$

The sampling distribution for this test is shown in Table 4.5. A depiction of the critical regions and half the p-value is shown in Figure 4.13.

Notice first that the left hand critical region has associated probability .00839. Were we to expand this region to include $\hat{p}$ of .1, the probability of $\hat{p}$ falling in this region under a true null hypothesis would be $.00839 + .05144 = .05983$ which is greater than $\alpha/2 = .05$. The right-hand region has probability $.00006 + .00102 + .00752 + .03272 = .04132$. Again, we cannot expand this region without exceeding $\alpha/2$. This means that the probability of a Type I error is not .10 as intended, but is only $.00839 + .04132 = .04971$. It also follows that critical $\hat{p}$ is .00 and .7.

---

[11]We will present a different method in conjunction with Fisher's exact test in Chapter 10. The method demonstrated there is also applicable to the current problem.

[12]The two regions will be balanced when π equals .5.

TABLE 4.6: Sampling distributions of $\hat{p}$ for $n = 8$ and $\pi = .35, .50,$ and .55.

Proportion $\hat{p}$	Number of Successes y	$\pi = .35$ $P(y)$	$\pi = .50$ $P(y)$	$\pi = .55$ $P(y)$
.000	0	.03186	.00391	.00168
.125	1	.13726	.03125	.01644
.250	2	.25869	.10937	.07033
.375	3	.27859	.21875	.17192
.500	4	.18751	.27344	.26266
.625	5	.08077	.21875	.25683
.750	6	.02175	.10937	.15695
.875	7	.00335	.03125	.05481
1.000	8	.00023	.00391	.00837

The p-value under this scheme is obtained by doubling the one-tailed p-value with the one-tailed p-value being defined as the smaller of the two-tail probabilities.[13] In this case the p-value would be

$$2\,(.00006 + .00102 + .00752 + .03272 + .09343) = .26950.$$

Note that any value of $\hat{p}$ that falls in a critical region will produce a p-value that is less than α while any $\hat{p}$ falling outside a critical region will produce a p-value greater than α.

The test of hypothesis is conducted via the p-value versus α method by noting that the p-value of .26950 is greater than $\alpha = .10$ so that the null hypothesis is not rejected. The same conclusion is reached via the obtained versus critical $\hat{p}$ method by noting that obtained $\hat{p}$ of .6 is between the critical values of 0.0 and .7. It follows that .6 is in neither critical region.

EXAMPLE 4.19

Use the binomial probabilities in Table 4.6 to perform the following hypothesis test. Report results for both the p-value versus alpha and obtained versus critical $\hat{p}$ methods.

$$H_0 : \pi = .35 \qquad \hat{p} = .25 \qquad n = 8$$
$$H_A : \pi \neq .35 \qquad \alpha = .05$$

Solution   As may be seen in Panel A of Figure 4.14, there is no critical region in the lower tail of the distribution. This comes about because the smallest possible value of $\hat{p}$ (i.e., 0) has probability .03186 which is greater than $\alpha/2 = .05/2 = .025$. The upper tail region consists of $\hat{p}$ values of .875 and 1.000. $\hat{p}$ of .750 cannot be included in the critical region because the summed probabilities $.00023 + .00335 + .02175 = .02533$ would exceed $\alpha/2 = .025$.

---

[13]That is, the smaller of the two probabilities obtained when the summed probabilities of $\hat{p}$ values that are less than or equal to obtained $\hat{p}$ are compared to those obtained from summing probabilities of $\hat{p}$ values that are greater than or equal to obtained $\hat{p}$.

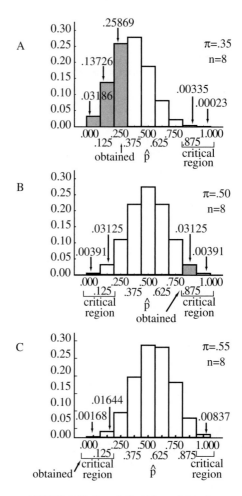

FIGURE 4.14: Two-tailed binomial tests.

There is only one critical value of $\hat{p}$ that being .875. Obtained $\hat{p}$ of .250 is not in the critical region so that the null hypothesis is not rejected. We compute the p-value as two times the one-tailed p-value or $2(.03186 + .13726 + .25869) = .85562$ which exceeds $\alpha = .05$ so that the null hypothesis is not rejected. ∎

EXAMPLE 4.20

Use the binomial probabilities in Table 4.6 to perform the following hypothesis test. Report results for both the p-value versus alpha and obtained versus critical $\hat{p}$ methods.

$$H_0 : \pi = .50 \qquad \hat{p} = .875 \qquad n = 8$$
$$H_A : \pi \neq .50 \qquad \alpha = .10$$

Solution As may be seen in Panel B of Figure 4.14, the lower tail critical region consists of the values .000 and .125 while the upper tail region consists of the values .875 and 1.000.

Notice that neither of these regions can be enhanced without the associated probabilities exceeding $\alpha/2 = .10/2 = .05$. Critical $\hat{p}$ values are then .125 and .875. Because obtained $\hat{p}$ of .875 is in the upper critical region, the null hypothesis is rejected.

The two-tailed p-value is twice the one-tailed p-value or

$$2\,(.00391 + .03125) = 0.07032.$$

Because this is less than $\alpha = .10$, the null hypothesis is rejected. The one-tailed p-value is depicted as the shaded portion of Panel B in Figure 4.14.   ■

EXAMPLE 4.21

Use the binomial probabilities in Table 4.6 to perform the following hypothesis test. Report results for both the p-value versus alpha and obtained versus critical $\hat{p}$ methods.

$$H_0 : \pi = .55 \qquad \hat{p} = .000 \qquad n = 8$$
$$H_A : \pi \neq .55 \qquad \alpha = .05$$

Solution   As may be seen in Panel C of Figure 4.14, the lower tail critical region consists of the values .000 and .125 while the upper tail region contains only 1.000. Notice that neither of these regions can be enhanced without the associated probabilities exceeding $\alpha/2 = .05/2 = .025$. Critical $\hat{p}$ values are then .125 and 1.000. Because obtained $\hat{p}$ of .000 is in the lower critical region, the null hypothesis is rejected.

The two-tailed p-value is twice the one-tailed p-value or $2\,(.00168) = .00336$. Because this is less than $\alpha = .05$, the null hypothesis is rejected. The one-tailed p-value is depicted as the shaded portion of Panel C in Figure 4.14.   ■

Approximate Test.   As you learned in Section 4.2.6, the normal curve can be used to approximate the sampling distribution of $\hat{p}$ provided that sample size is sufficiently large. It follows that in this circumstance the normal curve can be used as the basis for hypothesis tests. These tests are conducted in a manner similar to that used for the one mean Z test with the primary difference being that obtained Z is calculated by

$$Z = \frac{\hat{p} - \pi_0}{\sqrt{\frac{\pi_0(1-\pi_0)}{n}}} \qquad (4.9)$$

which is simply Equation 4.6 with π_0 substituted for π to indicate that an hypothesized value of π is being used rather than π. (You may wish to review Section 4.3.3 on page 89 before continuing.) The test is demonstrated below.

EXAMPLE 4.22

Use the information provided below to perform an approximate test of the stated null hypothesis. Justify your decision regarding the null hypothesis on the basis of both the p-value versus alpha and obtained Z versus critical Z methods.

$$H_0 : \pi = .20 \qquad \hat{p} = .217 \qquad n = 350$$
$$H_A : \pi > .20 \qquad \alpha = .01$$

Solution By Equation 4.9 obtained Z is

$$Z = \frac{\hat{p} - \pi_0}{\sqrt{\frac{\pi_0(1-\pi_0)}{n}}} = \frac{.217 - .200}{\sqrt{\frac{(.20)(.80)}{350}}} = .80.$$

Column three of Appendix A shows that the area above $Z = .80$ is .2119. Since this value is greater than $\alpha = .01$, the null hypothesis is not rejected. Column three of the normal curve table also shows that the closest value to .01 is .0099 which has an associated Z value of 2.33. Thus, the Z value at the leading edge of the critical region is 2.33. Since obtained Z of .80 is less than critical Z of 2.33, the null hypothesis is not rejected. ∎

EXAMPLE 4.23

Use the information provided below to perform an approximate test of the stated null hypothesis. Justify your decision regarding the null hypothesis on the basis of both the p-value versus alpha and obtained Z versus critical Z methods.

$$H_0 : \pi = .58 \qquad \hat{p} = .53 \qquad n = 400$$
$$H_A : \pi \neq .58 \qquad \alpha = .05$$

Solution By Equation 4.9 obtained Z is

$$Z = \frac{\hat{p} - \pi_0}{\sqrt{\frac{\pi_0(1-\pi_0)}{n}}} = \frac{.53 - .58}{\sqrt{\frac{(.58)(.42)}{400}}} = -2.03.$$

Column three of Appendix A shows that the area below $Z = -2.03$ is .0212. Multiplying this area by two yields a p-value of .0424. Since this value is less than $\alpha = .05$, the null hypothesis is rejected. Column three of the normal curve table also shows that the .025 tail area ($\alpha/2$) has an associated Z score of 1.96. The critical Z values for the two-tailed test are then ± 1.96. Since obtained Z of -2.03 is less than critical Z of -1.96, the null hypothesis is rejected. ∎

Assumptions and Consequences of Their Violations for the One Sample Test of a Proportion. The sampling distribution of $\hat{p}$ is appropriately modeled by the binomial distribution when the success/failure observations drawn from a dichotomous population are independent. The exact test of $\hat{p}$ is not generally robust against violations of the independence assumption so that violations may produce misleading results. Violations of the independence assumption may occur under circumstances similar to those discussed on page 98 in relation to the one mean Z test.

The approximate test is valid under the assumptions that observations are independent and that sampling is from a normally distributed population. Obviously, the normality assumption is always violated for this test since sampling is from a dichotomous population. (See Panel A of Figure 4.3 on page 81.) The approximate test is appropriately used, therefore, when sample size is sufficiently large so as to ensure the approximate normality of the sampling distribution of $\hat{p}$ via the central limit theorem. As we discussed in Section 4.2.6, a commonly invoked rule of thumb maintains that the normal curve model will be satisfactory so long as both $n\pi$ and $n(1-\pi)$ are greater than or equal to five. A more conservative rule states that the criterion should be ten rather than five. A continuity correction (see Section 4.2.6 on page 84) can be used in conjunction with the approximate test in an attempt to improve the approximation, but studies by Ramsey and Ramsey [39] discourage this practice.

The approximate test is not generally robust against violations of the independence assumption and is likely to produce misleading results in such instances.

4.3.6   Equivalence Tests

Equivalence testing is a method of testing rather than a specific statistical procedure. Thus, the tests you have studied to this point as well as many others can be used as tests for equivalence. It follows that equivalence testing can deal with population means, proportions, or other parameters. For the sake of efficiency we will restrict this discussion to means and will deal with proportions later. We will consider other parameters in subsequent chapters.

The rationale for equivalence testing stems from the fact that standard statistical tests are designed to establish what is *not* true rather than what *is* true. By this we mean that when a null hypothesis is rejected you can be confident that the null hypothesis is not true. By contrast, when the null hypothesis is not rejected you cannot be confident that the null hypothesis *is* true. You will learn more about why this is true later.

You will also learn about the usefulness of equivalence testing in Chapter 5, but for the moment let us consider a rather contrived example. Suppose that a particular drug, which we shall call drug A, is used to treat anxiety disorder. The object of the treatment is to reduce anxiety to a normal level but not so much that the patient becomes non-responsive to threats in his/her environment. Drug A is an effective treatment as shown by numerous studies based on the XYZ Anxiety Scale. These studies show that patients taking drug A produce a mean score of 80.4 on the XYZ Anxiety Scale which is considered quite satisfactory. An average too much above 80.4 would indicate that the drug leaves patients too anxious while an average too far below 80.4 would show too much desensitization.

While drug A is quite effective in the treatment of anxiety disorder, it has several serious side effects and for this reason has been replaced by drug B which does not produce the undesirable effects. A study is to be conducted to see if drug B controls anxiety as well as did drug A. More specifically, the study is designed to determine whether the population of patients taking drug B manifests a mean XYZ Anxiety Scale score of about 80.4.

Notice that this study attempts to show that the mean population value *is* 80.4 rather than showing that it *is not* 80.4. We will symbolize this value as μ_0. If the purpose of the study was to show that μ_0 is not the population mean, a standard test of significance such as the one mean Z or t test could be employed. Since the goal is to show that μ_0 is the mean, an equivalence test must be used.

The first step in equivalence testing is to define an **equivalence interval** (EI). The EI is a set of values around μ_0 that are sufficiently near μ_0 so as to produce essentially the same result as would be achieved if the mean were μ_0. For example, experts might decide that a mean population value that is less than three points from 80.4 would not be medically significant from 80.4. Therefore, it is decided that if the population mean is any value between 77.4 and 83.4, they will consider drug B *equivalent*, in terms of effectiveness, to drug A. This form of equivalence should not be confused **bioequivalence** which refers to equivalence in the rate and extent of absorption of drugs.

The strategy used to show that drug B produces results equivalent to those of drug A is to show that the mean value of the drug B treated population is less than 83.4 *and* greater than 77.4. That is, to show that the population mean is within EI. This can be done with two hypothesis tests of the following form.

Test One

$H_0 : \mu = 83.4$

$H_A : \mu < 83.4$

Test Two

$H_0 : \mu = 77.4$

$H_A : \mu > 77.4$

Notice that if the first test is significant the interpretation is that the population mean is less than 83.4. Rejection of the second test leads to the conclusion that the mean is greater than 77.4. Therefore, if *both* tests are significant it can be concluded that the population mean is in EI and that, from a practical standpoint, the two drugs are equivalent insofar as treatment of anxiety disorder is concerned.

Two-Tailed Test.   If we let EI_U represent the upper end of EI and EI_L the lower end, the null and alternative hypotheses for the (one mean) two-tailed equivalence test are as follows.

$$H_{0E} : \mu \leq EI_L \quad \text{or} \quad \mu \geq EI_U$$
$$H_{AE} : EI_L < \mu < EI_U$$

Notice that the null hypothesis states that the population mean is not in EI while the alternative states that the population mean is in EI. The null hypothesis, then, is an assertion of non-equivalence while the alternative asserts equivalence. In order to reject H_{0E} it is necessary to show that $\mu > EI_L$ *and* that $\mu < EI_U$. This is done by conducting two one-tailed tests with hypotheses as follows.

Test One

$H_{01} : \mu = EI_U$

$H_{A1} : \mu < EI_U$

Test Two

$H_{02} : \mu = EI_L$

$H_{A2} : \mu > EI_L$

We have used the notation H_{0E} and H_{AE} to indicate the null and alternative hypotheses for the equivalence test and H_{01}, H_{A1}, H_{02}, and H_{A2} for the null and alternative hypotheses of the two component tests. Using this notation, H_{0E} is rejected in favor of H_{AE}

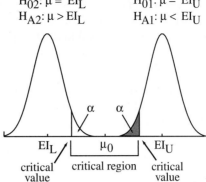

$H_{02}: \mu = EI_L$ $H_{01}: \mu = EI_U$
$H_{A2}: \mu > EI_L$ $H_{A1}: \mu < EI_U$

FIGURE 4.15: Two-tailed equivalence test for a population mean.

only if *both* H_{01} and H_{02} are rejected in favor of their respective alternatives. Figure 4.15 represents the two-tailed equivalence testing procedure. Notice that the critical region for the equivalence test is the region that encompasses the critical regions of both component tests.

Several differences between two-tailed equivalence and two-tailed standard hypothesis tests should be recognized. First, the equivalence test is performed by conducting two *one-tailed* tests. Thus, a two-tailed equivalence test at $\alpha = .05$ would be carried out by means of two one-tailed tests each conducted at $\alpha = .05$ *not* at $\alpha/2$ as would be done with the standard test. This stems from the fact that when H_{0E} is true, the population mean must be *either* EI_U *or* EI_L. If the population mean is EI_U, the probability of obtaining a significant result is α which is shown as the shaded area of the right hand curve in Figure 4.15. If the mean is EI_L the probability of a significant finding is the critical region of the left-hand curve. This means that the probability of falsely concluding equivalence is *either* the area in the critical region of the right hand curve *or* the area of the critical region in the left-hand curve. In either case, the probability of a significant finding in the face of non-equivalence will be α.[14]

A second difference, which follows from the first, is the manner in which the *p*-value is calculated. The *p*-value for the two-tailed equivalence test is obtained by calculating the *p*-value for each of the one-tailed component tests and choosing the larger value. For example, if the *p*-values for the two one-tailed tests were .07 and .001, the *p*-value for the two-tailed equivalence test would be .07. This method of obtaining the *p*-value follows from the fact that both component tests must be significant in order to reject the null hypothesis of non-equivalence in favor of a finding of equivalence. If the larger *p*-value is less than α the smaller value must also be less than α which means that both component tests are significant. On the other hand, if the larger *p*-value is greater than α then at least one of the two component tests was not significant which means that the non-equivalence hypothesis is not rejected.

---

[14]Actually, it may be less than α but we will not introduce that complication here.

Finally, when conducting the test via the obtained versus critical value method, you must compute two obtained values for comparison with the two critical values. H_{0E} is rejected when both obtained values fall in their respective critical regions.

EXAMPLE 4.24

Use one mean Z tests and the information provided below to conduct a two-tailed equivalence test of the null hypothesis that μ is not in the EI 77.4 to 83.4. Report results for both the p-value versus α and obtained versus critical Z methods.

$$\bar{x} = 82.2 \qquad n = 30$$
$$\sigma = 8 \qquad \alpha = .05$$

Solution   The equivalence test is carried out by conducting both of the following tests.

Test One

$H_{01} : \mu = 83.4$

$H_{A1} : \mu < 83.4$

Test Two

$H_{02} : \mu = 77.4$

$H_{A2} : \mu > 77.4$

For the first test

$$Z_1 = \frac{82.2 - 83.4}{\frac{8}{\sqrt{30}}} = -.82$$

and for the second

$$Z_2 = \frac{82.2 - 77.4}{\frac{8}{\sqrt{30}}} = 3.29.$$

Reference to column three of Appendix A shows the respective p-values to be .2061 and .0005. Because the larger of these is greater than $\alpha = .05$, the null hypothesis is not rejected.

Although Z_2 is greater than its associated critical Z of 1.65 and is therefore significant, Z_1 is greater than its associated critical value of -1.65 and is not significant. Since both tests must be significant in order to reject H_{0E}, the null hypothesis is not rejected. The interpretation of this result is that equivalence could not be demonstrated.   ■

EXAMPLE 4.25

Use one mean t tests and the information provided below to conduct a two-tailed equivalence test of the null hypothesis that μ is not in the EI 2 to -2. Use $\alpha = .05$.

Sample: 3 -1 0 -4 -2 2 1 -3 -1 0

Solution   The equivalence test is carried out by conducting both of the following tests.

Test One

$$H_{01} : \mu = 2$$
$$H_{A1} : \mu < 2$$

Test Two

$$H_{02} : \mu = -2$$
$$H_{A2} : \mu > -2$$

The values of $\bar{x}$ and s can be obtained from the following sums.

X	X^2
3	9
-1	1
0	0
-4	16
-2	4
2	4
1	1
-3	9
-1	1
0	0
-5	45

By Equations 2.1 and 2.16, $\bar{x}$ and s are

$$\bar{x} = \frac{\sum x}{n} = \frac{-5}{10} = -.5$$

and

$$s = \sqrt{\frac{\sum x^2 - \frac{(\sum x)^2}{n}}{n - 1}} = \sqrt{\frac{45 - \frac{-5^2}{10}}{9}} = 2.173.$$

By Equation 4.8, the two test statistics are computed as

$$t_1 = \frac{\bar{x} - \mu_0}{\frac{s}{\sqrt{n}}} = \frac{-.5 - 2}{\frac{2.173}{\sqrt{10}}} = -3.638$$

and

$$t_2 = \frac{-0.5 - (-2)}{\frac{2.173}{\sqrt{10}}} = 2.183.$$

Appendix B shows that a value of 1.833 cuts off .05 in the tail of a t distribution with 9 degrees of freedom. The critical values for the two one-tailed tests conducted at $\alpha = .05$ are, therefore, -1.833 and 1.833. Because $t_1 = -3.638 < -1.833$, and $t_2 = 2.183 > 1.833$, both component hypotheses are rejected. Since both component tests are significant, the null hypothesis of non-equivalence is rejected in favor of equivalence. ∎

TABLE 4.7: Sampling distributions of $\hat{p}$ for $n = 8$ and $\pi = .3$, and .7.

Proportion $\hat{p}$	Number of Successes y	$\pi = .30$ P (y)	$\pi = .70$ P (y)
.000	0	.05765	.00007
.125	1	.19765	.00122
.250	2	.29648	.01000
.375	3	.25412	.04668
.500	4	.13614	.13614
.625	5	.04668	.25412
.750	6	.01000	.29648
.875	7	.00122	.19765
1.000	8	.00007	.05765

EXAMPLE 4.26

Use exact one sample tests for a proportion to conduct a two-tailed equivalence test of the null hypothesis that π is not in the EI .3 to .7. Use $n = 8$, obtained $\hat{p}$ of .5, and $\alpha = .20$. Report results for both the p-value versus α and obtained versus critical $\hat{p}$ methods.

Solution The equivalence test is carried out by conducting both of the following tests.

Test One

$H_{01} : \pi = .7$

$H_{A1} : \pi < .7$

Test Two

$H_{02} : \pi = .3$

$H_{A2} : \pi > .3$

Table 4.7 shows binomial probabilities for $n = 8$ and $\pi = .3$ and .7. Using the probabilities for $\pi = .70$, the p-value for Test One is $P (0) + P (1) + P (2) + P (3) + P (4) = .00007 + .00122 + .01000 + .04668 + .13614 = .19411$. Likewise, using the probabilities for $\pi = .30$, the p-value for Test Two is $P (4) + P (5) + P (6) + P (7) + P (8) = .13614 + .04668 + .01000 + .00122 + .00007 = .19411$. Since both of these values are less than $\alpha = .20$, both tests are significant which leads to rejection of the null hypothesis of non-equivalence.

Because $P (0) + P (1) + P (2) + P (3) + P (4) = .19411$ which is less than $\alpha = .20$ and $P (0) + P (1) + P (2) + P (3) + P (4) + P (5) = .19411 + .25412 = .44823$ which is greater than $\alpha = .20$, the critical value for Test One is $\hat{p} = .50$. Likewise, because $P (8) + P (7) + P (6) + P (5) + P (4) = .19411$ which is less than $\alpha = .20$ and $P (8) + P (7) + P (6) + P (5) + P (4) + P (3) = .19411 + .25412 = .44823$ which is greater than $\alpha = .20$, the critical value for Test Two is also $\hat{p} = .50$. Since obtained $\hat{p} = .50$ is less

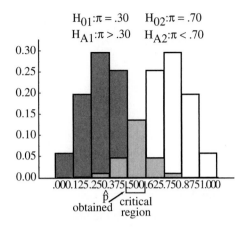

$H_{01}{:}\pi = .30$ $H_{02}{:}\pi = .70$
$H_{A1}{:}\pi > .30$ $H_{A2}{:}\pi < .70$

FIGURE 4.16: Two-tailed equivalence test for a population proportion.

than or equal to critical $\hat{p}$ of .50, the null hypothesis for Test One is rejected. Test Two is also significant because obtained $\hat{p}$ is greater than or equal to critical $\hat{p}$. Again, the null hypothesis of non-equivalence is rejected. The two binomial sampling distributions and the joint critical region are depicted in Figure 4.16.

   One-Tailed Test.   As you will learn in Chapters 5 and 6, one-tailed equivalence tests are more commonly employed in research than are two-tailed tests. The one-tailed test is carried out by conducting *either* Test One *or* Test Two but not both. The choice of test depends on the null hypothesis of interest. For example, suppose that a surgical procedure is known to restore visual acuity to near the normal level of 20/20 but is also known to have high risk of infection which may lead to total loss of vision. A new method has a much lower rate of infection but the question arises as to whether it is as effective as the older method in terms of vision restoration. A panel of vision experts decide that if the new procedure produces average acuities of 20/30 or less, the new method will be considered functionally equivalent to the older treatment method. Notice that neither method could be expected to improve vision to a level better than normal (i.e., 20/20).

   In this situation, Test One would be carried out with EI_U set at the numerical equivalent of 20/30. (The numerical equivalent is called a LogMar score and is .20 for an acuity of 20/30. The LogMar equivalent of 20/20 is 0.0) The equivalence null hypothesis would maintain that average acuity produced by the new surgical method is greater than or equal to 20/30 (.20 LogMar) while the alternative would hold that the average is less than 20/30. The one-tailed equivalence test is carried out by conducting the following test at the appropriate level of α.[15]

Test One

$H_{01} : \mu = .20$

$H_{A1} : \mu < .20$

---

[15]Visual acuity scores are typically skewed so that the robustness of the one mean Z or other normality assuming test would have to be relied upon for a valid result.

If the null hypothesis is rejected, the methods will be declared equivalent. In general, the null and alternative hypotheses for the one-tailed equivalence test are

$$H_{0E} : \mu \geq EI_U$$
$$H_{AE} : \mu < EI_U$$

or

$$H_{0E} : \mu \leq EI_L$$
$$H_{AE} : \mu > EI_L$$ ∎

EXAMPLE 4.27

Use a one mean Z test with the information provided below to conduct a one-tailed equivalence test of the null hypothesis that μ is greater than or equal to $EI_U = .20$. Report results for both the p-value versus α and obtained versus critical Z methods.

$$\bar{x} = .11 \qquad n = 30$$
$$\sigma = .16 \qquad \alpha = .05$$

Solution   The equivalence test is carried out by conducting Test One as follows.

Test One
$$H_{01} : \mu = .20$$
$$H_{A1} : \mu < .20$$

For this test
$$Z_1 = \frac{\bar{x} - \mu_0}{\frac{\sigma}{\sqrt{n}}} = \frac{.11 - .20}{\frac{.16}{\sqrt{30}}} = -3.08.$$

Reference to column three of Appendix A shows the associated p-value to be .0010. Because this value is less than $\alpha = .05$, the null hypothesis of non-equivalence is rejected in favor of the alternative of equivalence. Because obtained Z of -3.08 is less than critical Z of -1.65, the null hypothesis of non-equivalence is rejected. ∎

EXAMPLE 4.28

Persons with iron deficiency anemia often have hemoglobin levels in the range 8–10 g/dl whereas normal levels would be about 15 g/dl. Treatment with iron supplements usually brings levels to normal values but does not cause values to increase above normal levels. Suppose that a dietary supplement has been developed that can be easily and cheaply produced in developing countries. The supplement will be declared effective if it can be shown to be equivalent to more expensive treatments. It is decided that if iron deficient persons in developing countries who take the new supplement produce average hemoglobin values greater than 13, the new supplement will be declared equivalent to older methods.

 Use a one mean Z test with the information provided below to test the equivalence null hypothesis that persons taking the new supplement produce average hemoglobin levels that are less than or equal to 13.[16] Report results for both the *p*-value versus α and obtained versus critical Z methods.

$$\bar{x} = 13.8 \qquad n = 150$$
$$\sigma = 1.16 \qquad \alpha = .05$$

Solution The equivalence test is carried out by conducting Test Two as follows.

Test Two
$$H_0 : \mu = 13$$
$$H_A : \mu > 13$$

For this test

$$Z_2 = \frac{\bar{x} - \mu_0}{\frac{\sigma}{\sqrt{n}}} = \frac{13.8 - 13.0}{\frac{1.16}{\sqrt{150}}} = 8.45.$$

Reference to column three of Appendix A shows that the associated *p*-value for $Z = 3.5$ is .0002. It follows that *p* for $Z = 8.45$ will be less than this value. Because *p* is less than $\alpha = .05$, the null hypothesis of non-equivalence is rejected. Because obtained Z of 8.45 is greater than critical Z of 1.65, the null hypothesis of non-equivalence is rejected. ■

4.3.7 Errors and Correct Decisions in Hypothesis Testing

As you are now aware, it is possible to reject a null hypothesis even though it is true. You have also learned that the probability of such an occurrence is symbolized as α. It is also possible to err by failing to reject a false null hypothesis. In contrast to these two types of errors, a correct decision would be rendered when a true null hypothesis is not rejected or a false null hypothesis is rejected. In this section you will learn about each of these eventualities as well as the probability of their realization. The discussion that follows will be divided into two distinct components, namely, (1) events that occur when the null hypothesis is true and (2) events that occur when the null hypothesis is false. It is important that you keep in mind which of these eventualities is being addressed in the discussion that follows. The comments that follow apply to hypothesis testing in general, but for the sake of simplicity, will be presented in relation to the one mean Z test. Table 4.8 provides a summarization of the following discourse and should be consulted as you read.

Events That Occur When the Null Hypothesis Is True. A **Type I error** occurs when a *true* null hypothesis is rejected. The *probability* of making a Type I error is termed the **significance level** of the test or α. In order to avoid bias, α should be established before the analysis is performed. Note also that the probability of a Type I error is under the control of the researcher who establishes the level of significance for the test.

Failure to reject a *true* null hypothesis results in a **correct decision**. The probability of a correct decision when the null hypothesis is true is $1 - \alpha$. The probabilities of a Type I error and a correct decision for a one-tailed one mean Z test are depicted in Figure 4.17.

---

[16]Once again we must rely on the robustness of the one mean Z to departures from population normality.

TABLE 4.8: Outcomes associated with hypothesis testing.

	Null Hypothesis	
	True	False
Reject	Type I Error (α)	Correct Decision (Power)
Fail to Reject	Correct Decision $(1 - \alpha)$	Type II Error (β)

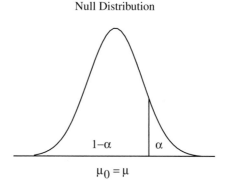

FIGURE 4.17: Probability of a Type I error and correct decision for a one-tailed one mean Z test.

Events That Occur When the Null Hypothesis Is False. A **Type II error** occurs when a *false* null hypothesis is not rejected. The *probability* of making a Type II error is termed **beta** (β).

Rejection of a *false* null hypothesis results in a **correct decision**. The probability of a correct decision when the null hypothesis is false is termed **power**.

Notice that when the null hypothesis is true, the hypothesized value of the population mean (μ_0) is the population mean (μ) (see Figure 4.17). In this circumstance the sampling distribution used to perform the hypothesis test is in fact the sampling distribution of the test statistic. In contrast, when the null hypothesis is false the mean of the population (μ) is different from the hypothesized mean (μ_0). This implies that the sampling distribution of the test statistic is centered around μ rather than μ_0 so that the model used to perform the hypothesis test is not an accurate depiction of the sampling distribution of the statistic. This situation is depicted in Figure 4.18.

It is important to understand that the null distribution is used to perform the hypothesis test but it is the **alternative** distribution that represents the sampling distribution of the statistic. This arises because the mean of the population is μ rather than μ_0. Suppose, for example, that a researcher wishes to test the null hypothesis $\mu = \mu_0 = 100$ and uses the null curve for this purpose. *Unknown* to the researcher is the fact that $\mu = 120$ so that

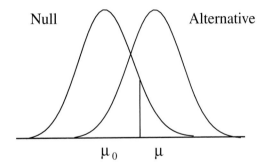

FIGURE 4.18: Null and alternative distributions.

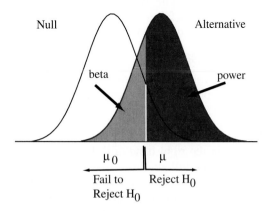

FIGURE 4.19: Power and beta for a one-tailed one mean Z test.

the sampling distribution is centered around this value rather than 100. It follows that the probability of obtaining a test statistic in any specified range is represented by an area under the alternative rather than the null curve.

Because power is the probability of rejecting a false null hypothesis, it can be depicted as the portion of the alternative curve that represents values of the test statistic that are large (or small) enough to cause rejection of the null hypothesis. Consider now Figure 4.19. In this figure the alternative curve is divided into two portions, namely, the portion representing power (darker shading) and the portion representing beta (lighter shading). (The leading edge of the critical region has been marked with a white line.) Notice that the null hypothesis is rejected when the value of the test statistic is greater than or equal to the value at the leading edge of the critical region. The probability of rejecting the null hypothesis is, then, the portion of the alternative curve that represents values of the test statistic that are equal to or exceed the leading edge of the critical region. The null hypothesis will not be rejected when the test statistic is below the leading edge of the critical region. This probability (β) is represented by the portion of the alternative curve that lies below the critical region. As can be seen in the figure, power $= 1 - \beta$ and is often represented as such in the statistics and research literature.

EXAMPLE 4.29

Answer the following questions as they relate to Panel A of Figure 4.20. (a) What is the alternative hypothesis for the test of significance? (b) What area(s) represent power? (c) What area(s) represent beta? (d) What does area "e" represent?

Solution (a) The alternative states $H_A : \mu < \mu_0$. This follows from the fact that the critical region is in the left hand tail of the null distribution. (b) Power is represented by areas a and c. This follows from the fact that areas a and c represent the probability that a test statistic will be small enough to meet the criterion for rejection of H_0 as established by the critical region of the null curve. (c) Beta is depicted by areas b and d. These areas are the portion of the alternative curve that falls outside the critical region and, therefore, represents the probability that the test will result in failure to reject the null hypothesis. (d) Because the null hypothesis is false, area e does not represent the probability of any outcome. All such probabilities are depicted as areas of the alternative curve. ∎

EXAMPLE 4.30

Answer the following questions as they relate to Panel B of Figure 4.20. (a) What is the alternative hypothesis for the test of significance? (b) What area(s) represent power? (c) What area(s) represent beta? (d) What does area "a" represent? (e) What does area d represent?

Solution (a) The alternative states $H_A : \mu > \mu_0$. This stems from the fact that the critical region is in the right hand tail of the null distribution. (b) Power is represented by areas b and c. This follows from the fact that areas b and c depict the probability that a test statistic will be large enough to meet the criterion for rejection of H_0 as established by the critical region of the null curve. (c) Beta is depicted as area e. This area is the portion of the alternative curve that falls outside the critical region and, therefore, represents the probability that the null hypothesis will not be rejected. (d) Because the null hypothesis is false, area a does not represent the probability of any outcome. All such probabilities are shown as areas of the alternative curve. (e) As with area a, area d does not represent any outcome. ∎

Factors That Determine Power and Beta. Power and beta are determined by (a) the level of significance of the test (α), (b) sample size (n), and (c) the form of the alternative distribution. We will now elaborate.

Level of Significance

Power increases as the level of significance is increased (e.g., .01 to .05). Because beta = 1 − power, beta decreases with increases in the level of significance. This relationship is shown in Figure 4.21. As can be seen in this figure, when $\alpha = .01$, power is represented by areas d and a with beta being depicted as areas b, c, f, and e. When α is increased by .04 to .05, power is increased by areas b and c with beta being decreased by this same amount. Naturally, the reverse is also true. Decreases in level of significance are associated with decreases in power and increases in beta. Thus, the price paid for decreasing the probability of a Type I error is an increase in the probability of a Type II error.

Sample Size (n)

From a practical point of view, perhaps the most important determinant of power and beta is sample size. As sample size increases, power increases. Thus, a researcher who wishes

A

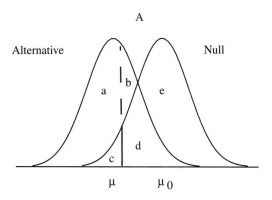

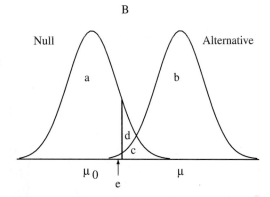

B

FIGURE 4.20: Two example depictions of power and beta for one-tailed one mean Z tests.

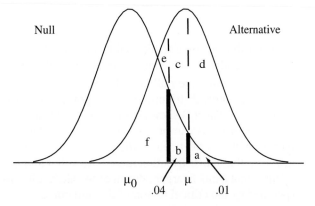

FIGURE 4.21: Relationship of level of significance to power and beta.

A

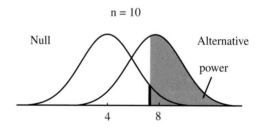

B

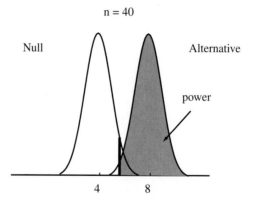

FIGURE 4.22: Relationship of sample size to power.

to establish α at some traditional level (e.g., .05) can increase power by increasing sample size. This is demonstrated in Figure 4.22 for the one mean Z test.

Panel A shows the null and alternative distributions when $n = 10$. The shaded portion of the alternative distribution depicts power and appears to represent slightly more than .5 of the curve. Panel B shows distributions with the same means and level of significance but with $n = 40$. As can be seen, the curves in Panel B have smaller standard errors of the mean which in effect causes the two distributions to draw away from each other. Notice, however, that the distribution means are still 4 and 8 as is true of the curves in Panel A. The result is that a much larger portion of the alternative distribution meets the rejection criterion specified by the critical region of the null curve.

Because level of significance is often set to some traditional level and the form of the alternative distribution is beyond the control of the researcher, it is usually sample size that is manipulated to address power issues in research. We will say more on this topic shortly.

A

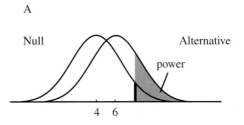

B

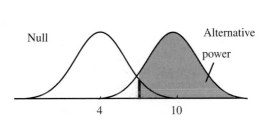

FIGURE 4.23: Power of the one mean Z test as a function of the difference between hypothesized and actual population means.

Form of the Alternative Distribution

The power of the one mean Z test is heavily dependent on the magnitude of the difference between μ_0 and μ. As this difference increases, power increases. As the difference decreases, power decreases until $\mu_0 = \mu$ at which point the null hypothesis is true and the probability of rejecting is α. This characteristic of the test has intuitive support. It seems reasonable that small differences from the null value would be more difficult to detect than would larger differences. Panel A of Figure 4.23 shows a situation where there is a two unit difference between the actual population mean ($\mu = 6$) and the hypothesized value ($\mu_0 = 4$). Panel B shows the increase in power that corresponds to a population mean of 10 which is six units from the hypothesized value.

In general, as the degree to which the null hypothesis is false increases, so does the power of the test. For the binomial test this is expressed as the difference between π_0 and π but may be expressed in very different ways for other types of tests which you will encounter in later chapters.

Calculating Power and Beta.   It is generally not possible to calculate power and beta in applied research situations. The reason is that in order to calculate these quantities you must have certain information about the alternative distribution that is typically not available. For example, for the one mean Z test you must know μ in order to make these calculations. But, if you knew μ, there would be no need for an hypothesis test. It is nevertheless helpful, insofar as understanding power and beta is concerned, to perform a few hypothetical calculations.

For example, what would be the power of a one mean Z test under the following conditions?

$$H_0 : \mu = 4 \qquad \sigma_{\bar{x}} = 2 \qquad \mu = 10$$
$$H_A : \mu > 4 \qquad \alpha = .05$$

Reference to Panels A and B of Figure 4.24 will help to clarify the following discussion. In order to find beta, it is necessary to find the portion of the alternative curve that lies below the leading edge of the critical region on the null curve. In order to find this area we first calculate

$$\frac{\mu_0 - \mu}{\sigma_{\bar{x}}} = \frac{4 - 10}{2} = -3.00.$$

This indicates that $\mu_0 = 4$ is three standard errors below $\mu = 10$. Because the critical Z value for a one-tailed Z test conducted at $\alpha = .05$ is 1.65, it follows that if this value is added to -3.00 we will obtain the number of standard errors that lie between the leading edge of the critical region and μ. For purposes of power calculations we will designate critical Z as Z_α and the Z score indicating the distance between μ and the leading edge of the critical region as Z_β. (See Panel B of Figure 4.24.) In this example, $Z_\beta = -3.00 + 1.65 = -1.35$. The portion of the alternative curve that lies below this value represents beta while the portion above this value is power. As can be seen in column three of the normal curve table, a Z score of -1.35 cuts off .0885 in the lower tail of the alternative curve. Power is $1 - .0885 = .9115$. The probability that the test will result in a Type II error is then .0885 while the probability of a correct decision is .9115.

The logic underlying this solution was to find the number of standard errors between the leading edge of the critical region and the mean of the alternative curve (i.e., Z_β). This was done by finding the number of standard errors between μ_0 and μ and adding the number of standard errors between μ_0 and the leading edge of the critical region (i.e., Z_α). Once this was accomplished it was a simple matter to find the appropriate area in the normal curve table. An expression for Z_β is given by

$$\boxed{Z_\beta = \frac{\mu_0 - \mu}{\sigma_{\bar{x}}} + Z_\alpha} \qquad (4.10)$$

EXAMPLE 4.31

Find power and beta for a one mean Z under the following conditions.

$$H_0 : \mu = 4 \qquad \sigma_{\bar{x}} = 2 \qquad \mu = -2$$
$$H_A : \mu < 4 \qquad \alpha = .05$$

Solution Substituting into equation 4.10 yields

$$Z_\beta = \frac{4 - (-2)}{2} + (-1.65) = 1.35.$$

The portion of the alternative curve that lies above this value represents beta while the portion below this value is power. Reference to Column three of the normal curve table shows that $Z = 1.35$ cuts off .0885 in the upper tail of the normal curve. Power is then $1 - .0885 = .9115$. Why were the same answers obtained here as were obtained in the above example? ∎

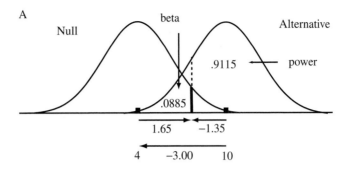

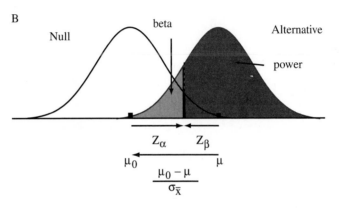

FIGURE 4.24: Power and beta calculation for a one-tailed one mean Z test. Distances are expressed in terms of standard errors.

EXAMPLE 4.32

Find power and beta for a one mean Z under the following conditions.

$$H_0 : \mu = 90 \qquad \sigma = 20 \qquad \alpha = .05$$
$$H_A : \mu \neq 90 \qquad n = 25 \qquad \mu = 88$$

Solution   Using

$$\sigma_{\bar{x}} = \frac{\sigma}{\sqrt{n}} = \frac{20}{\sqrt{25}} = 4$$

in equation 4.10 yields

$$Z_\beta = \frac{90 - 88}{4} + (-1.96) = -1.46.$$

The value for $Z_\alpha = -1.96$ was used because the test under consideration is two tailed with $\alpha = .05$ and $\mu < \mu_0$. This is depicted in Figure 4.25. Notice that the distance

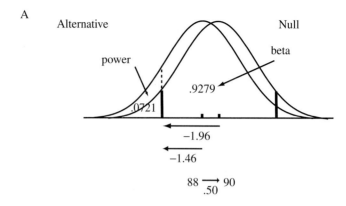

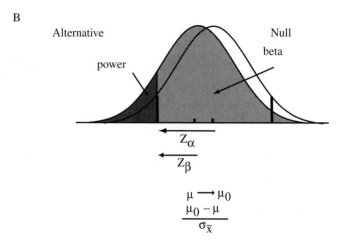

FIGURE 4.25: Power and beta calculation for a two-tailed one mean Z test. Distances are expressed in terms of standard errors.

from μ_0 to μ is .5 standard errors. Because the leading edge of the critical region is 1.96 standard errors below μ_0, the distance from μ to the leading edge of the critical region is $-1.96 + .50 = -1.46$ standard errors. Reference to column three of the normal curve table shows power to be .0721. Beta is $1 - .0721 = .9279$. ∎

EXAMPLE 4.33

Find power and beta for a one mean Z test under the following conditions.

$$H_0 : \mu = 90 \qquad \sigma = 20 \qquad \alpha = .05$$
$$H_A : \mu \neq 90 \qquad n = 25 \qquad \mu = 92$$

Solution   Using

$$\sigma_{\bar{x}} = \frac{\sigma}{\sqrt{n}} = \frac{20}{\sqrt{25}} = 4$$

in equation 4.10 yields

$$Z_\beta = \frac{90 - 92}{4} + (1.96) = 1.46$$

The value for $Z_\alpha = 1.96$ was used because the test under consideration is two-tailed with $\alpha = .05$ and $\mu > \mu_0$. Power is the portion of the normal curve that lies above $Z = 1.46$. Reference to column three of the normal curve table shows power to be .0721. Beta is $1 - .0721 = .9279$. Why are these the same values as were obtained in the above example? ∎

EXAMPLE 4.34

As another example, we will find power and beta for the following exact test for a proportion.

$$H_0 : \pi = .35 \qquad \alpha = .05 \qquad \pi = .50$$
$$H_A : \pi > .35 \qquad n = 8$$

Solution   The binomial probabilities for the null ($\pi = .35$) and alternative ($\pi = .50$) distributions are shown in Table 4.6 on page 113. Critical $\hat{p}$ can be determined by noting that $P(6) + P(7) + P(8) = .02175 + .00335 + .00023 = .02533 < \alpha = .05$ while $P(5) + P(6) + P(7) + P(8) = .08077 + .02533 = .10610 > \alpha = .05$. Critical $\hat{p}$ is then .750. Thus, the null hypothesis is rejected for any value of $\hat{p}$ that is greater than or equal to .750. But what is the probability that $\hat{p}$ will equal or exceed .750? Since $\pi = .50$, this probability is $P(6) + P(7) + P(8) = .10938 + .03125 + .00391 = .14454$. This, then, is the power of the test. By definition, beta is $1 - .14454 = .85546$. It, therefore, appears unlikely that a correct decision would be reached for this test. ∎

EXAMPLE 4.35

What would power and beta be if π were .55 rather than .50 in Example 4.34?

Solution   From Table 4.6 power is

$$.15695 + .05481 + .00837 = .22013.$$

This larger value than was calculated for $\pi = .50$ is expected since .55 is further from the null value of .35 than is .50. Beta is $1 - .22013 = .77987$. What do you think might be done to increase the power of these tests? ∎

Sample Size Calculation.   A question commonly faced by researchers is, "How many subjects should I have in my study?" This question is often resolved on the basis of financial or other considerations unrelated to statistics. In this section we will examine the statistical aspects of this question as they relate to tests of hypotheses. In general, the statistical solution to this question will require computer software or specially designed tables. We will simplify this process by considering sample size calculations for the one mean Z test which are quite straightforward. The goal of this section is to use the one mean Z test to present the logic underlying the solution. The same general logic applies to more complex tests.

Suppose that a researcher plans to use a one mean Z test as the primary analysis tool in his upcoming study. The null hypothesis to be tested is $H_0 : \mu = 100$ while the alternative is of the form $H_A : \mu > 100$. After some thought, the researcher decides that if the population mean is 104 or greater he would like to have a high probability (e.g., .90) of rejecting the null hypothesis. This implies that the researcher is less concerned about detecting means that differ by less than four units from the null value. This decision might be based on the opinion that differences of less than four units are of minor clinical importance while differences of four or more units are important to detect.

To further specify the problem, the researcher wants to have sufficient sample size (n) so as to have power of .90 to reject the null hypothesis if the population mean is 104. Notice that power will be greater than .90 if the population mean is greater than 104. The one-tailed test is to be conducted at $\alpha = .05$.

Solving for n in Equation 4.10 on page 132 yields

$$n = \frac{\sigma^2 \left(Z_\beta - Z_\alpha\right)^2}{(\mu_0 - \mu)^2} \tag{4.11}$$

We can substitute 100 and 104 for μ_0 and μ respectively, and 1.65 for Z_α. Because .90 of the normal curve lies above $Z = -1.28$ (see Panel A of Figure 4.26), this value can be substituted for Z_β. For purposes of this exercise we will assume $\sigma^2 = 400$. This then gives

$$n = \frac{400 \left(-1.28 - 1.65\right)^2}{(100 - 104)^2} = 214.6$$

This would be rounded up to 215. Thus, with 215 subjects in the study the researcher will reject the null hypothesis with probability .90 if the true mean is 104 and will realize a higher probability of rejection if the mean is greater than 104.

A problem in generating sample size estimates is finding a value for σ^2 for the expression. You might protest that we can simply use s^2 as was done with the t test but remember, this calculation is typically carried out *before* the study is conducted so that there may not be a sample from which to obtain the variance estimate. Usually the variance estimate is obtained from the research literature, a pilot study designed to generate the estimate or some other study that employed similar data. In any case, sample size estimates are just that—estimates.

EXAMPLE 4.36

Calculate the sample size required to attain power of .8 to detect a population mean of 8 for a two-tailed Z test of the null hypothesis $H_0 : \mu = 10$ conducted at $\alpha = .01$. Assume that $\sigma = 4$.

Solution   Twenty percent of the normal curve lies above, and 80 percent below, a Z value of .84. (See Panel B of Figure 4.26.) Substituting this value for Z_β, and -2.58 for Z_α, 10 for μ_0, 8 for μ and 16 for σ^2 gives

$$n = \frac{16 \left(.84 - (-2.58)\right)^2}{(10 - 8)^2} = 46.8$$ ■

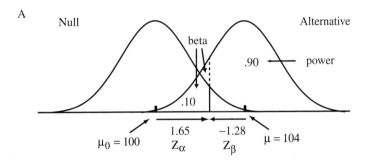

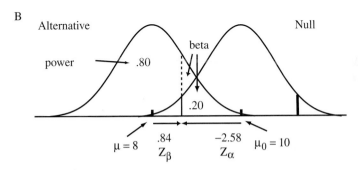

FIGURE 4.26: Sample size calculations for one- and two-tailed one mean z tests. Distances are expressed in terms of standard errors.

4.4 CONFIDENCE INTERVALS

4.4.1 Introduction

Earlier we indicated that there are two basic forms of inference, hypothesis testing, and confidence intervals. Now that you have completed an introduction to hypothesis testing we turn attention to the other form of inference. Perhaps the most important difference between the two inferential methods lies in the questions they address. For example, the one mean Z test inquires as to whether the population mean differs from the value specified by the null hypothesis while a comparable confidence interval asks the simpler question, "What is the population mean?" Notice that the latter question does not imply any form of hypothesis about the population mean but rather simply poses the question as to its value. As Figure 4.27 shows, both methods employ statistics to address questions posed about parameters.

After considering the rationale underlying confidence intervals, you will learn to construct such intervals for the population mean when σ is known and when σ is not known. You will then learn approximate and exact methods for estimating the population proportion. In the following section we will compare the two methods and show that confidence intervals are usually preferable to tests of hypotheses when both methods are applicable to a particular problem.

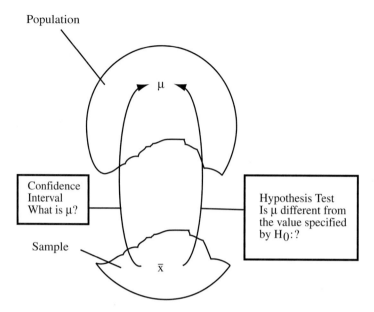

FIGURE 4.27: Questions regarding a population mean posed by an hypothesis test and confidence interval.

4.4.2   Rationale and Method

Two-Sided Confidence Intervals.   Consider the following. Suppose a researcher randomly selects a sample of size $n = 100$ from a normally distributed population with a standard deviation (σ) of 50. The sample mean ($\bar{x}$) is computed. What is the probability that $\bar{x}$ will take a value between μ and a point that is one standard error above μ? Column two of Appendix A gives this probability as .3413. This is also the probability that $\bar{x}$ will take a value between μ and a point that is one standard error below μ. Thus, the probability that $\bar{x}$ will take a value that is within one standard error of μ is $.3413 + .3413 = .6826$. We have marked this region on a curve representing the sampling distribution of $\bar{x}$ in Panel A of Figure 4.28.

Suppose now that one standard error, which in this example is

$$\frac{\sigma}{\sqrt{n}} = \frac{50}{\sqrt{100}} = 5,$$

is added to and subtracted from $\bar{x}$. We will designate $\bar{x} + \frac{\sigma}{\sqrt{n}}$ as U and $\bar{x} - \frac{\sigma}{\sqrt{n}}$ as L. We have depicted $\bar{x}$, L and U in Panel A of Figure 4.28. In the example shown, the randomly selected sample produced a value of $\bar{x}$ that is exactly one standard error above μ.

Several important points should be noted about the interval L, U. First, when $\bar{x}$ takes a value that is one standard error above μ, the interval L, U contains μ. That is, $L \leq \mu \leq U$. Second, if $\bar{x}$ had taken a value one standard error below μ (not depicted in the figure), the interval would again contain μ because μ would be equal to U. Third, any value of $\bar{x}$ that is within one standard error of μ would produce an interval such that μ would be a value

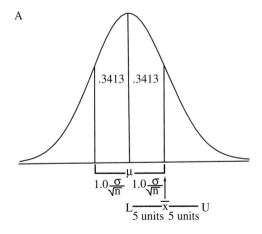

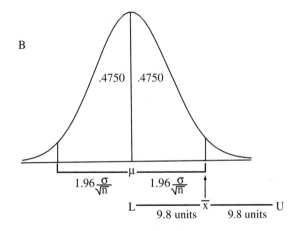

FIGURE 4.28: Rationale underlying two-sided confidence intervals for a population mean.

between L and U while any value of $\bar{x}$ that is not within one standard error of μ would produce values of L and U that do not contain μ. Fourth, the probability[17] that $\bar{x}$ will lie within one standard error of μ and therefore produce an interval L, U that includes the value of μ is approximately .68.

The interval L, U is termed a **confidence interval**. L is the **lower limit** of the confidence interval while U is the **upper limit**. The **level of confidence** or **coverage** of the confidence interval is .68. The statistic around which the interval is formed, $\bar{x}$ in this case, is termed the **point estimate**.

The implications of the above are as follows. A researcher who wishes to estimate the mean of a population may randomly select a sample of size n from the population, calculate

---

[17]We will comment on the use of the word probability in this context below.

the sample mean and generate the confidence interval as described above. The researcher can then assert with 68% confidence that the population mean is some value between L and U. But 68% does not seem a very high level of confidence that μ is in the interval. Suppose the researcher wants to achieve a higher level of confidence, say, a 95% level.

The level of confidence is determined by the *number* of standard errors added to and subtracted from the point estimate. How many standard errors would be added to and subtracted from $\bar{x}$ in order to form a 95% confidence interval? The interval will have to be of sufficient length so as to capture μ in the event that $\bar{x}$ takes any of the values that make up the central .95 of the normal curve. This area is shown in Panel B of Figure 4.28. As may be seen in this figure, .95 of the normal curve lies within 1.96 standard errors of the mean. It follows that adding and subtracting 1.96 standard errors will guarantee that the value of μ will lie between L and U when $\bar{x}$ is within 1.96 standard errors of μ. Of course, if $\bar{x}$ is not within 1.96 standard errors of μ the resulting interval will not contain μ. Thus, the researcher can assert with 95 percent confidence that μ is some value between L and U. Details of how to form specific confidence intervals will be given below.

One-Sided Confidence Intervals. In certain circumstances researchers may be more interested in one end of a confidence interval than the other. In such cases the researcher may choose to form a one-sided confidence interval. A **one-sided** confidence interval is a confidence statement that consists of *either L or U* but not both. For example, an environmental engineer may routinely sample discharge from a water treatment facility in order to estimate ammonia concentrations (measured as nitrogen) contained therein. If the concentration is too high, nearby rivers and ponds may be adversely affected. Ammonia concentrations that are low would be of no concern. In this case the one-sided confidence interval would consist only of U. The researcher could then assert with a specified level of confidenc that ammonia concentration in the discharge is *at most U*. If U is less than some acceptable level no action will be taken. Otherwise efforts must be made to reduce the concentration regardless of what L might be. The advantage of one-sided confidence intervals as compared to two-sided intervals will become clear in the following discussions. Before considering specific forms of one-sided confidence intervals, we will provide the rationale underlying their construction.

In Panel A of Figure 4.29 we have marked the point on the sampling distribution of $\bar{x}$ such that approximately .95 $(.5000 + .4505 = .9505)$ of the curve is below the designated point. As you can see, this point is 1.65 standard errors above the distribution mean. Notice that if L is formed by subtracting 1.65 standard errors from $\bar{x}$, L will always be less than or equal to μ so long as $\bar{x}$ is from the designated .95 region. If $\bar{x}$ is from a point more than 1.65 standard errors above μ, L will not be less than or equal to μ. This can be seen from the figure where $\bar{x}$ has been placed 1.65 standard errors above μ. Any value of $\bar{x}$ that is less than the value shown will produce an L that is less than μ while any $\bar{x}$ greater than the value shown will produce an L that is not less than or equal to μ. Because the probability is .95 that $\bar{x}$ will come from a point that is 1.65 standard errors above the mean of the distribution, or a lesser point, the researcher can declare with .95 confidence that the mean of the population is greater than or equal to L. As an example, if the standard error is 2.0, L would be equal to $\bar{x} - (1.65)(2.0)$ or $\bar{x} - 3.30$.

It may also happen that a researcher is interested in estimating the highest value μ might take without concern for a lower limit. That is, the researcher may wish to make a statement of the form "The mean of the population is no greater than U." In such circumstance a one-

A

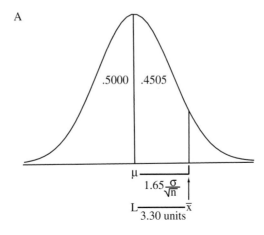

B

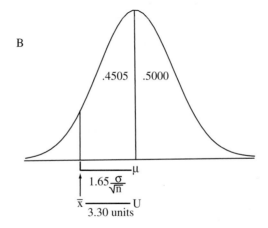

FIGURE 4.29: Rationale underlying one-sided confidence intervals for a population mean.

sided confidence interval can be formed that consists only of U. As panel B of Figure 4.29 shows, any value of $\bar{x}$ that is 1.65 standard errors below μ or higher in the distribution will produce a value of U that is greater than or equal to μ. Thus, the researcher can draw a random sample from the population, calculate U, and then assert with .95 confidence that μ is less than U.

4.4.3   A Note of Caution

When asked to state the meaning of a two-sided 95% confidence interval a student is likely to respond with something to the effect that, "The *probability* that μ is between L and U is .95." A statement of this sort is likely to bring a frown to the face of your instructor or, if committed to paper, the flash of a red pencil. The problem can best be explained by an analogy.

If a marble is randomly selected from *a box that contains seven white and three black marbles*, the probability that the chosen marble will be black is, by Equation 3.1 on page 52, simply the proportion of black marbles in the box or $\frac{3}{10} = .3$. Now suppose you have selected the marble but have not yet opened your hand. What is the probability that the unobserved marble *in your hand* is black? Again, it is the proportion of black marbles in your hand. But, there is only one marble in your hand. Thus, the probability is either one or zero depending on whether the marble in your hand is black or white.

The same logic applies to confidence intervals. *Before* selecting a sample from the population it can be asserted that the probability that the confidence interval yet to be constructed will contain μ is .95 (or some other specified level). The reason for this is that the proportion of possible values of $\bar{x}$ that will produce values of L and U that contain μ is .95. But once the sample has been selected and the confidence interval constructed, the question "What is the probability that *this* confidence interval contains μ" must be answered, as with the marble in your hand, one or zero. As with the marble, you have now limited the problem to one confidence interval so that the proportion containing μ is either one or zero.

Because of certain philosophical considerations that underlie this issue, some statistics instructors require very precise statements regarding the meaning of confidence intervals while others seem not to be so concerned. A statement acceptable to most statisticians is, "Ninety-five percent of all confidence intervals constructed in this fashion will capture μ."

4.4.4   Confidence Interval for μ When σ Is Known

In this and the two following sections we will address the specifics of constructing confidence intervals for μ and π. We begin with confidence intervals for μ when σ is known.

When σ is known, the normal curve may be used as a model of the sampling distribution of $\bar{x}$ so that L and U are obtained by

$$L = \bar{x} - Z\frac{\sigma}{\sqrt{n}}$$ (4.12)

and

$$U = \bar{x} + Z\frac{\sigma}{\sqrt{n}}$$ (4.13)

Two-Sided Intervals.   The level of confidence associated with an interval is determined by Z. For example, suppose that a two-sided .95 confidence interval is to be constructed. The number of standard errors to be subtracted from and added to $\bar{x}$ must be sufficient so as to have L and U capture μ when $\bar{x}$ takes any of the central 95 percent of the values about μ. This is demonstrated in panel B of Figure 4.28 on page 139. This figure shows that .95 of the area of the curve lies within 1.96 standard errors of μ. The appropriate Z value can be obtained by noting that $\frac{.95}{2} = .475$ of the curve lies between μ and the desired point. In order to find the appropriate Z score we must read the normal curve table in reverse order to that which we have previously used. That is, we must look up .4750 in column two of the table and then find the associated Z value. As you can see from the normal curve table, that score is 1.96. It follows that by adding and subtracting 1.96 standard errors to $\bar{x}$, we can be assured that μ will fall between L and U when $\bar{x}$ takes any value that is within 1.96 standard errors of μ. The probability that $\bar{x}$ will take such a value is .95. A few examples will help clarify this method of forming confidence intervals.

EXAMPLE 4.37

Given the information provided below, use two-sided 90, 95, and 99% confidence intervals to estimate the mean IQ of a population of children whose mothers received inadequate pre-natal care. What happens to the confidence interval as the level of confidence is increased?

$$\sigma = 16 \qquad n = 60 \qquad \bar{x} = 90.1$$

Solution   The standard error of the mean $(\frac{\sigma}{\sqrt{n}})$ is $\frac{16}{\sqrt{60}} = 2.066$. How many standard errors should be subtracted from and added to $\bar{x}$ in order to form a 90% confidence interval? This can be determined by noting that $.90/2 = .45$ of the curve lies between μ and the desired Z score. Column two of Appendix A shows that there is no value given for .4500 and that there are two values, .4495 and .4505 that are equally close to the area sought. We shall follow the convention of choosing the area in column two that is *closest* to the desired area when the desired area is not in the table, or choosing the larger area in column two when two areas are equally close to the area sought. In this case there are two areas equally close to .4500 so that we will choose the larger which is .4505. The associated Z score is 1.65. Notice that approximately .90 of the curve lies within 1.65 standard errors of μ. Thus, if 1.65 standard errors are subtracted from and added to $\bar{x}$, we can be 90% confident that μ will be captured. Thus by Equations 4.12 and 4.13,

$$L = \bar{x} - Z\frac{\sigma}{\sqrt{n}} = 90.1 - (1.65)(2.066) = 86.69$$

and

$$U = \bar{x} + Z\frac{\sigma}{\sqrt{n}} = 90.1 + (1.65)(2.066) = 93.51$$

As has been shown, a 95% confidence interval is formed by subtracting and adding 1.96 standard errors so that

$$L = 90.1 - (1.96)(2.066) = 86.05$$

and

$$U = 90.1 + (1.96)(2.066) = 94.15$$

The Z value to be used in the construction of a two-sided .99 confidence interval can be found by noting that $.99/2 = .4950$. Because there are two values equally close to .4950 in column two of Appendix A (i.e., .4949 and .4951), we follow the convention of using the larger area of .4951 which has an associated Z value of 2.58. The interval is then

$$L = 90.1 - (2.58)(2.066) = 84.77$$

and

$$U = 90.1 + (2.58)(2.066) = 95.43$$

As these results show, confidence is gained at the expense of longer intervals. ∎

EXAMPLE 4.38

Use the information provided below to form a two-sided 80% confidence interval.

$$\sigma = 22 \qquad n = 100 \qquad \bar{x} = 220.5$$

Solution   The standard error is $\frac{22}{\sqrt{100}} = 2.20$. The Z value to be used for the calculation must be of sufficient size so as to produce values of L and U that contain μ when $\bar{x}$ takes any of the values that makeup the middle .80 of the curve. To find this value of Z we look for the area $.80/2 = .40$ in column two of the normal curve table. The nearest area to .4000 is .3997 which has an associated Z value of 1.28. Substituting in Equations 4.12 and 4.13 gives

$$L = \bar{x} - Z\frac{\sigma}{\sqrt{n}} = 220.5 - (1.28)(2.2) = 217.68$$

and

$$U = \bar{x} + Z\frac{\sigma}{\sqrt{n}} = 220.5 + (1.28)(2.2) = 223.32.$$

We can thus be 80% confident that the population mean is some value in the range 217.68 to 223.32.   ■

One-Sided Intervals.   Suppose that a researcher wishes to form a one-sided 95% confidence interval in order to obtain a lower bound estimate of μ. That is, the researcher wishes to find a value L that, with 95% confidence, can be declared to be less than or equal to μ. As can be seen in Panel A of Figure 4.29, the appropriate Z value to be used for this purpose is the value that has $.95 - .50 = .45$ of the curve between that value and μ. Using the previously established rule regarding areas that are equally close to that sought, we identify 1.65 as the appropriate Z value.

If $\bar{x}$ is 10.2 and the standard error is 2.0 then by Equation 4.12

$$L = \bar{x} - Z\frac{\sigma}{\sqrt{n}} = 10.2 - (1.65)(2.0) = 6.9.$$

The researcher can assert with 95% confidence that the population mean is greater than or equal to 6.9.

Had the researcher been interested in an upper bound estimate of μ, then by Equation 4.13

$$U = \bar{x} + Z\frac{\sigma}{\sqrt{n}} = 10.2 + (1.65)(2.0) = 13.5.$$

Notice that, as with one- and two-tailed hypothesis tests, one- and two-sided confidence intervals do not employ the same value of Z when constructing intervals with equal levels of confidence. For example, 1.65 is used to construct a one-sided 95% interval but 1.96 is used for a two-sided 95 percent interval.

EXAMPLE 4.39

Use the information provided below to form lower bound estimates of μ at the 90 and 99% levels of confidence.

$$\sigma = 44 \qquad n = 150 \qquad \bar{x} = 105.8$$

Solution   One-sided confidence intervals will be constructed because only lower bound estimates of μ are required. The standard error of the mean is $\frac{44}{\sqrt{150}} = 3.593$. For the 90% interval, the appropriate Z value will have $.90 - .50 = .40$ of the curve between itself and the mean of the curve. Column two of Appendix A shows this value to be (approximately) 1.28. The lower bound estimate is then

$$L = \bar{x} - Z\frac{\sigma}{\sqrt{n}} = 105.8 - (1.28)(3.593) = 101.2.$$

For the .99 interval, the Z value will have $.99 - .50 = .49$ of the curve between itself and the distribution mean. Column two of Appendix A shows this value to be (approximately) 2.33. The lower bound estimate is then

$$L = 105.8 - (2.33)(3.593) = 97.4. \qquad \blacksquare$$

EXAMPLE 4.40

Use the information provided below to obtain 86 and 94% confidence intervals for the upper bound value of μ.

$$\sigma = 4.5 \qquad n = 80 \qquad \bar{x} = 12.0$$

Solution   The Z values for the two intervals are respectfully 1.08 and 1.56. Using these values with a standard error of $\frac{4.5}{\sqrt{80}} = .503$ yields

$$U = \bar{x} + Z\frac{\sigma}{\sqrt{n}} = 12.0 + (1.08)(.503) = 12.5$$

and

$$U = 12.0 + (1.56)(.503) = 12.8.$$

Assumptions.   The assumptions underlying the confidence interval for μ when σ is known are the same as those underlying the one mean Z test (see page 98). Violations of one or more of these assumptions may result in the level of confidence (coverage) actually realized being different from that intended. Table 4.2 on page 100 can be used to obtain the coverage properties for one such violation. This can be done by subtracting α_E from one. For example, when samples of size five are drawn from the nonnormal distribution considered in that table, the actual level of confidence is $1 - .046 = .954$ rather than the intended .950. Likewise, as seen in Table 4.3, under moderate violation of the assumption of independence, the 95% confidence interval for μ has actual coverage $1 - .11 = .89$. The confidence interval for μ when σ is known is robust or nonrobust under the same conditions that the one mean Z test is robust or nonrobust.

4.4.5   Confidence Interval for μ When σ Is Not Known

You will recall that the one mean t test is used in place of the one mean Z test in the commonly encountered situation where σ is not known. The same is true for confidence intervals. When σ is not known, s is used as an estimate of the unknown parameter σ which means that the relevant distribution is t rather than Z. The equations for L and U then become

$$L = \bar{x} - t\frac{s}{\sqrt{n}} \qquad (4.14)$$

and

$$U = \bar{x} + t\frac{s}{\sqrt{n}} \qquad (4.15)$$

where s is the sample standard deviation and t is the appropriate value from the t table with $n - 1$ degrees of freedom. For example, suppose we wish to construct a two-sided 95% confidence interval for μ using the data from Section 4.3.4 which we have reproduced here for convenience.

Sample: 6.0 8.0 5.5 4.5 8.5 4.0 3.5

The sample standard deviation can be obtained by application of Equation 2.16 on page 37 to the summed scores and the sum of the squared scores as given below.

X	X^2
6.0	36.00
8.0	64.00
5.5	30.25
4.5	20.25
8.5	72.25
4.0	16.00
3.5	12.25
40.0	251.00

$$s = \sqrt{\frac{\sum x^2 - \frac{(\sum x)^2}{n}}{n - 1}} = \sqrt{\frac{251 - \frac{(40)^2}{7}}{7 - 1}} = \sqrt{\frac{22.429}{6}} = 1.933$$

and

$$\bar{x} = \frac{\sum x}{n} = \frac{40}{7} = 5.714.$$

The appropriate value for t can be found by entering the table in Appendix B with $7 - 1 = 6$ degrees of freedom. Because we wish to construct a two-sided 95% confidence interval we use the row and column so designated. This gives a value for t of 2.447. By Equations 4.14 and 4.15

$$L = 5.714 - 2.447\frac{1.933}{\sqrt{7}} = 3.926$$

and

$$U = 5.714 + 2.447\frac{1.933}{\sqrt{7}} = 7.502.$$

Thus, the researcher can be 95% confident that this sample was drawn from a population whose mean was between 3.926 and 7.502. A one-sided 95% lower bound estimate of μ would be

$$L = 5.714 - 1.943\frac{1.933}{\sqrt{7}} = 4.294.$$ ■

EXAMPLE 4.41

The data provided below represent a (fictitious) sample of blood glucose values taken from 10 children aged 14–16 who report that they routinely eat fast food three or more times a week. Use this data to form a two-sided 99% confidence interval.

Sample: 100 99 97 104 124 120 89 122 118 101

Solution   The sum and sum of squared observations are as follows.

X	X^2
100	10000
99	9801
97	9409
104	10816
124	15376
120	14400
89	7921
122	14884
118	13924
101	10201
1074	116732

Application of Equations 2.1 and 2.16 yield

$$\bar{x} = \frac{\sum x}{n} = \frac{1074}{10} = 107.4$$

and

$$s = \sqrt{\frac{\sum x^2 - \frac{(\sum x)^2}{n}}{n-1}} = \sqrt{\frac{116732 - \frac{(1074)^2}{10}}{10-1}} = \sqrt{\frac{1384.4}{9}} = 12.403.$$

Appendix B shows that the appropriate t value for a two-sided 99% confidence interval based on $10 - 1 = 9$ degrees of freedom is 3.250. Then by Equations 4.14 and 4.15

$$L = \bar{x} - t\frac{s}{\sqrt{n}} = 107.4 - 3.250\frac{12.403}{\sqrt{10}} = 94.65$$

and

$$U = \bar{x} + t\frac{s}{\sqrt{n}} = 107.4 + 3.250\frac{12.403}{\sqrt{10}} = 120.15.$$ ■

Assumptions.　The assumptions underlying the confidence interval for μ when σ is not known are the same as those underlying the one mean t test which are the same as those underlying the one mean Z test (see page 98). Violations of one or more of these assumptions may result in the level of confidence (coverage) actually realized being different from that intended. Table 4.4 can be used to obtain the coverage properties for one such violation. This can be done by subtracting α_E from one. For example, when samples of size five are drawn from the nonnormal distribution considered in that table, the actual level of confidence is $1 - .069 = .931$ rather than the intended .950. The confidence interval for μ when σ is not known is robust or nonrobust under the same conditions that the one mean t test is robust or nonrobust.

4.4.6　Confidence Interval for π

There are a number of approximate methods for constructing confidence limits for a population proportion (π). However, these methods are often not sufficiently accurate for many applications especially when the sample size (n) is not large or when $\hat{p}$ is near zero or one. There is also an exact method that overcomes this difficulty.

Because of its common use by researchers, we will briefly describe and provide an example calculation for an approximate method based on the normal curve. However, for your own applications you should consider using the exact method described in a later section.

An Approximate Method.　An approximation for L and U is given by

$$L = \hat{p} - Z\sqrt{\frac{\hat{p}\hat{q}}{n}} \tag{4.16}$$

and

$$U = \hat{p} + Z\sqrt{\frac{\hat{p}\hat{q}}{n}} \tag{4.17}$$

where $\hat{p}$ is the proportion of successes in the sample of size n and $\hat{q}$ is the proportion of failures so that $\hat{q} = 1 - \hat{p}$. Z determines the level of confidence and is found in the manner described in Section 4.4.4.

As an example application, suppose that a health policy researcher wishes to estimate the proportion of adults living in a rural southern county who have some form of health insurance. To this end, a sample of 350 adults living in the county are interviewed. Of the 350 persons interviewed, 112 or $112/350 = .32$ report that they currently have some form of health insurance. A two-sided 95% confidence interval can be constructed as

$$L = \hat{p} - Z\sqrt{\frac{\hat{p}\hat{q}}{n}} = .32 - 1.96\sqrt{\frac{(.32)\,(.68)}{350}} = .27$$

and

$$U = \hat{p} + Z\sqrt{\frac{\hat{p}\hat{q}}{n}} = .32 + 1.96\sqrt{\frac{(.32)\,(.68)}{350}} = .37.$$

The researcher can then be 95% confident that the proportion of adults living in the county who have health insurance is between .27 and .37. It is instructive to note the result that would have been attained if $\hat{p}$ had been close to zero (e.g., .02) and n was much reduced (e.g., $n = 25$). In that case

$$L = .02 - 1.96\sqrt{\frac{(.02)\,(.98)}{25}} = -.03$$

and

$$U = .02 + 1.96\sqrt{\frac{(.02)\,(.98)}{25}} = .07.$$

But, of course, proportions cannot be negative. Rather than declaring that the population proportion is between $-.03$ and $.07$, the researcher would use the interval .00 to .07. The important point, however, is that this method of constructing confidence intervals is not accurate when the sample size is not sufficiently large and $\hat{p}$ is too near zero or one.

Assumptions.   See page 116. In addition to comments found there, you should bear in mind that confidence limits for proportions that are less than zero or greater than one should take these values as their legitimate lower or upper bound.

Exact Method.   Just as the binomial distribution can be used to form the basis of an exact hypothesis test regarding π (see page 108), this same distribution can form the basis of an exact confidence interval to estimate this parameter. The rationale underlying this exact interval can best be understood after reading the next section. For the moment we will simply state that it relies on a special relationship that exists between hypothesis tests and confidence intervals.

Having said this, however, we must add that the *method* by which these intervals are formed is not straightforward and relies on mathematical statistics concepts that are far beyond the scope of this book. For this reason, we will show you *how* to construct these intervals but will not attempt to tell you *why* this method is appropriate. These limits are calculated by

$$L = \frac{S}{S + (n - S + 1)\,F_L} \tag{4.18}$$

and

$$U = \frac{(S + 1)\,F_U}{n - S + (S + 1)\,F_U} \tag{4.19}$$

where S is the number of successes in the sample, n is the number of observations in the sample and F_L and F_U are the appropriate values from an F distribution. F_L and F_U can be obtained from Appendix C. Notice that, unlike the t table in Appendix B, the F table must be entered with *two* different degrees of freedom. The first of these, which we shall call the numerator degrees of freedom, is listed across the top of the table while the second, which we shall call the denominator degrees of freedom, is given along the edge of the table. Thus, for example, the appropriate F value to be used in the calculation of a two-sided 95% confidence interval assuming numerator degrees of freedom equal to 4 and denominator degrees of freedom of 20 would be 3.51. In order to facilitate calculations we will use the notation df_{LN}, df_{LD}, df_{UN}, and df_{UD} to represent respectively the numerator degrees of freedom for calculating L, the denominator degrees of freedom for calculating L, the numerator degrees of freedom for calculating U, and the denominator degrees of freedom for calculating U. The degrees of freedom used for calculation of the lower and upper limits are obtained by

$$\boxed{df_{LN} = 2\,(n - S + 1)} \tag{4.20}$$

$$\boxed{df_{LD} = 2S} \tag{4.21}$$

$$\boxed{df_{UN} = 2\,(S + 1)} \tag{4.22}$$

$$\boxed{df_{UD} = 2\,(n - S)} \tag{4.23}$$

EXAMPLE 4.42

A random sample of 10 children with normal blood glucose levels who have one or more siblings with diabetes are tested for antibodies associated with that disease. Forty percent of these children test positive for the specified antibody. Use the exact method to construct a two-sided 95% confidence interval to estimate the proportion of children of this type in the population who will test positive for the antibody. Compare this interval to one constructed by the approximate method discussed on page 148.

Use the same data to construct an exact one-sided 95% confidence interval for the lower bound of the population proportion. Compare this interval to one constructed by the approximate method discussed on page 148.

Solution   We begin by computing the degrees of freedom necessary to find F_L. By equations 4.20 and 4.21 and the fact that the number of successes (S) is $n\hat{p} = (10)\,(.4) = 4$,

$$df_{LN} = 2\,(n - S + 1) = 2\,(10 - 4 + 1) = 14$$

and

$$df_{LD} = 2S = (2)\,(4) = 8.$$

Appendix C shows that with numerator degrees of freedom of 14 and denominator degrees of freedom of 8, the appropriate F value to be used for the construction of a two-sided 95% confidence interval is 4.13. Using this value in Equation 4.18 gives

$$L = \frac{S}{S + (n - S + 1)\,F_L} = \frac{4}{4 + (10 - 4 + 1)\,4.13} = .122.$$

Calculating the degrees of freedom for F_U by means of Equations 4.22 and 4.23 gives

$$df_{UN} = 2\,(S + 1) = 2\,(4 + 1) = 10$$

and

$$df_{UD} = 2\,(n - S) = 2\,(10 - 4) = 12.$$

The F value with numerator and denominator degrees of freedom of 10 and 12 respectively to be used in the construction of a two-sided 95% confidence interval is, by Appendix C, 3.37. Then by Equation 4.19

$$U = \frac{(S + 1)\,F_U}{n - S + (S + 1)\,F_U} = \frac{(4 + 1)\,3.37}{10 - 4 + (4 + 1)\,3.37} = .737.$$

The exact two-sided 95% confidence interval is then .122 to .737. The sizable length of this interval is due to the small sample size.

The approximate interval as constructed by Equations 4.16 and 4.17 is

$$L = \hat{p} - Z\sqrt{\frac{\hat{p}\hat{q}}{n}} = .4 - 1.96\sqrt{\frac{(.4)\,(.6)}{10}} = .096$$

and

$$U = \hat{p} + Z\sqrt{\frac{\hat{p}\hat{q}}{n}} = .4 + 1.96\sqrt{\frac{(.4)\,(.6)}{10}} = .704.$$

The approximate interval of .096 to .704 is different from the exact interval calculated above. This is to be expected because the approximate method cannot generally be relied upon when sample size is small.

The exact one-sided interval for the lower bound estimate of π can be obtained by substituting the appropriate value from Appendix C into Equation 4.18. Entering Appendix C with the degrees of freedom previously computed by 4.20 and 4.21 (i.e., 14 and 8), the F value for a one-sided 95% interval is found to be 3.24. Substituting this value into Equation 4.18 gives

$$L = \frac{S}{S + (n - S + 1)\,F_L} = \frac{4}{4 + (10 - 4 + 1)\,3.24} = .150.$$

The approximate method yields

$$L = \hat{p} - Z\sqrt{\frac{\hat{p}\hat{q}}{n}} = .4 - 1.65\sqrt{\frac{(.4)\,(.6)}{10}} = .144. \qquad \blacksquare$$

EXAMPLE 4.43

Given 24 successes in a sample of size 28, form an exact two-sided 99% confidence interval for π. Form an exact one-sided 99% confidence interval for the upperbound of π.

Solution　The numerator and denominator degrees of freedom for F_L are by Equations 4.20 and 4.21

$$df_{LN} = 2\,(28 - 24 + 1) = 10$$

and

$$df_{LD} = (2)\,(24) = 48.$$

From Appendix C F_L is 3.01. The lower limit is by Equation 4.18 then

$$L = \frac{24}{24 + (28 - 24 + 1)\,3.01} = .615.$$

The numerator and denominator degrees of freedom for F_U are by Equations 4.22 and 4.23

$$df_{UN} = 2\,(24 + 1) = 50$$

and

$$df_{UD} = 2\,(28 - 24) = 8.$$

With F_U of 6.22, U is then

$$U = \frac{(24 + 1)\, 6.22}{28 - 24 + (24 + 1)\, 6.22} = .975.$$

The two-sided 99% confidence interval is then .615 to .975.
   The one-sided upperbound is

$$U = \frac{(24 + 1)\, 5.07}{28 - 24 + (24 + 1)\, 5.07} = .969.$$ ∎

Assumptions.   The assumptions underlying the exact method of formulating confidence intervals for the estimation of π are the same as those for the exact hypothesis test discussed on page 116. An additional explanatory note must be added for this method of forming exact confidence intervals. The accuracy of the calculation of L and U depend on the number of decimal places used in F_L and F_U. Because the F table in this book provides two decimal places, you should not rely on more than two digits of accuracy in your calculation of L and U. This may be particularly important when limits are close to zero or one. For example, if an actual lower limit is .0003, you cannot count on getting this result with the F table provided here. In such cases you can get more precise F values from other tables or from commonly available software.

4.5   COMPARISON OF HYPOTHESIS TESTS AND CONFIDENCE INTERVALS

Hypothesis tests and confidence intervals are more closely related than may be initially apparent. In fact, confidence intervals may be used to perform hypothesis tests. For this reason, and others to be discussed below, confidence intervals are usually preferable to hypothesis tests in situations where both might be employed. In this section we will show you why and how confidence intervals can be used to conduct hypothesis tests. We will end by comparing the information provided by each.

4.5.1   Two-Tailed Hypothesis Tests and Two-Sided Confidence Intervals

Suppose that a two-tailed one mean Z test is conducted at $\alpha = .05$. The null distribution used for this test is shown in Panels A and B of Figure 4.30. Because the test is two-tailed and α is .05, the distance from the hypothesized mean value (μ_0) to the leading edge of the critical regions is 1.96 standard errors (see Appendix A). In Panel A of this figure the test statistic ($\bar{x}$ in this case) is in a critical region which leads to rejection of the null hypothesis.

   Suppose now that a 95% confidence interval is formed around $\bar{x}$ as shown in Panel A. Note that a two-sided 95% confidence interval is formed by adding and subtracting 1.96 standard errors to and from $\bar{x}$. Because this is also the distance from μ_0 to the edge of the critical region, it follows that when $\bar{x}$ is in the critical region the null value (μ_0) will not be in the interval formed by L and U.

   By comparison, Panel B depicts the situation where the test statistic is not in the critical region leading to a failure to reject the null hypothesis. In this situation, μ_0 will be contained in the interval formed by L and U. From the results shown in Panels A and B, it

A

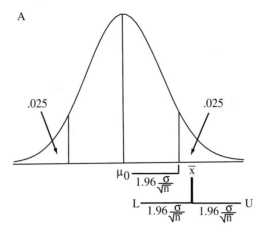

B

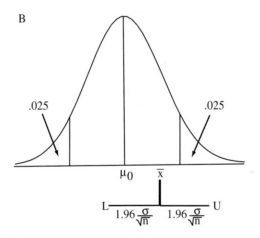

FIGURE 4.30: Relationship of two-tailed hypothesis test to two-sided confidence interval.

follows that a researcher wishing to use a two-tailed one mean Z with $\alpha = .05$ to test the null hypothesis that μ is equal to μ_0 could conduct the test by forming a two-sided 95% confidence interval and noting whether μ_0 is in the resultant interval. If μ_0 is not in the interval the null hypothesis will be rejected. If μ_0 is in the interval the null hypothesis will not be rejected.

If the researcher had wished to conduct the hypothesis test at $\alpha = .01$ a 99% confidence interval would be required. In general, if α and the level of confidence are expressed as decimals then the level of confidence used to perform a test at level α is $1 - \alpha$. Thus, a hypothesis test conducted at $\alpha = .10$ would require an interval with level of confidence $1 - .10 = .90$.

A

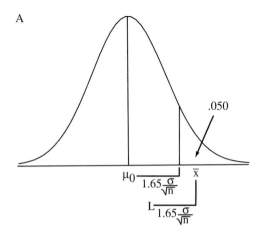

B

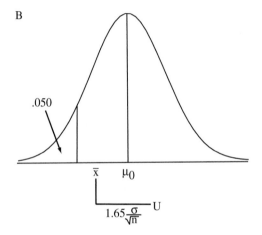

FIGURE 4.31: Relationship of one-tailed hypothesis test to one-sided confidence interval.

4.5.2   One-Tailed Hypothesis Tests and One-Sided Confidence Intervals

Just as two-sided confidence intervals can be used to perform two-tailed hypothesis tests,
so one-sided intervals can be used to conduct one-tailed tests. For example, a test of a null
hypothesis against an alternative of the form

$$H_A : \mu > \mu_0$$

can be conducted by forming a one-sided confidence interval for the lower bound of the
estimated parameter. If $\mu_0 \leq L$ the null hypothesis is rejected, otherwise it is not rejected.
Panel A of Figure 4.31 shows that when the test statistic is in the critical region, μ_0 will be
less than L.

A test with alternative of the form

$$H_A : \mu < \mu_0$$

is conducted by means of a one-sided interval for the upperbound of the parameter. This test is depicted in Panel B of Figure 4.31. In this example the test statistic is not in the critical region which implies that $\mu_0 < U$. The null hypothesis would be rejected if $\mu_0 \geq U$.

In a sense, then, numerous hypothesis tests can be conducted at a glance of a confidence interval. The researcher knows that any hypothesized value that lies between L and U of a two-sided interval would not be rejected while any value outside the interval would be rejected. Likewise, the researcher knows that any value less than or equal to L of a one-sided interval would lead to rejection while any value greater than L would not be rejected. It is also obvious that any value of μ_0 that is greater than or equal to U would be rejected while any value less than U would not lead to rejection.

4.5.3   Some Additional Comments

In addition to the above, two characteristics of confidence intervals make them generally preferable to hypothesis tests. First, confidence intervals usually answer more interesting questions than do hypothesis tests. For example, the question "What is μ?" is usually more interesting than the question "Is μ different from 12?" Second, if a sample is too small to provide adequate power, this fact is not signaled to the researcher by an hypothesis test. By contrast, an inadequate sample will produce an unduly long confidence interval thereby alerting the researcher to the fact that the sample is too small. In general, the larger the sample the shorter the confidence interval.

The question naturally arises as to why hypothesis tests should ever be used. The ancillary question is why you had to learn so much about such tests in this chapter. The primary reason is that many questions of interest to researchers cannot be answered by confidence intervals. The only answers to such questions are given by hypothesis tests. You will encounter such examples in the following chapters. It is also necessary to understand certain tests because they are still (arguably, inappropriately) commonly reported in the research literature. If you are to understand the research literature you must understand the hypothesis tests used therein.

4.6   A RE-ORIENTATION

You have now completed a rather elaborate introduction to inference. Most of the remainder of this book will deal with inferential methods commonly employed in health science research. If you understand the concepts in this chapter, you are in very good stead for the remaining material.

KEY WORDS AND PHRASES

After reading this chapter you should be able to demonstrate familiarity with the following words and phrases.

$1 - \alpha$ 125	$1 - \beta$ 127
$1 -$ power 128	alpha (α) 88

EXERCISES

4.1 Given $\sigma = 50$ and $n = 100$, find the variance of the sampling distribution of $\bar{x}$.

4.2 What is the standard deviation of the sampling distribution of $\bar{x}$ called?

4.3 What happens to the standard deviation of the sampling distribution of $\bar{x}$ as sample size increases?

4.4 Does the central limit theorem guarantee that the sampling distribution of $\bar{x}$ will be normally distributed when sample size is sufficiently large? Explain your answer.

4.5 Use the normal curve to find the approximate probability of randomly selecting 25 observations from a population whose mean is 50 and whose variance is 100 and finding that the sample mean is less than 45.

4.6 Suppose ten observations are randomly selected from a dichotomous population in which the proportion of successes is .40. Use the binomial distribution to find the probability that the proportion of successes in the sample will be less than or equal to .30.

4.7 Use the normal curve to approximate the probability described in Exercise 4.6

4.8 Given $\bar{x} = 135$, $\sigma = 40$ and $n = 90$, test the null hypothesis $H_0 : \mu = 125$ against the alternative $H_A : \mu \neq 125$ at

$\alpha = .10$. Report results for both th p-value versus alpha and critical versus obtained methods.

4.9 Use the data provided below to test the null hypothesis $H_0 : \mu = 5$ against the alternative $H_A : \mu < 5$ at the .05 level of significance.

$$\text{Sample}: 3 \quad 3 \quad 2 \quad 1 \quad 0 \quad 6 \quad 5 \quad 4$$

4.10 *Use a confidence interval* constructed from the data in Exercise 4.9 to perform a two-tailed test of the hypothesis $H_0 : \mu = 0$ at the .01 level of significance. Explain why you rejected or failed to reject the null hypothesis.

4.11 Given $\sigma = 24$, $n = 144$ and $\bar{x} = .52$, perform the following test at the .05 level of significance.

$$H_{0E} : \mu \leq -1.0 \quad \text{or} \quad \mu \geq 1.0$$
$$H_{AE} : -1.0 < \mu < 1.0$$

What are the terms used for tests of this sort? Is this a one- or two-tailed test? Why didn't you use t tests to conduct the test?

4.12 Use the data provided below to perform the following test at the .05 level of significance.

$$H_{0E} : \mu \leq 10$$
$$H_{AE} : \mu > 10$$

$$\text{Sample}: 14 \quad 10 \quad 12 \quad 16 \quad 11 \quad 19 \quad 15 \quad 14$$

Why didn't you use a Z test?

4.13 Given $n = 10$ and $\hat{p} = .1$, use an exact method to perform the following test at the .05 level of significance. Use both the p-value versus alpha and obtained versus critical methods.

$$H_0 : \pi = .7$$
$$H_A : \pi \neq .7$$

4.14 Use an approximate method to perform the test described in Exercise 4.13. How do the two p-values compare?

4.15 Given $n = 9$ and $\hat{p} = .11$, use an exact method to perform the following test at the .05 level of significance. Use both the p-value versus alpha and obtained versus critical methods.

$$H_0 : \pi = .5$$
$$H_A : \pi < .5$$

4.16 Define each of the following.

(a) Type I error.

(b) Type II error.

(c) Power

(d) Beta

(e) α

(f) $1 - \alpha$

4.17 What are the factors that determine the power of an inferential test?

4.18 Find power and beta for a one mean Z test under the following conditions.

$$H_0 : \mu = 40 \qquad \sigma = 20 \qquad \alpha = .05$$
$$H_A : \mu > 40 \qquad n = 25 \qquad \mu = 42$$

4.19 Recompute power and beta for the test described in Exercise 4.18 assuming that sample size has been increased to 100.

4.20 Find power and beta for the following test of significance.

$$H_0 : \pi = .40$$
$$H_A : \pi < .40$$

Where

$$\alpha = .05 \qquad n = 8 \qquad \pi = .10$$

4.21 Calculate the sample size required to attain power of .9 to detect a population mean of 6 for a two-tailed Z test of the null hypothesis $H_0 : \mu = 10$ conducted at $\alpha = .05$. Assume that $\sigma = 2$.

4.22 Use the data provided here to construct a 95% two-sided confidence interval for the estimation of μ.

$$\text{Sample}: 9 \quad 7 \quad 5 \quad 4 \cdot 8 \quad 9 \quad 5 \quad 8 \quad 7$$

4.23 Suppose in a random sample of 200 persons over age 65, it is found that 150 of the sampled subjects were vaccinated against smallpox at some time in their life. Use an approximate method to form a two-sided 95% confidence interval to estimate the proportion of persons in the population of interest who received such vaccinations.

4.24 Recalculate the confidence interval described in Exercise 4.23 under the condition that 300 of 400 sampled persons had the vaccination. What effect did the increase in sample size have on the confidence interval?

4.25 Given six successes in a sample of size 10, form an exact 95% confidence interval for a lower bound estimate of π.

4.26 Given 5 successes in a sample of size 10, use an exact method to construct a two-sided confidence interval for the estimation of π.

A. The following questions refer to Case Study A (page 469)

4.27 The mean age of the participants in this study is reported to be 31.5 years. This value can be taken as an estimate of the population mean age. What term is used for estimates of this sort? Use this value along with other information from the study to form a (two-sided) 95% confidence interval. Interpret this interval.

4.28 In summarizing their results, the study authors report, "At the beginning of wear $\cdots$ 33 subjects preferred the comfort with the conditioned lens versus 26 with the control, however, this was not statistically significant."

(a) Assuming the authors wanted to know whether the subjects preferred one type lens (treated) over the other (untreated), what sort of hypothesis test do you think they performed? State the null and alternative hypotheses, then indicate two methods they might have used for carrying out this test.[18]

(b) Can you propose another method of analysis to evaluate this question?

(c) Can the result reported by the authors be interpreted as meaning that the subjects had no lens preference?

(d) Would it be a good idea to combine the am and pm data for analysis purposes? After all, this would double sample size thereby increasing statistical power.

(e) Analyze the a.m. data for the "no" group alone and report your results. The authors report this result as significant ($P < .05$). Do you agree? What do you conclude as a result of this analysis? Do you believe an hypothesis test or confidence interval would be more appropriate for this analysis? (Note: F_U with 22 and 4 degrees of freedom for a 95% confidence interval is 8.53.)

(f) Analyze the p.m. data for the "no" group alone in order to determine whether subjects had a lens preference. Report your results. The authors report this result as not significant ($P > .05$). Do you agree? What do you conclude as a result of this analysis?

B. The following questions refer to Case Study B (page 470)

4.29 Use the data in Table J.3 to construct two-sided 95% confidence intervals for each of the three groups to estimate the proportion of subjects in each group who correctly identified the type of bracelet they wore. Note that a correct identification for the Weak group is "Dummy." Can you speculate as to how they were able to correctly identify the bracelet?

D. The following questions refer to Case Study D (page 471)

4.30 Use a 95% confidence interval to estimate the mean CD4 count for the HIV positive subjects.

4.31 Suppose you were asked to use the CD4 count data to test the null hypothesis $H_0 : \mu = 500$ against a two-tailed alternative by means of a one mean Z test. Does this test seem to provide a reasonable approach to this problem? Why or why not?

F. The following questions refer to Case Study F (page 473)

4.32 Construct exact and approximate confidence intervals for estimation of the proportion of recurrent cases of tuberculosis in the population that are attributable to new exogenous infection. How well would you say the approximate interval approximates the exact interval? (Note: F_U with 26 and 8 degrees of freedom for a 95 percent confidence interval is 3.93.)

O. The following questions refer to Case Study O (page 477)

4.33 Can you counter the assertion made by the author in point (e)?

---

[18]Don't be overly concerned if you didn't see this right away. It takes some experience to learn how to answer questions of this sort. You've now had your first experience.

Paired Samples Methods

5.1 INTRODUCTION

Paired data occur with some frequency in a variety of research contexts. For example, researchers may assess the relative effectiveness of two laser surgery techniques for the treatment of diabetic retinopathy by applying one technique to one eye of patients suffering from this disease with the second method being applied to the remaining eye. If, after some period of time, visual acuity is measured for each of the two eyes in order to determine which surgery technique resulted in better vision, the two acuities obtained for each patient would constitute a data pair.

As a second example, researchers wishing to determine whether an over-the-counter cold remedy raises systolic blood pressure as an undesirable side effect might take baseline blood pressure measurements for each patient, administer the cold remedy and, after some specified period of time, take a second blood pressure measurement. The mean blood pressures taken at the two points in time could then be compared to determine whether an increase had occurred. The two measurements taken on each subject would again constitute a data pair.

As a final example, sets of twins may be exposed to two forms of vaccine against some childhood illness. One twin from each pair is randomly assigned to receive the first vaccine with the remaining twin then receiving the second vaccine. The relative effectiveness of the two vaccines could then be compared by comparing the proportions of twins taking each vaccine type who develop the disease. The indicators as to whether each twin in the pair develops the disease would be a data pair.

This chapter deals with hypothesis tests and confidence intervals that are designed to compare means or proportions of paired data. As you will see, many of these methods are simply special cases of methods with which you are already familiar.

TABLE 5.1: Pre- and post-treatment systolic blood pressure measurements.

Pre-Treatment	Post-Treatment	(difference) d	d^2
95	99	4	16
111	120	9	81
97	97	0	0
132	130	−2	4
144	148	4	16
100	122	22	484
120	131	11	121
110	109	−1	1
131	140	9	81
154	153	−1	1
105	131	26	676
119	120	1	1
107	114	7	49
101	110	9	81
118	116	−2	4
$\sum$ 1744	1840	96	1616

5.2 METHODS RELATED TO MEAN DIFFERENCE

5.2.1 The Paired Samples (Difference) t Test

Rationale. As in the example alluded to above, suppose that a researcher is interested in determining whether an over-the-counter cold medication raises systolic blood pressure as an undesirable side effect. To this end, the researcher takes the blood pressures of 15 subjects of the type in which he/she is interested. Each subject is then given the recommended dose of the over the counter product. Thirty minutes after administration of the remedy, blood pressures are again taken on each subject. The question of interest is, "Are blood pressures higher after taking the medication than before?" Table 5.1 shows (fictitious) blood pressures for the 15 subjects taken before (pre-treatment) and after (post-treatment) administration of the medication.

We can determine the change in blood pressure for each subject by subtracting their pre-treatment value from their post-treatment value. These **difference scores** are labeled d in the table. The mean of these difference scores, which we shall designate as $\bar{d}$, is $96/15 = 6.40$. Notice that this is the same value as would be obtained if the pre-treatment mean were subtracted from the post-treatment mean or $122.67 - 116.27 = 6.40$.[1]

Thus, the average change in blood pressure as measured before and after taking the cold remedy is 6.40 units. But there are at least two explanations for this mean difference. First,

---

[1] You can verify this as a general result through application of the rules of summation outlined in Section 2.3.2.

we note that any time two blood pressure measurements are taken on the same person the resulting values are rarely the same even when no intervention has taken place between measurements. Therefore, it might well be that the medication had no effect on blood pressures. However, by random happenstance the later measurements were greater, on average, than the pre-treatment measurements. This implies that, if we were to repeat the experiment, we might obtain a mean difference of -6.40 or some other value fairly close to zero. A second explanation is that the medication does have the undesired side effect. As a result, individual post-treatment pressures tend to be elevated over what they would have been without the medication, thereby producing the mean difference of 6.40 units. Which of these explanations is to be believed? A test of significance might help decide the issue.

Suppose we conceive of the 15 d values as a random sample taken from a population of such values. We further conceive of the population of d values as those that would result if the cold medication had no impact on blood pressure. That is, some values will be positive and some negative, so that the population mean is zero. We will designate the mean of this difference score population as μ_d. Thus, if the medication has no impact we can conceive of our sample of ds as being a random sample from a difference score population whose mean is zero.

On the other hand, if the medication does have the undesired effect, the sample must be thought of as coming from a population whose mean is greater than zero. The following may help in conceptualizing the two populations.

Think of a large population of persons whose blood pressures are recorded at two points in time with no intervening treatment being administered. The statistical population is composed of the difference between the two measurements. Because no treatment was administered between measurements the ds are made up of positive and negative values whose mean is zero.

On the other hand, suppose a large population of persons had their blood pressures taken at two points in time with a treatment being administered between measurements that tends to raise blood pressures. In this case there would be more positive ds than negative ds in the population so that μ_d would be greater than zero.

In determining whether the medication raised blood pressures or not, it would be helpful if we could determine which of the two theoretical populations most plausibly represents the population from which our sample of ds was drawn.

The Test. Continuing the above discussion, information concerning which of the two theoretical populations best represents the population from which our data were sampled can be gained by conducting the following hypothesis test.

$$H_0 : \mu_d = 0$$
$$H_A : \mu_d > 0$$

In essence, the null hypothesis asserts that no change in mean blood pressure took place while the alternative maintains that a change took place that resulted in a higher post-treatment mean. But how is this test to be conducted? From your studies of Chapter 4, you know that a one mean t test carried out on the difference scores can be used to conduct the above significance test. Indeed, a **paired samples** or **paired difference** t test is nothing more than a one mean t test conducted on difference (d) scores. Obtained t is then

$$t = \frac{\bar{d} - \mu_{d0}}{\frac{s_d}{\sqrt{n}}} \tag{5.1}$$

where $\bar{d}$ is the mean of the sample difference scores, μ_{d0} is the hypothesized mean of the difference score population and s_d is the sample standard deviation of the difference scores. You will recognize Equation 5.1 as the equation for the one mean t statistic with $n - 1$ degrees of freedom. Subscripts have been added to remind you that the test is conducted on difference scores. In practice, μ_{d0} is almost always zero though it need not be. An exception occurs with equivalence testing where μ_{d0} is typically not zero.

EXAMPLE 5.1

Use a paired samples t test with $\alpha = .05$ to perform the above indicated test of hypothesis on the data in Table 5.1.

Solution   Using the sums from Table 5.1

$$s_d = \sqrt{\frac{\sum d^2 - \frac{(\sum d)^2}{n}}{n - 1}} = \sqrt{\frac{1616 - \frac{(96)^2}{15}}{15 - 1}} = \sqrt{\frac{1001.6}{14}} = 8.458$$

and

$$\bar{d} = \frac{\sum d}{n} = \frac{96}{15} = 6.40$$

By Equation 5.1 obtained t is then

$$t = \frac{6.40}{\frac{8.458}{\sqrt{15}}} = \frac{6.40}{2.184} = 2.930.$$

Appendix B shows that the critical value for a one-tailed test with 14 degrees of freedom conducted at $\alpha = .05$ is 1.761 so that the null hypothesis is rejected. We can conclude, therefore, that a change did take place and that the observed mean difference was not likely due to chance.   ■

Does this finding mean that the medication *caused* the elevation in blood pressure? Probably—but two facts must be kept in mind. First there is always the chance of a Type I error. Second and probably more important, we must bear in mind that the t test tells us that a change took place in blood pressures but does not tell us *why* the change took place. Suppose, for example, that the subjects became anxious about the experiment and suffered a mild elevation in blood pressure as a result of this anxiety. The t test cannot differentiate between the medication, anxiety or any other source as the causative agent; it simply asserts that the change was not due to chance. The discussion of cause and effect will become more important as you progress in this book.

TABLE 5.2: LogMar scores associated with two laser treatments.

Treatment One	Treatment Two
.0	.2
.8	1.1
.4	.9
1.0	.5
.5	.2
.4	.7
.5	.5

EXAMPLE 5.2

Researchers are interested in comparing the effectiveness of two laser surgery treatments for diabetic retinopathy. Patients who manifest the disease in both eyes have one eye treated by the first surgical method with the remaining eye then being treated by the second method. After a period of time, visual acuity is measured for each eye. The question of interest is "Does one treatment method produce better vision (as measured by acuity) than does the other?"

The determination as to which eye is treated with which method is made at random. The random assignment of eyes to treatments is important in order to avoid bias. For example, if this assignment were left to the surgeon, she/he might choose to use the first treatment method for more severely affected eyes. A later comparison of visual acuities might show that the first treatment method produced poorer vision than did the second. But this would not be a fair comparison because it was used with more severely affected eyes. The visual acuity scores, expressed in LogMar units,[2] obtained after the treatments are given in Table 5.2.

Use the data provided to perform a two-tailed paired samples t test with $\alpha = .05$ in order to assist in determining whether one treatment was more effective than the other.

Solution   The LogMar scores, their differences, the squares of their differences and the sum of the difference scores and their squares are given in Table 5.3.

Using these sums

$$s_d = \sqrt{\frac{.81 - \frac{(-.5)^2}{7}}{7 - 1}} = \sqrt{\frac{.774}{6}} = \sqrt{.129} = .359$$

and

$$\bar{d} = \frac{-.5}{7} = -.071.$$

By Equation 5.1 obtained t is

$$t = \frac{\bar{d} - \mu_{d0}}{\frac{s_d}{\sqrt{n}}} = \frac{-.071}{\frac{.359}{\sqrt{7}}} = \frac{-.071}{.136} = -.522.$$

---

[2]See the discussion of one-tailed equivalence tests on page 123 for an explanation of LogMar scores.

TABLE 5.3: LogMar scores, their differences, the squares of their differences and the sum of the difference scores and their squares.

Treatment One	Treatment Two	d	d^2
.0	.2	−.2	.04
.8	1.1	−.3	.09
.4	.9	−.5	.25
1.0	.5	.5	.25
.5	.2	.3	.09
.4	.7	−.3	.09
.5	.5	.0	.00
$\sum$ 3.6	4.1	−.5	.81

Because the critical values for a two-tailed t test with six degrees of freedom conducted at the .05 level are +2.447 and −2.447, the null hypothesis is not rejected. But what does this imply insofar as the question posed by the researcher is concerned? Does it mean that there is no difference in the effectiveness of the two treatments? If you answered "Yes" to this latter question you would do well to review the discussions of Type II error in Chapter 4. The point is that you do not know beta (the probability of a Type II error), so you do not know the probability of being wrong if you assert that the null hypothesis is true. Therefore, you should not use failure to reject the null hypothesis as evidence that the null hypothesis (which maintains that there is no difference in the average acuities) is true. An old maxim declares that "Absence of evidence is not evidence of absence." A more realistic interpretation of this result is that "There was insufficient evidence to allow for a conclusion that the treatments differed in their effectiveness." ■

A more mundane question that often plagues students is the issue of which set of scores should be subtracted from the other in order to obtain the d values. In the cold remedy example we subtracted the pre-treatment scores from the post-treatment scores. The result was $\bar{d} = 6.40$ with an associated obtained t of 2.930. The fact that $\bar{d}$ was positive indicates that the post-treatment average blood pressure was higher than the pre-treatment average blood pressure. Because the test was significant we rejected the null hypothesis in favor of the alternative that maintained that mean post-treatment blood pressure is higher than average pre-treatment blood pressure.

But what would have been the result if we had subtracted the post-treatment scores from the pre-treatment scores? In this case $\bar{d}$ would have been −6.40 and obtained t would be −2.930 leading to the conclusion that mean pre-treatment blood pressure was significantly lower than mean post-treatment blood pressure. It really doesn't matter whether we say post is significantly greater than pre or pre is significantly less than post. The conclusion is the same. Note, however, that if we had followed the second subtraction course we would have had to place the critical region in the left tail of the distribution which implies a negative critical value. The bottom line is that so long as you set up the test and interpret the result correctly, it doesn't matter which way the subtraction is carried out.

5.2.2   Establishing Equivalence by Means of Paired Samples t Tests

Rationale.   As we have pointed out in several places in this book, rejection of a null hypothesis provides good evidence (though not proof positive) that the null hypothesis is false. By contrast, failure to reject a null hypothesis does not, in general, provide good evidence that the null hypothesis is true. In situations where you wish to establish the validity of the null hypothesis, you must employ an equivalence test to show that the null hypothesis is (approximately) true. (See Section 4.3.6 on page 117.)

It sometimes happens that researchers would like to establish that there is *no* difference between treatments or between pre- and post-treatment means rather than showing that there *is* a difference. In such instances the paired samples t test can be used to establish equivalence.

As an example, suppose that a manufacturer seeks to gain FDA approval for a modification they have made to an automated blood pressure monitoring device. The modification is not meant to improve the performance of the device but does make it easier, and therefore cheaper, to manufacture. Before approving the modification, the FDA might require that a series of studies be carried out in order to establish that the modified device performs as well as the unmodified device. One of these studies might attempt to demonstrate that the two devices, when used with the same group of patients, produce equivalent mean blood pressure values. After appropriate consultations it might be decided that the mean blood pressures produced by the two devices will be deemed equivalent if the means differ by no more than four millimeters of mercury. Table 5.4 shows (fictitious) blood pressures for 16 subjects as measured by the modified and unmodified devices.

The relevant question is whether this sample of difference scores can be conceived of as coming from a difference score population whose mean is between 4 and -4. If this is the case, the two devices can be declared equivalent since the difference in mean blood pressures does not exceed four millimeters of mercury. If the mean population difference does not fall in this range, the devices cannot be declared equivalent.

The Test.   Because the paired samples t test is an application of the one mean t test to difference scores, an equivalence test for the means of paired data can be carried out via the methods related to the one mean t test discussed in Section 4.3.6 on page 117.[3] Using the notation of Section 4.3.6, the null and alternative hypotheses for a two-tailed equivalence test based on the paired samples t test can be stated as

$$H_{0E} : \mu_d \leq EI_L \quad \text{or} \quad \mu_d \geq EI_U$$
$$H_{AE} : EI_L < \mu_d < EI_U$$

where EI_L and EI_U are the lower and upper ends of the equivalence interval. In essence, the null hypothesis states that the difference between means does not lie in the equivalence interval while the alternative asserts that the difference is in the interval. You will recall that the null hypothesis is tested by conducting two one-tailed tests at level α. In order to reject the equivalence null hypothesis you must show that $\mu_d > EI_L$ *and* that $\mu_d < EI_U$ which

---

[3]You may wish to review this section before continuing.

TABLE 5.4: Blood pressure measurements obtained by means of modified and unmodified monitoring devices.

Modified	Unmodified	(difference) d	d^2
98	99	1	1
111	109	−2	4
97	100	3	9
132	133	1	1
144	148	4	16
100	100	0	0
120	119	−1	1
110	109	−1	1
131	136	5	25
154	153	−1	1
105	107	2	4
119	120	1	1
107	107	0	0
100	101	1	1
118	116	−2	4
122	127	5	25
$\sum$ 1868	1884	16	94

means that *both* of the following tests must be significant.

$$
\begin{array}{ll}
\text{Test One} & \text{Test Two} \\
H_{01} : \mu_d = EI_U & H_{02} : \mu_d = EI_L \\
H_{A1} : \mu_d < EI_U & H_{A2} : \mu_d > EI_L
\end{array}
$$

The null and alternative hypotheses for the one-tailed equivalence test are *one* of the following

$$
H_{0E} : \mu_d \geq EI_U
$$
$$
H_{AE} : \mu_d < EI_U
$$

or

$$
H_{0E} : \mu_d \leq EI_L
$$
$$
H_{AE} : \mu_d > EI_L
$$

The first one-tailed equivalence hypothesis given above is tested by Test One with the second being carried out by means of Test Two.

EXAMPLE 5.3

Carry out a two-tailed equivalence test as discussed in connection with the data in Table 5.4. State the null and alternative equivalence hypotheses before conducting the test.

Solution   The equivalence hypotheses are

$$H_{0E} : \mu_d \leq -4 \text{ or } \mu_d \geq 4$$
$$H_{AE} : -4 < \mu_d < 4$$

In order to test the equivalence null hypothesis the following paired samples t tests will be conducted at $\alpha = .05$.

Test One	Test Two
$H_{01} : \mu_d = 4$	$H_{02} : \mu_d = -4$
$H_{A1} : \mu_d < 4$	$H_{A2} : \mu_d > -4$

Using the sums from Table 5.4

$$s_d = \sqrt{\frac{\sum d^2 - \frac{(\sum d)^2}{n}}{n-1}} = \sqrt{\frac{94 - \frac{(16)^2}{16}}{16-1}} = \sqrt{\frac{78}{15}} = 2.280$$

and

$$\bar{d} = \frac{\sum d}{n} = \frac{16}{16} = 1.000.$$

Values of obtained t for Test One and Two are then

$$t_1 = \frac{1.0 - 4.0}{\frac{2.28}{\sqrt{16}}} = -5.263 \text{ and } t_2 = \frac{1.000 - (-4.0)}{\frac{2.28}{\sqrt{16}}} = 8.772.$$

Critical t values based on $16 - 1 = 15$ degrees of freedom for the two one-tailed tests are -1.753 and 1.753. It follows that both results are significant leading to rejection of the null hypothesis that the two devices are not equivalent insofar as mean blood pressures are concerned in favor of the alternative that maintains equivalence.

Panel A of Figure 5.1 depicts the logic underlying this test. Strictly speaking, equivalence implies a mean difference of zero but it has been decided, based on practical considerations, that a (population) mean difference between four and minus four will be sufficiently close to zero to be declared equivalent. Test One shows that the mean difference is less than four while Test Two shows that the difference is greater than minus four thereby demonstrating that the mean difference is in the range defined as equivalent.

What would you conclude if the equivalence null hypothesis had not been rejected? Would you conclude that the two devices do not produce equivalent means? No! You can only conclude that you were unable to show that they are equivalent. Why? Because you do not know beta. ∎

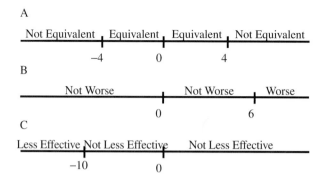

FIGURE 5.1: Logic underlying paired samples t equivalence tests.

EXAMPLE 5.4

Suppose a study is performed to determine whether the dosage of a commonly prescribed cholesterol lowering drug can be reduced in order to avoid liver damage that is sometimes observed with the current dose level. The question of interest is, "Will the reduced dose produce cholesterol levels that are not worse than those obtained with the higher dose?" In order to answer this question 18 subjects who have been taking the higher dose level are switched to the lower dosage. Cholesterol is measured while patients are still on the higher dose and again after they have been on the lower dose for a specified period of time. Cholesterol assessments (fictitious) for the 18 subjects are as shown in Table 5.5.

The researchers decide that if average cholesterol rises by less than six points the reduced dose can be declared to produce results that are "no worse" than those produced by the higher dose. Use a one-tailed equivalence test with $\alpha = .05$ to make this determination.

Solution Because the researchers wish to determine if the rise in cholesterol is less than six points, a one-tailed equivalence test of the form

$$H_{0E} : \mu_d \geq 6$$
$$H_{AE} : \mu_d < 6$$

can be used. The test is carried out by means of a paired-samples t test with hypotheses

$$H_{01} : \mu_d = 6$$
$$H_{A1} : \mu_d < 6$$

Using the sums from the above table

$$s_d = \sqrt{\frac{\sum d^2 - \frac{(\sum d)^2}{n}}{n-1}} = \sqrt{\frac{730 - \frac{(40)^2}{18}}{18-1}} = \sqrt{\frac{641.111}{17}} = 6.141$$

and

$$\bar{d} = \frac{\sum d}{n} = \frac{40}{18} = 2.222.$$

TABLE 5.5: Cholesterol assessments of 18 patients treated with full and reduced dosages of a cholesterol-lowering drug.

Higher Dose	Lower Dose	(difference) d	d^2
198	202	4	16
180	178	−2	4
165	175	10	100
152	158	6	36
211	214	3	9
261	264	3	9
140	140	0	0
200	206	6	36
188	180	−8	64
154	152	−2	4
204	206	2	4
144	148	4	16
155	160	5	25
199	194	−5	25
190	197	7	49
160	164	4	16
189	178	−11	121
190	204	14	196
$\sum$ 3280	3320	40	730

Therefore,

$$t_1 = \frac{\bar{d} - \mu_{d0}}{\frac{s_d}{\sqrt{n}}} = \frac{2.222 - 6.0}{\frac{6.141}{\sqrt{18}}} = -2.610.$$

Critical t for a one-tailed test conducted at $\alpha = .05$ with 17 degrees of freedom is -1.740. The null hypothesis is rejected leading to the conclusion that the reduced dose is not worse than the higher dose insofar as control of cholesterol is concerned.

Panel B of Figure 5.1 depicts the logic of this test. The researchers have decided that a mean increase of less than six units is sufficiently small so as to be considered "not worse." There is no possibility of better results so a two-tailed test is not needed. Once the null hypothesis is rejected in favor of the alternative that maintains the mean difference is less than six, the "not worse" proposition is supported. We won't ask you what the proper interpretation would be if the null hypothesis had not been rejected since we are sure you know this by now. (You do don't you?)   ■

EXAMPLE 5.5

A manufacturer of the generic form of a drug used for the treatment of AIDS wishes to demonstrate that the generic form is not less effective in increasing CD4 count than is the brand name form of the drug. To this end, 14 matched pairs of subjects are treated with the two forms of the drug. Each pair is matched on the basis of age (within two years) and initial CD4 count (within $10/\ mm^3$). One member of each pair is randomly chosen to receive the generic form with the remaining member of the pair then receiving the brand name form. After a specified period of treatment, the CD4 count of each subject is assessed. CD4 counts (fictitious) for the 14 pairs of subjects are provided in Table 5.6.

After consulting with a panel of experts, the researchers decide that the generic form can be declared equivalent to the brand name form if the mean CD4 count for patients taking the generic form is no more than $10/mm^3$ less than the mean level experienced by subjects taking the brand name form. Use a one-tailed equivalence test with $\alpha = .05$ to determine if the two drugs can be declared equivalent.

Solution   Because the researchers wish to determine if the average decline in CD4 level attributable to use of the generic form is less than $10/mm^3$, a one-tailed equivalence test of the form

$$H_{0E} : \mu_d \leq -10$$
$$H_{AE} : \mu_d > -10$$

can be used. The test is carried out by means of a paired-samples t test with hypotheses

$$H_{02} : \mu_d = -10$$
$$H_{A2} : \mu_d > -10$$

Using the sums from the above table

$$s_d = \sqrt{\frac{\sum d^2 - \frac{(\sum d)^2}{n}}{n-1}} = \sqrt{\frac{2673 - \frac{(-85)^2}{14}}{14-1}} = \sqrt{\frac{2156.929}{13}} = 12.881$$

and

$$\bar{d} = \frac{\sum d}{n} = \frac{-85}{14} = -6.071.$$

Therefore,

$$t_2 = \frac{\bar{d} - \mu_{d0}}{\frac{s_d}{\sqrt{n}}} = \frac{-6.071 - (-10.0)}{\frac{12.881}{\sqrt{14}}} = 1.141.$$

Critical t for a one-tailed test conducted at $\alpha = .05$ with 13 degrees of freedom is 1.771. Therefore, the null hypothesis is not rejected. This means that it cannot be asserted that the generic form is *not* less effective.

Panel C of Figure 5.1 depicts the logic of this test. The researchers have decided that a mean decrease of less than ten units is sufficiently small so as to be considered "not less effective." However, this was not demonstrated by the test.

As a final exercise, consider the implications for the above equivalence tests if d had been generated by subtracting the two variables in the opposite direction from that actually employed. Would any of the conclusions have changed? Try making your own sketch of Figure 5.1 as it would appear if the subtraction had been done in the opposite direction. ■

TABLE 5.6: CD4 counts of patients treated with brand name and generic drugs.

Name Brand	Generic	(difference) d	d^2
302	306	4	16
400	363	−37	1369
398	405	7	49
225	225	0	0
221	211	−10	100
261	245	−16	256
154	138	−16	256
300	280	−20	400
188	192	4	16
154	145	−9	81
204	206	2	4
190	200	10	100
210	211	1	1
199	194	−5	25
$\sum$ 3406	3321	−85	2673

5.2.3   Confidence Interval for Paired Samples Mean Difference

Rationale.   The paired samples t test attempts to determine whether there is a difference between the means of two paired variables. A related and usually more informative question is, "How large is the difference between the means of the paired variables?" This difference can be estimated with a confidence interval.

In Section 4.4.5 you learned how to estimate the mean of a population when σ is not known. This method is directly applicable to the problem of estimating the mean of a difference score population.

Earlier we argued that in the situation where there is no difference between the means of the paired variables, the appropriate model for the difference score population would have a mean of zero. In the situation where there is a treatment effect, change, or other mechanism to cause a difference in the means of the two paired variables, the difference score population is best modeled by a population whose mean is not zero. Further, the magnitude of this mean difference can be estimated by forming the appropriate confidence interval on the sample difference scores. By creating such intervals researchers can address such questions as "How much average change took place?" or "How large is the difference between the mean outcomes of the two treatment groups?" Thus, instead of asking the question "Does one drug create a larger mean response than does another" as would be done with a paired samples t test, the researcher can address the question, "How much larger is the mean response produced by one drug than that produced by some other drug," by means of a confidence interval.

The Confidence Interval.   The method of constructing confidence intervals for the mean of a difference score population is the same as that expressed in Equations 4.14 and 4.15 on page 146. We repeat those equations here with subscripts to indicate that difference scores are being used in the calculation.

The equations for L and U then become

$$L = \bar{d} - t\frac{s_d}{\sqrt{n}} \tag{5.2}$$

and

$$U = \bar{d} + t\frac{s_d}{\sqrt{n}} \tag{5.3}$$

where s_d is the sample standard deviation of the difference scores and t is the appropriate value from the t table with $n - 1$ degrees of freedom.

EXAMPLE 5.6

Use the difference scores in Table 5.1 on page 160 to form a one-sided 95% confidence interval for the lower bound of μ_d. Interpret the result.

Solution   As previously calculated, s_d and $\bar{d}$ are

$$s_d = \sqrt{\frac{\sum d^2 - \frac{(\sum d)^2}{n}}{n - 1}} = \sqrt{\frac{1616 - \frac{(96)^2}{15}}{15 - 1}} = \sqrt{\frac{1001.60}{14}} = 8.458$$

and

$$\bar{d} = \frac{\sum d}{n} = \frac{96}{15} = 6.40.$$

Noting from Appendix B that the appropriate t value (with 14 degrees of freedom) for forming a one-sided 95% confidence interval is 1.761, Equation 5.2 yields

$$L = 6.40 - 1.761\frac{8.458}{\sqrt{15}} = 2.554$$

A statistical interpretation of this interval would maintain that we can be 95% confident that μ_d is greater than or equal to 2.554. From the researcher's point of view, it can be stated with 95% confidence that the average increase in blood pressure as measured before and after taking the cold remedy was *at least* 2.554 millimeters of mercury.   ■

EXAMPLE 5.7

The data in Table 5.1 were previously used to conduct a one-tailed paired samples t test of the form

$$H_0 : \mu_d = 0$$
$$H_A : \mu_d > 0$$

at $\alpha = .05$. The result was statistically significant. Does the confidence interval you calculated support this result? Why?

Solution   It can be seen from the confidence interval that the null hypothesis was rejected. This follows from the fact that L is greater than zero.[4]   ∎

EXAMPLE 5.8

Use the difference scores in Table 5.3 on page 164 to form a two-sided 95% confidence interval. What is the meaning of this interval?

Solution   As previously calculated

$$s_d = \sqrt{\frac{.81 - \frac{(-.5)^2}{7}}{7 - 1}} = \sqrt{\frac{.774}{6}} = .359$$

and

$$\bar{d} = \frac{-.5}{7} = -.071.$$

Noting from Appendix B that the appropriate t value (with six degrees of freedom) for a two-sided 95% confidence interval is 2.447, Equations 5.2 and 5.3 yield

$$L = \bar{d} - t\frac{s_d}{\sqrt{n}} = -.071 - 2.447\frac{.359}{\sqrt{7}} = -.403$$

and

$$U = \bar{d} + t\frac{s_d}{\sqrt{n}} = -.071 + 2.447\frac{.359}{\sqrt{7}} = .261.$$

A statistical interpretation of this interval would maintain that we can be 95% confident that μ_d is between $-.403$ and $.261$. From the researcher's point of view, it can be stated with 95% confidence that the average difference in visual acuity between patients receiving laser treatment one and those receiving laser treatment two was between $-.403$ and $.261$ LogMar units. Based on this interval can you rule out a difference of zero?   ∎

EXAMPLE 5.9

The data in Table 5.2 were previously used to conduct a two-tailed paired samples t test of the form

$$H_0 : \mu_d = 0$$
$$H_A : \mu_d \neq 0$$

at $\alpha = .05$. The result was not statistically significant. Does the confidence interval you just calculated support this result? Why?

Solution   It can be seen from the confidence interval that the null hypothesis was not rejected. This follows from the fact that zero is in the interval $-.403$ to $.261$.[5]   ∎

---

[4] You should review Section 4.5 on page 152 if this explanation is not clear.

[5] Again, you should review Section 4.5 on page 152 if this explanation is not clear.

5.2.4 Assumptions

Assumptions underlying the paired samples t test, equivalence tests based on the paired samples t test, and the confidence interval for mean difference are the same as those for the one mean t test and consequently, the one mean Z test (see Section 4.3.3). These are the assumptions of normality and independence. You should note that these assumptions apply to the difference scores rather than to the two component distributions (e.g., pre- and post-treatment distributions) used to generate the difference scores.

Even when the two component distributions are radically nonnormal, difference scores are often symmetric and somewhat normal in shape. As a result, the statistical methods alluded to above tend to be quite robust in the presence of nonnormal data. As is true of the one mean t test, these methods cannot be relied upon to be robust against violations of the independence assumption.

5.3 METHODS RELATED TO PROPORTIONS

5.3.1 McNemar's Test of a Paired Samples Proportion

Rationale. Let us return to an example mentioned at the beginning of this chapter. Two vaccines are to be tested in an attempt to determine whether one is superior to the other in preventing some childhood disease. Because there is a suspicion that there may be a genetic predisposition to this disease, the study design calls for pairs of twins to be used in the study. One member of each pair is randomly designated to receive the first vaccine with the remaining member of the pair then being inoculated with the second vaccine. By using pairs of twins it is assured that genetic predispositions are equally distributed between the two vaccines thereby preventing bias that may result if such predisposition had not been equally represented in the two groups. After some period of time note is taken as to which children contracted the disease. Interest lies in determining whether one vaccine was more effective at preventing the disease than the other.

Table 5.7 shows the results of this (fictitious) study. Each member of the 16 pairs of twins is designated as D or $\overline{D}$ to indicate, respectively, that the twin contracted the disease or did not contract the disease. The last column of the table shows which vaccine was favored (disfavored) by the results obtained from each pair of twins. A one is awarded when the twin receiving vaccine one contracts the disease and the other twin does not. A zero indicates that the twin receiving vaccine two contracted the disease while the other did not. Thus, if vaccine one is more effective than vaccine two we would expect to see many zeros and few ones which would show more disease in the vaccine two group. Many ones and few zeros would indicate more disease in the vaccine one group. Notice that when the same result was obtained from both twins of a pair, no value is recorded since no information regarding superiority of one vaccine over the other was obtained from the pair. Data that provide no information regarding the question at hand (e.g., which vaccine is superior) are termed **noninformative**.

If the two vaccines are equal in their ability to prevent disease, we would expect to see approximately the same number of outcomes favoring one vaccine as favor the other. That is, we would expect to see approximately the same number of ones and zeros in the outcome column. This observation provides the basis for a test of significance.

TABLE 5.7: Presence or absence of a childhood disease among twins inoculated with two different vaccines.

Twin Pair	Vaccine One	Vaccine Two	Outcome
1	$\bar{D}$	D	0
2	D	D	—
3	$\bar{D}$	$\bar{D}$	—
4	D	$\bar{D}$	1
5	D	$\bar{D}$	1
6	$\bar{D}$	$\bar{D}$	—
7	D	$\bar{D}$	1
8	D	D	—
9	$\bar{D}$	$\bar{D}$	—
10	$\bar{D}$	D	0
11	D	$\bar{D}$	1
12	$\bar{D}$	$\bar{D}$	—
13	D	$\bar{D}$	1
14	$\bar{D}$	$\bar{D}$	—
15	$\bar{D}$	$\bar{D}$	—
16	D	$\bar{D}$	1

$$\sum = 6$$

The Test. If noninformative data are omitted, the eight outcomes in Table 5.7 can be conceived of as a sample from a dichotomous population such as is discussed in Section 4.2.4. Further, if there is no difference in the effectiveness of the two vaccines the population would be made up of an equal number of ones and zeros meaning that the proportion of successes in the population (π) would be .5. By contrast, if one vaccine were more effective than the other the appropriate population model would favor one vaccine over the other meaning it would have either more ones than zeros or more zeros than ones. This implies that π would be greater than .5 or less than .5 depending upon which vaccine was favored. It follows that a test of the null hypothesis

$$H_0 : \pi = .5$$

against an appropriate alternative is a test of the assertion that the vaccines have equal effect. Rejection of this hypothesis leads to the conclusion that the vaccines differ in their effectiveness.

In section 4.3.5 you learned to test hypotheses concerning the population proportion π. McNemar's test is simply an application of the methods you learned in Section 4.3.5 to a dichotomous outcome as illustrated in Table 5.7. The null hypothesis tested is as stated above, i.e., that $\pi = .5$. As you know, both approximate and exact methods for testing a population proportion are possible. We will illustrate both as they relate to McNemar's test.

TABLE 5.8: Two by two table for paired dichotomous data.

		First Variable	
		$+$	$-$
Second	$+$	a	b
Variable	$-$	c	d

Approximate Method

A test based on the normal curve approximation for the sampling distribution of $\hat{p}$ can be conducted by means of Equation 4.9. Because McNemar's test specifies the null value as .5, Equation 4.9 can be modified to produce the following

$$Z = \frac{\hat{p} - .5}{\frac{.5}{\sqrt{n}}} \qquad (5.4)$$

Although the number of observations in Table 5.7 are too few[6] to permit appropriate analysis via Equation 5.4, we will nevertheless use these data for illustrative purposes. Noting that $\hat{p} = \frac{6}{8} = .75$ and $n = 8$, Equation 5.4 yields

$$Z = \frac{.75 - .50}{\frac{.50}{\sqrt{8}}} = 1.414$$

If we assume a two-tailed test at $\alpha = .05$, Appendix A shows critical values to be $+1.96$ and -1.96 so that the null hypothesis is not rejected.

An alternative computational form preferred by epidemiologists and others who routinely use contingency tables is given by

$$\chi^2 = \frac{(b - c)^2}{b + c} \qquad (5.5)$$

where χ^2 represents a chi-square statistic (of which you will learn more later), and b and c are counts from the cells of a frequency table as shown in Table 5.8.

This table shows frequencies of two dichotomous variables. The $+$ and $-$ symbols are used to indicate the two conditions of the dichotomous variables. For the current example the frequencies from Table 5.7 would be

---

[6]See Section 4.2.6 on page 84 for a discussion of sample size requirements for this statistic.

		Vaccine One	
		D	$\overline{D}$
Vaccine	D	2	2
Two	$\overline{D}$	6	6

Applying Equation 5.5 gives

$$\chi^2 = \frac{(2-6)^2}{8} = 2.00$$

The critical value for this test is found by entering Appendix D with one degree of freedom. This value is 3.841 for $\alpha = .05$. Because the obtained χ^2 value must be greater than or equal to critical χ^2 in order to reject the null hypothesis, the result is not significant.

Interestingly, Equations 5.4 and 5.5 represent two forms of the same test so that the same result is obtained regardless of which equation is applied. For these tests, $Z^2 = \chi^2$ so that for this example $1.414^2 = 2.00$ (given rounding) for the two obtained values and $1.96^2 = 3.841$ (again, given rounding) for the critical values. As was stated above, Equation 5.5 is preferred by researchers who routinely use contingency tables, but Equation 5.4 may be more intuitive.

EXAMPLE 5.10

Two forms of sunscreen, based on different active ingredients, are to be compared with respect to the level of protection afforded against sun-induced skin damage. To this end, one form of sunscreen is applied to a randomly selected arm of each of 15 volunteer subjects with the second form then being applied to the remaining arm. After a specified period of sun exposure, the skin of each arm is microscopically examined and graded as exhibiting a satisfactory level of protection S or as not exhibiting a satisfactory level of protection $\overline{S}$. The (fictitious) results for each subject are shown in Table 5.9.

Use the two computational methods given above to perform a two-tailed McNemar's test at $\alpha = .05$ on the sunscreen data. What is your conclusion concerning the relative effectiveness of the two sunscreen products?

Solution A dichotomous outcome is constructed as shown in Table 5.10. Ignoring non-informative outcomes, we note that the proportion of outcomes favoring product one is $\frac{9}{10} = .90$ and $n = 10$. Equation 5.4 then gives

$$Z = \frac{.90 - .50}{\frac{.50}{\sqrt{10}}} = 2.530$$

Appendix A shows that the critical values for a two-tailed test conducted at $\alpha = .05$ are $+1.96$ and -1.96 thereby leading to rejection of the null hypothesis. Viewed from a statistical point of view the null and alternative hypotheses are

$$H_0 : \pi = .5 \quad \text{and} \quad H_A : \pi \neq .5$$

When viewed from the context of this study, the null hypothesis asserts that there is no difference in the effectiveness of the two products while the alternative maintains that such

TABLE 5.9: Satisfactory or unsatisfactory levels of protection against sun-induced skin damage rendered by two sunscreen products.

Subject	Product One	Product Two
1	$\overline{S}$	S
2	S	S
3	S	$\overline{S}$
4	S	$\overline{S}$
5	S	$\overline{S}$
6	S	$\overline{S}$
7	$\overline{S}$	$\overline{S}$
8	$\overline{S}$	$\overline{S}$
9	S	$\overline{S}$
10	S	$\overline{S}$
11	S	S
12	$\overline{S}$	$\overline{S}$
13	S	$\overline{S}$
14	S	$\overline{S}$
15	S	$\overline{S}$

TABLE 5.10: Satisfactory or unsatisfactory levels of protection against sun-induced skin damage with outcome variable.

Subject	Product One	Product Two	Outcome
1	$\overline{S}$	S	0
2	S	S	—
3	S	$\overline{S}$	1
4	S	$\overline{S}$	1
5	S	$\overline{S}$	1
6	S	$\overline{S}$	1
7	$\overline{S}$	$\overline{S}$	—
8	$\overline{S}$	$\overline{S}$	—
9	S	$\overline{S}$	1
10	S	$\overline{S}$	1
11	S	S	—
12	$\overline{S}$	$\overline{S}$	—
13	S	$\overline{S}$	1
14	S	$\overline{S}$	1
15	S	$\overline{S}$	1
			$\sum = 9$

a difference does exist. In this case, we rejected the claim of no difference in favor of a statement that the two products do differ in effectiveness. Further, we can conclude that product one is superior to product two in terms of protection provided.

The data can be arranged for analysis via the second computational method as follows.

Product One

		S	$\overline{S}$
Product	S	2	1
Two	$\overline{S}$	9	3

Applying Equation 5.5 gives

$$\chi^2 = \frac{(1-9)^2}{10} = 6.40$$

Because 6.40 exceeds the critical value of 3.841 obtained from Appendix D, the null hypothesis is rejected. Inspection of the summary table shows that there are nine instances in which product one prevents damage while product two does not and one instance in which product two provides protection and product one does not. The conclusion is that product one provides superior protection to that provided by product two. ∎

Exact Method

Previously (page 108) you learned to use the binomial distribution to perform exact tests of hypotheses of the form

$$H_0 : \pi = \pi_0$$

where π is the proportion of observations in a population that meet some specified criterion and π_0 is the hypothesized proportion for the same population. Because McNemar's test can be reduced to a test of the hypothesis

$$H_0 : \pi = .5$$

it follows that the exact method you learned previously (see page 108) can be used to perform an exact version of McNemar's test. Since the exact method was throughly covered in that section, we will not review it here.[7] Rather, we will apply the exact method to the two examples previously analyzed via the approximate method.

EXAMPLE 5.11

Use the data in Table 5.7 to perform an exact two-tailed McNemar's test at $\alpha = .05$. How does this result compare to that obtained from the approximate test?

Solution The sampling distribution of $\hat{p}$ under the condition $\pi = .5$ and $n = 8$ can be constructed by means of Equation 4.5 and is given in Table 5.11. Noting that $\hat{p} = \frac{6}{8} = .75$ and employing the previously learned method (page 108) for finding the p-value for a two-tailed exact test, this value is computed as $2(.10937 + .03125 + .00391) = .28906$.

[7]You may wish to review the discussion beginning on page 108 before continuing.

Because this value is greater than $\alpha = .05$, the null hypothesis is not rejected. This is the same decision reached with the approximate test. However, you should note that the obtained Z value of 1.41 calculated in connection with the approximate test has an associated p-value of $p = 2 \times .0793 = .1586$ which is substantially different from the value obtained from the exact test. ∎

EXAMPLE 5.12

Use the data in Table 5.9 to perform an exact McNemar test with alternative hypothesis

$$H_A : \pi > .5$$

Solution   The sampling distribution of $\hat{p}$ under the condition $\pi = .5$ and $n = 10$ can be constructed by means of Equation 4.5 and is given in Table 5.12. Noting that $\hat{p} = \frac{9}{10} = .90$ and employing the method outlined on page 108 for finding the p-value for a one-tailed exact test with alternative of the form $H_A : \pi > \pi_0$, this value is computed as $.00977 + .00098 = .01075$. Because this value is less than $\alpha = .05$, the null hypothesis is rejected. ∎

5.3.2   Establishing Equivalence for a Paired Samples Proportion

Rationale.   As you saw in Section 5.2.2, studies are sometimes conducted with the aim of showing that two treatments or methods produce equivalent (similar) results. When the data are paired and the outcome dichotomous, equivalence may be demonstrated by showing that the proportion of outcomes favoring one treatment is within some equivalence interval established around .5. For example, suppose that patients suffering from a particular mental disorder have historically been hospitalized for treatment. It is now hypothesized that this disorder might be just as successfully treated on an outpatient basis at significantly reduced cost. In order to test this hypothesis, patient pairs are formed based on age, gender and severity of condition. One member of each pair is randomly chosen for inpatient treatment with the remaining member then being treated as an outpatient. Notice that interest lies in demonstrating that the two treatment methods produce equivalent (similar) results rather than showing that one treatment is superior to the other. After a cost-benefit analysis and other considerations, it is decided that the two treatment options can be considered functionally equivalent if the proportion of paired outcomes favoring the outpatient protocol is greater than .40. (Notice that a proportion of .5 would indicate strict equivalence.) Table 5.13 shows (fictitious) outcomes for 18 patients treated by the two methods. The symbols S and $\overline{S}$ are used to indicate satisfactory and unsatisfactory treatments respectively.

The relevant question is whether π, the proportion of paired outcomes favoring the outpatient treatment is greater than .40. If this is the case, the treatments will be declared equivalent. If not, the treatments will not be deemed equivalent. Notice that the researchers are discounting the possibility that outpatient treatment may be superior to inpatient treatment and are, therefore, interested in one-sided equivalence.

The Test.   Because the equivalence intervals considered here are for population proportions, the methods outlined in Section 4.3.6 on page 117 for proportion equivalence are applicable here.[8] Using the notation of Section 4.3.6, the null and alternative hypotheses for

---

[8]You may wish to review this section before continuing.

TABLE 5.11: Sampling distribution of $\hat{p}$ for $n = 8$ and $\pi = .50$.

Proportion $\hat{p}$	Number of Successes y	Probability $P(y)$
.000	0	.00391
.125	1	.03125
.250	2	.10937
.375	3	.21875
.500	4	.27344
.625	5	.21875
.750	6	.10937
.875	7	.03125
1.00	8	.00391

TABLE 5.12: Sampling distribution of $\hat{p}$ for $n = 10$ and $\pi = .50$.

Proportion $\hat{p}$	Number of Successes y	Probability $P(y)$
.000	0	.00098
.100	1	.00977
.200	2	.04395
.300	3	.11719
.400	4	.20508
.500	5	.24609
.600	6	.20508
.700	7	.11719
.800	8	.04395
.900	9	.00977
1.00	10	.00098

TABLE 5.13: Satisfactory or unsatisfactory outcomes for pairs of patients suffering from a particular mental disorder when treated on an inpatient or outpatient bases.

Pair	Inpatient	Outpatient
1	$\overline{S}$	S
2	$\overline{S}$	$\overline{S}$
3	$\overline{S}$	S
4	$\overline{S}$	S
5	S	$\overline{S}$
6	S	$\overline{S}$
7	S	S
8	S	$\overline{S}$
9	$\overline{S}$	S
10	S	$\overline{S}$
11	S	S
12	S	S
13	$\overline{S}$	S
14	S	$\overline{S}$
15	S	$\overline{S}$
16	S	$\overline{S}$
17	S	$\overline{S}$
18	$\overline{S}$	S

a two-tailed equivalence test for paired proportions can be stated as

$$H_{0E} : \pi \leq EI_L \quad \text{or} \quad \pi \geq EI_U$$
$$H_{AE} : EI_L < \pi < EI_U$$

where EI_L and EI_U are the lower and upper ends of the equivalence interval. In essence, the null hypothesis states that the proportion of paired outcomes favoring one condition over the other does not lie in the equivalence interval while the alternative asserts that this proportion is in the interval. You will recall that the null hypothesis is tested by conducting two one-tailed tests at level α. In order to reject the null hypothesis you must show that $\pi > EI_L$ *and* that $\pi < EI_U$ which means that *both* of the following tests must be significant.

Test One	Test Two
$H_{01} : \pi = EI_U$	$H_{02} : \pi = EI_L$
$H_{A1} : \pi < EI_U$	$H_{A2} : \pi > EI_L$

The null and alternative hypotheses for the one-tailed equivalence test are *one* of the following

$$H_{0E} : \pi \geq EI_U$$
$$H_{AE} : \pi < EI_U$$

or

$$H_{0E} : \pi \le EI_L$$
$$H_{AE} : \pi > EI_L$$

The first one-tailed equivalence hypothesis given above is tested by Test One with the second being carried out by means of Test Two.

As you know from your study of Section 4.3.5, tests of the form represented by Test One and Two can be carried out by either approximate or exact means. Both methods will be illustrated in the examples that follow.

EXAMPLE 5.13

Use approximate and exact methods to carry out a one-tailed equivalence test at $\alpha = .05$ as discussed in connection with the data in Table 5.13. State the null and alternative equivalence hypotheses before conducting the tests.

Solution   An outcome variable that uses a one to indicate a patient pair whose result favors outpatient treatment and a zero for a result that favors inpatient treatment is constructed as shown in Table 5.14. Ignoring noninformative outcomes, it can be seen from this table that the proportion of outcomes favoring outpatient treatment is $\frac{6}{14} = .429$.

The equivalence hypotheses are

$$H_{0E} : \pi \le .40$$
$$H_{AE} : \pi > .40$$

The equivalence null hypothesis can be tested by Test Two as

$$H_{02} : \pi = .40$$
$$H_{A2} : \pi > .40$$

Obtained Z for the approximate method as given by Equation 4.9 is

$$Z_2 = \frac{\hat{p} - \pi_0}{\sqrt{\frac{\pi_0(1-\pi_0)}{n}}} = \frac{.429 - .40}{\sqrt{\frac{(.429)(.571)}{14}}} = .22$$

Because this value is less than critical Z of 1.65, the null hypothesis is not rejected. It follows that equivalence is not demonstrated.

The exact test is conducted by generating the sampling distribution of $\hat{p}$ under the condition $\pi = .40$ and $n = 14$. This is done by application of Equation 4.5 with the result being shown in Table 5.15.[9]

By the method outlined on page 108 for finding the p-value for a one-tailed exact test, $p = .20660 + .15741 + .09182 + .04081 + .01360 + .00330 + .00055 + .00006 + .00000 = .51415$. This value being larger than $\alpha = .05$, the null hypothesis is not rejected which is the same result obtained with the approximate test. Column two of Appendix A shows that the tail area associated with $Z = .22$ is .4129 so that the p-value for the approximate test is decidedly smaller than that for the exact test.   ■

---

[9]Is it necessary to generate all of the values in this table in order to conduct the test?

TABLE 5.14: Outcomes for pairs of patients suffering from a particular mental disorder.

Pair	Inpatient	Outpatient	Outcome
1	$\overline{S}$	S	1
2	$\overline{S}$	S	—
3	$\overline{S}$	S	1
4	$\overline{S}$	S	1
5	S	$\overline{S}$	0
6	S	$\overline{S}$	0
7	S	S	—
8	S	$\overline{S}$	0
9	$\overline{S}$	S	1
10	S	$\overline{S}$	0
11	S	S	—
12	S	S	—
13	$\overline{S}$	S	1
14	S	$\overline{S}$	0
15	S	$\overline{S}$	0
16	S	$\overline{S}$	0
17	$\overline{S}$	$\overline{S}$	0
18	$\overline{S}$	S	1
			$\sum = 6$

TABLE 5.15: Sampling distribution of $\hat{p}$ for $n = 14$ and $\pi = .40$.

Proportion $\hat{p}$	Number of Successes y	Probability $P(y)$
.000	0	.00078
.071	1	.00731
.143	2	.03169
.214	3	.08452
.286	4	.15495
.357	5	.20660
.429	6	.20660
.500	7	.15741
.571	8	.09182
.643	9	.04081
.714	10	.01360
.786	11	.00330
.857	12	.00055
.929	13	.00006
1.00	14	.00000

TABLE 5.16: Positive and negative results obtained under two conditions.

Pair	Condition One	Condition Two
1	+	−
2	+	−
3	−	+
4	−	−
5	+	−
6	−	+
7	+	+
8	+	−
9	−	+
10	−	+
11	+	−
12	+	−
13	−	+
14	−	+
15	+	−
16	−	+
17	−	+
18	+	−

EXAMPLE 5.14

Table 5.16 shows positive (+) and negative (−) results obtained under two conditions. Use approximate and exact methods to perform a two-tailed equivalence test of the null hypothesis that the proportion of outcomes favoring Condition One is not in the interval .3 to .7. Use $\alpha = .05$.

Solution   Table 5.17 shows the coded outcome for each pair. Because interest lies in the proportion of outcomes favoring Condition One, this condition is coded one while outcomes favoring Condition Two are coded zero. It can be seen from this table that $\hat{p} = \frac{8}{16} = .50$.

Because the test is two-tailed, both Test One and Test Two must be performed. The equivalence null hypothesis will be rejected only if both of these tests are significant. The hypotheses for these tests are as follows.

Test One	Test Two
$H_{01} : \pi = .7$	$H_{02} : \pi = .3$
$H_{A1} : \pi < .7$	$H_{A2} : \pi > .3$

The approximate test is carried out by application of Equation 4.9 which yields obtained Z values of

$$Z_1 = \frac{\hat{p} - \pi_0}{\sqrt{\frac{\pi_0(1-\pi_0)}{n}}} = \frac{.50 - .70}{\sqrt{\frac{(.70)(.30)}{16}}} = -1.75$$

and for the second

$$Z_2 = \frac{.50 - .30}{\sqrt{\frac{(.30)(.70)}{16}}} = 1.75.$$

As can be seen in Appendix A, the critical Z values for these two one-tailed tests are -1.65 and 1.65 respectively. It follows that both null hypotheses are rejected leading to rejection of the equivalence null hypothesis.

The exact equivalence test is carried out by use of Equation 4.5 to construct exact distributions of $\hat{p}$ under the conditions $\pi = .7$ and $\pi = .3$ for $n = 16$. These distributions are shown in Table 5.18. By the method outlined on page 108 for finding the p-value for a one-tailed exact test, the p-value for Test One is $.04868 + .01854 + .00556 + .00130 + .00023 + .00003 + .00000[10] + .00000 + .00000 = .07434$ with the same value being obtained for Test Two. As a result, neither hypothesis is rejected which means that the equivalence null hypothesis is not rejected. In this case the approximate test is significant while the exact test is not. ∎

5.3.3   Confidence Interval for a Paired Samples Proportion

Rationale.   While McNemar's Test addresses the question, "Is the proportion of outcomes favoring one condition over the other (π) different from .5," the comparable confidence interval asks the question, "What proportion of outcomes (π) favor one condition over the other?" As you saw in Section 4.5, confidence intervals are usually preferred to hypothesis tests in situations where both methods are applicable. In most situations where McNemar's Test is applicable, confidence intervals of the sort addressed in this section will also be applicable. When this is the case, the confidence interval approach is usually preferred.

The Confidence Interval.   In Section 4.4.6 you learned to construct confidence intervals for a population proportion (π). The approximate and exact methods you learned there are directly applicable to the problem at hand.[11]   For that reason, we will not review the mechanics of constructing confidence intervals for π here as such a course would be unnecessarily redundant, but rather, will provide a few sample applications.

EXAMPLE 5.15

Suppose that 14 paired observations are made in connection with a particular study. Of the 14 pairs, four outcomes favor treatment one, six favor treatment two, and four are noninformative. Use approximate and exact 95% confidence intervals as described in 4.4.6 to construct two-sided confidence intervals to estimate the proportion of outcomes favoring treatment one.

Solution   Equations 4.16 and 4.17 yield the following approximate interval

$$L = \hat{p} - Z\sqrt{\frac{\hat{p}\hat{q}}{n}} = .4 - 1.96\sqrt{\frac{(.4)(.6)}{10}} = .096$$

---

[10]These values are not actually zero but are reported as such because of the number of decimal places used in the approximation.

[11]You may wish to review Section 4.4.6 on page 148 before continuing.

TABLE 5.17: Outcomes with positive and negative results from two conditions.

Pair	Condition One	Condition Two	Outcome
1	+	−	1
2	+	−	1
3	−	+	0
4	−	−	−
5	+	−	1
6	−	+	0
7	+	+	−
8	+	−	1
9	−	+	0
10	−	+	0
11	+	−	1
12	+	−	1
13	−	+	0
14	−	+	0
15	+	−	1
16	−	+	0
17	−	+	0
18	+	−	1
			$\sum = 8$

TABLE 5.18: Sampling distributions of $\hat{p}$ for $n = 16$ and $\pi = .3$, and .7.

Proportion $\hat{p}$	Number of Successes y	$\pi = .30$ $P(y)$	$\pi = .70$ $P(y)$
.0000	0	.00332	.00000
.0625	1	.02279	.00000
.1250	2	.07325	.00000
.1875	3	.14650	.00003
.2500	4	.20405	.00023
.3125	5	.20988	.00130
.3750	6	.16490	.00556
.4375	7	.10096	.01854
.5000	8	.04868	.04868
.5625	9	.01854	.10096
.6250	10	.00556	.16490
.6875	11	.00130	.20988
.7500	12	.00023	.20405
.8125	13	.00003	.14650
.8750	14	.00000	.07325
.9375	15	.00000	.02279
1.0000	16	.00000	.00332

and

$$U = \hat{p} + Z\sqrt{\frac{\hat{p}\hat{q}}{n}} = .4 + 1.96\sqrt{\frac{(.4)(.6)}{10}} = .704.$$

Using $S = 4$ in equations 4.20 and 4.21, the degrees of freedom necessary for finding the F value required for computing the exact value of L are

$$df_{LN} = 2(n - S + 1) = 2(10 - 4 + 1) = 14$$

and

$$df_{LD} = 2S = (2)(4) = 8.$$

Appendix C shows that with numerator degrees of freedom of 14 and denominator degrees of freedom of 8, the appropriate F value to be used for construction of the lower limit of a 95% confidence interval is 4.13. Using this value in Equation 4.18 gives

$$L = \frac{S}{S + (n - S + 1) F_L} = \frac{4}{4 + (10 - 4 + 1)4.13} = .122.$$

The degrees of freedom necessary for finding the F value required for computing the exact value of U are by means of Equations 4.22 and 4.23

$$df_{UN} = 2(S + 1) = 2(4 + 1) = 10$$

and

$$df_{UD} = 2(n - S) = 2(10 - 4) = 12.$$

The F value with numerator and denominator degrees of freedom of 10 and 12 respectively to be used in the construction of the upper limit of a 95% confidence interval is, by Appendix C, 3.37. Then by Equation 4.19

$$U = \frac{(S + 1) F_U}{n - S + (S + 1) F_U} = \frac{(4 + 1) 3.37}{10 - 4 + (4 + 1) 3.37} = .737.$$

The exact two-sided 95% confidence interval is then .122 to .737. This compares with the approximate interval of .096 to .704. The discrepancy between these intervals is expected given the small sample size. The sizable length of both intervals is also due to the small sample size. ■

EXAMPLE 5.16

A (fictitious) study is to be conducted in order to determine whether the complex of symptoms reported by Gulf War veterans is more prevalent among this group than among a comparable group of veterans who did not participate in this action. To this end, 14,791 matched pairs of veterans are interviewed with one member of each pair having participated in the war while the other did not. Matching was done on the basis of age, gender, branch of military service, dates of service, and military occupation.

For 628 of these pairs, the Gulf War participant reported experiencing one or more of the symptoms that make up the alleged syndrome while the matched member of the pair did not. For 174 pairs, the non-participant reported one or more symptoms while the Gulf War participant did not. Thirteen thousand nine hundred eighty nine pairs were noninformative. Use these data to form two-sided 95% confidence intervals for the proportion of outcomes in which the participant reported one or more symptoms while the non-participant did not. Form both approximate and exact intervals.

Solution   Noting that $\hat{p} = \frac{628}{628+174} = .783$, Equations 4.16 and 4.17 yield the following approximate interval

$$L = \hat{p} - Z\sqrt{\frac{\hat{p}\hat{q}}{n}} = .783 - 1.96\sqrt{\frac{(.783)(.217)}{802}} = .754$$

and

$$U = \hat{p} + Z\sqrt{\frac{\hat{p}\hat{q}}{n}} = .783 + 1.96\sqrt{\frac{(.783)(.217)}{802}} = .812.$$

Using $S = 628$ in equations 4.20 and 4.21, the degrees of freedom necessary for finding the F value required for computing the exact value of L are

$$df_{LN} = 2(n - S + 1) = 2(802 - 628 + 1) = 350$$

and

$$df_{LD} = 2S = (2)(628) = 1256.$$

The calculated denominator degrees of freedom of 1256 is not shown in Appendix C. However, this value is so large as to permit use of the ∞ degrees of freedom entry. Thus, with numerator degrees of freedom of 350 and denominator degrees of freedom taken to be ∞, the appropriate F value to be used for construction of the lower limit of a 95% confidence interval is 1.15.[12] Using this value in Equation 4.18 gives

$$L = \frac{S}{S + (n - S + 1)F_L} = \frac{628}{628 + (802 - 628 + 1)1.15} = .757.$$

The degrees of freedom necessary for finding the F value required for computing the exact value of U are by means of Equations 4.22 and 4.23

$$df_{UN} = 2(S + 1) = 2(628 + 1) = 1258$$

and

$$df_{UD} = 2(n - S) = 2(802 - 628) = 348.$$

Since neither of these degrees of freedom are found in Appendix C, we will use ∞ for the numerator degrees of freedom and 350 for the denominator degrees of freedom[13] which yields an F value of 1.17. Then by Equation 4.19

$$U = \frac{(S + 1)F_U}{n - S + (S + 1)F_U} = \frac{(628 + 1)1.17}{802 - 628 + (628 + 1)1.17} = .809.$$

---

[12]In fact, the F value associated with 350 and 1256 degrees of freedom is 1.18 when reported to two decimal places.

[13]If this were an actual study rather than a textbook exercise, we would use computer software to find the F value with 1258 and 348 degrees of freedom.

Notice that the approximate interval of .754 to .812 is quite close to the exact interval of .757 to .809. Notice also that both intervals are fairly narrow thereby yielding estimates of π that would be satisfactory for many research applications. Both of these results are attributable to the large sample size.

Based on these intervals, what is your conclusion regarding the question of whether participants in the Gulf War report one or more symptoms more often than do non-participants? Would your conclusion change if the interval had been .470 to .530? Why? ∎

5.3.4 Assumptions

See page 116.

5.4 METHODS RELATED TO PAIRED SAMPLES RISK RATIOS

5.4.1 Background

The **risk** of some event to a particular group is the *probability* that the event will occur to a member of that group. For example, suppose that a group of industrial workers are routinely exposed (E) to a chemical solvent that is suspected of being related to bladder cancer (D). The risk of bladder cancer to this group is the probability that a member of the group will experience this disease. Using the notation of Chapter 3 the risk to the group can be characterized as $P(D \mid E)$ which is read, "The probability of disease given exposure." Likewise, we can characterize the risk to workers in the same factory who are not exposed to the solvent as $P(D \mid \overline{E})$ which is read, "The probability of disease given no exposure."

It seems reasonable that an industrial hygiene researcher may wish to compare the probability that an exposed worker will experience the disease to the probability that an unexposed worker will experience the disease. If the probability for exposed workers is greater than that for unexposed workers, the implication may be an association between exposure to the solvent and bladder cancer. A comparison between the two probabilities may be done in various ways including computation of the difference by simple subtraction. A popular way to compare the two probabilities is to form them into a ratio referred to as a **risk ratio** which is defined by Equation 5.6.

$$RR = \frac{P(D \mid E)}{P(D \mid \overline{E})} \qquad (5.6)$$

You will recognize this as Equation 3.11 from from page 59.[14] If the risk of disease[15] in the two groups is the same, the risk ratio is one which would imply that there is no relationship between the exposure and the disease. On the other hand, a risk ratio greater than one would imply that exposure is associated with greater risk of disease than is non-exposure. Ratios that are less than one are said to be **protective** because exposure implies reduced risk of disease. This might occur, for example, when persons are exposed to a

---

[14]You may wish to review Section 3.3.6 on page 59 at this point.

[15]While we refer to disease and exposure in Equation 5.6, the terms are generic and represent whatever variables are being examined.

vaccine. We will represent the parameter and statistic forms of the risk ratio as RR and $\widehat{RR}$ respectively.

The sample paired samples risk ratio is defined by Equation 5.7 where a, b, and c are as specified in Table 5.8 on page 176. Note that the value of the sample risk ratio will depend upon the manner in which you arrange the table. You will usually want to assess the risk of some exposure or condition to non-exposure or some other condition. In this case the exposure variable will be placed along the left side of the table with the non-exposure variable or the second condition to which comparison is to be made, along the top. The important point is that you should be aware of what $\widehat{RR}$ estimates. It is the probability that the variable placed along the left side of the table will occur divided by the probability that the variable placed along the top of the table will occur. The result derived from one arrangement of the table will be the reciprocal of the result obtained under the second arrangement. That is, if you calculate $\widehat{RR}$ to be 1.2 under one arrangement of the data, the value obtained under the other arrangement will be $1.0/1.2 = .833$. Take note of the manner in which the tables are arranged in the examples that follow.

$$\widehat{RR} = \frac{a+b}{a+c} \tag{5.7}$$

5.4.2   Test of the Hypothesis $RR = 1$ for Paired Samples

Rationale.   As in the situation alluded to above, suppose it is suspected that exposure to a particular chemical solvent commonly used in manufacturing increases the risk of bladder cancer. To test this suspicion, workers in a particular industry who are routinely exposed to the solvent might be followed across time to determine whether they develop the disease. For comparison purposes, each of these workers might be matched with a worker in the same industry who is not exposed to the solvent. After the period of follow-up, the sample risk ratio is computed by dividing the proportion of exposed workers developing the disease by the proportion of non-exposed workers who develop the disease. If the two risks were .005 and .001 respectively, the risk ratio would be $\frac{.005}{.001} = 5$. This indicates that the risk of bladder cancer for exposed workers is five times that of non-exposed workers. But is this simply a result of the particular sample used in the study? If the study were repeated with other workers might the result be markedly different? More importantly, is the risk ratio in the population one, indicating no relationship between exposure and disease (i.e., the risk in the two groups is the same), or is it different from one indicating that such a relationship does exist? A test of hypothesis can help answer these questions.

The Test.   A test of the null hypothesis $H_0 : RR = 1$ can be conducted by noting that whenever the risk ratio is 1.0, the proportion of outcomes favoring one condition over the other is .5. When the risk ratio is not 1.0, the proportion of outcomes favoring one condition over the other is not .5. This means that by using McNemar's test of the null hypothesis $H_0 : \pi = .5$ we can simultaneously test the null hypothesis $H_0 : RR = 1$. Thus, for this test both approximate and exact methods are available.

EXAMPLE 5.17

Use the data in Table 5.7 on page 175 to calculate the sample risk ratio showing the risk of disease for twins receiving vaccine two compared to that of twins receiving vaccine one, then test the null hypothesis $H_0 : RR = 1$.[16] Use both Equations 5.4 and 5.5 on page 176 for the hypothesis tests.

Solution   Forming the data in Table 5.7 into a two by two table yields

		Vaccine One	
		D	$\overline{D}$
Vaccine D		2	2
Two $\overline{D}$		6	6

By Equation 5.7 the sample risk ratio is

$$\widehat{RR} = \frac{(2+2)}{(2+6)} = .50$$

Thus, the probability of disease for a twin receiving vaccine two is only half that of a twin receiving vaccine one. The tests of hypotheses were previously conducted. (See the discussion beginning on page 175.) Obtained Z and χ^2 were respectively 1.414 and 2.0 indicating a nonsignificant result. Thus, we cannot demonstrate that the risk ratio differs from one thereby not allowing for the conclusion that a relationship exists between the type of vaccine used and disease.   ■

EXAMPLE 5.18

Use the data in Table 5.7 to perform an exact test of the null hypothesis $H_0 : RR = 1$.

Solution   This test was previously conducted (see the discussion beginning on page 175) with a resultant p-value of .28906. Because this value is greater than $\alpha = .05$, the null hypothesis $H_0 : RR = 1$ is not rejected.   ■

EXAMPLE 5.19

Evans and Frick [13] (as cited by Greenland [20]) provide results from the Fatal Accident Reporting System (FARS) dealing with fatal two-rider motorcycle accidents in which both driver and passenger were male and neither wore a helmet. They report that in 226 instances both died, in 546 instances only the driver died and in 378 cases only the passenger died. Use these data to calculate the risk ratio comparing risk of driver to passenger deaths. Use Equations 5.4 and 5.5 to test the null hypothesis $RR = 1$.

Solution   It is convenient to arrange the data into a two by two table in the manner represented by Table 5.8 on page 176.

		Passenger	
		died	lived
Driver	died	226	546
	lived	378	

---

[16]From this point on tests of hypotheses will be two-tailed at $\alpha = .05$ unless otherwise specified.

Notice that no information regarding the number of instances in which neither driver nor passenger died is provided. This is of no consequence for the problem at hand because this frequency does not enter into the calculation of $\widehat{RR}$ nor the hypothesis tests.

By Equation 5.7 the sample risk ratio is

$$\widehat{RR} = \frac{a + b}{a + c} = \frac{226 + 546}{226 + 378} = 1.278$$

By this data, then, the driver is at 1.278 times the risk of death of the passenger. In order to employ Equation 5.4 we note that cell a is noninformative insofar as the test of significance is concerned. Also, the proportion of instances in which the driver dies but the passenger does not die,[17] ($\hat{p}$) is

$$\hat{p} = \frac{b}{b + c} = \frac{546}{546 + 378} = .591$$

Then, by Equation 5.4.

$$Z = \frac{\hat{p} - .5}{\frac{.5}{\sqrt{n}}} = \frac{.591 - .50}{\frac{.50}{\sqrt{924}}} = 5.53$$

Because this value is greater than 1.96 the null hypothesis that asserts $RR = 1$ is rejected. We conclude, therefore, that drivers are at greater risk of death than are passengers.

The same conclusion is reached by application of Equation 5.5

$$\chi^2 = \frac{(b - c)^2}{b + c} = \frac{(546 - 378)^2}{546 + 378} = 30.55$$

This value is greater than the critical χ^2 from Appendix D of 3.841 so that the null hypothesis is rejected.   ■

5.4.3 Establishing Equivalence by Means of the Paired Samples Risk Ratio

Rationale.   As with mean differences and paired proportions, it is sometimes desirable to show that a risk associated with some condition is equivalent to a risk associated with some other condition. This would imply use of a two-tailed equivalence test based on risk ratios. Perhaps more commonly, it is sometimes desirable to show that a risk associated with one condition is *no greater than* a risk associated with some other condition. This implies a one-tailed test.

For example, suppose that a program designed to control the fruit fly is to be implemented by aerial spraying of large geographic areas with the chemical malathion. Because some environmental activists have charged that such spraying is harmful to humans, a study might be designed to show that the risk after spraying is no greater than the risk before spraying. To this end, certain subjects who live in the areas to be sprayed may undergo medical examination before and after spraying is carried out. Each condition examined is rated as either being present or not being present for each subject.

---

[17]In the table format used in conjunction with McNemar's test the entries in cell b were coded 1 while those in cell c were coded 0.

Examination of the collected data shows that for a specific condition (e.g., elevated heart rate), 126 subjects manifest the condition before and after spraying, 414 after spraying but not before, 390 before but not after, and 999 do not manifest the condition at either period. The researchers will test the null hypothesis that the risk after spraying is greater than the risk before spraying against the alternative that the risk after spraying is not greater than the risk before spraying.

What is the difference between the analysis outlined above and one that tests the null hypothesis of no difference in the risk against an alternative that the risk is increased after spraying? Where does the burden of proof lie in the two analyses? We'll leave these questions for you to ponder.

The Test. Equivalence intervals for risk ratios are usually, though not necessarily, symmetric about 1.0 in the sense that $EI_U = \frac{1}{EI_L}$ and vice versa. Thus, if EI_L is .8, EI_U would ordinarily be $\frac{1.0}{.8} = 1.25$.

Using the notation of Section 4.3.6, the null and alternative hypotheses for a two-tailed equivalence test for paired risk ratios can be stated as

$$H_{0E} : RR \leq EI_L \text{ or } RR \geq EI_U$$
$$H_{AE} : EI_L < RR < EI_U$$

where EI_L and EI_U are the lower and upper ends of the equivalence interval. In essence, the null hypothesis states that the risk ratio does not lie in the equivalence interval while the alternative asserts that the risk ratio is in the interval. You will recall that the null hypothesis is tested by conducting two one-tailed tests at level α. In order to reject the equivalence null hypothesis you must show that $RR > EI_L$ *and* that $RR < EI_U$ which means that *both* of the following tests must be significant.

Test One	Test Two
$H_{01} : RR = EI_U$	$H_{02} : RR = EI_L$
$H_{A1} : RR < EI_U$	$H_{A2} : RR > EI_L$

The null and alternative hypotheses for the one-tailed equivalence test is *one* of the following

$$H_{0E} : RR \geq EI_U$$
$$H_{AE} : RR < EI_U$$

or

$$H_{0E} : RR \leq EI_L$$
$$H_{AE} : RR > EI_L$$

The first one-tailed equivalence hypothesis given above is tested by Test One with the second being carried out by means of Test Two.

Because McNemar's test is constrained to a test of the hypothesis $H_0 : RR = 1$, it is not useful when testing for equivalence. However, a number of approximate tests are available for this purpose. One such test is conducted by the following.

$$Z = \frac{\ln\left(\widehat{RR}\right) - \ln\left(RR_0\right)}{\sqrt{\frac{b+c}{(a+b)(a+c)}}} \tag{5.8}$$

The symbols a, b, and c are shown in Table 5.8 on page 176, $\widehat{RR}$ is the sample risk ratio as defined in Equation 5.7 and RR_0 is the hypothesized population risk ratio. The symbol $\ln()$ indicates that the natural log is to be taken of the value in parentheses.

EXAMPLE 5.20

Compute the risk ratio for the malathion data discussed above. Assuming that an increase in risk of less than 1.1 is deemed acceptable, conduct a one-tailed equivalence test to show that the increase in risk is less than this value. State the null and alternative equivalence hypotheses and interpret the test results.

Solution   It is convenient to form the data into a two-by-two table as shown below.

		Before Spray	
		elevated	not elevated
After	elevated	126	414
Spray	not elevated	390	999

By Equation 5.7 the sample risk ratio is

$$\widehat{RR} = \frac{a+b}{a+c} = \frac{126+414}{126+390} = 1.047$$

Because the researchers are attempting to determine whether the risk ratio is less than 1.1, the test will be one-tailed with equivalence null hypothesis

$$H_{0E} : RR \geq 1.1$$

and alternative

$$H_{AE} : RR < 1.1$$

Conducting Test One by means of Equation 5.8 yields

$$Z_1 = \frac{\ln\left(\widehat{RR}\right) - \ln\left(RR_0\right)}{\sqrt{\frac{b+c}{(a+b)(a+c)}}} = \frac{\ln(1.047) - \ln(1.1)}{\sqrt{\frac{414+390}{(126+414)(126+390)}}} = -.919.$$

Because this value is greater than the critical value of -1.65 the test is not significant. Therefore, the researchers were not able to demonstrate that the increase in risk was in the acceptable range (i.e., less than 1.1). ∎

EXAMPLE 5.21

Use the data in the table below to compute $\widehat{RR}$. Then perform a two-tailed equivalence test using an EI of .833 to 1.2. Interpret your results.

		Variable Two	
		+	−
Variable	+	301	771
One	−	780	151

Solution By Equation 5.7 the sample risk ratio is

$$\widehat{RR} = \frac{301 + 771}{301 + 780} = .992.$$

This would suggest a slight protective effect for Variable One. Conducting Test One by means of Equation 5.8 yields

$$Z_1 = \frac{\ln\left(\widehat{RR}\right) - \ln\left(RR_0\right)}{\sqrt{\frac{b+c}{(a+b)(a+c)}}} = \frac{\ln\left(.992\right) - \ln\left(1.2\right)}{\sqrt{\frac{771+780}{(301+771)(301+780)}}} = -5.203.$$

Since this value is less than the critical value of -1.65, Test One is significant.
   Conducting Test Two by means of Equation 5.8 yields

$$Z_2 = \frac{\ln\left(.992\right) - \ln\left(.833\right)}{\sqrt{\frac{771+780}{(301+771)(301+780)}}} = 4.775.$$

Since this value is greater than the critical value of 1.65, Test Two is significant. Because both tests are significant, the equivalence null hypothesis is rejected. We conclude, therefore, that Variable One and Variable Two produce equivalent results as defined by the equivalence interval. ∎

5.4.4 Confidence Interval for a Paired Samples Risk Ratio

Rationale. We earlier indicated (Section 4.5) that confidence intervals are generally preferable to hypothesis tests. Thus, rather than asking the question, "Is the population risk ratio different from one?", a generally more informative question asks, "What is the population risk ratio?" While exact methods for constructing confidence intervals for RR are problematic, rather simple approximate methods are available.

   The Confidence Interval. Lower and upper limits for the paired samples risk ratio may be obtained by the following equations.

$$L = \exp\left[\ln\left(\widehat{RR}\right) - Z\sqrt{\frac{b+c}{(a+b)(a+c)}}\right] \tag{5.9}$$

and

$$U = \exp\left[\ln\left(\widehat{RR}\right) + Z\sqrt{\frac{b+c}{(a+b)(a+c)}}\right] \tag{5.10}$$

The symbols exp and ln indicate, respectively, that the natural exponential and logarithm of the enclosed expression is to be taken while a, b, and c are frequencies in a two by two table as previously discussed.

EXAMPLE 5.22

Use the Evans and Frick [13] data (reproduced for your convenience below) to construct a two-sided 95% confidence interval. What does this interval mean? Use *the interval* to perform a two-tailed test of the null hypothesis $RR = 1$. Explain how you reached your conclusion regarding this test.

		Passenger died	Passenger lived
Driver	died	226	546
	lived	378	

Solution   By Equation 5.7

$$\widehat{RR} = \frac{a+b}{a+c} = \frac{226+546}{226+378} = 1.278.$$

Then by Equation 5.9

$$L = \exp\left[\ln\left(\widehat{RR}\right) - Z\sqrt{\frac{b+c}{(a+b)(a+c)}}\right]$$

$$= \exp\left[\ln(1.278) - 1.96\sqrt{\frac{546+378}{(226+546)(226+378)}}\right]$$

$$= \exp\left[.245 - 1.96\sqrt{.002}\right]$$

$$= \exp[.157]$$

$$= 1.170.$$

And by Equation 5.10

$$U = \exp\left[\ln\left(\widehat{RR}\right) + Z\sqrt{\frac{b+c}{(a+b)(a+c)}}\right]$$

$$= \exp\left[.245 + 1.96\sqrt{.002}\right]$$

$$= \exp[.333]$$

$$= 1.395.$$

From this interval we can conclude that drivers of motorcycles (under the circumstances alluded to above) are at between 1.170 and 1.395 times the risk of death as compared to that of passengers. We can also conclude that if a two-tailed test of the null hypothesis $RR = 1$ were conducted at $\alpha = .05$, the result would be significant. This conclusion is reached by noting that the null value of 1.0 is not in the interval.[18] ■

EXAMPLE 5.23

Use the malathion data discussed previously (reproduced for your convenience below) to construct a one-sided 95% confidence interval to estimate the lower bound for increased heart rate. What does this interval mean? Use *the interval* to perform a one-tailed test of the null hypothesis $RR = 1$ against the alternative $RR > 1$. Explain how you reached your conclusion regarding this test.

| | | Before Spray ||
		elevated	not elevated
After	elevated	126	414
Spray	not elevated	390	999

Solution   By Equation 5.7

$$\widehat{RR} = \frac{a+b}{a+c} = \frac{126+414}{126+390} = 1.047.$$

Then by Equation 5.9

$$L = \exp\left[\ln\left(\widehat{RR}\right) - Z\sqrt{\frac{b+c}{(a+b)(a+c)}}\right]$$

$$= \exp\left[\ln\left(1.047\right) - 1.65\sqrt{\frac{414+390}{(126+414)(126+390)}}\right]$$

$$= \exp\left[.046 - 1.65\sqrt{.003}\right]$$

$$= \exp[-.044]$$

$$= .957.$$

From this interval we can conclude that the risk of elevated heartbeat after spraying as compared to before spraying is greater than or equal to .957. Notice that a value of .957 would be protective, a value of 1.0 would indicate no difference in risk and a value greater than 1.0 would imply increased risk. Since all of these values are possible, we are left with little information about the risk associated with spraying of this chemical. Not surprisingly, we can conclude that if a one-tailed test of the null hypothesis $RR = 1$ and alternative $RR > 1$ were conducted at $\alpha = .05$, the result would not be significant. This conclusion is reached by noting that the null value of 1.0 is in the interval.[19] ■

---

[18]See Section 4.5 on page 152 if this explanation is not clear.

[19]See Section 4.5 on page 152 if this explanation is not clear.

5.4.5 Assumptions

Because the method of testing $H_0 : RR = 1$ is based on McNemar's test, the assumptions underlying the test are those discussed on page 116.

The methods for conducting equivalence tests and constructing confidence intervals presented here depend on a normal curve approximation. Sample sizes should be large enough to ensure that the approximation provides sufficiently accurate results. Greenland [20] suggests a rule of thumb for sample size that requires both the number of exposed and unexposed subjects be greater than or equal to five. In addition, it is assumed that pairs of observations are mutually independent. That is, the results obtained from one pair neither influences nor is influenced by, results obtained from any other pair.

5.5 METHODS RELATED TO PAIRED SAMPLES ODDS RATIOS

5.5.1 Background

In certain circumstances, to be discussed in Chapter 6, the risk ratio is not an appropriate indicator of relative risk. In these circumstances, it is common practice to use the odds ratio as such a measure.

The **odds** associated with some event for a particular group is the *probability* that the event will occur to a member of that group divided by the probability that the event will not occur to a member of that group. For example, suppose that a group of industrial workers are identified who have been diagnosed as having bladder cancer. The odds that a member of this group was exposed to a particular industrial solvent suspected of causing bladder cancer is, using the notation of Chapter 3, $\frac{P(E|D)}{P(\overline{E}|D)}$ which is read, "The probability of exposure given disease divided by the probability of no exposure given disease." If, for example, this value were 2.0 we would say that the probability of exposure to the solvent for workers having bladder cancer is twice the probability that they were not exposed to the solvent.

Likewise, we can characterize the odds of exposure to the solvent for workers in the same factory who are free of bladder cancer as $\frac{P(E|\overline{D})}{P(\overline{E}|D)}$ which is read, "The probability of exposure given no disease divided by the probability of no exposure given no disease." If, for example, this value were .80 we would say that the probability of exposure to the solvent for workers not having bladder cancer is .8 the probability that they were not exposed to the solvent.

It seems reasonable that an industrial hygiene researcher may wish to compare the odds that a worker with bladder cancer was exposed to the solvent to the odds that a worker without bladder cancer was exposed to the solvent. If the odds of exposure for workers with bladder cancer is greater than the odds for workers without bladder cancer, the implication may be an association between exposure to the solvent and bladder cancer. A comparison between the two odds may be done in various ways including computation of the difference by simple subtraction. A popular way to compare the two odds is to form them into a ratio referred to as an **odds ratio** which is defined by Equation 5.11. Note that

$$OR = \frac{\frac{P(E|D)}{P(\overline{E}|D)}}{\frac{P(E|\overline{D})}{P(\overline{E}|\overline{D})}}$$

which simplifies to

$$OR = \frac{P(E \mid D)\, P\left(\overline{E} \mid \overline{D}\right)}{P\left(\overline{E} \mid D\right) P\left(E \mid \overline{D}\right)} \qquad (5.11)$$

Equation 3.12 on page 60[20] expresses the odds of disease for exposed and unexposed groups while Equation 5.11 expresses the odds of exposure for groups with and without disease. Both forms are used by researchers.

It will be helpful to distinguish between two general forms of research designs[21] with which the paired samples odds ratio is commonly employed. In Section 5.4.1 an example was given in which workers who were routinely exposed to a solvent along with workers who were not so exposed were followed across time in order to determine which developed bladder cancer. The odds ratio computed for this design would express the odds of disease (bladder cancer) for exposed (solvent) and unexposed groups of workers. Designs of this sort in which exposed and unexposed subjects are followed across time in order to determine which develop disease are termed **prospective cohort** designs. Data from this sort of design would typically be arranged for analysis as shown in Table 5.19 on the next page.

By contrast, in this section an example was given in which workers who had already developed disease (bladder cancer) were compared to workers who had not developed the disease. The odds ratio computed for this design would express the odds of exposure (solvent) for workers with and without disease (bladder cancer). Designs of this sort in which subjects who already have disease are compared to subjects who do not have the disease in order to determine which has greater exposure are termed **case-control** designs. Data from this sort of design would typically be arranged for analysis as shown in Table 5.20.

For purposes of the discussion that follows, we will assume that a case-control study is being discussed though, with proper substitutions of exposure for disease and vice versa, the remarks apply equally to prospective studies.

If the odds of exposure[22] in the two groups are the same, the odds ratio is one which would imply that there is no relationship between disease and exposure. On the other hand, an odds ratio greater than one would imply that disease is associated with greater exposure than is non-disease. Ratios that are less than one are said to be **protective** because disease implies less exposure than does no disease. This might occur, for example, when persons with disease have less exposure to a daily vitamin supplement than do persons without disease. We will represent the parameter and statistic forms of the odds ratio as OR and $\widehat{OR}$ respectively. The sample paired samples odds ratio is defined by Equation 5.12 where b, and c are as specified in Table 5.8 on page 176.

$$\widehat{OR} = \frac{b}{c} \qquad (5.12)$$

---

[20]You may wish to review the material on page 60 at this point.

[21]These are but two of many such designs.

[22]While we refer to exposure and disease in Equation 5.11, the terms are generic and represent whatever variables are being examined.

TABLE 5.19: Typical data arrangement for paired samples prospective study data.

		Not Exposed	
		Disease	No Disease
Exposed	Disease	a	b
	No Disease	c	d

TABLE 5.20: Typical data arrangement for paired samples case-control study data.

		No Disease	
		Exposed	Not Exposed
Disease	Exposed	a	b
	Not Exposed	c	d

5.5.2   Test of the Hypothesis $OR = 1$ for Paired Samples

Rationale.   As in the situation alluded to above, suppose it is suspected that exposure to a particular chemical solvent commonly used in manufacturing increases the risk of bladder cancer. To test this suspicion, workers in a particular industry who have been diagnosed with bladder cancer are matched with workers in the same industry who are free of bladder cancer. The work histories of these workers are then examined to determine which have been exposed to the solvent and which have not. The sample odds ratio is computed via Equation 5.12. If this ratio were 5.0, for example, the indication would be that the odds of exposure to the solvent for workers with bladder cancer is five times that of cancer-free workers. But is this simply a result of the particular sample used in the study? If the study were repeated with other workers might the result be markedly different? More importantly, is the odds ratio in the population one, indicating no relationship between disease and exposure (i.e., the odds in the two groups is the same), or is it different from one indicating that such a relationship does exist? A test of hypothesis can help answer these questions.

   The Test.   A test of the null hypothesis $H_0 : OR = 1$ can be conducted by noting that whenever the odds ratio is 1.0, the proportion of outcomes favoring one condition over the other is .5. When the odds ratio is not 1.0, the proportion of outcomes favoring one condition over the other is not .5. This means that by using McNemar's test of the null hypothesis $H_0 : \pi = .5$, we can simultaneously test the null hypothesis $H_0 : OR = 1$. You will recall that the same was true for the risk ratio. It follows that McNemar's test simultaneously tests the hypotheses $\pi = .5$, $RR = 1$ and $OR = 1$. Thus, for this test both approximate and exact methods are available.

EXAMPLE 5.24

Suppose that in the bladder cancer study discussed above, 13 pairs of workers were found in which both members of the pair had been exposed to the solvent, 25 pairs had the worker with cancer exposed and the cancer free member unexposed, 5 pairs had the worker with cancer unexposed and the worker with cancer exposed and 55 pairs had neither member exposed.

Use the above data to compute the sample odds ratio. What does this ratio mean? Use an approximate McNemar's test to test the null hypothesis $OR = 1$. Interpret your results.

Solution   For convenience, the data are arranged into a two by two table as shown in Table 5.21.

From Equation 5.12

$$\widehat{OR} = \frac{b}{c} = \frac{25}{5} = 5.0$$

Thus, the odds of exposure for persons with bladder cancer is five times that of persons without the disease.

By Equation 5.5

$$\chi^2 = \frac{(b-c)^2}{b+c} = \frac{(25-5)^2}{25+5} = \frac{400}{30} = 13.33$$

The critical value for this test is found by entering Appendix D with one degree of freedom. This value is 3.841 for $\alpha = .05$. Because the obtained value of 13.33 is greater than the critical value of 3.84, the null hypothesis that $H_0 : \pi = .5$ is rejected which means that $H_0 : \widehat{OR} = 1$ is also rejected. It can now be asserted that bladder cancer is related to exposure to the solvent in question. Do you think this result proves that the solvent *causes* bladder cancer?   ■

EXAMPLE 5.25

Suppose the data in Table 5.21 were modified as shown in Table 5.22. Use this data to calculate the $\widehat{OR}$. Then perform an exact two-tailed test of the null hypothesis $H_0 : OR = 1$.

Solution   By Equation 5.12

$$\widehat{OR} = \frac{b}{c} = \frac{6}{4} = 1.5$$

An exact test of the null hypothesis $H_0 : OR = 1$ can be conducted by testing the null hypothesis $H_0 : \pi = .5$ using McNemar's test. To this end, we note that the frequencies in cells a and d are noninformative[23] and that the proportion of pairs in which the worker with cancer was exposed to the solvent while the worker without cancer was not so exposed ($\hat{p}$) is $\frac{b}{b+c} = \frac{6}{6+4} = .6$.

The sampling distribution of $\hat{p}$ under the condition $\pi = .5$ and $n = 10$ can be constructed by means of Equation 4.5 and is given in Table 5.23. Noting that $\hat{p} = .60$ and employing the method outlined in the Section beginning on page 108 for finding the p-value for a two-tailed exact test, this value is computed as $.20508 + .11719 + .04395 + .00977 + .00098 = .37697$ for the right hand tail so that $p = (2)(.37697) = .75394$. Because this value is greater than $\alpha = .05$, the null hypothesis is not rejected. Thus, we have been unable to show that OR is different from one.   ■

---

[23] See page 174.

TABLE 5.21: Data from a case-control study relating bladder cancer to solvent exposure.

		No Cancer	
		Solvent	No Solvent
Cancer	Solvent	13	25
	No Solvent	5	55

TABLE 5.22: Data from case-control study relating bladder cancer to solvent exposure.

		No Cancer	
		Solvent	No Solvent
Cancer	Solvent	9	6
	No Solvent	4	29

TABLE 5.23: Sampling distribution of $\hat{p}$ for $n = 10$ and $\pi = .50$.

Proportion $\hat{p}$	Number of Successes y	Probability $P(y)$
.000	0	.00098
.100	1	.00977
.200	2	.04395
.300	3	.11719
.400	4	.20508
.500	5	.24609
.600	6	.20508
.700	7	.11719
.800	8	.04395
.900	9	.00977
1.00	10	.00098

5.5.3   Establishing Equivalence by Means of the Paired Samples Odds Ratio

Rationale.   As with risk ratios, it is sometimes desirable to show that the odds associated with some condition are equivalent to the odds associated with some other condition. This would imply use of a two-tailed equivalence test based on odds ratios. Perhaps more commonly, it is often desirable to show that the odds associated with one condition are *no greater than* the odds associated with some other condition. This implies a one-tailed test.

For example, reconsider the program discussed earlier that was designed to control the fruit fly by aerial spraying of large geographic areas with the chemical malathion. Because some environmental activists have charged that such spraying is harmful to humans, a study might be designed to show that the odds of some undesirable outcome after spraying are no greater than the odds before spraying. To this end, certain subjects who live in the areas to be sprayed may undergo medical examination before and after spraying is carried out. Each condition examined is rated as either being present or not being present for each subject.

Examination of the collected data shows that for a specific condition (e.g., elevated heart rate), 126 subjects manifest the condition before and after spraying, 414 after spraying but not before, 390 before but not after and 999 do not manifest the condition at either period. The researchers will test the null hypothesis that the odds after spraying are greater than the odds before spraying against the alternative that the odds after spraying are not greater than those before spraying.

The Test.   As with risk ratios, equivalence intervals for odds ratios are usually, though not necessarily, symmetric about 1.0 in the sense that $EI_U = \frac{1}{EI_L}$ and vice versa. Thus, if EI_L is .8, EI_U would ordinarily be $\frac{1.0}{.8} = 1.25$.

Using the notation of Section 4.3.6, the null and alternative hypotheses for a two-tailed equivalence test for paired odds ratios can be stated as

$$H_{0E} : OR \le EI_L \quad \text{or} \quad OR \ge EI_U$$
$$H_{AE} : EI_L < OR < EI_U$$

where EI_L and EI_U are the lower and upper ends of the equivalence interval. In essence, the null hypothesis states that the odds ratio does not lie in the equivalence interval while the alternative asserts that the odds ratio is in the interval. You will recall that the null hypothesis is tested by conducting two one-tailed tests at level α. In order to reject the equivalence null hypothesis you must show that $OR > EI_L$ *and* that $OR < EI_U$ which means that *both* of the following tests must be significant.

Test One	Test Two
$H_{01} : OR = EI_U$	$H_{02} : OR = EI_L$
$H_{A1} : OR < EI_U$	$H_{A2} : OR > EI_L$

The null and alternative hypotheses for the one-tailed equivalence test is *one* of the following

$$H_{0E} : OR \ge EI_U$$
$$H_{AE} : OR < EI_U$$

or

$$H_{0E} : OR \leq EI_L$$
$$H_{AE} : OR > EI_L$$

The first one-tailed equivalence hypothesis given above is tested by Test One with the second being carried out by means of Test Two.

Because McNemar's test is constrained to a test of the hypothesis $H_0 : OR = 1$, it is not useful when testing for equivalence. However, for paired samples, there is a relationship between π and OR on the one hand and $\hat{p}$ and $\widehat{OR}$ on the other, which allows tests concerning OR to be converted to tests of π thereby permitting testing by methods with which you are already familiar. This also means that both approximate and exact tests can be conducted.

The relationship between π and OR is given by

$$\pi = \frac{OR}{1 + OR} \qquad (5.13)$$

and between $\hat{p}$ and $\widehat{OR}$ by

$$\hat{p} = \frac{\widehat{OR}}{1 + \widehat{OR}} \qquad (5.14)$$

EXAMPLE 5.26

Compute the odds ratio for the malathion data contained in the following table. Assuming that an increase in the odds ratio of less than 1.1 is deemed acceptable, conduct an approximate one-tailed equivalence test to show that the increase in the odds ratio is less than this value. State the null and alternative equivalence hypotheses and interpret the test results.

		Before Spray	
		elevated	not elevated
After	elevated	126	414
Spray	not elevated	390	999

Solution   By Equation 5.12 the sample odds ratio is

$$\widehat{OR} = \frac{b}{c} = \frac{414}{390} = 1.062.$$

Because the researchers are attempting to determine whether the odds ratio is less than 1.1, the test will be one-tailed with equivalence null hypothesis

$$H_{0E} : OR \geq 1.1$$

and alternative

$$H_{AE} : OR < 1.1$$

In order to conduct Test One we first use Equation 5.13 to convert the null OR value of 1.1 to π. This is given as

$$\pi_0 = \frac{OR}{1 + OR} = \frac{1.1}{1.0 + 1.1} = .524.$$

Likewise

$$\hat{p} = \frac{\widehat{OR}}{1 + \widehat{OR}} = \frac{1.062}{1.0 + 1.062} = .515.$$

With these conversions we are now in a position to test the null hypothesis $H_0 : OR = 1.1$ against the alternative $H_A : OR < 1.1$ by testing $H_0 : \pi = .524$ against the alternative $H_A : \pi < .524$. As you are aware, this test can be conducted by means of a Z test for proportions as given in Equation 4.9.

Conducting Test One by means of Equation 4.9 on page 115 yields

$$Z_1 = \frac{\hat{p} - \pi_0}{\sqrt{\frac{\pi_0(1 - \pi_0)}{n}}} = \frac{.515 - .524}{\sqrt{\frac{.524(1 - .524)}{804}}} = -.51.$$

The value for $n = 804$ is obtained as $b + c = 414 + 390 = 804$. Because obtained Z of $-.51$ is greater than critical Z of -1.65, the test is not significant. Therefore, the researchers were not able to demonstrate that the increase in risk was in the acceptable range (i.e., less than 1.1). ∎

EXAMPLE 5.27

Use the data in the table below to compute the $\widehat{OR}$. Then perform an exact two-tailed equivalence test using the EI .833 to 1.2. Interpret your results.

		Variable Two	
		+	−
Variable	+	13	8
One	−	7	19

Solution By Equation 5.12

$$\widehat{OR} = \frac{b}{c} = \frac{8}{7} = 1.143.$$

Because paired samples odds ratios can be transformed into proportions, the methods for performing exact equivalence tests discussed in Sections 4.3.6 and 5.3.2 may be used to perform exact equivalence tests for the paired samples odds ratio.

We begin by transforming the equivalence interval expressed as odds ratios into an equivalence interval expressed as a proportion. Thus, by Equation 5.13 the upper and lower equivalence limits become

$$I_U = \frac{\widehat{OR}}{1 + \widehat{OR}} = \frac{1.2}{1.0 + 1.2} = .545$$

TABLE 5.24: Sampling distributions of $\hat{p}$ for $n = 15$ and $\pi = .545$, and .454.

Proportion $\hat{p}$	Number of Successes y	$\pi = .545$ $P(y)$	$\pi = .454$ $P(y)$
.000	0	.00001	.00011
.067	1	.00013	.00143
.133	2	.00112	.00829
.200	3	.00580	.02989
.267	4	.02084	.07455
.333	5	.05491	.13638
.400	6	.10962	.18900
.467	7	.16882	.20206
.533	8	.20221	.16801
.600	9	.18838	.10866
.667	10	.13539	.05421
.733	11	.07371	.02049
.800	12	.02943	.00568
.867	13	.00814	.00109
.933	14	.00139	.00013
1.000	15	.00011	.00001

and

$$I_L = \frac{\widehat{OR}}{1 + \widehat{OR}} = \frac{.833}{1.0 + .833} = .454.$$

By Equation 5.14

$$\hat{p} = \frac{1.143}{1.0 + 1.143} = .533.$$

Notice that this result can be obtained more simply as

$$\hat{p} = \frac{b}{b + c} = \frac{8}{8 + 7} = .533.$$

In order to perform the exact versions of Tests One and Two, it will be necessary to generate the exact distributions of $\hat{p}$ under the conditions $\pi = .545$ and $\pi = .454$ where $n = 15$. This is done by means of Equation 4.5 on page 81. These distributions are provided in Table 5.24.

Noting that $\hat{p} = .533$, the p-value for Test One is found by computing the probability that $\hat{p}$ will take a value of .533 or less under the condition $\pi = .545$ and $n = 15$. This value is computed as $.20221 + .16882 + .10962 + .05491 + .02084 + .00580 + .00112 + .00013 + .00001 = .56346$. The P-value for Test Two is found by computing the probability that $\hat{p}$ will take a value of .533 or greater under the condition $\pi = .454$ and $n = 15$. This value is computed as $.16801 + .10866 + .05421 + .02049 + .00568 + .00109 + .00013 +$

.00001 = .35828. In order to reject the equivalence null hypothesis both of these tests must be significant. Since neither test meets the .05 criterion the null hypothesis is not rejected so that equivalence is not demonstrated. ■

5.5.4   Confidence Interval for a Paired Samples Odds Ratio

Rationale.   We earlier indicated (Section 4.5) that confidence intervals are generally preferable to hypothesis tests. Thus, rather than asking the question, "Is the population odds ratio different from one?", a generally more informative question asks, "What is the population odds ratio?" Both approximate and exact methods for constructing confidence intervals for OR are available. We will present each in the section that follows.

The Confidence Interval.

Approximate Method

A number of methods are available for forming approximate confidence intervals for the paired samples odds ratio. We will outline two of these here.

The first uses Equations 4.16 and 4.17 (repeated here, with slight modification, as Equations 5.15 and 5.16) to form a confidence interval for π. The relationship between the paired samples odds ratio and proportion is then used to convert the endpoints of this interval (i.e. L and U) into an interval for the estimation of the population odds ratio. This conversion is carried out by means of Equation 5.17.

$$L = \hat{p} - Z\sqrt{\frac{\hat{p}\left(1 - \hat{p}\right)}{n}} \tag{5.15}$$

$$U = \hat{p} + Z\sqrt{\frac{\hat{p}\left(1 - \hat{p}\right)}{n}} \tag{5.16}$$

$$\widehat{OR} = \frac{\hat{p}}{1 - \hat{p}} \tag{5.17}$$

where $n = b + c$.

EXAMPLE 5.28

Use the data in Table 5.25 on the next page to calculate the $\widehat{OR}$. Form a two-sided 95% confidence interval to estimate the OR.

Solution   By Equation 5.12

$$\widehat{OR} = \frac{b}{c} = \frac{13}{11} = 1.182.$$

By Equation 5.14

$$\hat{p} = \frac{\widehat{OR}}{1 + \widehat{OR}} = \frac{1.182}{1 + 1.182} = .542.$$

TABLE 5.25: Data from case-control study relating bladder cancer to solvent exposure.

		Controls	
		Exposed	Not Exposed
Cases	Exposed	19	13
	Not Exposed	11	19

Then by Equation 5.16

$$U = \hat{p} + Z\sqrt{\frac{\hat{p}\,(1 - \hat{p})}{n}} = .542 + 1.96\sqrt{\frac{.542\,(1 - .542)}{24}} = .741$$

and by Equation 5.15

$$L = \hat{p} - Z\sqrt{\frac{\hat{p}\,(1 - \hat{p})}{n}} = .542 - 1.96\sqrt{\frac{.542\,(1 - .542)}{24}} = .343.$$

But .343 and .741 represent a confidence interval for the estimation of π. A confidence interval for the estimation of OR may be obtained by converting these endpoints to expressions for odds ratios. This is done by means of Equation 5.17 so that

$$U = \frac{\hat{p}}{1 - \hat{p}} = \frac{.741}{1 - .741} = 2.861$$

and

$$L = \frac{\hat{p}}{1 - \hat{p}} = \frac{.343}{1 - .343} = .522.$$

Thus, the two-sided 95% confidence interval for the estimation of OR is .522 to 2.861.

A second commonly used method employs Equations 5.18 and 5.19.

$$L = \exp\left(\ln\left(\widehat{OR}\right) - Z\sqrt{\frac{1}{b} + \frac{1}{c}}\right) \tag{5.18}$$

$$U = \exp\left(\ln\left(\widehat{OR}\right) + Z\sqrt{\frac{1}{b} + \frac{1}{c}}\right) \tag{5.19}$$

The b and c terms are cell frequencies as outlined in Table 5.19 or 5.20.   ∎

EXAMPLE 5.29

Use Equations 5.18 and 5.19 with the data in Table 5.25 to form a two-sided 95% confidence interval for the estimation of OR.

Solution   As was previously calculated, $\widehat{OR} = 1.182$. From Table 5.25 b and c are, respectively, 13 and 11. Then by Equations 5.19 and 5.18

$$U = \exp\left(\ln\left(\widehat{OR}\right) + Z\sqrt{\frac{1}{b} + \frac{1}{c}}\right) = \exp\left(\ln(1.182) + 1.96\sqrt{\frac{1}{13} + \frac{1}{11}}\right) = 2.638$$

and

$$L = \exp\left(\ln\left(\widehat{OR}\right) - Z\sqrt{\frac{1}{b} + \frac{1}{c}}\right) = \exp\left(\ln(1.182) - 1.96\sqrt{\frac{1}{13} + \frac{1}{11}}\right) = .530.$$

As may be seen, this is a rather wide interval. What would have made this interval shorter?

∎

Exact Method

Earlier you learned to construct an approximate confidence interval for the estimation of the RR by first employing a familiar method for the estimation of π and then converting the endpoints of the resulting interval to expressions that allow for the estimation of the RR. The same method can be employed to form an exact confidence interval for the estimation of the OR.

On page 149[24] you employed Equations 4.18, 4.19, 4.20, 4.21, 4.22, and 4.23 to form exact confidence intervals for the estimation of π. We provide these equations here with a slight change in notation to reflect the contingency table style notation we are using in this section. The endpoints of the confidence interval obtained by use of these equations can be converted via Equation 5.17 on page 208 to form an appropriate estimate of OR.

$$L = \frac{b}{b + (c + 1)\, F_L} \tag{5.20}$$

$$U = \frac{(b + 1)\, F_U}{c + (b + 1)\, F_U} \tag{5.21}$$

In these equations, the b and c terms are cell frequencies as illustrated in Table 5.19 or 5.20 on page 201 and F_L and F_U are the appropriate values from an F distribution. F_L and F_U can be obtained from Appendix C. Notice that, unlike the t table in Appendix B, the F table must be entered with *two* different degrees of freedom. The first of these, which we will call the numerator degrees of freedom, is listed across the top of the table while the second, which we shall call the denominator degrees of freedom, is given along the edge of the table. Thus, for example, the appropriate F value to be used in the calculation of a two-sided 95% confidence interval assuming numerator degrees of freedom equal to four and denominator degrees of freedom of 20 would be 3.51. In order to facilitate calculations ,we will use the notation $df_{LN}, df_{LD}, df_{UN}$, and df_{UD} to represent respectively the numerator degrees of freedom for calculating L, the denominator degrees of freedom for calculating L, the

---

[24]You may wish to review this section before continuing.

numerator degrees of freedom for calculating U, and the denominator degrees of freedom for calculating U.

$$\boxed{df_{LN} = 2\,(c+1)} \tag{5.22}$$

$$\boxed{df_{LD} = 2b} \tag{5.23}$$

$$\boxed{df_{UN} = 2\,(b+1)} \tag{5.24}$$

$$\boxed{df_{UD} = 2c} \tag{5.25}$$

EXAMPLE 5.30

Use the data in Table 5.25 to form an exact two-sided 95% confidence interval for the estimation of OR.

Solution   In order to find F_L we must first find df_{LN} and df_{LD} which are respectively,

$$df_{LN} = 2\,(c+1) = 2\,(11+1) = 24$$

and

$$df_{LD} = 2b = (2)\,(13) = 26.$$

Though not shown in Appendix C, the appropriate F value for a two-sided 95% confidence interval with numerator and denominator degrees of freedom of 24 and 26 respectively is 2.22. It follows from Equation 5.20 that

$$L = \frac{b}{b + (c+1)\,F_L} = \frac{13}{13 + (11+1)\,2.22} = .328$$

In order to find F_U we must first find df_{UN} and df_{UD} which are respectively,

$$df_{UN} = 2\,(b+1) = 2\,(13+1) = 28$$

and

$$df_{UD} = 2c = (2)\,(11) = 22.$$

Though not shown in Appendix C, the appropriate F value for a two-sided 95% confidence interval with numerator and denominator degrees of freedom of 28 and 22 respectively is 2.29. It follows from Equation 5.21 that

$$U = \frac{(b+1)\,F_U}{c + (b+1)\,F_U} = \frac{(13+1)\,2.29}{11 + (13+1)\,2.29} = .745$$

The confidence interval thus far calculated, $L = .328$ and $U = .745$, estimates π. In order to construct an interval for the estimation of OR we use Equation 5.17 to convert the end points of this interval to expressions for the odds ratio. Thus, we obtain

$$L = \frac{\hat{p}}{1 - \hat{p}} = \frac{.328}{1 - .328} = .488$$

and

$$U = \frac{.745}{1 - .745} = 2.922.$$

The confidence intervals obtained via the two approximate methods were .522 to 2.861, and .530 to 2.638 which show fairly good agreement. The exact method produced .488 to 2.922. While approximate methods are commonly employed in the research literature, the exact method is preferred. ■

5.5.5 Assumptions

The assumptions underlying the methods presented here for odds ratios are essentially the same as those discussed on page 116.

KEY WORDS AND PHRASES

After reading this chapter you should be able to demonstrate familiarity with the following words and phrases.

CI for paired odds ratio 208
CI for paired samples proportion 186
difference score population 165
equivalence for paired odds ratio 204
equivalence for paired risk ratio 193
McNemar's test 174
odds 199
paired data 159
paired samples odds ratio 199
paired samples t test 161
protective effect 190
risk ratio 190

CI for paired risk ratio 196
CI for paired samples mean difference 171
difference scores 160
equivalence for a paired samples proportion 180
equivalence via paired samples t tests 165
noninformative data 174
odds ratio 199
paired difference t test 161
paired samples risk ratio 190
prospective cohort 200
risk 190

EXERCISES

5.1 The effectiveness of a diet designed to lower serum cholesterol is to be evaluated by measuring the cholesterol of eight patients identified as being at risk of cardiovascular disease before and after being on the diet for 16 weeks. The data are provided below. Use these data to,

(a) conduct a two-tailed paired samples t test at $\alpha = .05$, and

(b) construct a two-sided 95% confidence interval for the estimation of μ_d.

(c) What is your conclusion as to the effectiveness of the diet?

Patient	Before Diet	After Diet
1	213	199
2	252	241
3	195	197
4	222	220
5	267	248
6	216	224
7	209	209
8	255	237

5.2 A study is conducted in which a gold standard invasive (i.e., blood drawn) method of monitoring glucose in diabetic patients is compared to a newly developed non-invasive (i.e., no blood drawn) monitoring method in order to determine whether the two systems produce equivalent results. To this end, each of the monitoring devices is simultaneously employed to assess glucose levels in ten diabetic patients. The two devices will be considered equivalent if the mean difference in measured glucose levels produced by the two methods is between plus and minus four points. The two measures for each patient are provided below.

Invasive	Non-invasive
140	144
84	82
200	200
249	247
71	64
131	138
140	132
122	123
139	146
119	117

(a) Would a one- or two-tailed equivalence test be appropriate for this study? Why?

(b) State the null and alternative equivalence hypotheses for the study.

(c) Conduct the equivalence test at $\alpha = .05$ and report the result.

(d) What is your conclusion regarding whether the two devices are equivalent?

5.3 Patients suffering from facial burns are asked to characterize their general appearance as "satisfactory" (S) or "unsatisfactory" (U) before and after undergoing a scar reduction treatment regimen.

(a) Suppose that five patients characterize their appearance as U before treatment and S afterward (U-S), two rate their appearance as S before and U after treatment (S-U), while one each represent their appearance as S before and S after treatment (S-S), and U before and U after treatment (U-U).

   i. Use an exact two-tailed version of McNemar's test with $\alpha = .05$ to determine whether the treatment is effective at improving patient's perception of their appearance.

   ii. Use an exact two-sided 95% confidence interval to estimate the proportion of patients who perceive their appearance as improved.

(b) Given U-S= 71, U-U= 22, S-S= 11, and S-U= 20,

   i. Use an approximate two-tailed version of McNemar's test with $\alpha = .05$ to determine whether the treatment is effective at improving patient's perception of their appearance.

   ii. Use an approximate two-sided 95% confidence interval to estimate the proportion of patients who perceive their appearance as improved.

5.4 Patrol officers on a large metropolitan police force are assigned to either motorcycles (traffic control) or patrol cars (other than traffic control). Suppose that officers from each category are matched on time in service, age and gender. Each pair is then screened for skin cancers on their face, neck or hands. The table provided below shows results for the paired officers. The category "cancer" indicates the presence of one or more skin cancers on the areas examined.

		Patrol Car	
		cancer	no cancer
Motor-	cancer	10	34
cycle	no cancer	18	444

(a) Calculate the risk ratio to compare the risk of skin cancer for motorcycle officers to that of patrol car officers. What does this ratio mean?

(b) Construct a 95% confidence interval for the estimation of RR.

5.5 Suppose that the skin cancer study in Exercise 5.4 had collected data retrospectively. That is, suppose that officers with skin cancer were matched with officers who are free of skin cancer on time in service, age and gender. The officer pairs are then categorized as being assigned to motorcycles or patrol cars. These case control data are provided below.

		No Skin Cancer	
		motorcycle	pat. car
Skin	motorcycle	10	24
Cancer	pat. car	9	444

(a) Calculate the odds ratio to compare the odds of being a motorcycle officer for officers with skin cancer to those without skin cancer. What does this odds ratio mean?

(b) Construct 95% confidence intervals for the estimation of OR by two different approximate methods. How well do the intervals agree?

(c) Use an exact method to construct a 95% confidence interval for the estimation of OR. How does the result compare to the two approximate results?

5.6 Use the data in the table provided below to calculate $\widehat{OR}$ then perform the following test. Interpret the result.

$H_{0E} : OR \leq .80$ or $OR \geq 1.25$
$H_{AE} : .8 < OR < 1.25$

		No Disease	
		Exposed	Not Exposed
Disease	Exposed	132	290
	Not Exposed	310	718

A. The following questions refer to Case Study A (page 469)

5.7 The authors report that "A total of 16 subjects changed preference between the beginning and end of wear, with eight switching preferences from the control to the conditioned lens, matched exactly by eight changing the opposite way $\cdots$." They then report that the result of a specific statistical test was not significant.

What test do you think they conducted? Conduct the same test. Do you agree with their result? What is your interpretation of this result?

B. The following questions refer to Case Study B (page 470)

5.8 Suppose the researchers wished to estimate the average change in WOMAC A scores that took place from baseline to the 12 week point. What technique could be used for this estimation? Specify equations to be used if any. Is there enough information provided in the case study to carry out the technique you recommend? Explain.

K. The following questions refer to Case Study K (page 474)

5.9 The authors performed a paired samples t test on the data reported in Table J.6. The question arises as to why they used school means for this purpose rather than the paired scores of individual children. After all, using scores of individual students would provide many more degrees of freedom. There were two reasons. First, logistical and confidentially difficulties made the pairing of student data difficult. The second reason was based on a strictly statistical consideration. What do you believe was the basis of this consideration? (Hint: see the discussion on page 99 regarding juvenile blood pressures.)

5.10 Conduct a two-tailed paired samples t test at the $\alpha = .01$ level on the school means. Interpret the result.

5.11 Can you think of a more informative way to analyze these data? Conduct your proposed analysis. Explain why this result is more informative.

M. The following questions refer to Case Study M (page 475)

5.12 Estimate the average change in oxygen level that took place after taking the sleep medication.

5.13 Estimate the average change in carbon dioxide level that took place after taking the sleep medication.

Two Independent Samples Methods

6.1 INTRODUCTION

It is not always practical to collect data in the paired samples configurations discussed in Chapter 5. For example, a researcher may wish to match subjects on weight, age and gender but find that there are few subjects in the two existing groups that are sufficiently similar in regards to these attributes to make such a strategy viable. The same problem might arise in studies where subjects are to be matched with subsequent random assignment to alternative treatment conditions. As a result, the majority of data collected in research contexts is unpaired.

Consider a study in which the efficacy of two drugs designed for the treatment of hypertension are to be compared. Researchers may find that it is not possible to form subject pairs that are sufficiently similar so as to be treated as paired samples. A simple strategy, in such a circumstance, would be to randomly assign subjects to the two treatments without regard to inherent underlying subject characteristics (e.g., weight, preexisting medical conditions etc.). The random assignment process guarantees that no systematic bias enters into the formation of the two groups.

The methods you studied in Chapter 5 are not appropriate for the analysis of unpaired data. This chapter, then, deals with hypothesis tests and confidence intervals that are designed to compare means or proportions of unpaired data.

6.2 METHODS RELATED TO DIFFERENCES BETWEEN MEANS

6.2.1 The Independent Samples t Test

Rationale. As alluded to above, suppose that a study is to be conducted in order to determine whether one drug is more efficacious than some other drug in the treatment of hypertension. To this end, researchers randomly assign 30 hypertensive subjects to one of two treatment groups. The subjects assigned to group one receive a treatment regimen based on the first drug while those assigned to group two receive a regimen based on the second drug.

TABLE 6.1: Systolic blood pressures of hypertensive patients after treatment via one of two drug regimens.

Group One	Group Two
129	138
111	120
140	137
139	154
144	148
120	122
131	131
129	128
131	140
154	145
119	131
138	120
142	144
110	129
140	141
$\bar{x}_1 = 131.80$	$\bar{x}_2 = 135.20$

The question of interest is, "Do the two treatment regimens produce different blood pressure levels in the hypertensive patients?" Table 6.1 shows (fictitious) systolic blood pressures for the 30 subjects taken after receiving treatment via the two drug regimens.

A reasonable strategy for answering the question as to whether the two drugs produce different levels of blood pressure would be to calculate the mean pressures for each of the two groups and then compare the two means to determine whether they differ. In the case at hand, the subjects taking drug one had average pressure of 131.80 while those taking drug two had average pressure of 135.20. Thus, the average blood pressure of the first group was 3.4 units less than that of the second group.

But, there are at least two explanations for this difference. First, we note that any time we randomly assign subjects to groups, the means of the two groups are almost sure to differ to some degree on almost any conceivable measure. Indeed, we would be surprised to find that the average height of two such groups was both 68.4 inches or that the two average weights were exactly 151.38 pounds.

Thus, even if the two treatments were equal in their impact on the variable of interest, we would still expect to see some difference in the averages of the two groups. The first explanation for the difference of 3.4 units observed in our fictitious study, then, is that this difference arose, not because of any difference in the impact of the medications, but rather by happenstance related to the manner in which the groups were formed.

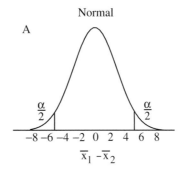

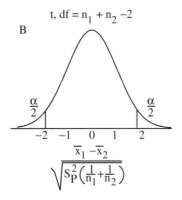

FIGURE 6.1: Distribution of $\bar{x}_1 - \bar{x}_2$ (Panel A) and a t statistic with $n_1 + n_2 - 2$ degrees of freedom (Panel B).

A second explanation is that the medication administered to group one is more effective at lowering the blood pressures of hypertensive patients than is the drug administered to the second group. This would explain why the pressures in group one tended to be lower than those in group two. But which of these explanations is to be believed? A test of significance might help decide the issue.

Conceive of a sampling distribution formed by repeatedly drawing a random sample from some population, randomly dividing each such sample into two component samples, and then calculating the quantity $\bar{x}_1 - \bar{x}_2$ where $\bar{x}_1$ and $\bar{x}_2$ are the means of the two component samples.[1] If this process were repeated many times the resultant values of $\bar{x}_1 - \bar{x}_2$ could then be formed into a sampling distribution as shown in Panel A of Figure 6.1. It is important to understand that this distribution represents the sampling distribution of $\bar{x}_1 - \bar{x}_2$ *when both samples come from populations that have the same mean.* Clearly, if both samples come from the same population then they come from populations with the same mean. We will designate the mean of the population from which the first sample was drawn as μ_1 and the

---

[1]Equivalently, we can conceive of the two component samples as simply two random samples taken from a common population.

mean of the population from which the second sample was drawn as μ_2. Because both samples come from the same population it is clear that $\mu_1 = \mu_2$.

Notice that the mean of this sampling distribution is zero and that the further values of $\bar{x}_1 - \bar{x}_2$ are from zero, the less likely is their occurrence. But how might this distribution be used to help answer the question posed above. That is, was the difference of 3.4 points between the means of the two groups treated with the two different hypertensive medications due to random happenstance as the groups were formed or rather was the drug administered to group one more effective than that administered to group two?

The two samples were originally taken from a common population so that $\mu_1 = \mu_2$. If the two drugs produced exactly the same result in the two samples then it would still be true that $\mu_1 = \mu_2$. In this circumstance we would not expect $\bar{x}_1 - \bar{x}_2$ to fall in a critical region. Why? Because the probability that $\bar{x}_1 - \bar{x}_2$ will fall in a critical region when $\mu_1 = \mu_2$ is only α.

But what would happen if drug one were more effective than drug two? In this case the blood pressures in group one would be lowered by taking the drug so that the absolute value of $\bar{x}_1 - \bar{x}_2$ would tend to be greater than would be the case where both drugs had equal impact. For example, in our hypothetical hypertensive study, $\bar{x}_1 - \bar{x}_2 = 131.8 - 135.2 = -3.4$. But suppose drug one had been more effective so that $\bar{x}_1 = 120.4$. Then $\bar{x}_1 - \bar{x}_2 = 120.4 - 135.2 = -14.8$. The more effective is drug one as compared to drug two, the further below the distribution mean of zero is $\bar{x}_1 - \bar{x}_2$ and the greater the chance that $\bar{x}_1 - \bar{x}_2$ will fall in the lower critical region. What would happen if it were drug two that was more effective at lowering blood pressure?

In short, we do not expect $\bar{x}_1 - \bar{x}_2$ to fall in a critical region when the treatments produce the same result. Indeed, the probability of this happening is only α. However, when treatments produce different results, we expect the value of $\bar{x}_1 - \bar{x}_2$ to move away from zero and toward a critical region. Further, when treatment effects differ, we can no longer claim that $\mu_1 = \mu_2$. Rather, it must be said that $\mu_1 \neq \mu_2$.

From the above, when $\bar{x}_1 - \bar{x}_2$ falls in the critical region, we can claim that the observed difference between these values was due to differential treatment effects rather than to the happenstance of how the groups were formed. We now see how this test is conducted.

The Test. The null hypothesis to be tested by the independent samples t test is most often (but not always)

$$H_0 : \mu_1 = \mu_2$$

or equivalently,

$$H_0 : \mu_1 - \mu_2 = 0$$

The two-tailed alternative is most often (but not always)

$$H_A : \mu_1 \neq \mu_2$$

or equivalently,

$$H_A : \mu_1 - \mu_2 \neq 0$$

One-tailed alternatives are most often (but not always) of the form

$$H_A : \mu_1 < \mu_2$$

or equivalently,

$$H_A : \mu_1 - \mu_2 < 0$$

or

$$H_A : \mu_1 > \mu_2$$

or equivalently,

$$H_A : \mu_1 - \mu_2 > 0$$

In essence, the null hypothesis asserts that there was no difference in the effects of the treatments administered the two groups while the two-tailed alternative maintains that a difference does exist. The one-tailed alternatives specify the form of that difference.

In order to carry out the hypothesis test, $\bar{x}_1 - \bar{x}_2$ cannot be directly referenced to the sampling distribution shown in Panel A of Figure 6.1 on page 217 but rather must be divided by an estimate of the standard error (see the denominator of Equation 6.1) which means that the test statistic must then be referenced to a t distribution as depicted in Panel B of Figure 6.1. Unlike the one mean t test with which you are already familiar, the degrees of freedom for the independent samples t test are $n_1 + n_2 - 2$. The test statistic is calculated by Equation 6.1.

$$t = \frac{\bar{x}_1 - \bar{x}_2 - \delta_0}{\sqrt{s_P^2 \left(\frac{1}{n_1} + \frac{1}{n_2} \right)}} \tag{6.1}$$

In Equation 6.1, $\bar{x}_1$ and $\bar{x}_2$ represent, respectively, the means of sample one and sample two, while n_1 and n_2 represent the number of observations in each of the two samples. The symbol δ_0 represents the hypothesized difference between μ_1 and μ_2 which is usually zero. However, as you will see in the discussion of equivalence tests based on the independent samples t test, this is not always the case.

The term s_P^2 requires some explanation. You recall that the one mean t test utilized an estimate of the standard deviation of the population from which the sample was drawn. We used the sample standard deviation (s) for this purpose. In the present case, we need to estimate the variance of the population from which the samples were drawn. We could use the sample variance (s^2) of one of the two samples for this purpose, but this would be wasteful since it employs an estimate based on only one of the two samples. A better strategy would be to base the variance estimate on *both* samples. This can be done through a particular form of averaging which statisticians refer to as **pooling** (hence the P subscript). Thus, s_P^2 is an estimate of the population variance based on an averaging or pooling of the information in the two samples. You can see that the two sample variances are used in the estimate from the following.

$$s_P^2 = \frac{(n_1 - 1)\, s_1^2 + (n_2 - 1)\, s_2^2}{n_1 + n_2 - 2}$$

s_1^2 and s_2^2 represent the variances of the first and second samples respectively.

The numerator of this expression is simply the sum of the sums of squares of the two samples. Therefore, a more efficient method of calculating s_P^2 is given by Equation 6.2.

We've placed parentheses in the numerator to help you identify the two sums of squares expressions. The subscripts 1 and 2 are used to identify the two samples.

$$s_P^2 = \frac{\left(\sum x_1^2 - \frac{\left(\sum x_1\right)^2}{n_1}\right) + \left(\sum x_2^2 - \frac{\left(\sum x_2\right)^2}{n_2}\right)}{n_1 + n_2 - 2} \tag{6.2}$$

EXAMPLE 6.1

Use the data in Table 6.1 on page 216 to perform a two-tailed independent samples t test at $\alpha = .05$. Interpret the result insofar as the two drug therapies are concerned.

Solution It will be convenient for computational purposes to arrange the data as in Table 6.2 on the facing page.

Using the sums from Table 6.2,

$$\bar{x}_1 = \frac{1977}{15} = 131.8$$

$$\bar{x}_2 = \frac{2028}{15} = 135.2$$

and

$$\begin{aligned}
s_P^2 &= \frac{\left(\sum x_1^2 - \frac{\left(\sum x_1\right)^2}{n_1}\right) + \left(\sum x_2^2 - \frac{\left(\sum x_2\right)^2}{n_2}\right)}{n_1 + n_2 - 2} \\
&= \frac{\left(262767 - \frac{(1977)^2}{15}\right) + \left(275706 - \frac{(2028)^2}{15}\right)}{15 + 15 - 2} \\
&= \frac{(2198.4) + (1520.4)}{28} \\
&= 132.814
\end{aligned}$$

Thus, our best estimate of the variance of the population from which the two samples were drawn is 132.814. By Equation 6.1 on the previous page, obtained t is

$$\begin{aligned}
t &= \frac{\bar{x}_1 - \bar{x}_2 - \delta_0}{\sqrt{s_P^2 \left(\frac{1}{n_1} + \frac{1}{n_2}\right)}} \\
&= \frac{131.8 - 135.2}{\sqrt{132.814 \left(\frac{1}{15} + \frac{1}{15}\right)}} \\
&= \frac{-3.4}{4.208} \\
&= -.808
\end{aligned}$$

TABLE 6.2: Blood pressures arranged for analysis via the independent samples t test.

Group One		Group Two	
X_1	X_1^2	X_2	X_2^2
129	16641	138	19044
111	12321	120	14400
140	19600	137	18769
139	19321	154	23716
144	20736	148	21904
120	14400	122	14884
131	17161	131	17161
129	16641	128	16384
131	17161	140	19600
154	23716	145	21025
119	14161	131	17161
138	19044	120	14400
142	20164	144	20736
110	12100	129	16641
140	19600	141	19881
Σ 1977	262767	2028	275706

Appendix B shows that the critical values for a two-tailed t test with

$$n_1 + n_2 - 2 = 15 + 15 - 2 = 28$$

degrees of freedom conducted at $\alpha = .05$ are -2.048 and 2.048 so that the null hypothesis is not rejected. Can we therefore conclude that the two drugs did not differ in their effect on blood pressure? No! Once again, you cannot use failure to reject the null hypothesis as evidence that the null hypothesis is true. Why not? Because you don't know the Type II error probability (Beta). Given the state of affairs, we must say that we did not find sufficient evidence of a differential drug impact to allow for such a conclusion. ■

EXAMPLE 6.2

A group of workers exposed to a toxic mold in the ventilating system of the building in which they worked are administered a scale designed to assess symptoms experienced before discovery of the mold. The scale rates symptom experience from zero (no symptoms) to 40 (numerous symptoms at least some of which were severe). Researchers believe that men tend to minimize such experiences and therefore will score significantly lower on the scale than will women.

Use the data in Table 6.3 on page 223 to perform a one-tailed independent samples t test at $\alpha = .05$ to assess the researcher's theory. Begin by clearly stating the null and alternative hypotheses.

Solution If men are designated as group one and women as group two, then the null and alternative hypotheses can be stated as

$$H_0 : \mu_1 = \mu_2$$
$$H_A : \mu_1 < \mu_2$$

How would these hypotheses be affected if men were designated as group two and women as group one?

It will be convenient for computational purposes to arrange the data as in Table 6.4. Using the sums from Table 6.4,

$$\bar{x}_1 = \frac{\sum x}{n} = \frac{147}{11} = 13.364$$

$$\bar{x}_2 = \frac{\sum x}{n} = \frac{372}{15} = 24.800$$

and

$$s_P^2 = \frac{\left(\sum x_1^2 - \frac{(\sum x_1)^2}{n_1} \right) + \left(\sum x_2^2 - \frac{(\sum x_2)^2}{n_2} \right)}{n_1 + n_2 - 2}$$

$$= \frac{\left(2287 - \frac{(147)^2}{11} \right) + \left(10222 - \frac{(372)^2}{15} \right)}{11 + 15 - 2}$$

$$= \frac{(322.545) + (996.400)}{24}$$

$$= 54.956$$

Thus, our best estimate of the variance of the population from which the two samples were drawn is 54.956. By Equation 6.1 on page 219, obtained t is

$$t = \frac{\bar{x}_1 - \bar{x}_2 - \delta_0}{\sqrt{s_P^2 \left(\frac{1}{n_1} + \frac{1}{n_2} \right)}}$$

$$= \frac{13.364 - 24.800}{\sqrt{54.956 \left(\frac{1}{11} + \frac{1}{15} \right)}}$$

$$= \frac{-11.436}{2.943}$$

$$= -3.886$$

Appendix B shows that the critical value for a one-tailed t test with $11+15-2 = 24$ degrees of freedom conducted at $\alpha = .05$ is -1.711 so that the null hypothesis is rejected. We can conclude, therefore, that the men who experienced the toxic mold scored significantly lower

TABLE 6.3: Symptom scale scores of men and women exposed to toxic mold.

Men	Women
14	28
9	20
16	22
7	31
10	13
20	10
13	32
23	29
5	30
11	9
19	38
	27
	27
	30
	26
$\bar{x}_1 = 13.364$	$\bar{x}_2 = 24.800$

TABLE 6.4: Symptom scale scores arranged for analysis via the independent samples t test.

Group One		Group Two	
X_1	X_1^2	X_2	X_2^2
14	196	28	784
9	81	20	400
16	256	22	484
7	49	31	961
10	100	13	169
20	400	10	100
13	169	32	1024
23	529	29	841
5	25	30	900
11	121	9	81
19	361	38	1444
		27	729
		27	729
		30	900
		26	676
$\sum$ 147	2287	372	10222

on the scale than did the women. Notice, however, that the design of this study did not permit the random assignment of subjects to groups. Indeed, we cannot randomly assign subjects to gender. We must be particularly cautious as to how results are interpreted in the absence of random assignment. In this case, we cannot be sure that the significant difference found between the two groups is due to gender. It may be, for example, that men tended to be assigned jobs in areas of the building where there was less mold.

In this case, the alternative hypothesis was of the form $H_A : \mu_1 < \mu_2$ which implies a critical region in the left-hand tail of the distribution. An alternative of the form $H_A : \mu_1 > \mu_2$ would require a critical region in the right-hand tail. ∎

6.2.2 Establishing Equivalence by Means of Independent Samples t Tests

Rationale. As we have pointed out several times, rejection of a null hypothesis provides good evidence (though not proof positive) that the null hypothesis is false. By contrast, failure to reject a null hypothesis does not, generally, provide good evidence that the null hypothesis is true. In situations where you wish to establish the validity of the null hypothesis, you must employ an equivalence test to show that the null hypothesis is (approximately) true. (See Section 4.3.6 on page 117.)

It sometimes happens that researchers would like to establish that there is *no* difference between treatments rather than showing that there *is* a difference. In such instances the independent samples t test can be used to establish equivalence.

As an example, a pharmaceutical company may wish to demonstrate that a new drug designed for the treatment of hypertension is just as effective as an older drug. It may be that the newer drug does not induce some undesirable side effect commonly associated with the older drug or is cheaper to manufacture than the older drug. It may also be that, because of the pharmaceutical properties of the new drug, it is not possible for this drug to be more effective than the older treatment.

The Test. Using the notation of Section 4.3.6 on page 117, the null and alternative hypotheses for a two-tailed equivalence test based on the independent samples t test can be stated as

$$H_{0E} : \mu_1 - \mu_2 \leq EI_L \text{ or } \mu_1 - \mu_2 \geq EI_U$$
$$H_{AE} : EI_L < \mu_1 - \mu_2 < EI_U$$

where EI_L and EI_U are the lower and upper ends of the equivalence interval. In essence, the null hypothesis states that the difference between means does not lie in the equivalence interval while the alternative asserts that the difference is in the interval. You will recall that the null hypothesis is tested by conducting two one-tailed tests at level α. In order to reject the equivalence null hypothesis you must show that $\mu_1 - \mu_2 > EI_L$ *and* that $\mu_1 - \mu_2 < EI_U$ which means that *both* of the following tests must be significant.

Test One	Test Two
$H_{01} : \mu_1 - \mu_2 = EI_U$	$H_{02} : \mu_1 - \mu_2 = EI_L$
$H_{A1} : \mu_1 - \mu_2 < EI_U$	$H_{A2} : \mu_1 - \mu_2 > EI_L$

The null and alternative hypotheses for the one-tailed equivalence test is *one* of the following

$$H_{0E} : \mu_1 - \mu_2 \geq EI_U$$
$$H_{AE} : \mu_1 - \mu_2 < EI_U$$

or

$$H_{0E} : \mu_1 - \mu_2 \leq EI_L$$
$$H_{AE} : \mu_1 - \mu_2 > EI_L$$

The first one-tailed equivalence hypothesis given above is tested by Test One with the second being carried out by means of Test Two.

EXAMPLE 6.3

Suppose that the data in Table 6.1 on page 216 was collected in connection with an equivalence study designed to show that the drug given group one is not more effective than the drug given group two insofar as control of hypertension is concerned. It is decided that drug one will be declared *not more effective* than drug two if the mean level of blood pressure achieved by drug one is less than five units below than that achieved by drug two.

Use the independent samples t test to conduct a one-tailed equivalence test on this data at $\alpha = .05$. State the equivalence null and alternative hypotheses before conducting the test.

Solution   The equivalence null hypothesis takes the form

$$H_{0E} : \mu_1 - \mu_2 \leq -5$$

while the alternative maintains

$$H_{AE} : \mu_1 - \mu_2 > -5$$

The test is carried out via Test Two with null and alternative hypotheses

$$H_{02} : \mu_1 - \mu_2 = -5$$

and

$$H_{A2} : \mu_1 - \mu_2 > -5$$

The values of $\bar{x}_1$, $\bar{x}_2$ and s_P^2 were found previously to be 131.8, 135.2, and 133.779 respectively. Because the null hypothesis specifies a nonzero difference between μ_1 and μ_2, the δ_0 term in Equation 6.1 is likewise nonzero and must be included in the calculation. Obtained t is then

$$t_2 = \frac{\bar{x}_1 - \bar{x}_2 - \delta_0}{\sqrt{s_P^2 \left(\frac{1}{n_1} + \frac{1}{n_2}\right)}} = \frac{131.8 - 135.2 - (-5)}{\sqrt{132.814 \left(\frac{1}{15} + \frac{1}{15}\right)}} = \frac{1.6}{4.208} = .380$$

Appendix B shows that the critical value for a one-tailed t test with $15 + 15 - 2 = 28$ degrees of freedom conducted at $\alpha = .05$ is 1.701.[2] Because obtained t of .380 is less than critical t of 1.701, the null hypothesis is not rejected. We have, therefore, not been able to demonstrate equivalence. ∎

---

[2]Notice that critical t is positive because the alternative specifies that $\mu_1 - \mu_2$ is greater than the value specified by the null hypothesis.

EXAMPLE 6.4

Use the data in Table 6.5 on the next page to perform a two-tailed equivalence test at $\alpha = .05$. Use equivalence interval $EI_L = -4$ and $EI_U = 4$. State the equivalence null and alternative hypotheses.

Solution The equivalence null and alternative hypotheses state that

$$H_{0E} : \mu_1 - \mu_2 \leq -4 \text{ or } \mu_1 - \mu_2 \geq 4$$

and

$$H_{AE} : -4 < \mu_1 - \mu_2 < 4.$$

It will be convenient for computational purposes to arrange the data as in Table 6.6 on the facing page.

Using the sums from Table 6.6,

$$\bar{x}_1 = \frac{\sum x}{n} = \frac{108}{16} = 6.750$$

$$\bar{x}_2 = \frac{\sum x}{n} = \frac{102}{16} = 6.375$$

and

$$s_P^2 = \frac{\left(\sum x_1^2 - \frac{(\sum x_1)^2}{n_1}\right) + \left(\sum x_2^2 - \frac{(\sum x_2)^2}{n_2}\right)}{n_1 + n_2 - 2}$$

$$= \frac{\left(800 - \frac{(108)^2}{16}\right) + \left(736 - \frac{(102)^2}{16}\right)}{16 + 16 - 2}$$

$$= \frac{(71.00) + (85.75)}{30}$$

$$= 5.225$$

Thus, our best estimate of the variance of the population from which the two samples were drawn is 5.225. By Test One and Equation 6.1 on page 219, obtained t_1 is

$$t_1 = \frac{\bar{x}_1 - \bar{x}_2 - \delta_0}{\sqrt{s_P^2 \left(\frac{1}{n_1} + \frac{1}{n_2}\right)}} = \frac{6.750 - 6.375 - 4.000}{\sqrt{5.225 \left(\frac{1}{16} + \frac{1}{16}\right)}} = \frac{-3.625}{.808} = -4.486$$

Appendix B shows that the critical value for a one-tailed t test with $16 + 16 - 2 = 30$ degrees of freedom conducted at $\alpha = .05$ is -1.697. It follows that the null hypothesis $H_0 : \mu_1 - \mu_2 = 4$ is rejected in favor of the alternative $H_A : \mu_1 - \mu_2 < 4$.

By Test Two and Equation 6.1, obtained t_2 is

$$t_2 = \frac{6.750 - 6.375 - (-4.0)}{\sqrt{5.225 \left(\frac{1}{16} + \frac{1}{16}\right)}} = \frac{4.375}{.808} = 5.415$$

TABLE 6.5: Practice data for two-tailed equivalence test based on the independent samples t test.

Group One	Group Two
8	8
9	3
5	4
7	10
4	8
4	3
10	5
8	3
3	7
10	9
9	5
7	6
6	6
5	8
6	10
7	7
$\bar{x}_1 = 6.750$	$\bar{x}_2 = 6.375$

TABLE 6.6: Practice data arranged for analysis via the two-tailed equivalence test based on the independent samples t test.

	Group One		Group Two
X_1	X_1^2	X_2	X_2^2
8	64	8	64
9	81	3	9
5	25	4	16
7	49	10	100
4	16	8	64
4	16	3	9
10	100	5	25
8	64	3	9
3	9	7	49
10	100	9	81
9	81	5	25
7	49	6	36
6	36	6	36
5	25	8	64
6	36	10	100
7	49	7	49
$\sum$ 108	800	102	736

Because obtained t_2 is greater than critical t of 1.697, the null hypothesis $H_0 : \mu_1 - \mu_2 = -4$ is rejected in favor of the alternative $H_A : \mu_1 - \mu_2 > -4$.

Because *both* Test One and Test Two are significant, the equivalence null hypothesis is rejected in favor of the alternative. Equivalence, as defined by the equivalence interval, is thereby established. ∎

6.2.3 Confidence Interval for the Difference Between Means of Two Independent Samples

Rationale. The independent samples t test attempts to determine whether there is a difference between the means of two populations (or whether the difference is of some specified value). A related and usually more informative question is, "How large is the difference between the population means?" This difference can be estimated with a confidence interval.

By way of application, if subjects are randomly assigned to treatment by one of two different drugs, the independent samples t test attempts to determine whether there is a difference in the impact produced by the two drugs. By contrast, a confidence interval for the difference between means based on independent samples addresses the question "How large is the difference between the impacts of the two drugs?"

The Confidence Interval. The confidence interval for the difference between means based on independent samples has the following forms for L and U.

$$L = (\bar{x}_1 - \bar{x}_2) - t\sqrt{s_P^2 \left(\frac{1}{n_1} + \frac{1}{n_2}\right)} \tag{6.3}$$

and

$$U = (\bar{x}_1 - \bar{x}_2) + t\sqrt{s_P^2 \left(\frac{1}{n_1} + \frac{1}{n_2}\right)} \tag{6.4}$$

where $\bar{x}_1$, $\bar{x}_2$ and s_P^2 are as previously defined for the independent samples t test, and t is the appropriate t value with $n_1 + n_2 - 2$ degrees of freedom.

EXAMPLE 6.5

Use the data in Table 6.1 on page 216 to form a two-sided 95% confidence interval for the estimation of $\mu_1 - \mu_2$. Interpret the result. Use the interval thus obtained to perform a two-tailed independent samples t test.

Solution As previously calculated on page 220, $\bar{x}_1 = 131.8$, $\bar{x}_2 = 135.2$, and $s_P^2 = 132.814$. Then, by Equations 6.3 and 6.4

$$L = (\bar{x}_1 - \bar{x}_2) - t\sqrt{s_P^2 \left(\frac{1}{n_1} + \frac{1}{n_2}\right)}$$

$$= (131.8 - 135.2) - 2.048\sqrt{132.814\left(\frac{1}{15} + \frac{1}{15}\right)}$$

$$= -12.018$$

and

$$U = (\bar{x}_1 - \bar{x}_2) + t\sqrt{s_P^2 \left(\frac{1}{n_1} + \frac{1}{n_2}\right)}$$

$$= (131.8 - 135.2) + 2.048\sqrt{132.814\left(\frac{1}{15} + \frac{1}{15}\right)}$$

$$= 5.218$$ ■

A strictly statistical interpretation of this interval would maintain that one can assert with 95% confidence that the difference $\mu_1 - \mu_2$ is between -12.018 and 5.218. From a research point of view, this interval maintains, with 95% confidence, that the average blood pressure level realized by patients treated with drug one minus the average blood pressure level of patients treated with drug two is between -12.018 and 5.218. Notice that this is an unsatisfactory result for a researcher trying to evaluate the relative effectiveness of the two drugs since the difference might be negative, indicating an advantage for drug one, positive, thereby indicating an advantage for drug two or zero indicating no difference.

This interval can be used to perform a two-tailed test of the null hypothesis $H_0 : \mu_1 - \mu_2 = 0$ by the method outlined in Section 4.5 on page 152. By this method, you simply observe the interval to note whether the value specified by the null hypothesis is contained therein. If the null value is between L and U, the null hypothesis is not rejected. Otherwise, the null hypothesis is rejected. In the present case, zero is between -12.018 and 5.218 so that the null hypothesis is not rejected. This is the same result obtained when an independent samples t test was conducted on this same data on page 220.

By way of comparison, suppose for a moment that the calculated interval had been -10.00 to -5.00. The conclusion in this case would be that drug one held the advantage over drug two in that the average blood pressure of patients treated with this drug would be between five and ten points below the average attained by patients treated with drug two. A significance test in this case would reject the null hypothesis because the value specified by the null hypothesis (zero) is not in the interval -10.00 to -5.00. It is important to note that while the independent samples t test would only assert the superiority of drug one, the confidence interval provides an estimate of the magnitude of that advantage.

EXAMPLE 6.6

Use the data in Table 6.3 on page 223 to form a one-sided 95% confidence interval to provide an upper bound estimate of the difference between mean symptom scale scores of males and females. Interpret the interval thus obtained. Use the interval to perform a one-tailed independent samples t test with null and alternative hypotheses $H_0 : \mu_1 - \mu_2 = 0$ and $H_A : \mu_1 - \mu_2 < 0$.

Solution   As previously calculated on page 222, $\bar{x}_1 = 13.364$, $\bar{x}_2 = 24.800$, and $s_P^2 = 54.956$. Then, by Equation 6.4 on the facing page

$$U = (13.364 - 24.800) + 1.711\sqrt{54.956\left(\frac{1}{11} + \frac{1}{15}\right)} = -6.401$$

Note that the degrees of freedom associated with $t = 1.711$ is $11 + 15 - 2 = 24$.

This interval indicates, with 95% confidence, that the mean symptom score of men minus the mean symptom score of women is *at most* -6.401. Because U is less than zero, the null hypothesis $H_0 : \mu_1 - \mu_2 = 0$ is rejected in favor of the alternative $H_A : \mu_1 - \mu_2 < 0$.[3] You will recall that this was the result obtained on page 222.   ∎

6.2.4   Assumptions

Assumptions underlying the independent samples t test, equivalence tests based on the independent samples t test and confidence intervals for the difference between means of independent samples are as follows. (1) The assumption of **normality** specifies that samples are from normally distributed populations. These procedures are generally (though not always) robust against violations of this assumption and are especially so when sample sizes are greater than or equal to 30 and are equal or approximately so. The assumption of **homogeneity of variance** specifies that the two samples are from populations that have the same variance. So long as heterogeneity of variance is not too extreme (e.g., ten to one), these procedures are generally (though not always) robust under the same conditions indicated above for the assumption of normality. The assumption of **independence** requires that each observation in the two samples be unrelated to every other observation in the two samples.[4] The procedures addressed here cannot be counted upon for robustness against violations of the independence assumption.

6.3   METHODS RELATED TO PROPORTIONS

6.3.1   An Independent Samples Test for the Difference Between Proportions

Rationale.   Suppose that a program designed to educate pregnant teens about proper nutrition during pregnancy is to be evaluated. As part of the evaluation, pregnant teens are randomly assigned to receive the educational program or to not receive the program. The outcome of interest is low birth weight. The question of interest is, "Do the pregnant teens who receive nutrition education produce a different proportion of low birth weight babies than do pregnant teens who do not receive such instruction?"

A reasonable strategy for answering the question as to whether the educational program produced a different proportion of low birth weight babies than did a no instruction strategy, would be to calculate the proportion of low birth weight babies realized in each of the two groups and then compare the two proportions to determine whether they differ. In the case at hand, suppose that 314 mothers received nutritional instruction of which 23 had low birth weight babies. By contrast, 39 of the 316 mothers in the non-instruction group had low birth weight babies. Thus, the proportion of low birth weight babies in the instructional group was $\hat{p}_1 = \frac{23}{314} = .073$ while that in the non-instructional group was $\hat{p}_2 = \frac{39}{316} = .123$. The difference between these two proportions is then $.073 - .123 = -.050$ which indicates that the instructional group had a lower proportion of low birth weight babies.

But, there are at least two explanations for this difference. First, we note that any time we randomly assign subjects to groups, the proportions of some outcome in the two groups are

---

[3] You should review Section 4.5 on page 152 if this result is not clear.

[4] See Section 4.3.3 on page 99 for examples and further details.

almost sure to differ to some degree on almost any conceivable measure. Indeed, we would be surprised to find that the proportion of randomly assigned persons with a history of heart disease was exactly .131 in the two groups or that the proportions of smokers was exactly the same.

Thus, even if the instructional program was no better than non-instruction, we would still expect to see some difference in the proportions of the two groups. The first explanation for the difference of −.050 observed in our (fictitious) study, then, is that this difference arose, not because of any difference in the impact of the treatments administered, but rather by happenstance related to the manner in which the groups were formed.

A second explanation is that the nutritional instruction afforded the one group was effective at lowering the incidence of low birth weight babies. This would explain why the proportion in group one was lower than that in group two. But which of these explanations is to be believed? A test of significance might help decide the issue.

Conceive of a sampling distribution formed by repeatedly drawing a random sample from some dichotomous population, randomly dividing each such sample into two component samples and then calculating the quantity $\hat{p}_1 - \hat{p}_2$ where $\hat{p}_1$ and $\hat{p}_2$ are the proportions of some outcome obtained from the two component samples.[5] If this process were repeated many times the resultant values of $\hat{p}_1 - \hat{p}_2$ could be formed into a sampling distribution. It is important to understand that this distribution represents the sampling distribution of $\hat{p}_1 - \hat{p}_2$ *when both samples come from populations that have the same proportion.* Clearly, if both samples come from the same population then they come from populations with the same proportion. We will designate the proportion in the population from which the first sample was drawn as π_1 and that of the population from which the second sample was drawn as π_2. Because both samples come from the same population it is clear that $\pi_1 = \pi_2$.

Notice that the mean of this sampling distribution would be zero and that the further values of $\hat{p}_1 - \hat{p}_2$ are from zero, the less likely is their occurrence. But how might this distribution be used to help answer the question posed above as to whether the difference of −.050 between the proportions of low birth weight babies in the two groups was due to random happenstance as the groups were formed or rather was due to the fact that the instructional program given group one was effective in reducing the incidence of low birth weight deliveries?

The two samples were originally taken from a common population so that $\pi_1 = \pi_2$. If the program of instruction produced exactly the same result as did non-instruction, then it would still be true that $\pi_1 = \pi_2$. In this circumstance we would not expect $\hat{p}_1 - \hat{p}_2$ to fall in a critical region. Why? Because the probability that $\hat{p}_1 - \hat{p}_2$ will fall in a critical region when $\pi_1 = \pi_2$ is only α.

But what would happen if nutritional instruction was more effective than non-instruction? In this case the incidence of low birth weights in group one would be lowered by having received the instruction so that the absolute value of $\hat{p}_1 - \hat{p}_2$ would tend to be greater (further from zero) than would be the case where both strategies had equal impact. For example, in our hypothetical study, $\hat{p}_1 - \hat{p}_2 = .073 - .123 = -.050$. But suppose the instruction had been more effective so that $\hat{p}_1 = .011$. Then $\hat{p}_1 - \hat{p}_2 = .011 - .123 = -.112$. The more

---

[5]Equivalently, we can conceive of the two component samples as simply two random samples taken from a common population.

effective is instruction as compared to non-instruction, the further below zero is $\hat{p}_1 - \hat{p}_2$ and the greater the chance that $\hat{p}_1 - \hat{p}_2$ will fall in the lower critical region. What would happen if non-instruction was superior to instruction?

In short, we do not expect $\hat{p}_1 - \hat{p}_2$ to fall in a critical region when the treatments produce the same result. Indeed, the probability of this happening is only α. However, when treatments produce different results, we expect the value of $\hat{p}_1 - \hat{p}_2$ to move away from zero and toward a critical region. Further, when treatment effects differ, we can no longer claim that $\pi_1 = \pi_2$. Rather, it must be said that $\pi_1 \neq \pi_2$.

From the above, when $\hat{p}_1 - \hat{p}_2$ falls in the critical region we can claim that the observed difference between these values was due to differential treatment effects rather than to the happenstance of how the groups were formed. We now see how this test is conducted.

The Test.   The null hypothesis to be tested by the independent samples Z test for the difference between proportions is most often (but not always)

$$H_0 : \pi_1 = \pi_2$$

or equivalently,

$$H_0 : \pi_1 - \pi_2 = 0$$

The two-tailed alternative is most often (but not always)

$$H_A : \pi_1 \neq \pi_2$$

or equivalently,

$$H_A : \pi_1 - \pi_2 \neq 0$$

One-tailed alternatives are most often (but not always) of the form

$$H_A : \pi_1 < \pi_2$$

or equivalently,

$$H_A : \pi_1 - \pi_2 < 0$$

or

$$H_A : \pi_1 > \pi_2$$

or equivalently,

$$H_A : \pi_1 - \pi_2 > 0$$

In essence, the null hypothesis asserts that there was no difference in the effects of the treatments administered the two groups while the two-tailed alternative maintains that a difference does exist. The one-tailed alternatives specify the form of that difference.

In order to carry out the hypothesis test, $\hat{p}_1 - \hat{p}_2$ cannot be directly referenced to a sampling distribution as described above but rather must be divided by the standard error (see the denominator of Equation 6.5) which means that the test statistic can then be referenced to a normal distribution. The test statistic is calculated by Equation 6.5.

$$Z = \frac{\hat{p}_1 - \hat{p}_2 - \delta_0}{\sqrt{\frac{\hat{p}_1 \hat{q}_1}{n_1} + \frac{\hat{p}_2 \hat{q}_2}{n_2}}} \qquad (6.5)$$

In this equation $\hat{p}_1$ and $\hat{p}_2$ represent the proportion of successes in the first and second samples respectively, $\hat{q}_1$ and $\hat{q}_2$ represent the proportion of failures in the two samples (i.e., $\hat{q}_1 = 1 - \hat{p}_1$ and $\hat{q}_2 = 1 - \hat{p}_2$) and n_1 and n_2 represent the two sample sizes. The symbol δ_0 represents the hypothesized difference between π_1 and π_2 which is usually zero. However, as you will see in the discussion of equivalence tests based on the independent samples Z test for the difference between proportions, this is not always the case. You should note that Equation 6.5 provides an *approximate* test. An exact test is available but will not be covered here.

EXAMPLE 6.7

Use the information provided above for the nutritional education study to conduct a one-tailed Z test of the null hypothesis $H_0 : \pi_1 = \pi_2$ against the alternative $H_A : \pi_1 < \pi_2$. What is your conclusion regarding the effectiveness of the instructional program?

Solution   As was provided above, 314 mothers received nutritional instruction of which 23 had low birth weight babies. By contrast, 39 of the 316 mothers in the non-instruction group had low birth weight babies. Thus, $\hat{p}_1 = \frac{23}{314} = .073$, $\hat{p}_2 = \frac{39}{316} = .123$, $\hat{q}_1 = 1 - .073 = .927$, $\hat{q}_2 = 1 - .123 = .877$, $n_1 = 314$ and $n_2 = 316$. Then by Equation 6.5 on the preceding page

$$Z = \frac{\hat{p}_1 - \hat{p}_2 - \delta_0}{\sqrt{\frac{\hat{p}_1\hat{q}_1}{n_1} + \frac{\hat{p}_2\hat{q}_2}{n_2}}} = \frac{.073 - .123}{\sqrt{\frac{(.073)(.927)}{314} + \frac{(.123)(.877)}{316}}} = \frac{-.050}{.0236} = -2.12$$

Appendix A shows that the critical value for a one-tailed Z test conducted at $\alpha = .05$ is -1.65 thereby leading to rejection of the null hypothesis. We can conclude, then, that the educational program was effective at lowering the incidence of low birth weight deliveries.

∎

EXAMPLE 6.8

A study shows that 61 of 414 adults who grew up in a single parent household report that they suffered at least one incident of sexual abuse during childhood. By contrast, 74 of 501 adults who grew up in two parent households report such abuse. Use an independent samples Z test for the difference between proportions to test the null hypothesis $H_0 : \pi_1 = \pi_2$ against the two-tailed alternative. What is your conclusion regarding whether there is a difference in proportions of abuse between one and two parent households?

Solution   From the above it follows that $\hat{p}_1 = \frac{61}{414} = .147$, $\hat{p}_2 = \frac{74}{501} = .148$, $\hat{q}_1 = 1 - .147 = .853$, $\hat{q}_2 = 1 - .148 = .852$, $n_1 = 414$, and $n_2 = 501$. Then by Equation 6.5 on the facing page

$$Z = \frac{\hat{p}_1 - \hat{p}_2 - \delta_0}{\sqrt{\frac{\hat{p}_1\hat{q}_1}{n_1} + \frac{\hat{p}_2\hat{q}_2}{n_2}}} = \frac{.147 - .148}{\sqrt{\frac{(.147)(.853)}{414} + \frac{(.148)(.852)}{501}}} = \frac{-.001}{.0235} = -.04$$

Appendix A shows that the critical values for a two-tailed Z test conducted at $\alpha = .05$ are -1.96 and 1.96 thereby leading to failure to reject the null hypothesis. It follows that we were unable to demonstrate a difference between the proportions of abuses in one and two parent households.

∎

6.3.2   Establishing Equivalence by Means of an Independent Samples Z Test for the Difference Between Proportions

Rationale.   As you now know, rejection of a null hypothesis provides good evidence that the null hypothesis is false. By contrast, failure to reject a null hypothesis does not, generally, provide good evidence that the null hypothesis is true. In situations where you wish to establish the validity of the null hypothesis, you must employ an equivalence test to show that the null hypothesis is (approximately) true. (See Section 4.3.6 on page 117.)

It sometimes happens that researchers would like to establish that there is *no* difference between treatments rather than showing that there *is* a difference. In such instances an independent samples Z test for the difference between proportions can be used to establish equivalence.

For example, a pharmaceutical company may wish to demonstrate that a new drug designed for the treatment of depression is similar to an older version of the drug insofar as the incidence of a particular undesirable side effect is concerned.

The Test.   Using the notation of Section 4.3.6 on page 117, the null and alternative hypotheses for a two-tailed equivalence test based on an independent samples Z test for the difference between proportions can be stated as

$$H_{0E} : \pi_1 - \pi_2 \leq EI_L \text{ or } \pi_1 - \pi_2 \geq EI_U$$
$$H_{AE} : EI_L < \pi_1 - \pi_2 < EI_U$$

where EI_L and EI_U are the lower and upper ends of the equivalence interval. In essence, the null hypothesis states that the difference between proportions does not lie in the equivalence interval while the alternative asserts that the difference is in the interval. You will recall that the null hypothesis is tested by conducting two one-tailed tests at level α. In order to reject the equivalence null hypothesis you must show that $\pi_1 - \pi_2 > EI_L$ *and* that $\pi_1 - \pi_2 < EI_U$ which means that *both* of the following tests must be significant.

Test One	Test Two
$H_{01} : \pi_1 - \pi_2 = EI_U$	$H_{02} : \pi_1 - \pi_2 = EI_L$
$H_{A1} : \pi_1 - \pi_2 < EI_U$	$H_{A2} : \pi_1 - \pi_2 > EI_L$

The null and alternative hypotheses for the one-tailed equivalence test are *one* of the following

$$H_{0E} : \pi_1 - \pi_2 \geq EI_U$$
$$H_{AE} : \pi_1 - \pi_2 < EI_U$$

or

$$H_{0E} : \pi_1 - \pi_2 \leq EI_L$$
$$H_{AE} : \pi_1 - \pi_2 > EI_L$$

The first one-tailed equivalence hypothesis given above is tested by Test One with the second being carried out by means of Test Two.

EXAMPLE 6.9

A clinical trial is conducted to determine whether the incidence of side effects associated with a new drug designed for the treatment of depression is similar to the incidence associated with an older version of the drug. It is decided that the two drugs will be declared equivalent insofar as the incidence of the indicated side effect is concerned if the difference between proportions of patients taking the new and old drug who experience the side effect is in the equivalence interval $-.04$ to $.04$. The proportion of patients taking the new drug who experience the side effect is $.07$ while that for the patients taking the older drug is $.06$. Each group was composed of 100 patients. Use a two-tailed equivalence test at $\alpha = .05$ to make this determination. Begin by stating the null and alternative equivalence hypotheses.

Solution   The null and alternative equivalence hypotheses are as follows.

$$H_{0E} : \pi_1 - \pi_2 \leq -.04 \text{ or } \pi_1 - \pi_2 \geq .04$$
$$H_{AE} : -.04 < \pi_1 - \pi_2 < .04$$

By Equation 6.5, obtained Z for Test One is

$$Z_1 = \frac{\hat{p}_1 - \hat{p}_2 - \delta_0}{\sqrt{\frac{\hat{p}_1 \hat{q}_1}{n_1} + \frac{\hat{p}_2 \hat{q}_2}{n_2}}} = \frac{(.07 - .06) - .040}{\sqrt{\frac{(.07)(.93)}{100} + \frac{(.06)(.94)}{100}}} = \frac{-.03}{.03486} = -.86$$

Obtained Z for Test Two is then

$$Z_2 = \frac{(.07 - .06) - (-.040)}{\sqrt{\frac{(.07)(.93)}{100} + \frac{(.06)(.94)}{100}}} = \frac{.05}{.03486} = 1.43$$

From Appendix A it can be determined that the critical value for Test One is -1.65 and that for Test Two is 1.65. In order to reject the equivalence null hypothesis, *both* tests must be significant. In this case neither test is significant leading to failure to reject the equivalence null hypothesis. Thus, we were unable to establish equivalence.   ∎

EXAMPLE 6.10

A study is conducted to determine whether an enhanced baby formula provides protection against a common childhood disease equivalent to that obtained by breast-fed babies. It is decided that if the proportion of formula fed babies contracting the disease is less than $.03$ greater than the proportion of breast-fed babies contracting the disease, equivalence will be declared. Fifty-two of the 390 formula fed babies contracted the disease while 94 of the 750 breast-fed babies contracted the disease. Use a one-tailed test at $\alpha = .05$ to test for equivalence. State the null and alternative equivalence hypotheses.

Solution   If we designate the formula fed babies as group one, the null and alternative equivalence hypotheses can be stated as

$$H_{0E} : \pi_1 - \pi_2 \geq .03$$
$$H_{AE} : \pi_1 - \pi_2 < .03$$

From the information provided above, $\hat{p}_1 = \frac{52}{390} = .133$, $\hat{p}_2 = \frac{94}{750} = .125$, $\hat{q}_1 = 1 - .133 = .867$, $\hat{q}_2 = 1 - .125 = .875$, $n_1 = 390$, and $n_2 = 750$. By Equation 6.5 on page 232, obtained Z for Test One is

$$Z_1 = \frac{\hat{p}_1 - \hat{p}_2 - \delta_0}{\sqrt{\frac{\hat{p}_1 \hat{q}_1}{n_1} + \frac{\hat{p}_2 \hat{q}_2}{n_2}}} = \frac{(.133 - .125) - .030}{\sqrt{\frac{(.133)(.867)}{390} + \frac{(.125)(.875)}{750}}} = \frac{-.022}{.0210} = -1.05$$

From Appendix A it can be determined that the critical value for Test One is -1.65 thereby resulting in a failure to reject the null hypothesis. As a result, equivalence of the baby formula to breast milk, insofar as prevention of the specified childhood disease is concerned, was not established. ■

6.3.3   Confidence Interval for a Difference Between Proportions Based on Two Independent Samples

Rationale.   The independent samples Z test for a difference between proportions attempts to determine whether there is a difference between two population proportions. A related and usually more informative question is, "How large is the difference between the population proportions?" This difference can be estimated with a confidence interval.

By way of application, if subjects are randomly assigned to treatment by one of two different drugs, the independent samples Z test for a difference between proportions attempts to determine whether there is a difference in the impact produced by the two drugs as expressed by some dichotomous outcome. By contrast, a confidence interval for the difference between proportions based on two independent samples addresses the question "What is the difference between the impact of the two drugs?"

The Confidence Interval.   The confidence interval for the difference between proportions based on independent samples has the following forms for L and U.

$$L = (\hat{p}_1 - \hat{p}_2) - \left(Z \sqrt{\frac{\hat{p}_1 \hat{q}_1}{n_1 - 1} + \frac{\hat{p}_2 \hat{q}_2}{n_2 - 1} + \frac{1}{2}\left(\frac{1}{n_1} + \frac{1}{n_2}\right)} \right) \tag{6.6}$$

and

$$U = (\hat{p}_1 - \hat{p}_2) + \left(Z \sqrt{\frac{\hat{p}_1 \hat{q}_1}{n_1 - 1} + \frac{\hat{p}_2 \hat{q}_2}{n_2 - 1} + \frac{1}{2}\left(\frac{1}{n_1} + \frac{1}{n_2}\right)} \right) \tag{6.7}$$

where $\hat{p}_1$ and $\hat{p}_2$ are the proportions of successes in the two samples, $\hat{q}_1$ and $\hat{q}_2$ are the proportions of failures in the two samples (i.e., $\hat{q}_1 = 1 - \hat{p}_1$, $\hat{q}_2 = 1 - \hat{p}_2$), n_1 and n_2, are the respective sample sizes and Z is the appropriate Z value for the specified interval.

EXAMPLE 6.11

A (fictitious) survey of teens aged 12 to 16 reports that 106 of 299 males and 66 of 313 females surveyed indicate that they routinely smoke three or more cigarettes a day. Form a two-sided 95% confidence interval to estimate the difference between proportions of males and females in the population who smoke three or more cigarettes per day. Use this confidence interval to test the null hypothesis $H_0 : \pi_1 = \pi_2$ against the alternative $\pi_1 \neq \pi_2$ at the .05 level of significance. State the reason for your decision regarding the null hypothesis.

Solution   By Equations 6.6 and 6.7 on the facing page

$$L = (\hat{p}_1 - \hat{p}_2) - \left(Z\sqrt{\frac{\hat{p}_1 \hat{q}_1}{n_1 - 1} + \frac{\hat{p}_2 \hat{q}_2}{n_2 - 1}} + \frac{1}{2}\left(\frac{1}{n_1} + \frac{1}{n_2} \right) \right)$$

$$= (.355 - .211) - \left(1.96\sqrt{\frac{(.355)(.645)}{298} + \frac{(.211)(.789)}{312}} + \frac{1}{2}\left(\frac{1}{299} + \frac{1}{313} \right) \right)$$

$$= .070$$

and

$$U = (\hat{p}_1 - \hat{p}_2) + \left(Z\sqrt{\frac{\hat{p}_1 \hat{q}_1}{n_1 - 1} + \frac{\hat{p}_2 \hat{q}_2}{n_2 - 1}} + \frac{1}{2}\left(\frac{1}{n_1} + \frac{1}{n_2} \right) \right)$$

$$= (.355 - .211) + \left(1.96\sqrt{\frac{(.355)(.645)}{298} + \frac{(.211)(.789)}{312}} + \frac{1}{2}\left(\frac{1}{299} + \frac{1}{313} \right) \right)$$

$$= .218$$

Thus, the 95% confidence interval is .070 to .218.

This interval can be used to perform a two-tailed test of the null hypothesis $H_0 : \pi_1 - \pi_2 = 0$ by the method outlined in Section 4.5 on page 152. By this method, you simply observe the interval to note whether the value specified by the null hypothesis is contained therein. If the null value is between L and U, the null hypothesis is not rejected. Otherwise, the null hypothesis is rejected. In the present case, zero is not between .070 and .218 so that the null hypothesis is rejected. Performing hypothesis tests in this manner by use of Equations 6.6 and 6.7 may provide somewhat superior results to those obtained via Equation 6.5 on page 232 due to the fact that the former equations may better approximate the normal curve model [18].   ∎

EXAMPLE 6.12

In Example 6.7 on page 233 you performed a one-tailed independent samples Z test for a difference between proportions in connection with the evaluation of a (fictitious) nutrition education program. You tested the null hypothesis $H_0 : \pi_1 = \pi_2$ against the alternative $H_A : \pi_1 < \pi_2$. In the course of conducting this test you found that $\hat{p}_1 = .073$, $\hat{p}_2 = .123$, $\hat{q}_1 = .927$, $\hat{q}_2 = .877$, $n_1 = 314$, and $n_2 = 316$. Use this information to construct a one-sided confidence interval for the upper bound of $\pi_1 - \pi_2$. Interpret this interval and use it to perform a similar test of significance to the one you conducted earlier. Explain how you conducted the test.

Solution By Equation 6.7 on page 236

$$U = (.073 - .123) + \left(1.65 \sqrt{\frac{(.073)\,(.927)}{313} + \frac{(.123)\,(.877)}{315} + \frac{1}{2} \left(\frac{1}{314} + \frac{1}{316} \right)} \right)$$

$$= -.008$$

From a research point of view this interval maintains, with 95% confidence, that the proportion of low birth weight babies in the educated group was *at least* .008 below that of the non-educated group. The null hypothesis stated above is rejected because the upper end of the interval is below zero.[6] ∎

6.3.4 Assumptions

Assumptions underlying the independent samples Z test for the difference between proportions, the test for equivalence based on the independent samples Z test for the difference between proportions, and the confidence interval for a difference between proportions based on two independent samples are essentially the same as those underlying the approximate test for a proportion discussed on page 116.

6.4 METHODS RELATED TO INDEPENDENT SAMPLES RISK RATIOS

6.4.1 Background

As discussed in Section 5.4.1 on page 190,[7] the **risk** of some event to a particular group is the *probability* that the event will occur to a member of that group. Using the notation of Chapter 3, the risk to some group that has been exposed to a potentially harmful (or beneficial) agent can be characterized as $P\,(D \mid E)$ which is read, "The probability of disease given exposure." Likewise, we can characterize the risk to subjects in an unexposed group as $P\left(D \mid \overline{E}\right)$ which is read, "The probability of disease given no exposure." These two probabilities can be compared by forming them into a **risk ratio** as follows.

$$RR = \frac{P\,(D \mid E)}{P\left(D \mid \overline{E}\right)} \tag{6.8}$$

You will recognize this as Equation 3.11 from page 59.[8] If the risk of disease[9] in the two groups is the same, the risk ratio is one which would imply that there is no relationship between exposure and the disease. On the other hand, a risk ratio greater than one would imply that exposure is associated with greater risk of disease than is non-exposure. Ratios less than one are said to be **protective** because exposure implies reduced risk of disease. This might occur, for example, when persons are exposed to a vaccine. As noted previously, we will represent the parameter and statistic forms of the risk ratio as RR and $\widehat{RR}$ respectively.

---

[6]See Section 4.5 on page 152 if this explanation is not clear.

[7]You may wish to review this section before continuing.

[8]You may wish to review the discussion beginning on page 59 at this point.

[9]While we refer to disease and exposure in Equation 6.8, the terms are generic and represent whatever variables are being examined.

TABLE 6.7: Two by two table for unpaired dichotomous data.

		Disease yes	no
Exposed	yes	a	b
	no	c	d

It will be helpful to refer to Table 6.7 for the discussions that follow. While we will use the generic terms "exposed" and "disease" for this table, you should understand that other variables such as "gender" and "access to medical care" might be used. A typical study in which risk ratios might be employed would follow two groups of subjects, one of which had experienced some form of exposure with the other not having been so exposed, in order to determine whether the risk of some outcome is different in the two groups. Studies of this type are termed **prospective cohort** studies.

The sample independent samples risk ratio is defined by Equation 6.9 where a, b, c, and d are the cell frequencies as specified in Table 6.7. Note that the risk to subjects in the first row is being compared to that of subjects in the second row. Thus, for example, a calculated risk ratio of 2.0 would indicate that subjects in the first row are at twice the risk of the specified outcome as subjects in the second row. You should bear this in mind when constructing tables for analysis.

$$\widehat{RR} = \frac{a/(a+b)}{c/(c+d)} \tag{6.9}$$

6.4.2   Test of the Hypothesis RR = 1 for Independent Samples

Rationale.   Returning to an example presented in Chapter 5, suppose it is suspected that exposure to a particular chemical solvent commonly used in manufacturing increases the risk of bladder cancer. To test this suspicion, workers in a particular industry who are routinely exposed to the solvent might be followed across time to determine whether they develop the disease. For comparison purposes, a group of workers in the same industry who are not exposed to the solvent might also be followed. After the period of followup, the sample risk ratio is computed by dividing the proportion of exposed workers developing the disease by the proportion of non-exposed workers who develop the disease. If the two risks were .005 and .001 respectively, the risk ratio would be $\frac{.005}{.001} = 5$. This indicates that the risk of bladder cancer for exposed workers is five times that of non-exposed workers. But is this simply a result of the particular sample used in the study? If the study were repeated with other workers might the result be markedly different? More importantly, is the risk ratio in the population one, indicating no relationship between exposure and disease (i.e., the risk in the two groups is the same), or is it different from one indicating that such a relationship does exist? A test of hypothesis can help answer this question.

The Test.   A test of the null hypothesis $RR = 1$ against one- or two-tailed alternatives can be conducted via Equation 6.10 on the next page. The frequencies a, b, c, and d are as shown in Table 6.7. The hypothesized population risk ratio is represented by RR_0. Because the log of 1 is zero, the expression $\ln(RR_0)$ may be omitted when testing $H_0 : RR = 1$.

$$Z = \frac{\ln\left(\widehat{RR}\right) - \ln(RR_0)}{\sqrt{\frac{b/a}{a+b} + \frac{d/c}{c+d}}}$$ (6.10)

EXAMPLE 6.13

Suppose that the results of the industrial solvent example given above had been as follows; 42 of the workers exposed to the solvent developed bladder cancer while 2,981 did not. Of the unexposed workers, 21 developed the disease while 4,088 did not. Use this data to calculate and interpret the sample risk ratio. Then test the null hypothesis $H_0 : RR = 1$ against a two-tailed alternative.

Solution   Forming the data into a two-by-two table yields

		Bladder Cancer	
		yes	no
Exposed to	yes	42	2,981
Solvent	no	21	4,088

By Equation 6.9 on the previous page the sample risk ratio is

$$\widehat{RR} = \frac{a/(a+b)}{c/(c+d)} = \frac{42/(42+2981)}{21/(21+4088)} = 2.718$$

This means that the risk of bladder cancer among workers exposed to the solvent is 2.718 times that of workers who were not exposed.
   By Equation 6.10

$$Z = \frac{\ln\left(\widehat{RR}\right) - \ln(RR_0)}{\sqrt{\frac{b/a}{a+b} + \frac{d/c}{c+d}}} = \frac{\ln(2.718)}{\sqrt{\frac{2981/42}{42+2981} + \frac{4088/21}{21+4088}}} = \frac{1.00}{\sqrt{.023 + .047}} = 3.78$$

Appendix A shows that the critical values for a two-tailed Z test conducted at $\alpha = .05$ are -1.96 and 1.96. Because obtained Z of 3.78 exceeds critical Z of 1.96, the null hypothesis is rejected. It is therefore concluded that the risk for the exposed group is greater than that for the unexposed group.   ∎

EXAMPLE 6.14

A large-scale study is conducted to determine whether regular ingestion of folic acid (vitamin B) by women of child bearing age who are at elevated risk for having babies affected by spina bifida will lower the risk of this disease. To this end, a large group of women assessed as being at high risk in this regard are encouraged to take a daily folic acid supplement. A similar group of women are not provided such encouragement. Of the live births recorded for the supplement encouraged group, 10 are diagnosed as having spina bifida while 12,344 are deemed not to have the disease. By contrast, 17 of the babies born to the nonencouraged group had the disease while 11,202 did not.

Calculate and interpret the risk ratio for the two groups. Test the null hypothesis $H_0 : RR = 1$ against the one-tailed alternative $H_A : RR < 1$ at the .05 level.

Solution   Forming the data into a two-by-two table yields

		Spina Bifida yes	no
Encouraged	yes	10	12,344
	no	17	11,202

By Equation 6.9 on page 239 the sample risk ratio is

$$\widehat{RR} = \frac{a/(a+b)}{c/(c+d)} = \frac{10/(10+12344)}{17/(17+11202)} = .534$$

This means that the risk of a baby being born with spina bifida to a member of the "encouraged" group is only .534 of the risk for the "nonencouraged" group. By Equation 6.10 on the preceding page

$$Z = \frac{\ln\left(\widehat{RR}\right) - \ln\left(RR_0\right)}{\sqrt{\frac{b/a}{a+b} + \frac{d/c}{c+d}}} = \frac{\ln(.534)}{\sqrt{\frac{12344/10}{10+12344} + \frac{11202/17}{17+11202}}} = \frac{-.627}{\sqrt{.100 + .059}} = -1.57$$

Appendix A shows that the critical value for a one-tailed Z test with the specified alternative conducted at $\alpha = .05$ is -1.65. Because obtained Z of -1.57 is not in the critical region defined by Z of -1.65, the null hypothesis is not rejected. As a result, the researchers in this fictitious study were unable to demonstrate a protective effect for mothers encouraged to include folic acid in their diet. ∎

6.4.3   Establishing Equivalence by Means of the Independent Samples Risk Ratio

Rationale.   As with differences between means and proportions, it is sometimes desirable to show that the risk associated with some condition is equivalent to the risk associated with some other condition. This would imply use of a two-tailed equivalence test based on risk ratios. Perhaps more commonly, it is sometimes desirable to show that the risk associated with one condition is *no greater than* the risk associated with some other condition. This implies a one-tailed test.

Recalling an example from Chapter 5, suppose that a program designed to control the fruit fly is to be implemented by aerial spraying of large geographic areas with the chemical malathion. Because some environmental activists have charged that such spraying is harmful to humans, a study might be designed to show that the risk associated with spraying is no greater than the risk for persons not exposed to the spray. To this end, subjects who live in a recently sprayed area may undergo medical examination as do a similar group of subjects who live in an unsprayed area. Each condition examined is rated as either being present or not being present for each subject.

Examination of the collected data shows that for a specific condition (e.g., elevated heart rate), 126 subjects in the sprayed area manifest the condition while 819 do not. In the unsprayed area, 119 residents manifest the condition while 811 do not. The researchers will test the null hypothesis that risk in the sprayed area is greater than risk in the unsprayed area against the alternative that risk in the sprayed area is not greater than risk in the unsprayed area.

What is the difference between the analysis outlined here and one that tests the null hypothesis of no difference in risk against an alternative that risk is increased in sprayed areas? Where does the burden of proof lie in the two analyses? We will leave these question for you to ponder.[10]

The Test.   Equivalence intervals for risk ratios are usually, though not necessarily, symmetric about 1.0 in the sense that $EI_U = \frac{1}{EI_L}$ and visa versa. Thus, if EI_L is .8, EI_U would ordinarily be $\frac{1.0}{.8} = 1.25$.

Using the notation of Section 4.3.6 on page 117, the null and alternative hypotheses for a two-tailed equivalence test for independent risk ratios can be stated as

$$H_{0E} : RR \le EI_L \text{ or } RR \ge EI_U$$
$$H_{AE} : EI_L < RR < EI_U$$

where EI_L and EI_U are the lower and upper ends of the equivalence interval. In essence, the null hypothesis states that the risk ratio does not lie in the equivalence interval while the alternative asserts that the risk ratio is in the interval. You will recall that the null hypothesis is tested by conducting two one-tailed tests at level α. In order to reject the equivalence null hypothesis you must show that $RR > EI_L$ *and* that $RR < EI_U$ which means that *both* of the following tests must be significant.

Test One	Test Two
$H_{01} : RR = EI_U$	$H_{02} : RR = EI_L$
$H_{A1} : RR < EI_U$	$H_{A2} : RR > EI_L$

The null and alternative hypotheses for the one-tailed equivalence test are *one* of the following

$$H_{0E} : RR \ge EI_U$$
$$H_{AE} : RR < EI_U$$

or

$$H_{0E} : RR \le EI_L$$
$$H_{AE} : RR > EI_L$$

---

[10]Don't get discouraged if you can't quite get all this figured out. Equivalence testing can be a bit difficult to get ones mind around. Just keep working at it.

The first one-tailed equivalence hypothesis given above is tested by Test One with the second being carried out by means of Test Two.

Test One and Two can be conducted by use of Equation 6.10 on page 240. An example will show how this is accomplished.

EXAMPLE 6.15

Compute the risk ratio for the malathion data discussed above. Assuming that an increase in risk of less than 1.1 is deemed acceptable, conduct a one-tailed equivalence test to show that the increase in risk is less than this value. State the null and alternative equivalence hypotheses and interpret the test results.

Solution   It is convenient to form the data into a two-by-two table as shown below.

		Elevated Heart Rate yes	no
Exposed to	yes	126	819
Malathion	no	119	811

By Equation 6.9 on page 239 the sample risk ratio is

$$\widehat{RR} = \frac{a/(a+b)}{c/(c+d)} = \frac{126/(126+819)}{119/(119+811)} = 1.042$$

Because the researchers are attempting to determine whether the risk ratio is less than 1.1, the test will be one-tailed with equivalence null hypothesis

$$H_{0E} : RR \geq 1.1$$

and alternative

$$H_{AE} : RR < 1.1$$

Conducting Test One by means of Equation 6.10 yields

$$Z_1 = \frac{\ln\left(\widehat{RR}\right) - \ln\left(RR_0\right)}{\sqrt{\frac{b/a}{a+b} + \frac{d/c}{c+d}}} = \frac{\ln(1.042) - \ln(1.100)}{\sqrt{\frac{819/126}{126+819} + \frac{811/119}{119+811}}} = \frac{.041 - .095}{\sqrt{.007 + .007}} = -.456$$

Because this value is greater than the critical value of -1.65 the test is not significant. Therefore, the researchers were not able to demonstrate that the increase in risk was in the acceptable range (i.e., less than 1.1). ∎

EXAMPLE 6.16

Use the data in the table below to compute $\widehat{RR}$. Then perform a two-tailed equivalence test using an EI of .833 to 1.200. Interpret your results.

		Result Positive yes	no
Exposed	yes	301	771
	no	345	820

Solution By Equation 6.9 on page 239 the sample risk ratio is

$$\widehat{RR} = \frac{a/(a+b)}{c/(c+d)} = \frac{301/(301+771)}{345/(345+820)} = .948$$

This would suggest that exposure provides a small protective effect. Conducting Test One by means of Equation 6.10 on page 240 yields

$$Z_1 = \frac{\ln\left(\widehat{RR}\right) - \ln\left(RR_0\right)}{\sqrt{\frac{b/a}{a+b} + \frac{d/c}{c+d}}} = \frac{\ln(.948) - \ln(1.200)}{\sqrt{\frac{771/301}{301+771} + \frac{820/345}{345+820}}} = \frac{-.053 - .182}{\sqrt{.002 + .002}} = -3.72$$

Since this value is less than the critical value of -1.65, Test One is significant.

Conducting Test Two by means of Equation 6.10 yields

$$Z_2 = \frac{\ln(.948) - \ln(.833)}{\sqrt{\frac{771/301}{301+771} + \frac{820/345}{345+820}}} = \frac{-.053 - (-.183)}{\sqrt{.002 + .002}} = 2.06$$

Since this value is greater than the critical value of 1.65, Test Two is significant. Because both tests are significant, the equivalence null hypothesis is rejected. We conclude, therefore, that exposure and nonexposure produce equivalent results as defined by the equivalence interval. ∎

6.4.4 Confidence Interval for the Independent Samples Risk Ratio

Rationale. We earlier indicated (Section 4.5 on page 152) that confidence intervals are generally preferable to hypothesis tests. Thus, rather than asking the question, "Is the population risk ratio different from one?", a generally more informative question asks, "What is the population risk ratio?" While exact methods for constructing confidence intervals for RR are problematic, rather simple approximate methods are available.

The Confidence Interval. Lower and upper limits for the independent samples risk ratio may be obtained by the following equations.

$$L = \exp\left[\ln\left(\widehat{RR}\right) - Z\sqrt{\frac{b/a}{a+b} + \frac{d/c}{c+d}}\right] \qquad (6.11)$$

and

$$U = \exp\left[\ln\left(\widehat{RR}\right) + Z\sqrt{\frac{b/a}{a+b} + \frac{d/c}{c+d}}\right] \qquad (6.12)$$

The symbols exp and ln indicate, respectively, that the natural exponential and logarithm of the enclosed expression is to be taken while a, b, and c are frequencies in a two-by-two table as previously discussed in connection with Table 6.7 on page 239. $\widehat{RR}$ is the sample risk ratio as defined by Equation 6.9 on page 239.

EXAMPLE 6.17

In Example 6.13 on page 240 you performed a test of the null hypothesis $H_0 : RR = 1$ using data from the table provided here. Use this data to construct a 95% confidence interval. What does this interval mean? Does this interval imply that the null hypothesis $H_0 : RR = 1$ was rejected? What is the reason for your answerer?

		Bladder Cancer yes	no
Exposed to	yes	42	2,981
Solvent	no	21	4,088

Solution   By Equation 6.9 the sample risk ratio is

$$\widehat{RR} = \frac{a/(a+b)}{c/(c+d)} = \frac{42/(42+2981)}{21/(21+4088)} = 2.718$$

Then by Equation 6.11 on the facing page

$$L = \exp\left[\ln\left(\widehat{RR}\right) - Z\sqrt{\frac{b/a}{a+b} + \frac{d/c}{c+d}}\right]$$

$$= \exp\left[\ln(2.718) - 1.96\sqrt{\frac{2981/42}{42+2981} + \frac{4088/21}{21+4088}}\right]$$

$$= \exp\left[1.000 - 1.96\sqrt{.023 + .047}\right]$$

$$= \exp[.481]$$

$$= 1.618$$

And by Equation 6.12 on the preceding page

$$U = \exp\left[\ln\left(\widehat{RR}\right) + Z\sqrt{\frac{b/a}{a+b} + \frac{d/c}{c+d}}\right]$$

$$= \exp\left[\ln(2.718) + 1.96\sqrt{\frac{2981/42}{42+2981} + \frac{4088/21}{21+4088}}\right]$$

$$= \exp\left[1.000 + 1.96\sqrt{.023 + .047}\right]$$

$$= \exp[1.519]$$

$$= 4.568$$

From this interval we can conclude that persons exposed to the solvent are at between 1.618 and 4.568 times the risk of bladder cancer as compared to persons not exposed. We can also conclude that a two-tailed test of the null hypothesis $RR = 1$ conducted at $\alpha = .05$

would be significant. This conclusion is reached by noting that the null value of 1.0 is not in the interval.[11] ∎

EXAMPLE 6.18

In Example 6.14 on page 240 you used fictitious data from a study evaluating the role of folic acid in preventing spina bifida to perform a test of the null hypothesis $H_0 : RR = 1$ against the alternative $H_A : RR < 1$. These data are provided in the table given below. Use these data to construct a one-sided 95% confidence interval for the upperbound of the population risk ratio. Use this interval to conduct the test mentioned above. How does this interval justify your conclusion?

		Spina Bifida yes	no
Encouraged	yes	10	12,344
	no	17	11,202

Solution   By Equation 6.9 on page 239 the sample risk ratio is

$$\widehat{RR} = \frac{a/(a+b)}{c/(c+d)} = \frac{10/(10+12344)}{17/(17+11202)} = .534$$

And by Equation 6.12 on page 244

$$U = \exp\left[\ln\left(\widehat{RR}\right) + Z\sqrt{\frac{b/a}{a+b} + \frac{d/c}{c+d}}\right]$$

$$= \exp\left[\ln(.534) + 1.65\sqrt{\frac{12344/10}{10+12344} + \frac{11202/17}{17+11202}}\right]$$

$$= \exp\left[-.627 + 1.65\sqrt{.100 + .059}\right]$$

$$= \exp[.031]$$

$$= 1.031$$

Because the upper bound of 1.031 is greater than one, the null hypothesis is not rejected. This is the same result obtained previously with the hypothesis test. ∎

6.4.5   Assumptions

The methods for conducting hypothesis tests, equivalence tests and constructing confidence intervals presented here depend on a normal curve approximation. Sample sizes should be large enough to ensure that the approximation provides sufficiently accurate results. These procedures also require the assumption of, and are generally not robust against violations of, independence.

---

[11] See Section 4.5 on page 152 if this explanation is not clear.

6.5 METHODS RELATED TO INDEPENDENT SAMPLES ODDS RATIOS

6.5.1 Background

Another, perhaps more commonly employed, statistic used to relate disease and exposure is the odds ratio. The **odds** associated with some event for a particular group is the *probability* that the event will occur to a member of that group divided by the probability that the event will not occur to a member of that group. For example, suppose that a group of industrial workers are identified who have been diagnosed as having bladder cancer. The odds that a member of this group was exposed to a particular industrial solvent suspected of causing bladder cancer is, using the notation of Chapter 3, $\frac{P(E|D)}{P(\overline{E}|D)}$ which is read, "The probability of exposure given disease divided by the probability of no exposure given disease." If, for example, this value were 2.0 we would say that the probability of exposure to the solvent for workers having bladder cancer is twice the probability that they were not exposed to the solvent.

Likewise, we can characterize the odds of exposure to the solvent for workers in the same factory who are free of bladder cancer as $\frac{P(E|\overline{D})}{P(\overline{E}|\overline{D})}$ which is read, "The probability of exposure given no disease divided by the probability of no exposure given no disease." If, for example, this value were .80 we would say that the probability of exposure to the solvent for workers not having bladder cancer is .8 the probability that they were not exposed to the solvent.

It seems reasonable that an industrial hygiene researcher may wish to compare the odds that a worker with bladder cancer was exposed to the solvent to the odds that a worker without bladder cancer was exposed to the solvent. If the odds of exposure for workers with bladder cancer is greater than the odds for workers without bladder cancer, the implication may be an association between exposure to the solvent and bladder cancer. A comparison between the two odds may be done in various ways including computation of the difference by simple subtraction. A popular way to compare the two odds is to form them into a ratio referred to as an **odds ratio** which is defined by Equation 6.13. Note that

$$OR = \frac{\frac{P(E|D)}{P(\overline{E}|D)}}{\frac{P(E|\overline{D})}{P(\overline{E}|\overline{D})}}$$

which simplifies to

$$OR = \frac{P(E\mid D)\,P\left(\overline{E}\mid\overline{D}\right)}{P\left(\overline{E}\mid D\right)P\left(E\mid\overline{D}\right)} \tag{6.13}$$

Equation 3.12 on page 60[12] expresses the odds of disease for exposed and unexposed groups while Equation 6.13 expresses the odds of exposure for groups with and without disease. Both forms are used by researchers.

It will be helpful to distinguish between two general forms of research designs[13] with which the independent samples odds ratio is commonly employed. In Section 6.4.2 an example was given in which workers who were routinely exposed to a solvent along with

---

[12]You may wish to review the information on page 60 at this point.

[13]These are but two of many such designs.

TABLE 6.8: Typical data arrangement for independent samples prospective cohort study data.

		Disease yes	Disease no
Exposed	yes	a	b
	no	c	d

TABLE 6.9: Typical data arrangement for independent samples case-control study data.

		Exposed yes	Exposed no
Disease	yes	a	b
	no	c	d

workers who were not exposed were followed across time in order to determine which developed bladder cancer. The odds ratio computed for this design would express the odds of disease (bladder cancer) for exposed (solvent) and unexposed groups of workers. Designs of this sort in which exposed and unexposed subjects are followed across time in order to determine who develops the disease are termed **prospective cohort** designs. Data from this sort of design would typically be arranged for analysis as shown in Table 6.8.

By contrast, in this section an example was given in which workers who had already developed disease (bladder cancer) were compared to workers who had not developed the disease. The odds ratio computed for this design would express the odds of exposure (solvent) for workers with and without disease (bladder cancer). Designs of this sort in which subjects who already have disease are compared to subjects who do not have the disease in order to determine which has greater exposure are termed **case-control** designs. Data from this sort of design would typically be arranged for analysis as shown in Table 6.9.

For purposes of the discussion that follows, we will assume that a case-control study is being discussed though, with proper substitutions of exposure for disease and visa versa, the remarks apply equally to prospective cohort studies.

If the odds of exposure[14] in the two groups are the same, the odds ratio is one which would imply that there is no relationship between disease and exposure. On the other hand, an odds ratio greater than one would imply that disease is associated with greater exposure than is non-disease. Ratios that are less than one are said to be **protective** because disease implies less exposure than does no disease. This might occur, for example, when persons with disease have less exposure to a daily vitamin supplement than do persons without disease. We will represent the parameter and statistic forms of the odds ratio as OR and $\widehat{OR}$ respectively. The sample independent samples odds ratio is defined by Equation 6.14. Note

---

[14]While we refer to exposure and disease in Equation 6.13, the terms are generic and represent whatever variables are being examined.

that

$$\widehat{OR} = \frac{\frac{a}{b}}{\frac{c}{d}}$$

which simplifies to

$$\boxed{\widehat{OR} = \frac{ad}{bc}} \tag{6.14}$$

where a, b, c, and d are as shown in Tables 6.8 and 6.9 on the facing page.

6.5.2   Test of the Hypothesis OR $= 1$ for Independent Samples

Rationale.   As in the situation alluded to above, suppose it is suspected that exposure to a particular chemical solvent commonly used in manufacturing increases the risk of bladder cancer. To test this suspicion, workers in a particular industry who have been diagnosed with bladder cancer are queried as to whether their job involved exposure to the solvent. Likewise, a group of workers in the same industry who do not have the disease are also queried as to exposure to this chemical.

Given this data, the sample odds ratio can be computed via Equation 6.14. If this ratio were 5.0, for example, the indication would be that the odds of exposure to the solvent for workers with bladder cancer is five times that of cancer free workers. But is this simply a result of the particular sample used in the study? If the study were repeated with other workers might the result be markedly different? More importantly, is the odds ratio in the population one, indicating no relationship between disease and exposure (i.e., the odds in the two groups is the same), or is it different from one indicating that such a relationship does exist? A test of hypothesis can help answer this question.

The Test.   A test of the null hypothesis $OR = 1$ against one- or two-tailed alternatives can be conducted via Equation 6.15. The frequencies a, b, c, and d in this equation are as shown in Tables 6.8 and 6.9 on the preceding page. The hypothesized population odds ratio is represented by OR_0. Because the log of 1 is zero, the expression $\ln(OR_0)$ may be omitted when testing $H_0 : OR = 1$.

$$\boxed{Z = \frac{\ln\left(\widehat{OR}\right) - \ln(OR_0)}{\sqrt{\frac{1}{a} + \frac{1}{b} + \frac{1}{c} + \frac{1}{d}}}} \tag{6.15}$$

EXAMPLE 6.19

Suppose that in the bladder cancer study discussed above, 13 of the workers who had bladder cancer were exposed to the solvent while 56 were not. Of the workers without bladder cancer, four had been exposed to the solvent while 65 had not. Use this data to compute the sample odds ratio. What does this ratio mean? Use Equation 6.15 to test the null hypothesis $OR = 1$. Interpret your results.

Solution   For convenience, the data are arranged into a two by two table as shown in Table 6.10.

TABLE 6.10: Data from case-control study relating bladder cancer to solvent exposure.

		Exposed to Solvent	
		yes	no
Bladder	yes	13	56
Cancer	no	4	65

From Equation 6.14 on the previous page

$$\widehat{OR} = \frac{ad}{bc} = \frac{(13)(65)}{(56)(4)} = 3.772$$

This means that the odds of exposure to the solvent for workers with bladder cancer was 3.772 times that of workers who do not suffer from the disease.

By Equation 6.15 on the preceding page

$$Z = \frac{\ln\left(\widehat{OR}\right) - \ln\left(OR_0\right)}{\sqrt{\frac{1}{a} + \frac{1}{b} + \frac{1}{c} + \frac{1}{d}}} = \frac{\ln(3.772)}{\sqrt{\frac{1}{13} + \frac{1}{56} + \frac{1}{4} + \frac{1}{65}}} = \frac{1.328}{.600} = 2.21$$

The critical values for a two-tailed Z test conducted at $\alpha = .05$ as reported in Appendix A are 1.96 and -1.96. Because the obtained Z value of 2.21 exceeds 1.96, the null hypothesis is rejected. It is thus concluded that the population odds ratio is greater than one thereby indicating that workers with bladder cancer have greater odds of having been exposed to the solvent than do workers without the disease. ■

EXAMPLE 6.20

In Example 6.14 on page 240 you used the data in the table provided below to assess the relationship between ingestion of folic acid and spina bifida. You found that the sample risk ratio was .534. A subsequent test of the null hypothesis $H_0 : RR = 1$ was not significant. Use this data to calculate and interpret the sample odds ratio. Conduct a test of the null hypothesis $H_0 : OR = 1$ against the alternative $H_A : OR < 1$ at the .05 level. Interpret the results of this test.

		Spina Bifida	
		yes	no
Encouraged	yes	10	12,344
	no	17	11,202

Solution   By Equation 6.14 on the previous page

$$\widehat{OR} = \frac{ad}{bc} = \frac{(10)(11202)}{(12344)(17)} = .534$$

Notice that this is quite close to the value obtained for $\widehat{RR}$. This is usually the case when dealing with a rare disease. Then by Equation 6.15 on page 249

$$Z = \frac{\ln\left(\widehat{OR}\right) - \ln\left(OR_0\right)}{\sqrt{\frac{1}{a} + \frac{1}{b} + \frac{1}{c} + \frac{1}{d}}} = \frac{\ln\left(.534\right)}{\sqrt{\frac{1}{10} + \frac{1}{12344} + \frac{1}{17} + \frac{1}{11202}}} = \frac{-.627}{.399} = -1.57$$

The critical value for a one-tailed Z test conducted at $\alpha = .05$ as reported in Appendix A is -1.65. Because the obtained Z value of -1.57 is greater than this critical value, the null hypothesis is not rejected. We were unable, therefore, to demonstrate a protective effect for mothers encouraged to include folic acid in their diet.   ∎

6.5.3 Establishing Equivalence by Means of the Independent Samples Odds Ratio

Rationale.   As with differences between means, differences between proportions and risk ratios, it is sometimes desirable to show that the odds associated with some condition is equivalent to the odds associated with some other condition. This would imply use of a two-tailed equivalence test based on odds ratios. Perhaps more commonly, it is sometimes desirable to show that the odds associated with one condition are *not greater than* the odds associated with some other condition. This implies a one-tailed test.

Suppose, for example, that military epidemiologists are directed to conduct a study to assess the relationship between a particular skin disorder that afflicts certain personnel on a military base and exposure to a particular munition. After due consideration, it is decided that the study should attempt to demonstrate that the odds of exposure to the munition for personnel with the skin disorder is not greater than the odds for other base personnel who do not suffer from the disease. It is also decided that an odds ratio less than 1.1 will suffice to show that the odds of exposure for persons with the disease was not greater than the odds for persons without the disease.

Notice that in a standard (i.e., non-equivalence) study the burden of proof would be to show that exposure *was* greater while in this study the burden is to show that exposure *was not* greater. In a sense, the standard study would assume the munition safe until proven unsafe while the equivalence study assumes the munition unsafe until shown to be safe. The difference between standard and equivalence studies is not always easy to grasp but the distinction is sufficiently important so as to justify the effort needed to make the distinction clear.

The Test.   As with risk ratios, equivalence intervals for odds ratios are usually, though not necessarily, symmetric about 1.0 in the sense that $EI_U = \frac{1}{EI_L}$ and visa versa. Thus, if EI_L is .8, EI_U would ordinarily be $\frac{1.0}{.8} = 1.25$.

Using the notation of Section 4.3.6 on page 117, the null and alternative hypotheses for a two-tailed equivalence test for independent samples odds ratios can be stated as

$$H_{0E} : OR \leq EI_L \quad \text{or} \quad OR \geq EI_U$$
$$H_{AE} : EI_L < OR < EI_U$$

where EI_L and EI_U are the lower and upper ends of the equivalence interval. In essence, the null hypothesis states that the odds ratio does not lie in the equivalence interval while the alternative asserts that the odds ratio is in the interval. You will recall that the null hypothesis is tested by conducting two one-tailed tests at level α. In order to reject the equivalence null hypothesis you must show that $OR > EI_L$ *and* that $OR < EI_U$ which means that *both* of the following tests must be significant.

Test One	Test Two
$H_{01} : OR = EI_U$	$H_{02} : OR = EI_L$
$H_{A1} : OR < EI_U$	$H_{A2} : OR > EI_L$

The null and alternative hypotheses for the one-tailed equivalence test are *one* of the following

$$H_{0E} : OR \geq EI_U$$
$$H_{AE} : OR < EI_U$$

or

$$H_{0E} : OR \leq EI_L$$
$$H_{AE} : OR > EI_L$$

The first one-tailed equivalence hypothesis given above is tested by Test One with the second being carried out by means of Test Two.

EXAMPLE 6.21

Compute the odds ratio for the skin disease data contained in the following table. Assuming that an odds ratio of less than 1.1 would be accepted as demonstrating that the odds of exposure for personnel with the disease was not greater than the odds of exposure for personnel without the disease, conduct a one-tailed equivalence test to show that the odds ratio is less than this value. State the null and alternative equivalence hypotheses and interpret the test results.

		Exposed to Munition	
		yes	no
Skin	yes	126	819
Disease	no	119	811

Solution By Equation 6.14 on page 249 the sample odds ratio is

$$\widehat{OR} = \frac{ad}{bc} = \frac{(126)(811)}{(819)(119)} = 1.048$$

The null and alternative hypotheses for the one-tailed equivalence test are

$$H_{0E} : OR \geq 1.1$$
$$H_{AE} : OR < 1.1$$

Then, conducting Test One by means of Equation 6.15 on page 249

$$Z_1 = \frac{\ln\left(\widehat{OR}\right) - \ln\left(OR_0\right)}{\sqrt{\frac{1}{a} + \frac{1}{b} + \frac{1}{c} + \frac{1}{d}}} = \frac{\ln\left(1.048\right) - \ln\left(1.100\right)}{\sqrt{\frac{1}{126} + \frac{1}{819} + \frac{1}{119} + \frac{1}{811}}} = \frac{-.048}{.137} = -.350$$

Because this value is greater than the critical value of -1.65, the null hypothesis is not rejected. Therefore, the researchers were unable to demonstrate that the odds of exposure for personnel with the skin disease is not greater than the odds of exposure for personnel without the disease. ∎

EXAMPLE 6.22

Use the data in the table below to compute the $\widehat{OR}$. What does this ratio mean? Then perform a two-tailed equivalence test using the EI .833 to 1.200. Interpret your results.

		Variable Two	
		yes	no
Variable	yes	48	12
One	no	24	6

Solution   By Equation 6.14 on page 249 the sample odds ratio is

$$\widehat{OR} = \frac{ad}{bc} = \frac{(48)(6)}{(12)(24)} = 1.000$$

This ratio indicates that the odds of a "yes" for variable two is the same for those with a "yes" for variable one as for those with a "no" for variable one. We can move the example from abstract to a more concrete form by substituting "disease" for variable one and "exposure" for variable two. We can now interpret the ratio as indicating that the odds of exposure are the same for those with and without disease.

The two-tailed equivalence test is conducted via Test One *and* Test Two as implemented by Equation 6.15 as

$$Z_1 = \frac{\ln\left(\widehat{OR}\right) - \ln\left(OR_0\right)}{\sqrt{\frac{1}{a} + \frac{1}{b} + \frac{1}{c} + \frac{1}{d}}} = \frac{\ln\left(1.000\right) - \ln\left(1.200\right)}{\sqrt{\frac{1}{48} + \frac{1}{12} + \frac{1}{24} + \frac{1}{6}}} = \frac{-.183}{.559} = -.327$$

and

$$Z_2 = \frac{\ln\left(1.000\right) - \ln\left(.833\right)}{\sqrt{\frac{1}{48} + \frac{1}{12} + \frac{1}{24} + \frac{1}{6}}} = \frac{.182}{.559} = .326$$

From Appendix A it can be seen that the critical values for Test One and Test Two are -1.65 and 1.65 respectively. Because neither of these tests is significant, the equivalence null hypothesis is not rejected. You will recall that *both* tests must be significant in order for the equivalence null hypothesis to be rejected. Therefore, we were unable to establish equivalence in this case. ∎

6.5.4   Confidence Interval for the Independent Samples Odds Ratio

Rationale.   We earlier indicated (Section 4.5 on page 152) that confidence intervals are generally preferable to hypothesis tests. Thus, rather than asking the question, "Is the population odds ratio different from one?", a generally more informative question asks, "What is the population odds ratio?" While exact methods for constructing confidence intervals for OR are problematic, rather simple approximate methods are available.

The Confidence Interval.   Lower and upper limits for the independent samples odds ratio may be obtained by the following equations.

$$L = \exp\left[\ln\left(\widehat{OR}\right) - Z\sqrt{\frac{1}{a} + \frac{1}{b} + \frac{1}{c} + \frac{1}{d}}\right] \qquad (6.16)$$

and

$$U = \exp\left[\ln\left(\widehat{OR}\right) + Z\sqrt{\frac{1}{a} + \frac{1}{b} + \frac{1}{c} + \frac{1}{d}}\right] \qquad (6.17)$$

The symbols exp and ln indicate, respectively, that the natural exponential and logarithm of the enclosed expression is to be taken while a, b, c, and d are frequencies in a two-by-two table as previously discussed in connection with Tables 6.8 and 6.9 on page 248. $\widehat{OR}$ is the sample odds ratio as defined by Equation 6.14 (page 249).

EXAMPLE 6.23

In Example 6.19 on page 249 you used the data from Table 6.11 on the next page to perform a two-tailed test of the null hypothesis $H_0 : OR = 1$. Use this data to construct a two-sided 95% confidence interval for the estimation of OR. What does this interval mean? Does this interval imply that the null hypothesis $H_0 : OR = 1$ was rejected? What is the reason for your answer?

Solution   By Equation 6.14 on page 249 the sample odds ratio is

$$\widehat{OR} = \frac{ad}{bc} = \frac{(13)(65)}{(56)(4)} = 3.772$$

Then by Equations 6.16 and 6.17

$$L = \exp\left[\ln\left(\widehat{OR}\right) - Z\sqrt{\frac{1}{a} + \frac{1}{b} + \frac{1}{c} + \frac{1}{d}}\right]$$

$$= \exp\left[\ln(3.772) - 1.96\sqrt{\frac{1}{13} + \frac{1}{56} + \frac{1}{4} + \frac{1}{65}}\right]$$

$$= \exp[1.328 - 1.96(.600)]$$

$$= \exp[.152]$$

$$= 1.164$$

TABLE 6.11: Data from case-control study relating bladder cancer to solvent exposure.

		Exposed to Solvent yes	Exposed to Solvent no
Bladder	yes	13	56
Cancer	no	4	65

and

$$U = \exp\left[\ln\left(\widehat{OR}\right) + Z\sqrt{\frac{1}{a} + \frac{1}{b} + \frac{1}{c} + \frac{1}{d}}\right]$$

$$= \exp\left[\ln\left(3.772\right) + 1.96\sqrt{\frac{1}{13} + \frac{1}{56} + \frac{1}{4} + \frac{1}{65}}\right]$$

$$= \exp\left[1.328 + 1.96\left(.600\right)\right]$$

$$= \exp\left[2.504\right]$$

$$= 12.231$$

Thus, we can be 95% confident that the odds of disease for persons exposed to the solvent are between 1.164 and 12.231 times that of persons not exposed.

From this interval we can assert that a two-tailed test of the null hypothesis $H_0 : OR = 1$ would be rejected. This follows from the fact that the value specified by the null hypothesis, one in this case, is not in the interval. ∎

EXAMPLE 6.24

In Example 6.20 on page 250 you used fictitious data from a study evaluating the role of folic acid in preventing spina bifida to perform a test of the null hypothesis $H_0 : OR = 1$ against the alternative $H_A : OR < 1$. These data are provided in the table given below. Use these data to construct a one-sided confidence interval for the upperbound of the population odds ratio. Use this interval to conduct the test alluded to here. How does this interval justify your conclusion?

		Spina Bifida yes	Spina Bifida no
Encouraged	yes	10	12,344
	no	17	11,202

Solution   By Equation 6.14 on page 249 the sample odds ratio is

$$\widehat{OR} = \frac{ad}{bc} = \frac{(10)\,(11202)}{(12344)\,(17)} = .534$$

Then by Equation 6.17 on page 254

$$U = \exp\left[\ln\left(\widehat{OR}\right) + Z\sqrt{\frac{1}{a} + \frac{1}{b} + \frac{1}{c} + \frac{1}{d}}\right]$$

$$= \exp\left[\ln(.534) + 1.65\sqrt{\frac{1}{10} + \frac{1}{12344} + \frac{1}{17} + \frac{1}{11202}}\right]$$

$$= \exp\left[-.627 + 1.65(.399)\right]$$

$$= \exp\left[.031\right]$$

$$= 1.031$$

This upper bound indicates that the odds of a spina bifida afflicted baby for mothers encouraged to include folic acid in their diet is 1.031 *or less* times that of mothers not so encouraged. Because the upper bound of 1.031 is greater than one, the null hypothesis is not rejected. This is the same result obtained previously with the hypothesis test. ■

6.5.5 Assumptions

The methods for conducting hypothesis tests, equivalence tests and constructing confidence intervals presented here depend on a normal curve approximation. Sample sizes should be large enough to ensure that the approximation provides sufficiently accurate results. These procedures also require the assumption of, and are generally not robust against violations of, independence.

6.5.6 Estimating Risk of Disease from Case-Control Data

On pages 239 and 248 we briefly discussed two common study designs—the cohort and case-control studies. Subsequently, we used these designs to demonstrate calculation and interpretation of risk and odds ratios. In this section we provide a few additional comments regarding use of these statistics in connection with data collected in the context of case-control studies.

You will recall that in case-control studies, subjects with disease are compared to subjects without disease regarding their exposure to some risk factor. These types of studies are usually more easily carried out than are cohort studies that require following subjects with and without exposure to some risk factor across time to determine which manifest some disease. One shortcoming of case-control studies, however, is that while risk or odds of exposure may be assessed, risk of disease cannot be obtained. The risk of disease is often of primary concern for researchers in the health sciences. In this section we will show why risk of disease cannot be directly obtained from case-control studies and will further show that, under certain conditions, an *estimate* of the risk ratio for disease can be obtained in case-control studies through use of the odds ratio.

In order to facilitate discussion of the above, let us assume that the cells of Table 6.12 on the facing page contain *population* frequencies for exposure and disease. To distinguish these population cell frequencies from the sample frequencies a, b, c, and d, we will use the corresponding designations A, B, C, and D. Using this notation, the risk of disease

TABLE 6.12: Population contingency table relating exposure to disease.

		Disease yes	no
Exposed	yes	A	B
	no	C	D

for exposed subjects is the proportion of exposed subjects in the population who have the disease or

$$\frac{A}{A+B}$$

Likewise, the risk of disease in the unexposed group is

$$\frac{C}{C+D}$$

It follows that

$$RR = \frac{\frac{A}{A+B}}{\frac{C}{C+D}}$$

Clearly, in order to estimate RR, we must be able to estimate the proportion of subjects in the exposed and unexposed groups who have the disease. This implies that the sample data must be representative of the exposed and unexposed groups in the population. But in a case-control study the researcher *chooses* the number of persons who have and do not have, the disease. This means that the proportion of persons with and without disease in the exposed and unexposed groups is fixed by the researcher. This is certainly unacceptable for estimating the odds/risk of disease. The most effective method for estimating the odds/risk of disease is to obtain random samples of exposed and unexposed members of the population.

Ideally then, we would want to randomly sample exposed and unexposed subjects from the population in order to estimate RR for disease. But what would happen in a case-control study? You will recall that in this instance subjects are selected not on the basis of exposure/nonexposure, but on the basis of disease/nondisease status. While a random sample of subjects with and without disease would allow for estimates of the risk of exposure in these two groups, the resultant samples would provide no information concerning risk of disease in the exposed and unexposed groups. Because these data allow for estimation of risk in the disease/nondisease groups, we can estimate the risk of exposure but not of disease.

Note, however, that when disease is rare, A and C are small as compared to B and D. This implies that

$$\frac{A}{A+B} \approx \frac{A}{B}$$

and

$$\frac{C}{C+D} \approx \frac{C}{D}$$

(The symbol $\approx$ is read "is approximately equal to.") Thus, *when disease is rare,*

$$RR = \frac{\frac{A}{A+B}}{\frac{c}{c+D}} \approx \frac{\frac{A}{B}}{\frac{C}{D}} = \frac{AD}{BC}$$

The expression

$$\frac{AD}{BC}$$

is the population odds ratio which is estimated by the sample odds ratio. All this implies that, when disease is rare, the sample odds ratio for disease can be used to estimate the population risk ratio for disease. For example, if $A = 100$, $B = 5,000$, $C = 75$, and $D = 4,900$,

$$RR = \frac{\frac{100}{100+5000}}{\frac{75}{75+4900}} = 1.30 \approx \frac{(100)\,(4900)}{(5000)\,(75)} = 1.31$$

Many of the issues related to use of risk and odds ratios for the analysis of data collected in various research contexts are beyond the scope of this book. Standard epidemiologic research texts such as Breslow and Day [5] or Lilienfeld and Lilienfeld [31] should be consulted for further details.

KEY WORDS AND PHRASES

After reading this chapter you should be able to demonstrate familiarity with the following words and phrases.

$\mu_1 - \mu_2$ 215
s_P^2 219
assumption of normality 230
CI for $\pi_1 - \pi_2$ 236
CI for OR 254
equivalence of π_1 and π_2 234
homogeneity of variance 230
odds 247
odds ratio 247
protective exposure 238, 248
risk equivalence 241
RR and OR when disease is rare 258
test of $\mu_1 - \mu_2$ 215
test of RR = 1 239

$\hat{p}_1 - \hat{p}_2$ 231
$\bar{x}_1 - \bar{x}_2$ 217
case-control study 248
CI for RR 244
CI for $\mu_1 - \mu_2$ 228
equivalence of μ_1 and μ_1 224
independence assumption 99
odds equivalence 251
prospective cohort study 239, 248
risk 238
risk ratio 238
test for $\pi_1 - \pi_2$ 230
test of OR = 1 249

EXERCISES

6.1 A study is conducted to determine whether children who live within one mile of a fast food restaurant have higher average body mass index (BMI) then do children who do not live in such proximity. As a rule of thumb, BMI categories represent the following; 18.5 or less = underweight, 18.5 to 24.9 = normal, 25.0 to 29.9 = overweight while an index of 30.0 or greater is characterized as obese. The data are provided below.

Conduct a test of the hypothesis $\mu_1 = \mu_2$ against the alternative $\mu_1 > \mu_2$ at the .05 level of significance. Let the "within one mile" group be designated group one. What is your conclusion regarding the question at hand?

Group One	Group Two
26.2	25.9
24.5	20.1
20.0	22.2
30.2	29.7
28.4	28.0
18.6	29.4
21.5	20.2
21.7	20.7
29.9	26.3
18.3	18.2

6.2 Suppose it is discovered that an elementary school has been built in close proximity to the site of a long defunct chemical factory. The chemicals produced by the factory are now suspected of causing a complex of respiratory problems. The school board is charged with showing that the school grounds are safe insofar as the inducement of respiratory problems is concerned.

As part of their study, the school board mails a check list of 30 respiratory symptoms to all students who attended the elementary school for four or more years and are now 30 years of age or older. The same check list is sent to a group of former students who meet the same criteria insofar as attendance at an elementary school not near the chemical plant site is concerned. It is decided that if it can be shown that the mean number of symptoms checked by the group attending the school in question is less than two more than the mean number checked by attendees of the control school, the suspect school will be deemed safe insofar as respiratory symptoms are concerned.

The responses for attendees of the two schools are provided below. Designate the attendees of the suspect school as group one.

Group One	Group Two
6	12
4	1
20	22
17	19
9	3
2	0
8	5
14	9
12	11
0	15
24	5
8	17

(a) Perform an equivalence test to carry out the charge given the school board. Would this most likely be a one- or two-tailed test? Why?

(b) State the null and alternative equivalence hypotheses.

(c) How might the analysis be different if the school board were charged with showing that the school grounds are unsafe?

6.3 Use the data in Exercise 6.1 to form a two-sided 95% confidence interval. What is being estimated by this interval?

6.4 Use the data in Exercise 6.1 to form a one-sided 95% confidence interval for the lower bound of $\mu_1 - \mu_2$. Does this interval support the result of your hypothesis test for these data? How so?

6.5 Suppose that a survey conducted by an anti-smoking education group shows that .312 of high school senior males and .288 of females report that they routinely smoke one or more cigarettes per day. Assuming that 1,777 males and 1,821 females were surveyed, perform a two-tailed test of the null hypothesis $\pi_1 = \pi_2$ at the .05 level of significance. Interpret the result.

6.6 Researchers suspect that a rather costly, commonly employed medical treatment is of no more benefit to patients than is a placebo. In order to investigate this impression, a large scale study is carried out in which 1,400 patients are randomly assigned to receive the standard of care treatment or a placebo. Of the 700 patients receiving active treatment, 313 report a beneficial effect. By contrast, 317 of the placebo treated patients report a benefit. The researchers decide it is reasonable to declare the active treatment "not better than placebo" if the proportion of patients benefiting from the treatment is less than .04 greater than the proportion benefiting from placebo.

Use an equivalence test with $\alpha = .05$ to analyze the study result. State the null and alternative hypotheses before conducting the test.

6.7 Use the data in Exercise 6.5 to construct a two-sided 95% confidence interval. If this interval were used to conduct a two-tailed test of $H_0 : \pi_1 = \pi_2$ at the .05 level of significance, would the result be significant? Why or why not? What does the interval estimate?

6.8 It is believed that patients suffering from optic neuritis (characterized by inflammation of the optic nerve) who have 3 or more brain stem lesions (demonstrated by MRI) are at greater risk of developing multiple sclerosis (MS) in the succeeding ten years than are optic neuritis patients who do not manifest such lesions. To examine this belief, 719 optic neuritis patients, 291 of whom have 3 or more lesions and 428 of whom do not, are followed for ten years. Of those patients with lesions, 196 developed MS while 191 of those without lesions developed the disease.

(a) Calculate a risk ratio comparing the risk of MS for patients with and without brain stem lesions. Interpret this ratio.

(b) Test the hypothesis $H_0 : RR = 1$ against the alternative $H_A : RR \neq 1$ at the .05 level of significance.

6.9 It has long been claimed that persons residing near high voltage power lines are at greater risk of developing certain forms of cancer than are persons not living near such lines. Many studies have failed to demonstrate a relationship between proximity to these lines and risk of cancer. A power company seeking permission to construct a power grid near a housing development has met with resistance from the County Planning Board (CPB). The consulting statistician for the CPB has pointed out that while previous studies have failed to show a relationship between residence near such lines and the cancers in question, there have been no studies to demonstrate the safety of close proximity to such lines.

A study commissioned by the power company shows that of 9,848 persons residing within 500 yards of high voltage lines, 590 have developed one of the cancers in question. Of 13,112 residents living more than 500 yards from such lines, 577 contracted one of the cancers.

(a) Using risk ratios, perform a test for equivalence using the equivalence interval .91 to 1.1 and $\alpha = .05$.

(b) State the null and alternative hypotheses for the equivalence test.

(c) State the null hypothesis you believe was used for the previous studies that failed to show a relationship between the variables in question.

(d) Discuss the difference between the previous (standard) tests and the equivalence test used here insofar as objective of the analyses is concerned.

6.10 Use the data in Exercise 6.8 to form a two-sided 95% confidence interval for the estimation of RR. Interpret this interval. Does this interval support the result of your hypothesis test? Why or why not?

6.11 Reanalyze the problem described in Exercise 6.8 using odds ratios in place of risk ratios.

6.12 Answer the questions in Exercise 6.9 again using odds ratios in place of risk ratios.

6.13 Use the data in Exercise 6.8 to form a two-sided 95% confidence interval for the estimation of OR. Interpret this interval. Does this interval support the result of your hypothesis test? Why or why not?

A. The following questions refer to Case Study A (page 469)

6.14 Use the a.m. data to form a two-sided 95% confidence for the estimation of $\pi_n - \pi_y$ where π_n represents the proportion of subjects answering no to the duration question who chose the treated lens and π_y represents the same proportion for the group answering yes to the duration question. Interpret this interval.

6.15 Use the CI constructed in 6.14 to perform a two-tailed test of the null hypothesis $H_0 : \pi_n = \pi_y$ at the .05 level. Explain why you rejected or failed to reject the null hypothesis.

6.16 Form the confidence interval described in 6.14 using the p.m. data. Carry out the associated hypothesis test. What can you say about the results of the two tests insofar as the study is concerned?

B. The following questions refer to Case Study B (page 470)

6.17 Would you expect that a two-tailed independent samples t test using baseline WOMAC A data to compare the means of the Standard and Placebo groups conducted at $\alpha = .05$ would produce a significant result? Explain your answer. Suppose the result *were* significant, how would you explain this fact?

6.18 Use the baseline WOMAC A data to construct a two-sided 95% confidence interval for the estimation of the difference between the means of the Standard and Placebo groups. (Hint: think about how knowledge of the sample standard deviation and sample size can be used to obtain the sum of squares for a particular group.) Does this interval support your contention in 6.17 ?

6.19 Use the 12 week WOMAC A data to construct a two-sided 95% confidence interval for the estimation of the difference between the means of the Standard and Placebo groups. From the researcher's point of view, what is estimated by this interval?

6.20 Use the interval constructed in 6.19 to perform a two-tailed test of the hypothesis $H_0 : \mu_s = \mu_p$ where μ_s and μ_p represent the means of the Standard and Placebo groups respectively. Explain how you arrived at your conclusion regarding whether to reject the null hypothesis.

6.21 Use the data in Table J.3 to test the following null hypotheses against two-tailed alternatives at the .05 level of significance. Note that π_s, π_w and π_p represent the proportions in each group who correctly identified the type of bracelet worn. A correct identification for the Weak group is "Dummy."

(a) $H_0 : \pi_s = \pi_w$

(b) $H_0 : \pi_s = \pi_p$

(c) $H_0 : \pi_w = \pi_p$

6.22 Using μ_w and μ_p to represent the mean WOMAC A score for the Weak and Placebo groups at 12 weeks respectively, carry out the following test at the .05 level of significance. Does the result justify treating the Weak and Placebo groups as equivalent?

$$H_{0E} : \mu_w - \mu_p \leq -1.0 \quad \text{or} \quad \mu_w - \mu_p \geq 1.0$$
$$H_{AE} : -1.0 < \mu_w - \mu_p < 1.0$$

6.23 Using π_w and π_p to represent the proportion of subjects in the Weak and Placebo groups respectively who indicate that they are wearing a Dummy bracelet, carry out the following test at the .05 level of significance. Does

the result justify treating the Weak and Placebo groups as equivalent?

$$H_{0E} : \pi_w - \pi_p \leq -.02 \quad \text{or} \quad \pi_w - \pi_p \geq .02$$
$$H_{AE} : -.02 < \pi_w - \pi_p < .02$$

C. The following questions refer to Case Study C (page 471)

6.24 Upon what statistics (e.g., difference between means etc.) might an evaluation of the two treatments be based?

6.25 Would you suggest that a test of hypothesis or confidence interval be used for this purpose?

6.26 Calculate a risk ratio comparing the risk of death for the invasive group to that of the non-invasive group for each of the time periods. Interpret each of these ratios.

6.27 Form 95% confidence intervals for the estimation of RR for each of the time periods.

6.28 What is your conclusion about the effectiveness of the two treatments?

D. The following questions refer to Case Study D (page 471)

6.29 The researchers report that "HIV-infected patients had significantly lower NPZ-8 scores (t[18]=2.26, $P < .05$) and lower PBV (t[18]=1.79, $P < .01$) than those of healthy control participants. " The notation t[18]=2.26, $P < .05$ means that a test was conducted with 18 degrees of freedom that produced an obtained t of 2.26 which had an associated p-value less than .05. A similar interpretation applies to the test on the PBV scores.

(a) What kind of t test do you think the researchers used? What makes you believe that this is the case?

(b) Do you see anything peculiar about the reported p-values?

(c) Conduct the same two tests. Do you get the same obtained t value? Assuming two-tailed tests, indicate whether each test is significant at the .05 or .01 level or is not significant at either.

(d) Did you notice anything regarding these tests that might lead you question the validity of either test?

6.30 Form a two-sided 95% confidence interval to estimate the difference between the mean NPZ-8 scores of the HIV-infected subjects who have positive ADC Stage assessments and HIV-infected subjects who have negative ADC Stage assessments.

E. The following questions refer to Case Study E (page 472)

6.31 Form an odds ratio to express the odds of a person testing positive actually having the disease to the odds of a person testing negative having the disease. Construct a 95% confidence interval for this estimate of OR.

G. The following questions refer to Case Study G (page 473)

6.32 Estimate the magnitude of the difference between the proportion of dizygotic twins and nontwin siblings with MS.

H. The following questions refer to Case Study H (page 473)

6.33 Do you agree with the conclusion reached by the authors of this study? What are the reasons for your answer?

6.34 Do you believe that the authors of this study were privileged to study biostatistics via this text?

6.35 Markello et al. [34] when discussing treatment of cystinosis (a disease that can impair kidney function) state the following (citations omitted).

> By 1987, oral cysteamine had proved efficacious in preserving renal function and improving growth. Since then, *bioequivalence has been demonstrated between cysteamine and phosphocysteamine*, a more palatable phosphothioester of cysteamine... [Italics added]

These authors cite Smolin et al. [43] as the authority for their bioequivalence[15] statement. Do you believe bioequivalence was established in the Smolin et al. study? Explain your answer.

6.36 What modifications to the Smolin et al. study would you make in order to establish bioequivalence for the two treatments?

I. The following questions refer to Case Study I (page 473)

6.37 Do you believe the analysis should be conducted by means of a one- or two-tailed equivalence test?

6.38 Use the notation of equivalence hypotheses to state the null and alternative hypotheses for the study.

6.39 Conduct the test at the .05 level. Were the researchers able to declare equivalence?

6.40 The researchers stated that they would conduct the equivalence test by constructing a one-sided 95% confidence interval and noting whether the upper end of the interval was above or below .004. Is this a reasonable way to conduct this test? Explain. Cite a page or pages in the text that support your contention.

6.41 The researchers note that other similar clinical trials had used .01 as the criterion for equivalence. Conduct the test again using the less stringent .01 value. Would equivalence have been declared if this criterion had been used?

J. The following questions refer to Case Study J (page 474)

6.42 State the equivalence null and alternative hypotheses for this study.

6.43 Conduct the equivalence test at $\alpha = .05$. What is your conclusion insofar as oral and intravenous therapies are concerned?

N. The following questions refer to Case Study N (page 476)

6.44 Evaluate each of the potential risk factors listed in Table 1 by calculating the odds ratio comparing the odds of being an abuser for those with the potential risk factor to those without the factor. Form a 95% confidence interval for OR. What are your conclusions?

O. The following questions refer to Case Study O (page 477)

6.45 Comment on the importance of point (i).

---

[15]See page 117

Multi-Sample Methods

7.1 INTRODUCTION

In Chapter 6 you studied various statistical methods designed for the analysis of data collected from research designs in which two independent groups are employed. As an example, subjects might be randomly assigned to one of two groups with a treatment condition being administered to the subjects in one group while a placebo is administered to the other. If the outcome is continuous, methods related to the independent samples t test or the related confidence interval might be employed in the analysis. If the outcome were dichotomous, various methods based on proportions might be useful.

But suppose a researcher wishes to compare three treatment conditions such as three different drug therapies? Or, perhaps the researcher may wish to study two active treatments and a placebo. In this chapter you will encounter methods that can be employed for the analysis of data collected from two *or more* groups. The first method you will study is the one-way Analysis Of Variance (ANOVA) which can be viewed as an extension of the independent samples t test. The second is a form of chi-square that can be viewed as an extension of the Z test for a difference between proportions. After you have mastered these two statistics, we will broach the subject of Multiple Comparison Procedures (MCPs).

As we begin study of the statistics outlined above, you will notice that the ANOVA and chi-square methods are presented as hypothesis tests and that no complementary confidence interval or equivalence test is provided. This is a result of the fact that the logic underlying these methods does not endow them with such statistical forms. Although confidence intervals are commonly calculated in connection with MCPs, they are generally problematic without use of computers and will not be covered in this text. Equivalence tests for MCPs will not be presented because they are not commonly used in current research practice. In fact, you have completed your study of equivalence tests insofar as this text is concerned.

7.2 THE ONE-WAY ANALYSIS OF VARIANCE (ANOVA) F TEST

7.2.1 Hypotheses

You will recall that the null hypothesis tested by the independent samples t test is of the form $H_0 : \mu_1 = \mu_2$ which essentially states that treatments provided the two groups produce equivalent results insofar as means are concerned. The null hypothesis of the one-way ANOVA extends this concept to multiple groups and is stated as

$$H_0 : \mu_1 = \mu_2 = \cdots = \mu_k \tag{7.1}$$

which asserts that all population means are equal. The notation indicates that the equality extends to any number of groups with the last group characterized as group k. Notice that, in the case of two groups, 7.1 would reduce to $H_0 : \mu_1 = \mu_2$ which is the hypothesis tested by the independent samples t test.

The alternative hypothesis is any condition that renders the null hypothesis false. Thus, given three groups, any of the following alternative conditions, barring a Type II error, would cause rejection of the null hypothesis.

1. $\mu_1 = \mu_2 \neq \mu_3$

2. $\mu_1 \neq \mu_2 = \mu_3$

3. $\mu_1 = \mu_3 \neq \mu_2$

4. $\mu_1 \neq \mu_2 \neq \mu_3$

It is important to understand that, for $k = 3$, when 7.1 is rejected, there is no way to know which of the four conditions listed above caused the rejection. As you will see, it is this fact that gives rise to the MCPs to be presented later in this chapter. To put this into some perspective, suppose that a study is conducted to compare three surgical methods. The outcome variable is the amount of patient blood loss experienced during the procedure. The null hypothesis states that mean blood loss for the three surgical methods is the same. If the null hypothesis is rejected, the researcher will know that the null hypothesis is false. That is, mean blood loss for the three techniques is not the same. But why is the null hypothesis false? Is it because method one produces the same result as method two but these methods do not produce the same result as method three as is stated in alternative 1? Or is it because all three methods produce different mean blood loss as is asserted in alternative 4? The one-way ANOVA cannot answer this question. It can only declare the null hypothesis false or fail to declare it false.

7.2.2 Obtained F

Just as you used obtained and critical Z and t statistics to conduct certain tests of significance, so the one-way ANOVA uses obtained and critical F statistics for this purpose. Obtained F is given by the following equation.

$$F = \frac{MS_b}{MS_w} \tag{7.2}$$

As may be seen, obtained F is a ratio of two quantities—the mean square between symbolized by MS_b[1] and the mean square within symbolized by MS_w.[2] It will be instructive to examine each of these quantities in turn.

The Mean Square Within (MS_w).   The mean square within is also a ratio and is defined as follows

$$MS_w = \frac{SS_w}{N-k} \tag{7.3}$$

where SS_w is the sum of squares within, N is the total number of observations, and k is the number of groups. The quantity $N - k$ is termed the **denominator degrees of freedom**. For example, if there are three groups with five subjects in each, $N = 15$, $k = 3$ and the denominator degrees of freedom is $15 - 3 = 12$.

The **sum of squares within** is defined as

$$SS_w = SS_1 + SS_2 + \cdots + SS_k \tag{7.4}$$

where SS_1 is the sum of squares[3] of the first group, SS_2 is the sum of squares of the second group and SS_k is the sum of squares of the last group. You may recall from Chapter 2 that the sum of squares for a data set is computed by

$$SS = \sum (x - \bar{x})^2$$

or equivalently

$$SS = \sum x^2 - \frac{\left(\sum x\right)^2}{n}$$

Thus, Equation 7.4 can be written

$$SS_w = \left[\sum x_1^2 - \frac{\left(\sum x_1\right)^2}{n_1}\right] + \left[\sum x_2^2 - \frac{\left(\sum x_2\right)^2}{n_2}\right] + \cdots + \left[\sum x_k^2 - \frac{\left(\sum x_k\right)^2}{n_k}\right] \tag{7.5}$$

The subscripts $1, 2, \ldots, k$ refer to the group represented by the data. Thus, x_1 represents observations from the first group, x_2 the second group, and so on.

EXAMPLE 7.1

The (fictitious) data in Table 7.1 on the next page represent the weights of subjects who have been engaged in three different dieting regimens. Use these data to calculate MS_w.

Solution   It will be convenient to arrange the data for analysis as shown in Table 7.2 on the following page. In this table each observation is shown along with its square. The sums of

---

[1] The mean square between is also referred to as the mean square for treatments.

[2] The mean square within is also referred to as the mean square for error.

[3] You first encountered the sum of squares for a data set on page 36.

TABLE 7.1: Weights of subjects engaged in three different dieting regimens.

Diet One	Diet Two	Diet Three
198	214	174
211	200	176
240	259	213
189	194	201
178	188	158

TABLE 7.2: Weights of subjects engaged in three different dieting regimens arranged for analysis.

	Diet One		Diet Two		Diet Three
X_1	X_1^2	X_2	X_2^2	X_3	X_3^2
198	39204	214	45796	174	30276
211	44521	200	40000	176	30976
240	57600	259	67081	213	45369
189	35721	194	37636	201	40401
178	31684	188	35344	158	24964
$\sum$ 1016	208730	1055	225857	922	171986

the observations and their squares are shown at the bottom of the table. Using these results, the sums of squares for the individual groups are calculated as follows.

$$SS_1 = \sum x_1^2 - \frac{\left(\sum x_1\right)^2}{n_1} = 208730 - \frac{(1016)^2}{5} = 2278.8$$

$$SS_2 = \sum x_2^2 - \frac{\left(\sum x_2\right)^2}{n_2} = 225857 - \frac{(1055)^2}{5} = 3252.0$$

$$SS_3 = \sum x_3^2 - \frac{\left(\sum x_3\right)^2}{n_3} = 171986 - \frac{(922)^2}{5} = 1969.2$$

Notice that the sums of squares calculated in this manner are obtained from *within* the groups. This contrasts with the sum of squares between as you will see shortly.

By Equation 7.4

$$SS_w = SS_1 + SS_2 + SS_3 = 2278.8 + 3252.0 + 1969.2 = 7500.0$$

Then by Equation 7.3

$$MS_w = \frac{SS_w}{N - k} = \frac{7500.0}{15 - 3} = 625$$

The denominator degrees of freedom is $15 - 3 = 12$.   ∎

The Mean Square Between (MS_b).   As with the mean square within, the mean square between is a ratio of a sum of squares to a degrees of freedom. More precisely,

$$MS_b = \frac{SS_b}{k-1} \qquad (7.6)$$

where SS_b is the sum of squares between and k is the number of groups. The quantity $k - 1$ is termed the **numerator degrees of freedom**. For example, if there are three groups the numerator degrees of freedom is $3 - 1 = 2$. The **sum of squares between** is defined as

$$SS_b = n \left[\sum_{j=1}^{k} \bar{x}_j^2 - \frac{\left(\sum_{j=1}^{k} \bar{x}_j \right)^2}{k} \right] \qquad (7.7)$$

where n is the number of observations in *each* group, and $\bar{x}_j$ are the group means. This equation implies that an equal number of observations are associated with each group. We will shortly present a sum of squares equation that relaxes this restriction. Of particular interest in Equation 7.7 is the expression contained between the brackets []. For this calculation, the mean of each group is obtained with the sum of squares *between* group means then being calculated. The expression between the brackets is simply the familiar sum of squares formula with x being replaced by $\bar{x}$. Because there are k means, the familiar n in the equation is replaced by k. Notice also that we use the subscript j rather than the more familiar i because we have previously used the letter i to designate observations within samples.

EXAMPLE 7.2

Use the information in Table 7.2 to calculate MS_b.

Solution   Using the sums from Table 7.2, the means of the three groups are respectively,

$$\bar{x}_1 = \frac{\sum x_1}{n_1} = \frac{1016}{5} = 203.2$$

$$\bar{x}_2 = \frac{\sum x_2}{n_2} = \frac{1055}{5} = 211.0$$

$$\bar{x}_3 = \frac{\sum x_3}{n_3} = \frac{922}{5} = 184.4$$

The sum of the three means as well as the sum of the three squared means are as follows.

$\bar{x}$	$\bar{x}^2$
203.2	41290.24
211.0	44521.00
184.4	34003.36
$\sum$ 598.6	119814.60

By Equation 7.7

$$SS_b = n \left[\sum_{j=1}^{k} \bar{x}_j^2 - \frac{\left(\sum_{j=1}^{k} \bar{x}_j \right)^2}{k} \right]$$

$$= 5 \left[119814.60 - \frac{(598.6)^2}{3} \right]$$

$$= (5)(373.95)$$

$$= 1869.75$$

Then by Equation 7.6

$$MS_b = \frac{SS_b}{k-1} = \frac{1869.75}{3-1} = 934.88$$

The form of Equation 7.7 was chosen so as to emphasize that this sum of squares is obtained from the sum of squares calculated on the sample means so that this sum of squares is obtained *between* groups. The reason for this emphasis will become clear shortly. While pedagogically appealing, this form of the equation has the disadvantage of being applicable only to groups with equal numbers of observations. A more useful, albeit less intuitive, form of the sum of squares between equation is given by

$$SS_b = \frac{\left(\sum_{i=1}^{n_1} x_{i1} \right)^2}{n_1} + \frac{\left(\sum_{i=1}^{n_2} x_{i2} \right)^2}{n_2} + \cdots + \frac{\left(\sum_{i=1}^{n_k} x_{ik} \right)^2}{n_k} - \frac{\left(\sum_{All} x_{..} \right)^2}{N} \qquad (7.8)$$

The terms before the minus sign in Equation 7.8 indicate that the observations in each group are to be summed with the sum then being squared and the result then being divided by the number of observations in the group. This calculation is carried out for each group with the results then being summed. The term after the minus sign indicates that *all* observations are to be summed and the result squared. The division of this term is by N which represents the total number of observations—i.e., $n_1 + n_2 + \cdots + n_k$. Using the sums in Table 7.2 on page 266 with Equation 7.8 yields

$$SS_b = \frac{\left(\sum_{i=1}^{n_1} x_{i1} \right)^2}{n_1} + \frac{\left(\sum_{i=1}^{n_2} x_{i2} \right)^2}{n_2} + \frac{\left(\sum_{i=1}^{n_3} x_{i3} \right)^2}{n_3} - \frac{\left(\sum_{All} x_{..} \right)^2}{N}$$

$$= \frac{(1016)^2}{5} + \frac{(1055)^2}{5} + \frac{(922)^2}{5} - \frac{(2993)^2}{15}$$

$$= 599073 - 597203.267$$

$$= 1869.73$$

which, within rounding, is the same result obtained via Equation 7.7.   ∎

TABLE 7.3: One-way ANOVA Table.

Source of Variation	Sum of Squares	df	Mean Squares	F Ratio	Critical F	p-value
Between	SS_b	$k-1$	$SS_b/k-1$	MS_b/MS_w	(table)	(computer)
Within	SS_b	$N-k$	$SS_w/N-k$			
Total	SS_t	$N-1$				

7.2.3 The Test of Significance

By Equation 7.2 obtained F is

$$F = \frac{MS_b}{MS_w} = \frac{934.88}{625.00} = 1.50$$

Critical F is obtained by first noting that the numerator degrees of freedom for the analysis are $k - 1 = 3 - 1 = 2$ and the denominator degrees of freedom are $N - k = 15 - 3 = 12$. To use Appendix C, the numerator degrees of freedom are located across the top of the table and the denominator degrees of freedom down the side. For $\alpha = .05$ with 2 and 12 degrees of freedom, Appendix C shows that critical F is 3.89. The null hypothesis is rejected when obtained F is greater than, or equal to, critical F. In this case 1.50 is not greater than or equal to 3.89 so the null hypothesis is not rejected.

How should this failure to reject H_0 be interpreted? Does it mean that there are no differences between the weight loss systems insofar as weight loss is concerned? NO! It simply means that we could not show that there *are* differences. (Have you heard this theme before?)

7.2.4 The ANOVA Table

The results of an ANOVA analysis are traditionally summarized in an **ANOVA table**. Table 7.3 shows how such a table is constructed. The total sum of squares (SS_t) is the sum of squares that would be obtained if the sum of squares for all the data were computed without regard to group membership. It is also obtained as $SS_b + SS_w$. This means that the total sum of squares for the outcome variable can be partitioned into components that are related to the treatment effect (SS_b) and a random error component (SS_w). The p-value associated with the F statistic requires a computer for calculation. The ANOVA table for the dieting regimen analysis is shown in Table 7.4 on the next page.

EXAMPLE 7.3

In Example 6.1 on page 220 you used the data in Table 6.2 on page 221 to perform a two-tailed independent samples t test at $\alpha = .05$. Use this same data to perform a one-way ANOVA F test.

Solution   From Table 6.2, $\sum x_1 = 1977, \sum x_1^2 = 262767, \sum x_2 = 2028, \sum x_2^2 = 275706$

TABLE 7.4: One-way ANOVA Table for dieting regimen analysis.

Source of Variation	Sum of Squares	df	Mean Squares	F Ratio	Critical F	p-value
Between	1869.75	2	934.88	1.50	3.89	(computer)
Within	7500.00	12	625.00			
Total	9369.75	14				

and $n_1 = n_2 = 15$. Using these values

$$SS_1 = \sum x_1^2 - \frac{\left(\sum x_1\right)^2}{n_1} = 262767 - \frac{(1977)^2}{15} = 2198.4$$

and

$$SS_2 = \sum x_2^2 - \frac{\left(\sum x_2\right)^2}{n_2} = 275706 - \frac{(2028)^2}{15} = 1520.4.$$

Then by Equation 7.4

$$SS_w = SS_1 + SS_2 = 2198.4 + 1520.4 = 3718.8.$$

By Equation 7.3

$$MS_w = \frac{SS_w}{N - k} = \frac{3718.8}{30 - 2} = 132.814.$$

Using Equation 7.8

$$SS_b = \frac{\left(\sum_{i=1}^{n_1} x_{i1}\right)^2}{n_1} + \frac{\left(\sum_{i=1}^{n_2} x_{i2}\right)^2}{n_2} - \frac{\left(\sum_{All} x_{..}\right)^2}{N} = \frac{(1977)^2}{15} + \frac{(2028)^2}{15} - \frac{(4005)^2}{30} = 86.7.$$

Dividing this quantity by the numerator degrees of freedom $k - 1 = 2 - 1 = 1$ gives $MS_b = 86.7$. Obtained F is then

$$F = \frac{MS_b}{MS_w} = \frac{86.7}{132.814} = .653.$$

Table C gives critical F with 1 and 28 degrees of freedom as 4.20. Because obtained F is less than this value, the null hypothesis is not rejected.

As this example demonstrates, both the independent samples t test and the one-way ANOVA may be used for analysis when testing for the difference between two means. Note that when only two groups are involved the null hypothesis for the ANOVA test becomes $H_0 : \mu_1 = \mu_2$ which is the same hypothesis tested by the t test. In fact, when only two groups are involved, obtained F is equal to the square of obtained t with the same relationship being true for (two-tailed) critical values.[4] Because of this relationship, the two tests will always produce the same result regarding the fail to reject or reject decision of the null hypothesis when two groups are involved. The ANOVA test has the advantage of being applicable to more than two groups as well as to two groups. ∎

---

[4]Two-tailed in this context refers to the t test. With trivial exceptions, F tests are always one-tailed.

TABLE 7.5: Quality of care ratings of emergency care in four metropolitan hospitals.

Hospital One	Hospital Two	Hospital Three	Hospital Four
10	18	14	9
12	15	16	8
16	14	16	10
9	18	14	11
	12	17	6
	12	18	
	13	20	
	14		

EXAMPLE 7.4

As part of a quality control study, nurses employed in the emergency care departments of four hospitals located in a given metropolitan area are asked to anonymously rate the quality of care provided by their facility along a number of dimensions. Each dimension is rated on a summative scale that ranges from zero (worst) to 20 (best). The ratings for one such dimension are provided in Table 7.5. Use these data to perform a one-way ANOVA F test at $\alpha = .05$. Summarize your results in an ANOVA table.

Solution   It is convenient to arrange the data for analysis as shown in Table 7.6 on the next page. In this table each observation is shown along with its square. The sums of the observations and their squares are given at the bottom of the table. Using these results, the sums of squares for the individual groups are calculated as follows.

$$SS_1 = \sum x_1^2 - \frac{\left(\sum x_1\right)^2}{n_1} = 581 - \frac{(47)^2}{4} = 28.75$$

$$SS_2 = \sum x_2^2 - \frac{\left(\sum x_2\right)^2}{n_2} = 1722 - \frac{(116)^2}{8} = 40.00$$

$$SS_3 = \sum x_3^2 - \frac{\left(\sum x_3\right)^2}{n_3} = 1917 - \frac{(115)^2}{7} = 27.71$$

$$SS_4 = \sum x_4^2 - \frac{\left(\sum x_4\right)^2}{n_4} = 402 - \frac{(44)^2}{5} = 14.8$$

By Equation 7.4

$$SS_w = SS_1 + SS_2 + SS_3 + SS_4 = 28.75 + 40.00 + 27.71 + 14.80 = 111.26.$$

Then by Equation 7.3

$$MS_w = \frac{SS_w}{N-k} = \frac{111.26}{24-4} = 5.56.$$

TABLE 7.6: Quality of care ratings of emergency care in four metropolitan hospitals arranged for analysis.

Hospital One		Hospital Two		Hospital Three		Hospital Four	
X_1	X_1^2	X_2	X_2^2	X_3	X_3^2	X_4	X_4^2
10	100	18	324	14	196	9	81
12	144	15	225	16	256	8	64
16	256	14	196	16	256	10	100
9	81	18	324	14	196	11	121
		12	144	17	289	6	36
		12	144	18	324		
		13	169	20	400		
		14	196				
$\sum$ 47	581	116	1722	115	1917	44	402

By Equation 7.8

$$SS_b = \frac{\left(\sum_{i=1}^{n_1} x_{i1}\right)^2}{n_1} + \frac{\left(\sum_{i=1}^{n_2} x_{i2}\right)^2}{n_2} + \frac{\left(\sum_{i=1}^{n_3} x_{i3}\right)^2}{n_3} + \frac{\left(\sum_{i=1}^{n_4} x_{i4}\right)^2}{n_4} - \frac{\left(\sum_{All} x_{..}\right)^2}{N}$$

$$= \frac{(47)^2}{4} + \frac{(116)^2}{8} + \frac{(115)^2}{7} + \frac{(44)^2}{5} - \frac{(322)^2}{24}$$

$$= 190.57.$$

Dividing SS_b by the numerator degrees of freedom gives

$$MS_b = \frac{SS_b}{k-1} = \frac{190.57}{4-1} = 63.52.$$

Obtained F is then

$$F = \frac{MS_b}{MS_w} = \frac{63.52}{5.56} = 11.42.$$

Entering Table C with 3 and 20 degrees of freedom gives critical F as 3.10. Because obtained F is greater than critical F, the null hypothesis is rejected. We can conclude, therefore, that the data from the hospitals did not arise from a common population mean. These results are summarized in Table 7.7 on the facing page.   ■

7.2.5   Two Important Characteristics of MS_b and MS_w

As you now know, the null hypothesis is rejected when obtained F is equal to or exceeds critical F. This implies that the value of obtained F is somehow increased when the null hypothesis is false. An understanding of how this comes about will enhance your understanding of ANOVA and is the topic of this section.

TABLE 7.7: One-way ANOVA table for quality of care ratings of four hospital emergency departments.

Source of Variation	Sum of Squares	df	Mean Squares	F Ratio	Critical F	p-value
Between	190.57	3	63.52	11.42	3.10	(computer)
Within	111.26	20	5.56			
Total	301.83	23				

When the null hypothesis is true, MS_w and MS_b estimate the variance of the population(s) from which the samples were drawn. Since MS_w and MS_b are simply two different methods for estimating σ^2 when the null hypothesis is true, we would expect obtained F to take a value in the neighborhood of one in this circumstance. But what happens to obtained F when the null hypothesis is false? A demonstration will clarify the issue.

Let us reconsider the data in Table 7.2 on page 266. If we assume these subject weights were obtained under a true null hypothesis, we can create a facsimile of a false null hypothesis by a simple modification. For example, assume that diet three was more effective than the other two so that subjects following this diet lost 20 pounds more than would have otherwise been the case. The weights and squared weights of the group three dieters are given below. These weights were created by subtracting 20 pounds from the weight of each group three member as reported in Table 7.2.

X_3	X_3^2
154	23716
156	24336
193	37249
181	32761
138	19044
$\sum$ 822	137106

What impact will this dramatic slimming effect have on obtained F? Let us first see what happens to MS_w. As previously calculated,

$$SS_1 = 2278.8$$
$$SS_2 = 3252.0$$
$$SS_3 = 1969.2$$

Since we did not modify the weights of the subjects in groups one or two, we can use the sums of squares calculated previously for these dieters. Using the sums from the above table, we re-calculate SS_3 as

$$SS_3 = \sum x_3^2 - \frac{\left(\sum x_3\right)^2}{n_3} = 137106 - \frac{(822)^2}{5} = 1969.2$$

But this is the same sum of squares obtained without the dramatic weight loss. It follows that MS_w is unchanged by the effect experienced by group three. We now examine the impact (if any) on SS_b.

As previously calculated (page 267)

$$\bar{x}_1 = 203.2$$
$$\bar{x}_2 = 211.0$$
$$\bar{x}_3 = 184.4$$

Since we did not alter the observations for diets one or two, the means for these groups will remain as previously calculated. The new mean for group three will be

$$\bar{x}_3 = \frac{\sum x_3}{n_3} = \frac{822}{5} = 164.4.$$

The sum of the two original means and the new mean for diet three as well as the sum of their squared values are as follows.

$\bar{x}$	$\bar{x}^2$
203.2	41290.24
211.0	44521.00
164.4	27027.36
$\sum$ 578.6	112838.60

By Equation 7.7

$$SS_b = n \left[\sum_{j=1}^{k} \bar{x}_j^2 - \frac{\left(\sum_{j=1}^{k} \bar{x}_j \right)^2}{k} \right]$$

$$= 5 \left[112838.60 - \frac{(578.6)^2}{3} \right]$$

$$= (5)(1245.95)$$

$$= 6229.75$$

which is considerably larger than the value of 1869.75 obtained under the true null hypothesis (page 268).

By Equation 7.6

$$MS_b = \frac{SS_b}{k-1} = \frac{6229.75}{3-1} = 3114.875.$$

Obtained F is now

$$F = \frac{MS_b}{MS_w} = \frac{3114.875}{625.000} = 4.98.$$

H_0 : True

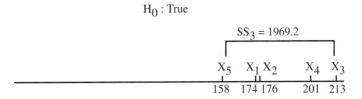

H_0 : False

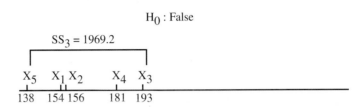

FIGURE 7.1: Depiction of SS_3 under true and false null hypotheses.

which is greater than critical F of 3.89 so that the null hypothesis of equal diet effects is rejected.

As previously stated, when the null hypothesis is true, both MS_w and MS_b estimate the population variance so that obtained F is typically near one. But, as you can see from the above demonstration, when the null hypothesis is false, MS_w is unchanged by this condition and continues as an estimate of the population variance. By contrast, MS_b inflates under a false null hypothesis thereby increasing the magnitude of obtained F.

The relationship between the two sums of squares and the condition (i.e., true or false) of the null hypothesis are shown in Figures 7.1 and 7.2 on the following page. You can see from Figure 7.1 that the weight loss experienced by the third group simply shifted their weights down the number line but did not alter the relative positions of the weights in the group. As a consequence, the sums of squares calculated from these values did not change.

By contrast, as may be seen in Figure 7.2 the weight loss in group three caused the mean of that group to be shifted away from the means of groups one and two. The effect is to spread the group means further apart and hence to increase the sum of squares obtained from these values. What do you think would happen to the F ratio if, in addition to the 20-pound weight loss experienced by group three, each member of group two had gained 20 pounds? Are you curious enough to check your answer by doing the calculations?[5]

A caveat must be expressed at this point. We modeled the treatment effect by subtracting a constant (i.e., 20 pounds) from the weight of each member of a selected group. When all members of a group respond to treatment in the same *additive*[6] manner which is different from the response in other groups as was the case here, the result is termed a **shift alternative**. Contrast this form of response to one in which some members of a group do

---

[5]Caution, if so, you may be in danger of becoming a biostatistician.

[6]That is, by adding or subtracting the same constant.

H_0 : True

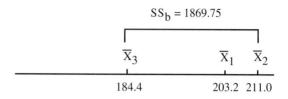

H_0 : False

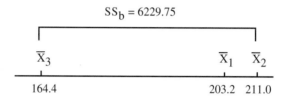

FIGURE 7.2: Depiction of SS_b under true and false null hypotheses.

not respond at all while others benefit greatly from the treatment. This form may cause an increase in the sum of squares of the group as well as a change in mean. This can have a deleterious effect on the power of the F test. Although the F test is designed primarily to detect shift alternatives, it has served well in a great variety of disciplines where other than shift alternatives are sometimes encountered.

7.2.6   Assumptions

The assumptions of the ANOVA F test are the same as those for the independent samples t test discussed in Section 6.2.4 on page 230.

7.3   THE 2 BY K CHI-SQUARE TEST

7.3.1   Hypotheses

In Chapter 6 on page 230 you learned to test the null hypothesis $H_0 : \pi_1 = \pi_2$ by means of an independent samples Z test. The 2 by k chi-square test extends this concept to test for equality of any number of proportions. This null hypothesis is stated as

$$\boxed{H_0 : \pi_1 = \pi_2 = \cdots = \pi_k} \tag{7.9}$$

which asserts that all population proportions are equal. The notation indicates that the equality extends to any number of groups with the last group characterized as group k. Notice that, in the case of two groups, 7.9 would reduce to $H_0 : \pi_1 = \pi_2$ which is the hypothesis tested by the independent samples Z test.

The alternative hypothesis is any condition that renders the null hypothesis false. Thus, given three groups, any of the following conditions, baring a Type II error, would cause rejection of the null hypothesis.

1. $\pi_1 = \pi_2 \neq \pi_3$

2. $\pi_1 \neq \pi_2 = \pi_3$

3. $\pi_1 = \pi_3 \neq \pi_2$

4. $\pi_1 \neq \pi_2 \neq \pi_3$

It is important to understand that when 7.9 is rejected, there is no way to know which of the four conditions listed above caused the rejection. As you will see, it is this fact that gives rise to the MCPs to be presented later in this chapter. To put this into some perspective, suppose that a study is conducted to compare three treatments for a terminal condition. The outcome variable is an indicator as to whether the patient is alive five years after treatment. The null hypothesis states that the proportions surviving five years are the same for the three treatments. If the null hypothesis is rejected, the researcher will know that the null hypothesis is false. That is, the three treatments do not produce the same proportions of five year survivors. But why is the null hypothesis false? Is it because treatment one produces the same result as treatment two but these treatments do not produce the same result as treatment three as is stated in alternative 1? Or is it because all three treatments produce different proportions of survivors as is asserted in alternative 4? The chi-square test cannot answer this question. It can only declare the null hypothesis false or fail to declare it false.

7.3.2   Obtained χ^2

As with other statistics with which you are now familiar, the hypothesis test is carried out by calculating an obtained value with a subsequent comparison to a critical value. For the chi-square test the obtained value is calculated by

$$\chi^2 = \sum_{\text{all cells}} \left[\frac{(f_o - f_e)^2}{f_e} \right] \tag{7.10}$$

where f_o and f_e are referred to respectively as the observed and expected frequencies. The **observed frequency** is simply the number of outcomes occurring in the given cell as shown in Table 7.8 on the next page. In this table we have used double subscripts to indicate the row and column of each cell entry. For example, if outcome one indicates "dead after five years" and outcome two represents "alive after five years," then f_{o11} would be the number of persons in group one who were dead at the five year mark and f_{o21} would be the number of persons in group one who were still alive at this point in time. The entries f_{o12} and f_{o22} would represent the same counts for group two. This table can be extended to represent data for any number of groups.

The **expected frequency** represents the expected number of persons to be found in each cell *if the null hypothesis is true*. This concept will require some explanation.[7] As expressed

---

[7]Don't be overly concerned if you don't understand all that follows. You can lead a *fairly* full and productive life without full mastery of this concept.

TABLE 7.8: Depiction of a 2 by 3 chi-square table.

	Group One	Group Two	Group Three
Outcome 1	f_{o11} f_{e11}	f_{o12} f_{e12}	f_{o13} f_{e13}
Outcome 2	f_{o21} f_{e21}	f_{o22} f_{e22}	f_{o23} f_{e23}

by Equation 3.5 on page 56, two events are independent if the product of their probabilities is equal to their joint probability. Expressed in terms of the problem at hand, we can say that if the treatment received is independent of whether or not the patient lives then

$$P\,(G_1\,D) = P\,(G_1)\,P\,(D)$$

where G_1 represents membership in group one and D indicates being dead. Thus, the above expression says that if group membership and outcome are independent, the probability of being dead and being a member of group one is simply equal to the product of the two individual probabilities. The same statement form can be applied to each cell of the table. The key concept here is recognition that a claim of independence between group membership and outcome is equivalent to a claim that the null hypothesis (see Equation 7.9 on page 276) is true. After all, if the particular treatment received has nothing to do with whether or not the patient dies, then the proportion living (or dying) in each group is the same.

To continue, if the null hypothesis is true, the probability of being in group one *and* being dead is equal to the probability of being in group one times the probability of being dead. The probability of being in group one is just the proportion of persons in group one or

$$\frac{N_{G_1}}{N}$$

where N_{G_1} is the total number of patients in group one and N is the total number of patients in the table. Likewise, the probability of being dead is simply the proportion of persons who are dead or

$$\frac{N_D}{N}$$

where N_D is the total number of persons who are dead. Thus, the probability of being dead and being in group one, if the null hypothesis is true, is simply

$$\left(\frac{N_{G_1}}{N}\right)\left(\frac{N_D}{N}\right)$$

Because this expression is a probability, we can think of it as a proportion—the proportion of the total number of patients who will be found in this cell *if the null hypothesis is true*.

TABLE 7.9: Data from treatment study arranged for chi-square analysis.

	Group One	Group Two	Group Three	
Dead	[17] (20.81)	[29] (25.03)	[11] (11.16)	57
Alive	[52] (48.19)	[54] (57.97)	[26] (25.84)	132
	69	83	37	$N = 189$

The *number* of persons we would expect to find in this cell would then be the proportion to be found times the total number of patients or

$$\left(\frac{N_{G_1}}{N}\right)\left(\frac{N_D}{N}\right)N = \frac{(N_{G_1})(N_D)}{N}$$

To make this applicable to any cell of the table we will write the expected frequency (f_e) for any given cell as

$$f_e = \frac{(N_R)(N_C)}{N} \tag{7.11}$$

where N_R is the row total for the cell whose expected frequency is being calculated and N_C is the column total for the same cell.

Once f_o and f_e are obtained for each cell, the quantity

$$\frac{(f_o - f_e)^2}{f_e}$$

is computed for each cell with the results then being summed to generate the obtained chi-square value. An example will be instructive.

EXAMPLE 7.5

Suppose that in the treatment of a terminal illness example discussed above, the following results are obtained. Of the patients receiving treatment one, 17 are dead at the end of five years while 52 are still alive. For treatment two, 29 are dead while 54 remain alive and for group three 11 are dead and 26 remain alive. Use these data to calculate obtained chi-square.

Solution   The observed (brackets) and expected (parentheses) frequencies as well as row and column totals are arranged as shown in Table 7.9. Expected frequencies were calculated via Equation 7.11 as follows.

$$f_{e11} = \frac{(N_D)(N_{G1})}{N} = \frac{(57)(69)}{189} = 20.81$$

$$f_{e12} = \frac{(N_D)(N_{G2})}{N} = \frac{(57)(83)}{189} = 25.03$$

$$f_{e13} = \frac{(N_D)(N_{G3})}{N} = \frac{(57)(37)}{189} = 11.16$$

$$f_{e21} = \frac{(N_A)(N_{G1})}{N} = \frac{(132)(69)}{189} = 48.19$$

$$f_{e22} = \frac{(N_A)(N_{G2})}{N} = \frac{(132)(83)}{189} = 57.97$$

$$f_{e23} = \frac{(N_A)(N_{G3})}{N} = \frac{(132)(37)}{189} = 25.84$$

Then by Equation 7.10 obtained chi-square is

$$\chi^2 = \sum_{\text{all cells}} \left[\frac{(f_o - f_e)^2}{f_e} \right]$$

$$= \frac{(17\text{-}20.81)^2}{20.81} + \frac{(29\text{-}25.03)^2}{25.03} + \frac{(11\text{-}11.16)^2}{11.16} + \frac{(52\text{-}48.19)^2}{48.19} + \frac{(54\text{-}57.97)^2}{57.97} + \frac{(26\text{-}25.84)^2}{25.84}$$

$$= .70 + .63 + .00 + .30 + .27 + .00$$

$$= 1.9$$

The critical value is obtained by entering Appendix D with $k-1$ degrees of freedom where k is the number of groups. In the present case, Appendix D shows that for $\alpha = .05$ and $3-1 = 2$ degrees of freedom, critical χ^2 is 5.991. The null hypothesis is rejected when obtained chi-square is greater than or equal to critical chi-square. Because 1.9 is less than 5.991, the null hypothesis is not rejected. We conclude, therefore, that a difference between population proportions cannot be demonstrated. In research terms, we conclude that we could not show a difference in the effectiveness of the three treatments. ∎

EXAMPLE 7.6

In Example 6.7 on page 233 you used a Z statistic to test $H_0 : \pi_1 = \pi_2$ using data from a (fictitious) study regarding nutritional instruction and low birth weight babies. In that study, 314 mothers received nutritional instruction of which 23 had low birth weight babies. By contrast, 39 of the 316 mothers in the non-instruction group had low birth weight babies. Use a chi-square test to test the above hypothesis. That is, test the hypothesis that the proportion of low birth weight babies born to the mothers receiving nutritional instruction is the same as the proportion born to mothers who did not receive instruction. Do you see a relationship between the obtained chi-square and the obtained Z of -2.12 you calculated previously?

Solution   Observed (in brackets) and expected (in parentheses) frequencies are shown in Table 7.10 on the facing page. The observed frequencies were obtained by noting that 23 of the 314 instructed mothers had low birth weight babies so that $314 - 23 = 291$ had babies that were not low birth weight. Likewise, since 39 of the 316 uninstructed mothers had low birth weight babies, $316 - 39 = 277$ of these mothers did not have low birth weight babies. Expected frequencies were calculated via Equation 7.11 as follows.

$$f_{e11} = \frac{(N_L)(N_I)}{N} = \frac{(62)(314)}{630} = 30.90$$

$$f_{e12} = \frac{(N_L)(N_{NI})}{N} = \frac{(62)(316)}{630} = 31.10$$

TABLE 7.10: Data from low birth weight study arranged for chi-square analysis.

	Instruction	No Instruction	
Low Birth Weight	[23] (30.90)	[39] (31.10)	62
Not Low Birth Weight	[291] (283.10)	[277] (284.90)	568
	314	316	N=630

$$f_{e21} = \frac{(N_{NL})(N_I)}{N} = \frac{(568)(314)}{630} = 283.10$$

$$f_{e22} = \frac{(N_{NL})(N_{NI})}{N} = \frac{(568)(316)}{630} = 284.90$$

The notation N_L, N_{NL}, N_I, and N_{NI} represents the number of low birth weight babies, the number of not low birth weight babies, the number of mothers receiving nutritional instruction and the number not receiving such instruction respectively.

Then by Equation 7.10 obtained chi-square is

$$\chi^2 = \sum_{\text{all cells}} \left[\frac{(f_o - f_e)^2}{f_e} \right]$$

$$= \frac{(23\text{-}30.90)^2}{30.90} + \frac{(39\text{-}31.10)^2}{31.10} + \frac{(291\text{-}283.10)^2}{283.10} + \frac{(277\text{-}284.90)^2}{284.90}$$

$$= 2.02 + 2.01 + .22 + .22$$

$$= 4.47$$

The critical value is obtained by entering Appendix D with $k - 1 = 2 - 1 = 1$ degree of freedom. For $\alpha = .05$ critical χ^2 is 3.841. The null hypothesis is rejected when obtained chi-square is greater than or equal to critical chi-square. Because 4.47 is greater than 3.841, the null hypothesis is rejected. We conclude, therefore, that a difference between population proportions does exist. In research terms, we conclude that there is a difference in the proportions of low birth weight babies produced by the mothers receiving nutritional information and those who did not receive this information.

Notice that obtained chi-square of 4.47 is the square (given rounding) of the obtained Z value of -2.12 calculated on page 233. Because the independent samples Z test for a difference between proportions tests the same null hypothesis as does the chi-square test with 1 degree of freedom, these tests produce the same result in terms of the reject or fail to reject decision. Just as the one-way ANOVA F test can be thought of as a generalization of the independent samples t test, so the 2 by k chi-square can be considered a generalization of the independent samples Z test for a difference between proportions.

As a final comment, recall that while f_e is the number of observations to be expected in a particular cell if the null hypothesis is true, f_o is the number actually found there. It is reasonable that the greater the difference between these two values, the greater the evidence against the null hypothesis. The numerator of the chi-square statistic (Equation 7.10) causes obtained chi-square to increase in size as a function of this difference. When this difference is large, obtained chi-square is large. If f_e and f_o were always the same value, obtained chi-square would be zero. Thus, rejection occurs when obtained chi-square is equal to or greater than the critical value.   ■

7.3.3   Assumptions

The 2 by k chi-square test presented here is an approximate rather than exact test.[8] In order for the approximation to be sufficiently accurate, a rule of thumb states that at least 80% of the f_e in the table should be greater than or equal to five and no cell should have f_e less than one.

It is also assumed that observations are independent. For example, one outcome can neither influence nor be influenced by another outcome. In the last example, this would mean that whether one mother has a low birth weight baby should be unrelated to whether some other mother has a low birth weight baby. A violation would likely occured if twins born to the same mother were entered into the analysis. The chi-square test cannot be counted upon to be robust against violation of the independence assumption.

7.4   MULTIPLE COMPARISON PROCEDURES

7.4.1   Introduction

You were first introduced to Type I errors on page 125. You learned there that such errors occur when a true null hypothesis is rejected and that the probability of such an event is termed α. In this section you will learn about a second form of Type I error. In order to differentiate the two we will refer to the now familiar form as **Per Comparison Error** or **PCE** and will designate the probability of a PCE as α_{PCE}. We will refer to the new form of Type I error as **Familywise Error** or **FWE** and will use the symbol α_{FWE} to represent the probability of its occurrence.

If you were to perform a single test at level α_{PCE}, the probability of rejecting a true null hypothesis would be α_{PCE}. But suppose you performed 100 such tests at level α_{PCE}. If all 100 null hypotheses were true, what would the probability be that you would reject one or more of these true null hypotheses? It stands to reason that if you performed each test at, for example, $\alpha_{PCE} = .05$, the chance would be very great that you would reject *at least* one true null hypothesis. To make the example even more extreme, what would you think the chance would be of rejecting one or more true null hypotheses if you did 1,000,000 such tests? Certainly, this probability would be very near one.

When at least one[9] true null hypothesis is rejected in a series (or "family") of tests, it is said that a **Familywise Error** has occurred. FWE becomes a concern in at least two contexts. The first of these we will call "multiple comparison analysis" and the second "multiple endpoint analysis."[10]

---

[8]An exact test is possible but requires special computer software and is, therefore, not discussed here.

[9]"At least one" means the same thing as "one or more."

[10]Many authors do not distinguish between the two forms and simply lump them together as "multiple

Multiple comparison analysis refers to the situation where multiple groups are being compared on a single outcome variable. For example, Table 7.1 on page 266 gives the weights of subjects participating in three different dieting strategies. As you now know, a one-way ANOVA F test only deals with the question of whether all population means are equal. But a researcher is likely to have more specific questions. For example, is there a difference in the effectiveness of diets one and two? Is there a difference in the effectiveness of diets one and three? In general, there are $\frac{k(k-1)}{2}$ pairs of comparisons that might be made.[11] In the case of three groups this would mean that $\frac{(3)(3-1)}{2} = 3$ pairwise comparisons can be made. The null hypotheses for these comparisons are

1. $\mu_1 = \mu_2$

2. $\mu_1 = \mu_3$

3. $\mu_2 = \mu_3$

Clearly, we could address the questions implied by these hypotheses by conducting three independent samples t tests. If there were four groups we might wish to test all $\frac{(4)(3)}{2} = 6$ pairwise hypotheses. But what are the implications for FWE when we conduct a series of tests of this sort? As we conduct more tests the probability that we will obtain *some* significant results increases. If we conduct six such tests and find that one is significant we do not know if this significant result came about because the specific null hypothesis addressed by the test is false or if we simply conducted so many tests that we were bound to get a significant result sooner or later. This is another way of saying that the result may be attributable to an increase in α_{FWE}.

Multiple endpoint analysis refers to the situation where two groups are being compared on multiple outcome measures. For example, if we wished to compare two groups of patients on five blood chemistries, we would likely conduct five t tests—one for each blood chemistry. Because you are conducting multiple tests you will once again be confronted by the problem of FWE.

The relationship between α_{PCE} and α_{FWE} is important and is illustrated in Table 7.11 on the following page. The entries in this table were generated via computer simulation methods whereby random samples from normally distributed populations were analyzed by means of independent samples t tests. The first column shows the level of significance at which each test was conducted while the second shows the number of samples (i.e., groups) involved in the analyses. The third column shows the number of tests conducted, the fourth the familywise error rate that the multiple tests generated and the last shows the **per family error rate** which we will explain below.

To further elucidate, the first column of the first row shows that the first series of tests were conducted at pre-specified level $\alpha_{PCE} = .05$. The second column shows that three groups were involved which, as the third column shows, would produce $\frac{(3)(2)}{2} = 3$ pairwise tests. The fourth column shows the three t tests had an associated familywise error rate of .122. This means that the probability of rejecting at least one of the three true null

comparisons."

---

[11]Often, researchers are interested in only a subset of these comparisons so that not all are carried out. There are power advantages to be gained by omitting comparisons that are of no interest.

TABLE 7.11: Per comparison and familywise error rates for specified numbers of comparisons.

α_{PCE}	Number of Groups	Number of Comparisons	α_{FWE}	PFE
.05	3	3	.122	0.150
	5	10	.286	0.499
	10	45	.630	2.249
	20	190	.920	9.508
.01	3	3	.027	0.030
	5	10	.075	0.100
	10	45	.231	0.451
	20	190	.528	1.898

hypotheses was .122. The last column shows the **per family error rate** which is defined as the average number of rejections occurring in the set of comparisons. Notice that PFE is not a probability. This is evident by observing that the average number of erroneous rejections is sometimes greater than one.

Three important points should be gleaned from this table. First, α_{FWE} increases as the number of tests conducted increases. Indeed, when 10 tests were conducted at $\alpha_{PCE} = .05$, the probability of obtaining a significant result was greater than one in four as compared to .122 when three tests were conducted. The second notable point is that α_{FWE} can be dramatically decreased by lowering α_{PCE}. For the case where 10 tests were conducted, the familywise error rate of .286 obtained when the per comparison level was .05 was lowered to .075 when α_{PCE} was lowered to .01. Third, α_{FWE} is always larger than α_{PCE}.

7.4.2   Controlling Familywise Errors

When you reject a single null hypothesis the interpretation is clear. You have an α_{PCE} probability that you did so incorrectly. Because this probability is small, you can be confident that the null hypothesis is false. When you perform a *series* of tests and reject one or more null hypotheses, the interpretation is not so clear. Did you reject these hypotheses because they are false or because the familywise Type I error rate is so high that rejections were highly likely even in the face of true null hypotheses? You were confident in your result for the single test because you were able to control the probability of a false rejection at α_{PCE}. You could gain this same confidence in your results for multiple tests if you could control α_{FWE} to some specified level—.05 for example.

Many methods for control of familywise error have been developed. Some of these are useful while others appear to be flawed and, therefore, not so useful. A few of these tests are designed for use in specific research contexts while others are designed for more general use. We will not attempt to discuss or even list these techniques here, but rather will explain the use of three such methods. We do not recommend use of the first of these, known as the Bonferroni method, for reasons to be outlined below. We include it here because of its simplicity and hence the insight it provides into the multiple testing problem. The step-down

TABLE 7.12: Effects of Bonferroni adjustments on familywise error rates for specified numbers of comparisons.

α_{PCE}	Number of Groups	Number of Comparisons	α_{FWE}	PFE
.0167	3	3	.044	0.050
.0050	5	10	.040	0.050
.0011	10	45	.037	0.050
.0003	20	190	.036	0.050

Bonferroni is an improved version of the older method and is generally superior to that form of the test. While the Bonferroni and step-down Bonferroni techniques are applicable to a variety of multiple endpoint and multiple comparison testing situations, Tukey's HSD test is restricted to multiple comparison analyses but is very useful in that setting.

The Bonferroni Method of Controlling Familywise Errors.   As shown in Table 7.11, α_{FWE} can be reduced by reducing α_{PCE}. But suppose you wish to establish α_{FWE} at some specified value—for example .05. How low must you set α_{PCE} in order to have α_{FWE} be .05? One of the oldest, simplest, and most widely used methods for finding this level is known as the **Bonferroni**[12] **adjustment**. By this adjustment

$$\alpha_{PCE} = \frac{\alpha_{FWE}}{NT} \qquad (7.12)$$

where N_T represents the number of tests to be performed. Thus, for example, if we wish to control α_{FWE} at .05 while we perform three tests, each test would be carried out at the $\frac{.05}{3} = .017$ level of significance. For 10 tests the appropriate level would be $\frac{.05}{10} = .005$. As you can see, as the number of tests increases we must lower α_{PCE} in order to maintain the desired Familywise Error level.

Table 7.12 shows the results of applying a Bonferroni correction on α_{FWE}. As this table demonstrates, the Bonferroni technique does not establish α_{FWE} at the specified level, but rather guarantees that α_{FWE} will not rise above the specified level. Thus, the entries in Table 7.12 for α_{FWE} are always below the specified level of .05 but are not equal to .05. We will now present a technique that, though not as popular as the Bonferroni adjustment, is generally superior.

The Step-Down Bonferroni Method of Controlling Familywise Errors.   In 1979, Holm [24] proposed a modification to the Bonferroni procedure that is usually more powerful than, is never less powerful than, and maintains familywise error at the same level as, the classical procedure. This modified Bonferroni, or more properly, step-down Bonferroni procedure is illustrated in Figure 7.3 on the following page and is carried out as follows.

1. The multiple test statistics are calculated.

2. The p-value for each statistic calculated in 1 is obtained.

---

[12]Named for Carlo Emilio Bonferroni, 1892–1960.

	Step One	Step Two	Step Three	...	Step NT
P-value	P (1)	P (2)	P (3)	...	P (NT)
	↓	↓	↓		↓
Step-down	$\dfrac{FWE}{NT}$	$\dfrac{FWE}{NT\text{-}1}$	$\dfrac{FWE}{NT\text{-}2}$	...	$\dfrac{FWE}{1}$
Classical	$\dfrac{FWE}{NT}$	$\dfrac{FWE}{NT}$	$\dfrac{FWE}{NT}$	...	$\dfrac{FWE}{NT}$

FIGURE 7.3: An illustration of the step-down Bonferroni multiple comparison procedure.

3. The p-values are ordered from smallest to largest with the smallest being designated $p_{(1)}$, the second smallest $p_{(2)}$ and so forth with the largest being $p_{(NT)}$ where NT is the number of tests.

4. At the first step, $p_{(1)}$ is compared to $\frac{\alpha_{FWE}}{NT}$. If $p_{(1)} \leq \frac{\alpha_{FWE}}{NT}$, the test is declared significant and the second step is carried out. If $p_{(1)} > \frac{\alpha_{FWE}}{NT}$, the test is declared nonsignificant and testing ceases with all remaining comparisons being declared not significant.

5. If the first step is significant, step two is carried out by comparing $p_{(2)}$ with $\frac{\alpha_{FWE}}{NT-1}$. If $p_{(2)} \leq \frac{\alpha_{FWE}}{NT-1}$, the result is declared significant and testing continues to the next step. Otherwise, the test is declared nonsignificant and testing ceases with all remaining tests being declared nonsignificant.

6. The steps are continued as shown in Figure 7.3 until a nonsignificant result is obtained or until the last step is completed.

It is important that testing cease when the first nonsignificant result is obtained. If this "stopping rule" is not adhered to, α_{FWE} will not be constrained to the desired level.

Notice that at the first step, the test is identical to the classical Bonferroni adjustment as defined in Equation 7.12. Subsequent steps, if they are carried out, are less stringent than the classical testing procedure. So, for example, if five tests are to be conducted with α_{FWE} being constrained to .05, α_{PCE} for the first step is $\frac{.05}{5} = .010$, for the second step is $\frac{.05}{4} = .0125$ and for subsequent steps are $\frac{.05}{3} = .0167$, $\frac{.05}{2} = .025$ and for the last step is $\frac{.05}{1} = .05$. By contrast, the classical method employs the most stringent level of .01 for all tests.

EXAMPLE 7.7

Suppose a clinical trial is conducted to compare results of cornea transplant surgery in which donor corneas are obtained from older donors (≥ 60 years of age) and younger donors (< 60 years of age). To this end patients are randomly assigned to one of two groups with the first group receiving corneas from older donors and the second from younger donors. Four outcomes are of interest to the researchers: (A) whether the patient experiences rejection of the cornea within five years after surgery, (B) whether the cornea becomes opaque within five years after surgery, (C) whether further medical intervention is required within one year after surgery, and (D) whether vision is measured as 20/30 or better within 30 days of surgery.

The results are as follows. Of the 60 patients in each of the two groups, .12 of the patients in group one and .10 of the patients in group two experienced rejection, .21 of the patients in group one and .03 of the patients in group two developed opaque corneas, .06 of the patients in group one and .07 in group two required further medical treatment and .68 of the patients in group one and .89 of the patients in group two achieved vision of 20/30 or better.

Perform multiple endpoint analyses on these outcomes while assuring that familywise error does not exceed .05. What form of statistical test did you choose for the analyses? Why?

Solution   Because the outcomes are dichotomous and two groups are being compared, we will employ the independent samples Z test for a difference between proportions as described in Section 6.3.1 on page 230. The obtained Z statistics (with appropriate subscripts to identify the outcome on which the test is being conducted) are calculated via Equation 6.5 on page 232 as follows.

$$Z_A = \frac{\hat{p}_1 - \hat{p}_2 - \delta_0}{\sqrt{\frac{\hat{p}_1 \hat{q}_1}{n_1} + \frac{\hat{p}_2 \hat{q}_2}{n_2}}} = \frac{.12 - .10}{\sqrt{\frac{(.12)(.88)}{60} + \frac{(.10)(.90)}{60}}} = .35$$

$$Z_B = \frac{.21 - .03}{\sqrt{\frac{(.21)(.79)}{60} + \frac{(.03)(.97)}{60}}} = 3.16$$

$$Z_C = \frac{.06 - .07}{\sqrt{\frac{(.06)(.94)}{60} + \frac{(.07)(.93)}{60}}} = -.22$$

$$Z_D = \frac{.68 - .89}{\sqrt{\frac{(.68)(.32)}{60} + \frac{(.89)(.11)}{60}}} = -2.90$$

By the method outlined on page 96 for finding the two-tailed p-value for a Z test, the p-values associated with the above statistics are as follows.

$$p_A = .7264$$
$$p_B = .0016$$
$$p_C = .8258$$
$$p_D = .0038$$

The four p values, along with a designation of the test from which each was derived, are listed in ascending order below. Also shown are the step-down (S-D) values of α_{PCE} for each test of significance. As can be seen here, the test for B is significant (S) because the p-value 0f .0016 is less than the α_{PCE} of .0125 which was calculated by dividing .05 by 4. Likewise, the test for D is significant because the p-value 0f .0038 is less than the α_{PCE} of .0167 which was calculated by dividing .05 by 3. The test for A is not significant (NS) because .7264 is greater than .0250. It is important to understand that C is automatically declared nonsignificant at this point due to the stopping rule. The researcher conducting these tests can be assured that α_{FWE} is not greater than .05.

Test	B	D	A	C
p-value	.0016	.0038	.7264	.8258
S-D α_{PCE}	.0125	.0167	.0250	.0500
	S	S	NS	NS

∎

EXAMPLE 7.8

A researcher involved in a study employing multiple groups of subjects wishes to test a series of null hypotheses by means of independent samples t tests. The null hypotheses with accompanying p-values associated with each test are given below. Use these results to perform a step-down Bonferroni procedure with α_{FWE} not to exceed .05. How do these results compare to results that would be obtained from classical Bonferroni tests?

H_0 :	p-value
$\mu_1 = \mu_3$	.0111
$\mu_2 = \mu_4$	.0419
$\mu_2 = \mu_5$	.0090
$\mu_3 = \mu_4$	.0200
$\mu_4 = \mu_5$	.0181

Solution   The five p values, along with the hypothesis test from which each was derived, are listed in ascending order below. Also shown are the step-down values of α_{PCE} (S-D) and the classical Bonferroni values of α_{PCE} (CB) for each test of significance.

As can be seen here, the test of $\mu_2 - \mu_5$ is significant because the p-value 0f .0090 is less than the α_{PCE} of .0100 which was calculated by dividing .05 by 5. Likewise, the test of $\mu_1 - \mu_3$ is significant because the p-value of .0111 is less than the α_{PCE} of .0125 which was calculated by dividing .05 by 4. The test of $\mu_4 - \mu_5$ is not significant because .0181 is greater than .0167. It is important to understand that the tests of $\mu_3 - \mu_4$ and $\mu_2 - \mu_4$ are automatically declared nonsignificant at this point due to the stopping rule. These last two tests are nonsignificant in spite of the fact that their p-values are less than their associated value of α_{FWE}. You may be tempted to declare these tests significant but you should bear in mind that violation of the stopping rule invalidates the procedure. The researcher conducting these tests can be assured that α_{FWE} is not greater than .05.

Notice that had the researcher employed the classical Bonferroni method, which unfortunately is still common practice, only $\mu_2 - \mu_5$ would have been significant.

Test	$\mu_2 - \mu_5$	$\mu_1 - \mu_3$	$\mu_4 - \mu_5$	$\mu_3 - \mu_4$	$\mu_2 - \mu_4$
p-value	.0090	.0111	.0181	.0200	.0419
S-D α_{PCE}	.0100	.0125	.0167	.0250	.0500
CB α_{PCE}	.0100	.0100	.0100	.0100	.0100
	S	S	NS	NS	NS

∎

Tukey's HSD Method of Controlling Familywise Errors.   As stated previously, the Bonferroni and step-down Bonferroni methods of controlling Familywise error are broad in application to include both multiple comparison and multiple endpoint analysis. By contrast, Tukey's HSD (Honestly Significant Difference) test [47] is designed for use in multiple comparison settings where all pairwise comparisons of group means are to be carried out. These tests are conducted by computing the test statistic, commonly symbolized as q, for each of the $\frac{k(k-1)}{2}$ comparisons with the resultant q statistics then being referenced to an appropriate table of critical values. The test statistic is defined as follows.

$$q_{ij} = \frac{\bar{x}_i - \bar{x}_j}{\sqrt{\frac{MS_w}{n_h}}} \tag{7.13}$$

The subscripts i and j denote the two groups being compared so that $\bar{x}_i$ and $\bar{x}_j$ are the means of groups i and j respectively. MS_w is the mean square within as computed for a one-way ANOVA via Equations 7.3 on page 265 and 7.4 on page 265. The symbol n_h represents the harmonic mean of the two sample sizes and is computed as

$$n_h = \frac{2}{\frac{1}{n_i} + \frac{1}{n_j}}$$

When $n_i = n_j$, $n_h = n$ which is the sample size of either group.

EXAMPLE 7.9

Use the data from the dieting study depicted in Table 7.1 on page 266 to perform Tukey's HSD test. Begin by stating the null hypotheses to be tested, then perform the tests and finally, state you conclusions. Maintain α_{FWE} at .05.

Solution   Because there are three groups and we wish to make all pairwise comparisons, we will have $\frac{3(2)}{2} = 3$ hypotheses to test. They are

$$H_0 : \mu_1 = \mu_2$$
$$H_0 : \mu_1 = \mu_3$$
$$H_0 : \mu_2 = \mu_3$$

Previous calculations (see page 267) obtained when performing a one-way ANOVA on these data provide the following.

$$\bar{x}_1 = 203.2$$
$$\bar{x}_2 = 211.0$$
$$\bar{x}_3 = 184.4$$

From page 266 we obtain

$$MS_w = 625$$

Because sample sizes are the same for all groups, n_h will be

$$n_h = \frac{2}{\frac{1}{n_i} + \frac{1}{n_j}} = \frac{2}{\frac{1}{5} + \frac{1}{5}} = 5$$

for all comparisons. The test statistics for the three comparisons are by Equation 7.13

$$q_{12} = \frac{\bar{x}_1 - \bar{x}_2}{\sqrt{\frac{MS_w}{n_h}}} = \frac{203.2 - 211.0}{\sqrt{\frac{625}{5}}} = -.698$$

$$q_{13} = \frac{\bar{x}_1 - \bar{x}_3}{\sqrt{\frac{MS_w}{n_h}}} = \frac{203.2 - 184.4}{\sqrt{\frac{625}{5}}} = 1.682$$

$$q_{23} = \frac{\bar{x}_2 - \bar{x}_3}{\sqrt{\frac{MS_w}{n_h}}} = \frac{211.0 - 184.4}{\sqrt{\frac{625}{5}}} = 2.379$$

Critical values of q [13] are obtained from Appendix E. The table is entered with the number of means in the analysis and the appropriate degrees of freedom. The degrees of freedom for Tukey's test are the same as the denominator degrees of freedom for the one-way ANOVA, namely, $N - k$. Because there are a total of 15 subjects and three groups, the degrees of freedom for the present tests are $15 - 3 = 12$. Referencing Appendix E for 3 means and 12 degrees of freedom yields a critical value of 3.773. The decision to reject or fail to reject the null hypothesis is made in the same manner as for a t test. That is, for a two-tailed test, the null hypothesis will be rejected if obtained q is greater than or equal to the table value or is less than or equal to the negative of the table value.

As may be seen, none of the hypotheses are rejected so that no differences between group means can be demonstrated. This result is not surprising since the one-way ANOVA conducted on these data was not significant. ∎

EXAMPLE 7.10

Use the quality of care ratings in Table 7.5 on page 271 to perform Tukey's HSD test. Maintain familywise error at .05. How many comparisons will have to be made? What are your conclusions?

Solution   Using the sums from Table 7.6 on page 272 we calculate

$$\bar{x}_1 = \frac{47}{4} = 11.75$$

$$\bar{x}_2 = \frac{116}{8} = 14.50$$

$$\bar{x}_3 = \frac{115}{7} = 16.43$$

$$\bar{x}_4 = \frac{44}{5} = 8.80$$

---

[13]The distribution of q is known as the **studentized range distribution.**

From page 271 we obtain

$$MS_w = 5.56$$

Because sample sizes are not equal,[14] n_h will differ for the six comparisons. For q_{12}

$$n_h = \frac{2}{\frac{1}{n_i} + \frac{1}{n_j}} = \frac{2}{\frac{1}{4} + \frac{1}{8}} = 5.33$$

We leave the remaining calculations of n_h to you.

The $\frac{4(3)}{2} = 6$ comparisons are then

$$q_{12} = \frac{\bar{x}_1 - \bar{x}_2}{\sqrt{\frac{MS_w}{n_h}}} = \frac{11.75 - 14.50}{\sqrt{\frac{5.56}{5.33}}} = -2.693$$

$$q_{13} = \frac{\bar{x}_1 - \bar{x}_3}{\sqrt{\frac{MS_w}{n_h}}} = \frac{11.75 - 16.43}{\sqrt{\frac{5.56}{5.09}}} = -4.478$$

$$q_{14} = \frac{\bar{x}_1 - \bar{x}_4}{\sqrt{\frac{MS_w}{n_h}}} = \frac{11.75 - 8.80}{\sqrt{\frac{5.56}{4.44}}} = 2.636$$

$$q_{23} = \frac{\bar{x}_2 - \bar{x}_3}{\sqrt{\frac{MS_w}{n_h}}} = \frac{14.50 - 16.43}{\sqrt{\frac{5.56}{7.47}}} = -2.237$$

$$q_{24} = \frac{\bar{x}_2 - \bar{x}_4}{\sqrt{\frac{MS_w}{n_h}}} = \frac{14.50 - 8.80}{\sqrt{\frac{5.56}{6.15}}} = 5.995$$

$$q_{34} = \frac{\bar{x}_3 - \bar{x}_4}{\sqrt{\frac{MS_w}{n_h}}} = \frac{16.43 - 8.80}{\sqrt{\frac{5.56}{5.83}}} = 7.813$$

Entering Appendix E with four means and $N - k = 24 - 4 = 20$ degrees of freedom we find critical q of 3.958. Because $q_{13} = -4.478$ is less than -3.958 and $q_{24} = 5.995$ and $q_{34} = 7.813$ are greater than 3.958, we declare these comparisons significant. We conclude, therefore, that quality of care ratings in hospital one are significantly lower than those in hospital three, and that ratings in hospitals two and three are greater than those in hospital four. We can be confident in the validity of these conclusions because familywise error has been maintained at .05. ∎

7.4.3   Further Comments Regarding Multiple Comparison Procedures

In statistics, it is often true that in order to gain some advantage, you must pay some price. This is true of MCPs. In order to gain control of α_{FWE}, you must reduce the level of significance at which each test is conducted. As you know from previous studies, reducing α

---

[14]Tukey's HSD test was originally devised for equal sample sizes so that n was used in place of n_h in Equation 7.13. When sample sizes are not equal i.e., $n_h \neq n$, the test is sometimes referred to as the Tukey-Kramer test.

also reduces power. As a result, it sometimes happens that a one-way ANOVA test or 2 by k chi-square test demonstrates that not all parameter values are equal but all subsequent multiple comparisons are nonsignificant. Among other reasons, this may result from a lack of power at the individual comparison level.

You should also know that the use of MCPs is a controversial subject among statisticians. For philosophical reasons, some statisticians believe that such methods should not be used. In making your own decision on this issue in a particular research context, we suggest that you consider the following question. Which concerns me more, (a) rejecting one or more true null hypotheses in this group of comparisons or, (b) failing to reject one or more false null hypotheses in this set of tests. If (a) is of prime concern, then you should probably make the adjustment necessary to control α_{FWE}. If (b) is more important, then you may consider not making such adjustments. You may also choose a compromise position in that you may decide to control α_{FWE} at the .10 rather than the more traditional .05 level. While this elevates the probability of a FWE somewhat, it also allows for less stringent adjustments to α_{PCE} thereby not reducing power as much as would be the case with use of .05.

Finally, as noted previously, there are many methods for controlling FWE. We have demonstrated only three of them here and have discouraged use of one of these. You would do well in your own research efforts to read more about such methods. A particularly lucid treatment of the subject is provided by Kirk [27] while a more technical discussion in found in Hochberg and Tamhane [22].

KEY WORDS AND PHRASES

After reading this chapter you should be able to demonstrate familiarity with the following words and phrases.

2 by k chi-square 276

additive treatment effect 275

Bonferroni adjustment 285

denominator degrees of freedom 265

expected frequency (f_e) 277

familywise error 282

$H_0 : \mu_1 = \mu_2 = \cdots = \mu_k$ 264

$H_0 : \pi_1 = \pi_2 = \cdots = \pi_k$ 276

mean square between (MS_b) 267

mean square within (MS_w) 265

multiple comparison analysis 283

multiple comparison procedures (MCP) 282

multiple endpoint analysis 283

numerator degrees of freedom 267

observed frequency (f_o) 277

one-way Analysis of Variance (ANOVA) 264

per comparison error 282

shift alternative 275

step-down Bonferroni 285

sum of square between (SS_b) 267

sum of square within (SS_w) 265

Tukey's HSD 289

EXERCISES

7.1 Researchers interested in stress as it relates to work assignments in industrial settings, conduct a study in which pulse rates of three groups of workers are compared. The first group consists of office staff who perform routine clerical duties, the second work with hazardous materials while the third performs tasks similar to those performed by the second group but do not come in contact with hazardous materials. Pulse rates are taken at the midpoint of the one hour lunch period for all three groups. The pulse rates thus obtained are provided here.

Office Workers	Hazardous Materials	Non- Hazardous Materials
58	88	65
64	59	70
71	74	79
66	80	66
79	81	74
74	69	79
70	90	60

(a) Calculate SS_w and MS_w.
(b) Calculate SS_b and MS_b.
(c) Test the hypothesis $\mu_1 = \mu_2 = \mu_3$ at the .05 level of significance. What is your conclusion regarding the question at hand?

7.2 Inmates in juvenile correctional facilities are routinely encouraged to voluntarily undergo HIV testing. In order to increase the rate of participation in the testing program, authorities have developed three different instructional programs that emphasize respectively, knowledge of the virus, obligation to potential partners and importance of early treatment. In order to determine whether the three approaches are differentially effective at encouraging participation in the testing program, each is implemented at a different juvenile facility. No instruction is provided at a fourth facility which acts as a control.

The results are as follows. Facility one, 46 participate, 51 do not, facility two 52 participate, 44 do not, facility three 36 participate, 37 do not, facility four (control) 24 participate, 64 do not. Use these data to test the hypothesis $H_0 : \pi_1 = \pi_2 = \pi_3 = \pi_4$ at the .05 level of significance. Does this analysis really answer the question posed by the researchers? What research question is addressed by this analysis?

7.3 Perform an analysis of the data in Exercise 7.2 that addresses the question of interest to the researchers.

7.4 Use the data in Exercise 7.2 to test the following hypotheses while not allowing FWE to exceed .05. Use two-tailed tests.

(A) $H_0 : \pi_1 = \pi_4$
(B) $H_0 : \pi_2 = \pi_4$
(C) $H_0 : \pi_3 = \pi_4$

7.5 The following quotation regarding familywise error is taken from a popular biostatistics text. Do you agree with this statement? Provide the rationale for your answer.

> To perform this **multiple comparison test**, we select an overall significance level, which denotes *the probability that one or more of the null hypotheses is false.*[Italics added]

7.6 Use the data in Exercise 7.1 to perform Tukey's HSD test. Maintain familywise error at .05. What are your conclusions? Given the result of the ANOVA obtained in Exercise 7.1 are you surprised by this result? Explain your answer.

B. The following questions refer to Case Study B (page 470)

7.7 What do you believe is the probability that an obtained F from a one-way ANOVA conducted on the baseline WOMAC A data for the three groups would be significant? Explain your answer.

7.8 Suppose that a one-way ANOVA conducted on the baseline WOMAC A data were to be significant. What would you conclude from this result?

7.9 Use the 12 week WOMAC A data to test the hypothesis $H_0 : \mu_s = \mu_w = \mu_p$ at the .05 level where the subscripts represent the Standard, Weak, and Placebo groups respectively. Interpret the result.

7.10 Conduct Tukey's HSD test on the 12 week WOMAC A data. Maintain familywise error at .05.

7.11 Use the data in Table J.3 to test the hypothesis $H_0 : \pi_s = \pi_w = \pi_p$ at the .05 level of significance where π represents the proportion of subjects who correctly identify their bracelet as Real or Dummy. The correct answer for the Weak group is Dummy. Interpret the result.

7.12 Test the hypotheses

$$H_0 : \pi_s = \pi_w$$
$$H_0 : \pi_s = \pi_p$$
$$H_0 : \pi_w = \pi_p$$

while not permitting familywise error to rise above .05. Interpret the results.

C. The following questions refer to Case Study C (page 471)

7.13 Test the hypothesis $H_0 : RR = 1$ for each of the time periods. Don't let familywise error rise above .05 for these tests.

D. The following questions refer to Case Study D (page 471)

7.14 Test the null hypothesis $H_0 : \mu_1 = \mu_2 = \mu_3$ where μ_1 is the mean NPZ-8 score for HIV-infected subjects who have positive ADC Stage assessments, μ_2 is the mean for HIV-infected subjects who have negative ADC Stage assessments, and μ_3 is the mean for subjects who are negative for HIV. Conduct the test at $\alpha = .05$. *If* this test is significant, conduct Tukey's HSD test with familywise error set to .05.

G. The following questions refer to Case Study G (page 473)

7.15 Test the null hypothesis that the proportions of monozygotic twins, dizygotic twins, and nontwin siblings with multiple sclerosis are the same. Use $\alpha = .05$. Interpret the result.

7.16 Test the difference in proportions of multiple sclerosis cases for all pairs of groups. Do not let familywise error rise above .05. Interpret your results.

The Assessment of Relationships

C H A P T E R

8

8.1 BACKGROUND

Many of the statistical procedures with which you are now familiar are designed to assess relationships. For example, the one-way ANOVA and 2 by k chi-square tests can be used to assess whether a relationship exists between treatments afforded subjects in a study and some outcome.

In this chapter we consider the Pearson product-moment correlation coefficient which is designed to assess a specific form of relationship between two continuous variables—namely, the degree to which they are linearly related. We will also briefly consider the chi-square test for independence whose function is to determine whether two categorical variables are independent. Because the chi-square test for independence is a straight forward extension of the 2 by k chi-square test with which you are already familiar, we will take only a cursory look at this test.

8.2 THE PEARSON PRODUCT-MOMENT CORRELATION COEFFICIENT

The Pearson product-moment correlation coefficient (P-M) takes values between -1 and 1 inclusive. In order to understand the P-M, you must understand this value. Two pieces of information are provided by the P-M which we shall term the nature and strength of the relationship between two continuous variables. We will discuss each in turn. Before doing so, however, it will be helpful to learn how the P-M is calculated.

8.2.1 Calculation of the Product-Moment Correlation Coefficient

Equations for calculating P-M are many and varied but are algebraically identical. In this section we shall present conceptual and computational forms[1] of these equations as well as one based on z scores.[2] The conceptual and z score forms are given to enhance your

---

[1] See page 36.
[2] See page 42.

understanding of the P-M while the computational form will be used for calculations. The conceptual form is given by

$$r = \frac{\sum (x - \bar{x})(y - \bar{y})}{\sqrt{\left[\sum (x - \bar{x})^2\right]\left[\sum (y - \bar{y})^2\right]}} \tag{8.1}$$

where r is the sample correlation coefficient, x and y are the two variables to be correlated and n is the number of paired observations. Note that if the two expressions in the denominator (i.e., the sums of squares of x and y) were divided by $n - 1$, the result would be the sample variances of x and y. Likewise, if the numerator of the correlation coefficient were divided by $n - 1$ the result is what is termed the **covariance** of x and y. Covariance is a measure of the linear relationship between two variable but is largely uninterpretable because it depends on the scale upon which the two variables are measured and hence is not bounded by 1 and -1 as is P-M. We could write r with the $n - 1$ divisors alluded to here, but these terms cancel from the expression and are, therefore, not necessary.

The computational form of r is given by

$$r = \frac{\sum xy - \frac{(\sum x)(\sum y)}{n}}{\sqrt{\left[\sum x^2 - \frac{(\sum x)^2}{n}\right]\left[\sum y^2 - \frac{(\sum y)^2}{n}\right]}} \tag{8.2}$$

Equation 8.2 is algebraically equivalent to Equation 8.1 but is designed for convenience of computation. You will recognize the sums of squares forms in the denominator. You will also note that the only term in this equation that you have not previously used in a calculation is $\sum xy$. This is obtained by multiplying each x value by each y value and summing the products thus obtained.

A rarely used but conceptually revealing form is given by

$$r = \frac{\sum z_x z_y}{n - 1} \tag{8.3}$$

where z_x and z_y are the x and y variables expressed in z score form. We will return to Equation 8.3 in a later section.

EXAMPLE 8.1

Use Equation 8.2 to compute P-M for the data in Table 8.1 on the next page. You need not be concerned as to the meaning of the data in this table at this time.

Solution   For computational purposes, it will be convenient to arrange the data in Table 8.1 as shown in Table 8.2 on the facing page. We will designate the access scores as variable x and the wellness scores as variable y.

TABLE 8.1: Wellness and access to medical care scores for 15 subjects.

Subject Number	Access Score	Wellness Score
1	3	2
2	6	6
3	13	9
4	1	1
5	7	5
6	8	7
7	13	10
8	10	8
9	2	2
10	4	3
11	5	4
12	11	9
13	4	5
14	3	4
15	9	8

TABLE 8.2:   Wellness and access to medical care scores arranged for computation of P-M.

Subject Number	x	y	xy	x^2	y^2
1	3	2	6	9	4
2	6	6	36	36	36
3	13	9	117	169	81
4	1	1	1	1	1
5	7	5	35	49	25
6	8	7	56	64	49
7	13	10	130	169	100
8	10	8	80	100	64
9	2	2	4	4	4
10	4	3	12	16	9
11	5	4	20	25	16
12	11	9	99	121	81
13	4	5	20	16	25
14	3	4	12	9	16
15	9	8	72	81	64
Σ	99	83	700	869	575

Using the sums in Table 8.2 with Equation 8.2 gives

$$r = \frac{\sum xy - \frac{(\sum x)(\sum y)}{n}}{\sqrt{\left[\sum x^2 - \frac{(\sum x)^2}{n}\right]\left[\sum y^2 - \frac{(\sum y)^2}{n}\right]}}$$

$$= \frac{700 - \frac{(99)(83)}{15}}{\sqrt{\left[869 - \frac{(99)^2}{15}\right]\left[575 - \frac{(83)^2}{15}\right]}}$$

$$= \frac{152.20}{\sqrt{[215.60][115.73]}}$$

$$= .964 \qquad\blacksquare$$

EXAMPLE 8.2

Use Equation 8.2 to compute P-M for the data in Table 8.3 on the next page. You need not be concerned as to the meaning of the data in this table at this time.

Solution For computational purposes, it will be convenient to arrange the data in Table 8.3 as shown in Table 8.4 on the facing page. We will designate percent of free lunch as variable x and percent of helmet wear as variable y.

Using the sums in Table 8.4 with Equation 8.2 gives

$$r = \frac{\sum xy - \frac{(\sum x)(\sum y)}{n}}{\sqrt{\left[\sum x^2 - \frac{(\sum x)^2}{n}\right]\left[\sum y^2 - \frac{(\sum y)^2}{n}\right]}}$$

$$= \frac{2703 - \frac{(194)(161)}{9}}{\sqrt{\left[5356 - \frac{(194)^2}{9}\right]\left[3409 - \frac{(161)^2}{9}\right]}}$$

$$= \frac{-767.44}{\sqrt{[1174.22][528.89]}}$$

$$= -.974 \qquad\blacksquare$$

Now that you can calculate P-M, we return to the subject of the meaning of the result thus obtained. More specifically, we examine the nature and strength of the relationship between x and y. $\qquad\blacksquare$

8.2.2 The Nature of the Relationship

Consider the data in Table 8.1 on the previous page. These fictitious data represent two measures taken on 15 subjects. The first is a questionnaire score that represents the degree to which the subject has access to medical care. The range of this scale is one to 13 with one representing no access and 13 indicating that full access is available to all facets of

TABLE 8.3: Percent of students on free or reduced lunch and percent of students who routinely wear a helmet when riding their bicycle in 9 schools.

School Number	Percent On Free or Reduced Lunch	Percent Wearing Helmet
1	17	18
2	15	25
3	25	15
4	4	29
5	36	8
6	40	8
7	20	19
8	8	28
9	29	11

TABLE 8.4: Percent of students on free or reduced lunch and percent of students who routinely wear a helmet when riding their bicycle arranged for computation of P-M.

School Number	x	y	xy	x^2	y^2
1	17	18	306	289	324
2	15	25	375	225	625
3	25	15	375	625	225
4	4	29	116	16	841
5	36	8	288	1296	64
6	40	8	320	1600	64
7	20	19	380	400	361
8	8	28	224	64	784
9	29	11	319	841	121
Σ	194	161	2703	5356	3409

medical care (i.e., medical, dental etc.). The second measure is an overall wellness index score that ranges from one to 10. One would represent a very poor overall state of health while 10 indicates very good overall health. Notice the pattern in these data. Persons with high access scores tend to also have high wellness scores and visa versa. At the same time, persons with low access scores tend to have low wellness scores.

This pattern is more clearly discerned in the bivariate plot depicted in Figure 8.1 on page 301. A **bivariate plot** uses dots to represent pairs of scores for each subject. For example, by reading down to the X (access) axis and across to the Y (wellness) axis, you can see that one subject had access and wellness scores of 11 and 9 respectively. A second subject had scores of 2 and 2. A distinct pattern can be seen in this figure. As access scores increase, wellness scores also tend to increase. When relationships of this sort exist—that is, when high values of one variable tend to be associated with high values of the other variable and low values of one variable tend to be associated with low values of the other variable, the data are said to be **positively correlated**. In this circumstance the P-M will be a positive number. This is the case for the data in Table 8.1 whose correlation we computed on page 298 and found to be .964.

Consider now the data in Table 8.3 on the preceding page and the bivariate plot in Figure 8.2 on the next page. A crude but commonly used measure of the overall socioeconomic status of a school is the percent of students in the school who qualify for free or reduced lunch. The percents of children qualifying for free or reduced lunch and the percents who report that they routinely wear a helmet when riding their bicycle are shown in Table 8.3 and Figure 8.2 for nine schools. From this table and more clearly, from this figure, you can see that schools with low percents of children on free or reduced lunch tend to have higher percents reporting routine use of bicycle helmets while schools with higher percents of free or reduced lunch students typically have less helmet wear. When relationships of this sort exist—that is, when high values of one variable tend to be associated with low values of the other variable and low values of one variable tend to be associated with high values of the other variable, the data are said to be **negatively correlated**. In this circumstance the P-M will be a negative number. This is the case for the data in Table 8.3 whose correlation we computed on page 298 and found to be $-.974$.

8.2.3   The Strength of the Relationship

In describing the nature of the relationship between two variables, we said that when two variables are positively correlated, high values on one variable *tend* to be associated with high values on the second variable and that low values on one variable *tend* to be associated with low values on the second variable. The word *tend* was also used in describing negative correlations. But how strong is this tendency? Does every subject who scores high on one variable also score high on the other when data are positively correlated or do some subjects violate the pattern? Does a subject who scores high on one variable score as high on the other? These questions allude to the strength of the relationship between two variable.

You will recall that P-M can take values between -1 and 1. The P-M is at maximum strength at either of these values and loses strength as it recedes toward zero. At zero,[3] the correlation coefficient is at minimum strength. Let us begin by looking at examples in which $r = 1.0$ and $r = -1.0$.

Consider the data in Table 8.5 on page 302. It appears that the x and y variables in this table are positively correlated, but how strong is this correlation? Application of Equation 8.2 on page 296[4] would show that P-M$= 1.0$ for these data. But what exactly does $r = 1.0$ imply?

A correlation coefficient of 1.0 means that each subject made exactly the same score on the two variables when scaling differences are eliminated by expressing the two variables in terms of z scores.[5] The data in Table 8.6 show the variables in Table 8.5 expressed as z

---

[3]We will deal with the case of zero correlation in a separate section.

[4]We'll leave it to you to carry out this computation.

[5]You would do well to review page 42 before continuing.

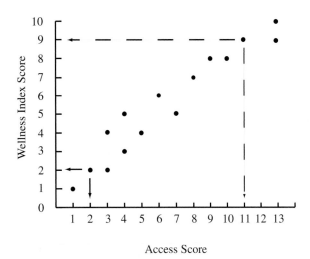

FIGURE 8.1: Bivariate plot of positively correlated health care access scores and wellness index scores.

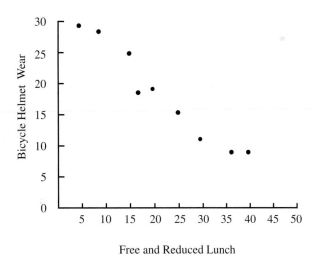

FIGURE 8.2: Bivariate plot of negatively correlated percents of students on free or reduced lunch and percents using bicycle helmets in nine schools.

TABLE 8.5: Data set where $r = 1.0$.

Subject Number	x	y	x^2	y^2
1	4	11	16	121
2	6	15	36	225
3	11	25	121	625
4	4	11	16	121
5	9	21	81	441
6	5	13	25	169
7	8	19	64	361
8	13	29	169	841
Σ	60	144	528	2904

scores. We will show you how the first two x and y values were converted to z scores but leave it to you to confirm the other conversions.

Sample data are converted to z scores through application of Equation 2.23 on page 42 which is repeated here for your convenience.

$$z = \frac{x - \bar{x}}{s}$$

Here x is the score to be converted, and $\bar{x}$ and s are the sample mean and standard deviation respectively. Using the sums from Table 8.5 and computing the mean and standard deviation of x by means of Equation 2.1 on page 25 and Equation 2.16 on page 37 gives

$$\bar{x} = \frac{\sum x}{n} = \frac{60}{8} = 7.5$$

and

$$s = \sqrt{\frac{\sum x^2 - \frac{(\sum x)^2}{n}}{n-1}} = \sqrt{\frac{528 - \frac{(60)^2}{8}}{7}} = 3.338$$

Then for $x = 4$ and $x = 6$,

$$z_4 = \frac{x - \bar{x}}{s} = \frac{4.0 - 7.5}{3.338} = -1.049$$

and

$$z_6 = \frac{x - \bar{x}}{s} = \frac{6.0 - 7.5}{3.338} = -.449$$

Using Equations 2.1 and 2.16 for y gives

$$\bar{y} = \frac{\sum y}{n} = \frac{144}{8} = 18.0$$

TABLE 8.6: x and y values from Table 8.5 expressed as z scores.

Subject Number	z_x	z_y	$z_x z_y$
1	−1.049	−1.049	1.100
2	−.449	−.449	.202
3	1.049	1.049	1.100
4	−1.049	−1.049	1.100
5	.449	.449	.202
6	−.749	−.749	.561
7	.150	.150	.023
8	1.648	1.648	2.716
$\sum$	00.000[a]	00.000	7.000[b]

[a] The sum of z_x and z_y are zero when calculations are carried out a sufficient number of decimal places.

[b] The sum of $z_x z_y$ is 7.000 when calculations are carried out a sufficient number of decimal places.

and

$$s = \sqrt{\frac{\sum y^2 - \frac{(\sum y)^2}{n}}{n-1}} = \sqrt{\frac{2904 - \frac{(144)^2}{8}}{7}} = 6.676.$$

Then for $y = 11$ and $y = 15$,

$$z_{11} = \frac{y - \bar{y}}{s} = \frac{11.0 - 18.0}{6.676} = -1.049$$

and

$$z_{15} = \frac{y - \bar{y}}{s} = \frac{15.0 - 18.0}{6.676} = -.449.$$

Notice that subjects one and two have exactly the same score on x and y when these variables are converted to the common z score scale. This is also true for the remaining subjects as can be seen in Table 8.6. We can confirm that $r = 1$ by application of Equation 8.3 which yields

$$r = \frac{\sum z_x z_y}{n-1} = \frac{7}{7} = 1.000.$$

Because P-M assesses the degree to which x and y are linearly related, when $r = 1.0$ the bivariate plot of x and y shows that the individual points fall on a line. This can be seen in Figure 8.3 on the next page. By contrast, the plot in Figure 8.1 of data for which $r = .964$ suggests a linear relationship but does not constitute a line showing that there is not a perfect linear relationship.

Contrast the z scores from Table 8.6 where $r = 1.0$ to those in Table 8.7 which were generated from the data in Table 8.1 where $r = .964$. Though each subject had similar z

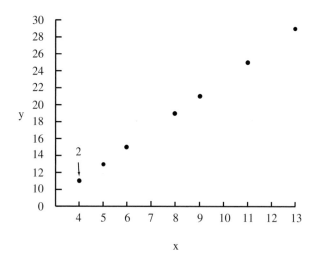

FIGURE 8.3: Bivariate plot of data from Table 8.5 for which $r = 1.0$.

scores on the access and wellness variables, their scores were not identical resulting in a P-M of less than 1.00. Nevertheless, the similarity of the two sets of z scores was sufficient to produce a *near* perfect correlation of .964. The value of the correlation coefficient computed earlier can be confirmed through application of Equation 8.3 which, using results from Table 8.7, yields

$$r = \frac{\sum z_x z_y}{n-1}$$
$$= \frac{13.490}{14}$$
$$= .964.$$

We now turn to the case in which $r = -1.0$. Application of Equation 8.2 to the data in Table 8.8 on the facing page would yield $r = -1.0$. These data are converted to z scores in Table 8.9.

Notice that when $r = -1.0$, the magnitude of each subject's score on the two variables is the same but is always opposite in sign. Thus, if a subject is 1.5 standard deviations above the mean on variable x she will be 1.5 standard deviation below the mean on variable y. As when $r = 1.0$, the bivariate plot of x and y falls on a line when $r = -1.0$ but as can be seen in Figure 8.4, the line has negative slope. Notice that the data depicted in Figure 8.2 on page 301 where $r = -.974$ tend toward a negatively sloped line but do not constitute a line. This is because the data are negatively correlated but not perfectly so.

In general then, for positively correlated data, the closer r is to one the more similar are the scores on variables x and y when both variables are expressed on the common z score scale. When $r = 1.0$ the z scores for the two variables are identical. For negatively correlated data, the closer r is to negative one the stronger is the tendency for the two variables to have equal magnitude but opposite signs. We now examine what happens when r is zero.

TABLE 8.7: Access to medical care and wellness scores from Table 8.1 expressed as z scores.

Subject Number	Access Score	Wellness Score	Access X Wellness
1	−.917	−1.229	1.127
2	−.153	.162	−.025
3	1.631	1.206	1.967
4	−1.427	−1.577	2.250
5	.102	−.185	−.019
6	.357	.510	.182
7	1.631	1.554	2.535
8	.866	.858	.743
9	−1.172	−1.229	1.440
10	−.663	−.881	.584
11	−.408	−.533	.217
12	1.121	1.206	1.352
13	−.663	−.185	.123
14	−.917	−.533	.489
15	.612	.858	.525
Σ	00.000	00.000[a]	13.490

[a] The sum of the wellness z scores is 0.000 when calculations are carried out a sufficient number of decimal places.

TABLE 8.8: Data set where $r = -1.0$.

Subject Number	x	y
1	4	29
2	6	25
3	11	15
4	4	29
5	9	19
6	5	27
7	8	21
8	13	11
Σ	60	176

TABLE 8.9: x and y values from Table 8.8 expressed as z scores.

Subject Number	z_x	z_y	$z_x z_y$
1	−1.049	1.049	−1.100
2	−.449	.449	−.202
3	1.049	−1.049	−1.100
4	−1.049	1.049	−1.100
5	.449	−.449	−.202
6	−.749	.749	−.561
7	.150	−.150	−.023
8	1.648	−1.648	−2.716
$\sum$	00.000	00.000	−7.000[a]

[a] The sum of $z_x z_y$ is −7.000 when calculations are carried out a sufficient number of decimal places.

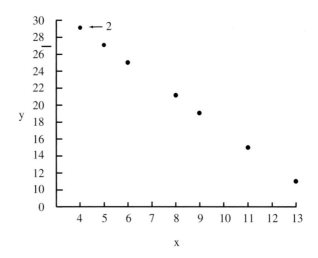

FIGURE 8.4: Bivariate plot of data from Table 8.8 for which $r = -1.0$.

8.2.4   Zero Correlation

Naive researchers sometimes interpret zero correlation between x and y as indicating that there is no relationship between the two variables. This is not necessarily the case. A zero correlation will exist when there is no relationship between x and y but may also exist when certain *nonlinear* relationships are expressed in the data. Consider the bivariate plot depicted in Figure 8.5.

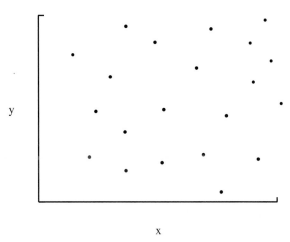

FIGURE 8.5: Bivariate plot in which there is no relationship between x and y.

This plot is meant to represent a situation in which there is no relationship between the two variables. Indeed, no pattern is discernable in these data. Such a data set will generate zero correlation because x and y are unrelated. But consider now the data in Table 8.10.

Application of Equation 8.2 to these data will show that $r = 0.0$. But does this mean that there is no relationship between x and y? The answer is clearly no as can be seen in the plot depicted in Figure 8.6 on the next page.

TABLE 8.10: Data set for which $r = 0.0$.

Subject Number	x	y
1	1	1
2	2	2
3	3	3
4	4	4
5	5	4
6	6	3
7	7	2
8	8	1

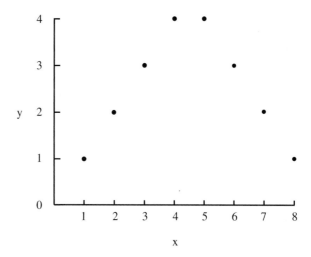

FIGURE 8.6: Bivariate plot of data for which $r = 0.0$ but a relationship between x and y is apparent.

It is obvious that for values of x between 1 and 4 y increases as x increases but for values of x between 5 and 8 y decreases with increases in x. P-M is simply not designed to detect relationships of this sort. P-M is designed to detect *linear* relationships. That is, relationships that tend to be characterized by a line. Many other nonlinear forms are not detectable by P-M.

To summarize, zero correlation does not necessarily indicate that there is no relationship between x and y.[6] It may be that there is no relationship but it may also be that there is a relationship but it is of a form not detectable by P-M.

8.2.5 Cause-Effect Relationships

It is sometimes tempting to interpret strong correlations between variables as an indication that one variable causes the other. It is the traditional burden of authors of introductory statistics texts to warn students against such conclusions. Two variables may be correlated because one causes the other but may also be correlated when no such causal relationship exists. Thus, you must be extremely cautious when claiming a causal relationship between two variables based on an observed correlation between those variables. Indeed, it would not be imprudent to simply say "Don't do it." Consider the following examples.

Early researchers noted a strong positive correlation between estimated number of cigarettes smoked in a lifetime and amount of lung damage observed at autopsy. Does cigarette smoking harm the human lung? Numerous studies have shown that it does. However, no such conclusion should be reached based on the high positive correlation between number of cigarettes smoked and observed pathology. A trivial but revealing example will show why.

---

[6]There are theoretical exceptions such as the case of bivariate normality but these are more of theoretical than practical interest.

Suppose that for each of the 365 days in a given year we were to record (1) the amount of ice cream consumed on that day in Florida and (2) the number of persons who drown in pools or at the beach in Florida on that day. Do you think these two variables would be correlated? The answer is obviously yes. On days when large amounts of ice cream are consumed there will also be a tendency to note more drownings. When little ice cream is consumed few drownings occur. Does eating ice cream cause people to drown? More ludicrously, does drowning cause people to eat ice cream?[7]

The reason for the correlation between these two variables is obvious—both are related to temperature. On warm days more ice cream is eaten and more people go to pools or the beach. On cool days less ice cream is consumed and fewer people go to beaches and pools. Thus, ice cream consumption and drowning are substantially related because of their relationship to a third variable and not because of any direct relationship between themselves. This type situation can develop in almost any research context. Our duty has now been discharged. You have been duly warned.

8.2.6   Test of Hypothesis and Confidence Interval

To this point we have characterized P-M as a descriptive statistic. That is, as a statistic that describes the relationship between two variables. But just as we used the sample mean ($\bar{x}$) to make inferences about the population mean (μ), so we can use the sample correlation coefficient (r) to make inferences about the population correlation coefficient (ρ). We will do so through the mechanisms of hypothesis testing and confidence intervals.

A Test of $H_0 : \rho = 0$.   In most cases the researcher will be interested in determining whether a correlation between x and y exists in the population. In such circumstance the natural null hypothesis to be tested is

$$H_0 : \rho = 0$$

where ρ is the population correlation. This test can be carried out by means of a t test whose form is as follows.

$$t = \frac{r}{\sqrt{\frac{1-r^2}{n-2}}} \tag{8.4}$$

Degrees of freedom for this test are $n - 2$ where n is the number of pairs. The two-tailed alternative hypothesis is

$$H_A : \rho \neq 0$$

with one-tailed alternatives taking one of the following forms.

$$H_A : \rho < 0$$

or

$$H_A : \rho > 0$$

---

[7]As far as is known, there is no recorded claim by witnesses to a drowning that the victim was heard calling for a double scoop chocolate cone as he/she went down for the third time.

EXAMPLE 8.3

On page 298 you found that the correlation between the 15 pairs of wellness and access to medical care scores was .964. Test this coefficient for significance with the following null and alternative hypotheses.

$$H_0 : \rho = 0$$
$$H_A : \rho > 0$$

Solution   By Equation 8.4

$$t = \frac{r}{\sqrt{\frac{1-r^2}{n-2}}} = \frac{.964}{\sqrt{\frac{1-.964^2}{15-2}}} = 13.072$$

Reference to Appendix B shows that critical t for a one-tailed test conducted at $\alpha = .05$ with 13 degrees of freedom is 1.771. Because obtained t exceeds this value, the null hypothesis is rejected. As a result, the researcher can be assured that the population correlation is greater than 0.   ∎

A Test of $H_0 : \rho = \rho_0$.   In most circumstances the researcher will want to know if the population correlation differs from zero. Occasionally, however, the researcher will want to know if the population correlation differs from some value other than zero. Because the sampling distribution of r is symmetric when $\rho = 0$, the t test is robust for tests of $H_0 : \rho = 0$. When the specified null value is other than zero, the sampling distribution of r is skewed under the null hypothesis so that the t test may produce misleading results due to a failure of the normality assumption. As a result, Equation 8.4 will generally not be appropriate for the test. Fisher [15, 16] proposed a method of testing that largely circumvents this problem. The following test statistic is based on Fisher's method.

$$Z = \frac{.5 \ln\left(\frac{1+r}{1-r}\right) - .5 \ln\left(\frac{1+\rho_0}{1-\rho_0}\right)}{\sqrt{\frac{1}{\sqrt{n-3}}}} \tag{8.5}$$

Here, ln is the natural log, ρ_0 is the hypothesized value of the population correlation coefficient and n is the number of pairs. This statistic is approximately normally distributed so that the test may be conducted by reference to the normal curve.

EXAMPLE 8.4

From the literature, a researcher knows that the correlation between a shortened form of the Attitude Toward Risky Sexual Behaviors Assessment Scale and a more elaborate form of the scale is .57. In an attempt to improve the correlation between the short and long forms of the scale, the researcher replaces several items on the shortened form. Both instruments are then administered to a group of 18 college students. The researcher finds a .71 correlation between the new version of the short form and the long form. Use this information to perform a two-tailed test of the null hypothesis

$$H_0 : \rho = .57$$

Solution   By Equation 8.5

$$Z = \frac{.5 \ln\left(\frac{1+r}{1-r}\right) - .5 \ln\left(\frac{1+\rho_0}{1-\rho_0}\right)}{\sqrt{\frac{1}{\sqrt{n-3}}}} = \frac{.5 \ln\left(\frac{1+.71}{1-.71}\right) - .5 \ln\left(\frac{1+.57}{1-.57}\right)}{\sqrt{\frac{1}{18-3}}} = \frac{.887 - .648}{.258} = .926$$

From Appendix A it can be seen that the critical Z values for a two-tailed test conducted at $\alpha = .05$ are -1.96 and 1.96. Because obtained Z is between these two values, the null hypothesis is not rejected. This means that the researcher has been unable to demonstrate a change in correlation after modifying the short form of the scale.   ■

A Confidence Interval for the Estimation of ρ.   A confidence interval based on the rationale underlying Equation 8.5 is possible and is probably more commonly employed than the method presented here. However, the method given here is generally superior and is therefore recommended.

Lower and upper bound estimates for ρ can be obtained from the following.

$$L = \frac{(1+F)r + (1-F)}{(1+F) + (1-F)r} \tag{8.6}$$

$$U = \frac{(1+F)r - (1-F)}{(1+F) - (1-F)r} \tag{8.7}$$

In these equations r is the sample correlation coefficient and F is the appropriate value from Appendix C. The degrees of freedom for F are $n - 2$ for both numerator and denominator degrees of freedom. One and two-sided intervals are possible as indicated at the top left portion of the F table.

EXAMPLE 8.5

Use the Risky Sexual Behaviors Assessment Scale data alluded to above to form a two-sided 95% confidence interval for the estimation of ρ. The observed correlation in that study was .71 which was obtained from data collected on 18 students. Use the resulting interval to perform a two-tailed test of $H_0 : \rho = .57$ at $\alpha = .05$. How did you obtain your result?

Solution   By Equation 8.6

$$L = \frac{(1+2.76).71 + (1-2.76)}{(1+2.76) + (1-2.76).71} = \frac{.910}{2.510} = .363$$

and

$$U = \frac{(1+2.76).71 - (1-2.76)}{(1+2.76) - (1-2.76).71} = \frac{4.430}{5.010} = .884$$

$F = 2.76$ was obtained by entering Appendix C for a two-sided confidence interval with numerator and denominator degrees of freedom of $18 - 2 = 16$. The researcher can, therefore, be 95 percent confident that the population correlation coefficient lies between .363 and .884. The two-tailed hypothesis test would not be rejected because the hypothesized value of .57 lies between the two limits.   ■

EXAMPLE 8.6

Given $r = .82$ and $n = 52$, find a lower bound estimate for ρ in which you can be 99% confident.

Solution   By Equation 8.6

$$
\begin{aligned}
L &= \frac{(1 + F)\, r + (1 - F)}{(1 + F) + (1 - F)\, r} \\
&= \frac{(1 + 1.95)\,.82 + (1 - 1.95)}{(1 + 1.95) + (1 - 1.95)\,.82} \\
&= \frac{1.469}{2.171} \\
&= .68
\end{aligned}
$$

$F = 1.95$ was obtained by entering Appendix C for a one-sided 99% confidence interval with 50 numerator and denominator degrees of freedom. ∎

8.2.7   Assumptions

Contrary to the belief of some researchers, there are no statistical assumptions underlying the calculation of r and its subsequent use as a descriptive statistic. However, certain assumptions do apply to hypothesis tests and confidence intervals associated with this statistic. Two primary assumptions are of concern. First, it is assumed that data pairs are sampled from a bivariate normal distribution. This distribution may be thought of as a three dimensional version of the normal curve. Tests of hypotheses and confidence intervals may or may not be robust to violations of this assumption depending on the shape of the population distribution and the value of ρ. In general, researchers should be cautious when using the inferential methods presented here and should be particularly so when either x and/or y appear to depart from normality to a significant degree.[8]

It is also assumed that each data pair is independent of all other data pairs. A violation of this assumption might occur, for example, if the same subject were tested for x and y at two different points in time. Inferences concerning ρ cannot be depended upon when this assumption is violated.

8.3   THE CHI-SQUARE TEST FOR INDEPENDENCE

The chi-square test for independence is used to test the null hypothesis that two discrete variables are independent against the alternative that they are not independent. You have already dealt with this test, albeit in a restricted form, in Section 7.3 on page 276. This is because the chi-square test for independence is simply a more general form of the 2 by k chi-square test you dealt with in that section.[9]

---

[8]Having normally distributed x and y does not guarantee bivariate normality but having nonnormal x and/or y does guarantee that bivariate normality has not been achieved.

[9]We strongly recommend you re-read Section 7.3 on page 276 before continuing.

TABLE 8.11: Depiction of a j by k chi-square table.

| | | Variable One | | | |
		Category One	Category Two	Category $\cdots$	Category k
	Category One	f_{o11} f_{e11}	f_{o12} f_{e12}	$\cdots$ $\cdots$	f_{o1k} f_{e1k}
Variable	Category Two	f_{o21} f_{e21}	f_{o22} f_{e22}	$\cdots$ $\cdots$	f_{o2k} f_{e2k}
Two	Category $\vdots$	$\vdots$	$\vdots$	$\vdots \cdots$	$\vdots$
	Category j	f_{oj1} f_{ej1}	f_{oj2} f_{ej2}	$\cdots$ $\cdots$	f_{ojk} f_{ejk}

Obtained chi-square is calculated by Equation 7.10 on page 277 which is repeated here for your convenience.

$$\chi^2 = \sum_{\text{all cells}} \left[\frac{(f_o - f_e)^2}{f_e} \right]$$

Here f_o and f_e are the observed and expected frequencies. The **observed frequency** is simply the number of outcomes occurring in the given cell as shown in Table 8.11. In this table we have used double subscripts to indicate the row and column of each cell entry. For example, if variable one represents vaccination against hepatitis B status with category one representing "not vaccinated," category two representing "vaccinated," and category three representing "not known" and variable two indicates county of residency, then f_{o11} would be the number of persons in county one who have been vaccinated and f_{o21} would be the number of persons in county two who have been vaccinated. The entries f_{o13} and f_{o23} would represent the numbers in these counties whose status is unknown.

The f_e are the numbers to be **expected** in each cell *if the null hypothesis is true* and are computed via Equation 7.11 on page 279 which is repeated here for your convenience.

$$f_e = \frac{(N_R)(N_C)}{N}$$

Here N_R is the row total for the cell whose expected frequency is being calculated and N_C is the column total for the same cell.[10]

The notation "$\cdots$" used in Table 8.11 indicates that the number of columns continues to the last column which is designated column k. Likewise, the notation "$\vdots$" indicates that the number of rows continues to the last row which is designated row j. All this means is that the table can have any number of rows and columns. The degrees of freedom for the test

---

[10]See Section 7.3 on page 276 for further details.

statistic are computed by

$$\boxed{\chi^2_{df} = (j-1)(k-1)} \tag{8.8}$$

where j and k are the number of rows and columns in the table respectively.

EXAMPLE 8.7

Suppose a survey is conducted in three rural counties to determine vaccination status against hepatitis B. It is found that in county one, 41 persons have been vaccinated, 126 have not been vaccinated and 452 do not know their status. In county two, 202 have been vaccinated, 210 have not been vaccinated and 440 do not know their status. In the last county 330 had been vaccinated, 614 had not been vaccinated, and 680 did not know their status.

Use this data to perform a chi-square analysis. Interpret the results.

Solution   The observed (brackets) and expected (parentheses) frequencies as well as row and column totals are arranged as shown in Table 8.12 on the facing page. Expected frequencies were calculated via Equation 7.11 as follows.

$$f_{e11} = \frac{(N_{One})(N_V)}{N} = \frac{(619)(573)}{3095} = 114.60$$

$$f_{e12} = \frac{(N_{One})(N_{NV})}{N} = \frac{(619)(950)}{3095} = 190.00$$

$$f_{e13} = \frac{(N_{One})(N_U)}{N} = \frac{(619)(1572)}{3095} = 314.40$$

$$f_{e21} = \frac{(N_{Two})(N_V)}{N} = \frac{(852)(573)}{3095} = 157.74$$

$$f_{e22} = \frac{(N_{Two})(N_{NV})}{N} = \frac{(852)(950)}{3095} = 261.52$$

$$f_{e23} = \frac{(N_{Two})(N_U)}{N} = \frac{(852)(1572)}{3095} = 432.74$$

$$f_{e31} = \frac{(N_{Three})(N_V)}{N} = \frac{(1624)(573)}{3095} = 300.66$$

$$f_{e32} = \frac{(N_{Three})(N_{NV})}{N} = \frac{(1624)(950)}{3095} = 498.48$$

$$f_{e33} = \frac{(N_{Three})(N_U)}{N} = \frac{(1624)(1572)}{3095} = 824.86$$

TABLE 8.12: Data from survey arranged for chi-square analysis.

	Vaccinated	Not Vaccinated	Unknown	
County One	[41] (114.60)	[126] (190.00)	[452] (314.40)	619
County Two	[202] (157.74)	[210] (261.52)	[440] (432.74)	852
County Three	[330] (300.66)	[614] (498.48)	[680] (824.86)	1624
	573	950	1572	$N = 3095$

Then by Equation 7.10 obtained chi-square is

$$\chi^2 = \sum_{\text{all cells}} \left[\frac{(f_o - f_e)^2}{f_e} \right]$$

$$= \frac{(41 - 114.60)^2}{114.60} + \frac{(126 - 190.00)^2}{190.00} + \frac{(452 - 314.40)^2}{314.40} + \frac{(202 - 157.74)^2}{157.74}$$

$$+ \frac{(210 - 261.52)^2}{261.52} + \frac{(440 - 432.74)^2}{432.74} + \frac{(330 - 300.66)^2}{300.66} + \frac{(614 - 498.48)^2}{498.48}$$

$$+ \frac{(680 - 824.86)^2}{824.86}$$

$$= 47.27 + 21.56 + 60.22 + 12.42 + 10.15 + .12 + 2.86 + 26.77 + 25.44$$

$$= 206.81$$

The critical value is obtained by entering Appendix D with

$$(j - 1)(k - 1) = (3 - 1)(3 - 1) = 4 \text{ degrees of freedom}$$

where j is the number of rows and k the number of columns in the table. In the present case, Appendix D shows that for $\alpha = .05$ and four degrees of freedom, critical χ^2 is 9.488. The null hypothesis is rejected when obtained chi-square is greater than or equal to critical chi-square. Because 206.81 is greater than 9.488, the null hypothesis is rejected. We conclude, therefore, that vaccination status and county of residence are not independent. This means that vaccination status depends on the county of residence. Said yet another way, the proportions of vaccinated, not vaccinated and unknown status, persons in the three counties are not the same. ∎

8.3.1 Assumptions

The chi-square test for independence presented here is an approximate rather than exact test.[11] In order for the approximation to be sufficiently accurate, a rule of thumb states that

---

[11] An exact test is possible but requires special computer software and is, therefore, not discussed here.

at least 80% of the f_e in the table should be greater than or equal to five and no cell should have f_e less than one.

It is also assumed that observations are independent. For example, one outcome can neither influence nor be influenced by another outcome. In the last example, a violation would likely have occured if some individuals had accidently been queried multiple times with their responses then being entered into the analysis. The chi-square test cannot be counted upon to be robust against violation of the independence assumption.

KEY WORDS AND PHRASES

After reading this chapter you should be able to demonstrate familiarity with the following words and phrases.

bivariate plot 299	cause-effect relationship 308
CI for ρ 311	chi-square test for independence 312
covariance 296	expected frequency 313
linear relationship 308	nature of the relationship 298
negative correlation 300	nonlinear relationship 307
observed frequency 313	Pearson product-moment correlation coefficient 295
positive correlation 300	strength of the relationship 300
test of $\rho = 0$ 309	test of $\rho = \rho_0$ 310

EXERCISES

8.1 Given variables x and y, define or explain the following terms.

(a) Positive correlation.

(b) Negative correlation.

(c) Zero correlation.

(d) Strength of relationship.

(e) Linear relationship.

(f) Correlation of 1.0 and -1.0.

8.2 APGAR scores are assigned to newborns one and five minutes after birth and indicate the general state of well being of the baby. Scores of seven to ten are typical and indicate that the baby requires only routine post-natal care. Scores of four to six indicate that the baby may require assistance while scores of three or less indicate that the baby requires immediate assistance if life is to be sustained.

Suppose a study is conducted to determine the consistency with which APGAR scores are assigned. To this end, two doctors observe 12 births and independently assign AP-GAR scores to the babies. Use the tabled APGAR scores given here to perform each of the following tasks. We will designate scores assigned by the first doctor as x and those assigned by the second doctor as y.

Baby Number	x	y
1	9	7
2	8	9
3	7	8
4	8	8
5	6	7
6	4	4
7	9	8
8	7	7
9	2	3
10	8	7
11	7	8
12	9	9

(a) From visual inspection, do you believe there is a positive, negative, or zero correlation between the two variables. Why?

(b) Calculate the Pearson product-moment correlation coefficient for the data.

(c) Perform a two-tailed test of the hypothesis $H_0 : \rho = 0$ at the .05 level of significance.

(d) Form a two-sided 95% CI for the estimation of ρ.

8.3 As part of a public safety pilot study, motorists arrested for driving under the influence of alcohol are also tested for controlled substances. The results of such tests may be positive (+), negative (−), or inconclusive (I). In order to utilize resources efficiently, researchers wish to determine whether test results are related to time of arrest. To this end, times of arrests are placed in one of three categories—midnight to 8:00 AM (category one), 8:00 AM to 4:00 PM (category two), and 4:00 PM to midnight (category three).

Results of the pilot study are as follows; category one, 19 (+), 44 (−), 6 (I), category two 7 (+), 29 (−), 3 (I), category three 13 (+), 51 (−), 9 (I). Test the hypothesis that time of day and test result are independent against the alternative that they are related. Use $\alpha = .05$.

D. The following questions refer to Case Study D (page 471)

8.4 The researchers report a −.50 Spearman rank order correlation between PBV and NPZ-8 for the 20 study participants. Calculate a Pearson correlation for the same data. Would you characterize the result as being markedly different from that produced by the Spearman?

8.5 Given the skewed nature of the NPZ-8 variable, do you believe the correlation coefficient you calculated in 8.4 is a valid expression of the linear relationship between PBV and NPZ-8. Explain your answer.

8.6 Would you have any reservations about conducting a test of the hypothesis $H_0 : \rho = 0$ by means of Equation 8.4 on page 309 for the Pearson you calculated? Explain.

8.7 Does the *nature* of the relationship expressed by the Pearson correlation you calculated in 8.4 make sense? Explain.

8.8 Calculate the Pearson correlation again using only the 15 HIV infected subjects. Does removing the five healthy subjects make a difference?

8.9 Given that the researchers' stated goal for this study was to determine whether neuropsychological function in HIV-infected persons is correlated with loss of brain volume, would you say their goal was achieved?

8.10 Calculate the Pearson correlation between NPZ-8 and CD4 for the 15 infected subjects. Construct a 95% CI to estimate ρ. Do you have any reservations concerning this interval?

L. The following questions refer to Case Study L (page 475)

8.11 Perform a test of significance to determine whether severity of injury is related to age. Use $\alpha = .05$. What is your conclusion?

M. The following questions refer to Case Study M (page 475)

8.12 From visual inspection, does there appear to be a positive, negative or no correlation between the before and after oxygen measures?

8.13 Compute P-M for the before and after oxygen values. Estimate ρ with a 95% confidence interval.

8.14 As originally published, the "after" oxygen value for subject #4 was incorrectly entered into the table as 4.9 rather than the correct value of 8.3. Repeat M.8.13 using the incorrect value of 4.9 in place of 8.3. Would entering this one incorrect value into the analysis have made much difference insofar as the correlation is concerned?

Linear Regression

9.1 BACKGROUND

In Chapter 8 you learned that for pairs of observations taken on a number of subjects or objects, a positive correlation between the two measured variables implies that high values on one variable tend to be associated with high values on the other variable and that low values on one variable tend to be associated with low values on the other variable. Likewise, if the two variables are negatively correlated, an inverse relationship exists between the two variables.

This information would prove helpful if, for some reason, you were asked to observe one of the two variables, i.e., x and predict the value of the other variable—y. Reasonably, if a particular subject had a high x value and you were asked to predict (or guess) his y value, you would choose some high y value as your guess if you knew the data were positively correlated. By the same token, you would choose a low y value as your guess if you knew the data were negatively correlated. Your guesses might not be too accurate, but they would likely be more accurate than would be the case if you had no knowledge concerning the relationship between x and y.

Using the relationship between two variables to predict the value of one from the value of the other can be formalized in such a way as to provide the best possible predictions.[1] Further, and perhaps more importantly, this methodology can be used to "explain" the variation in one variable as a consequence of its relationship with one or more other variables. The models used to generate these predictions and explanations are the subjects of this chapter.

The models to be developed in this chapter can be classified as either simple or multiple linear regression models. Simple linear regression (SLR) models use only one variable to predict the value of some other variable while multiple linear regression (MLR) models use one or more predictor variables for this purpose. Thus, SLR models constitute the simplest form of MLR. For this reason, we will begin with the simpler model.

---

[1] The term "best possible" will be defined shortly.

9.2   SIMPLE LINEAR REGRESSION

The simple linear regression model is formulated as follows.

$$\hat{y} = a + bx \tag{9.1}$$

where $\hat{y}$ (pronounced "y hat") is the *predicted* value of y, a and b are constants to be defined below and x is the variable from which predictions are to be made. For example, if $a = 2$ and $b = 4$ then the predicted value of y for a subject whose value on the x variable is 10 would be

$$\hat{y} = 2 + 4\,(10) = 42$$

But how good is this prediction? A simple way to characterize the accuracy of this prediction is to find the difference between the subject's actual y value and her/his predicted value of y. This quantity, $y - \hat{y}$, is called a **residual**. A natural question to ask at this point is, "How were a and b chosen?" Before answering this question it will be helpful to calculate a and b for some data set and assess the predictions made from the model thus formulated.

9.2.1   Calculation of a and b

The equations for the calculation of a and b are as follows.

$$a = \bar{y} - (b)\,(\bar{x}) \tag{9.2}$$

$$b = \frac{\sum xy - \frac{(\sum x)(\sum y)}{n}}{\sum x^2 - \frac{(\sum x)^2}{n}} \tag{9.3}$$

Two points about these equations are notable. First, a is a function of b so that b must be calculated before a. Second, the equation for b is similar to the equation used for the computation of the Pearson product-moment correlation coefficient.[2] This similarity leads to a second equation for the calculation of b which is useful when the correlation coefficient is already known.

$$b = r\sqrt{\frac{SS_y}{SS_x}} \tag{9.4}$$

Here, r is the Pearson product-moment correlation coefficient, SS_y is the y sum of squares and SS_x is the x sum of squares. It can be seen by Equation 9.4 that b will be zero when r is zero and that b will be positive or negative whenever r is positive or negative. This follows because $\sqrt{\frac{SS_y}{SS_x}}$ is the square root of the ratio of two sums of squares and will, therefore, always be positive.

---

[2]See Equation 8.2 on page 296.

EXAMPLE 9.1

Use the data in Table 8.2 on page 297 to form a SLR model for the prediction of wellness scores from access to medical care scores.

Solution   Designating the access scores as x and the wellness scores as y and using the sums from Table 8.2 with Equation 9.3 yields

$$b = \frac{\sum xy - \frac{(\sum x)(\sum y)}{n}}{\sum x^2 - \frac{(\sum x)^2}{n}} = \frac{700 - \frac{(99)(83)}{15}}{869 - \frac{(99)^2}{15}} = \frac{152.20}{215.60} = .7059$$

Notice that in general, b represents the change in $\hat{y}$ that occurs for each unit change in x. For the current case, an increase of one point in access score results in a .7059 increase in the predicted wellnes score. Likewise, a one unit decrease in the access score results in a .7059 decrease in the predicted wellness score.

Because we previously calculated r (on page 298) to be .9635,[3] we obtain the same result via Equation 9.4 by computing

$$SS_x = \sum x^2 - \frac{(\sum x)^2}{n} = 869 - \frac{(99)^2}{15} = 215.6000$$

and

$$SS_y = \sum y^2 - \frac{(\sum y)^2}{n} = 575 - \frac{(83)^2}{15} = 115.7333$$

Then

$$b = r\sqrt{\frac{SS_y}{SS_x}} = .9635\sqrt{\frac{115.7333}{215.6000}} = .7059$$

which is the same result achieved above.

We can obtain a by computing

$$\bar{y} = \frac{\sum y}{n} = \frac{83}{15} = 5.5333$$

and

$$\bar{x} = \frac{\sum x}{n} = \frac{99}{15} = 6.6000$$

Then by Equation 9.2

$$a = \bar{y} - b\bar{x} = 5.5333 - (.7059)(6.6000) = .8744$$

The prediction model is then

$$\hat{y} = a + bx = .8744 + .7059x$$

The predicted wellness scores obtained from application of this model are shown in Table 9.1.

---

[3]We carry this and other calculations out additional decimal places in order to enhance accuracy of results.

9.2.2 The Residual and Regression Sums of Squares and the Coefficients of Determination and Nondetermination

Table 9.1 shows wellness (y) and access to medical care (x) scores taken from Table 8.2 on page 297. Also shown are predicted wellness scores ($\hat{y}$), squared residuals ($(y-\hat{y})^2$) and the quantity $(\hat{y}-\bar{y})^2$ which will be explained below. Predicted values were obtained through application of the model constructed above. For example, predicted values for the first two subjects were calculated as follows.

$$\hat{y}_1 = .8744 + .7059x_1 = .8744 + .7059\,(3) = 2.9921$$

and

$$\hat{y}_2 = .8744 + .7059x_2 = .8744 + .7059\,(6) = 5.1098.$$

Squared residuals for the first two subjects are then

$$\left(y_1 - \hat{y}_1\right)^2 = (2 - 2.9921)^2 = .9843$$

and

$$\left(y_2 - \hat{y}_2\right)^2 = (6 - 5.1098)^2 = .7925.$$

The first two observations in the column headed $\left(\hat{y}-\bar{y}\right)^2$ were obtained as follows. Noting that $\bar{y} = 83/15 = 5.5333$,

$$\left(\hat{y}_1 - \bar{y}\right) = (2.9921 - 5.5333)^2 = 6.4577$$

and

$$\left(\hat{y}_2 - \bar{y}\right) = (5.1098 - 5.5333)^2 = .1794.$$

The y sum of squares (SS_y) represents the variation on the outcome variable (y). That is, it represents the fact that some subjects have wellness scores well above the average wellness score value (5.5333) while other subjects score well below this value. What explains this variation? To answer this question first note that

$$\boxed{SS_y = SS_{reg} + SS_{res}} \tag{9.5}$$

where SS_{reg} is the **regression sum of squares** and is calculated by

$$\boxed{SS_{reg} = \sum \left(\hat{y} - \bar{y}\right)^2} \tag{9.6}$$

and SS_{res} is the **residual sum of squares** and is calculated by

$$\boxed{SS_{res} = \sum \left(y - \hat{y}\right)^2} \tag{9.7}$$

When these two sums from Table 9.1 on the facing page are added together the result is

$$SS_y = 107.4322 + 8.2898 = 115.7720$$

TABLE 9.1: Predicted wellness scores with residual and regression sums of squares.

Subject Number	Wellness y	$\hat{y}$	Access x	$(y - \hat{y})^2$	$(\hat{y} - \bar{y})^2$
1	2	2.9921	3	.9843	6.4577
2	6	5.1098	6	.7925	.1794
3	9	10.0511	13	1.1048	20.4105
4	1	1.5803	1	.3367	15.6262
5	5	5.8157	7	.6654	.0797
6	7	6.5216	8	.2289	.9767
7	10	10.0511	13	.0026	20.4105
8	8	7.9334	10	.0044	5.7605
9	2	2.2862	2	.0819	10.5437
10	3	3.6980	4	.4872	3.3683
11	4	4.4039	5	.1631	1.2755
12	9	8.6393	11	.1301	9.6472
13	5	3.6980	4	1.6952	3.3683
14	4	2.9921	3	1.0159	6.4577
15	8	7.2275	9	.5968	2.8703
$\sum$	83	83.000	99	8.2898	107.4322

which, except for rounding errors, is the same result obtained above when SS_y was computed directly. ∎

Let us now examine these two components of SS_y more closely. First note that $y_i - \hat{y}_i$ is a measure of the amount by which the model erred in predicting subject i's y value. If we wish to characterize the total error involved in predicting the scores of a group of n subjects you might be tempted to take $\sum (y - \hat{y})$ as such but this would not be a satisfactory measure because this sum is always zero. Rather than summing the difference between y and $\hat{y}$, the squared differences are summed as indicated in Equation 9.7.

Earlier we posed the question as to how the model values of a and b were chosen. The answer is that they are the values that minimize SS_{res}. It is in this sense that a and b produce the best possible model.[4] For this reason, models of this type are referred to as **least squares models**.

The **coefficient of nondetermination** is symbolized by $1 - \widehat{R}^2$ and is the ratio of SS_{res} to SS_y or more formally

$$1 - \widehat{R}^2 = \frac{SS_{res}}{SS_y}$$

(9.8)

---

[4]Other "best possible" models are possible. For example, a model that minimizes $\sum |y - \hat{y}|$ might be constructed.

The **coefficient of nondetermination** is then, the proportion of the y sum of squares that is not accounted for or explained by the model.

But what of the other component of SS_y? If $SS_y = SS_{reg} + SS_{res}$ and we define SS_{res}/SS_y to be the proportion of SS_y not explained by x then it is reasonable to define the proportion that *is* explained by x as the ratio of SS_{reg} to SS_y. The **coefficient of determination** is symbolized by $\widehat{R}^2$ and is formally defined as

$$\widehat{R}^2 = \frac{SS_{reg}}{SS_y}$$

(9.9)

The **coefficient of determination** is then the proportion of the y sum of squares that is accounted for or explained by the model.

EXAMPLE 9.2

Find and interpret the coefficients of nondetermination and determination for the wellness and access data in Table 9.1 on the previous page.

Solution Using $SS_y = 115.722$, $SS_{reg} = 107.4322$ and $SS_{res} = 8.2898$ as calculated above,

$$1 - \widehat{R}^2 = \frac{SS_{res}}{SS_y} = \frac{8.2898}{115.722} = .072$$

and

$$\widehat{R}^2 = \frac{SS_{reg}}{SS_y} = \frac{107.4322}{115.722} = .928$$

This means that approximately 93% of the variation in wellness scores was related to the subject's access to medical care scores.[5] This also means that approximately 7% of this variation is not accounted for by access to medical care and must, therefore, be attributed to other unknown factors. ∎

9.2.3 A Note on the Calculation of SS_{res} and SS_{reg}

In the computations carried out above we computed all values of $\hat{y}$ in order to find SS_{res} and SS_{reg}. We used this method in order to help you understand the coefficients of determination and nondetermination. However, it is not necessary to calculate the $\hat{y}$ in order to obtain these quantities. We will demonstrate these alternative methods in this section. They will prove useful when we later study multiple linear regression.

If we substitute $a + bx$ for $\hat{y}$ in Equation 9.7 on page 322, use some basic algebra and the summation rules outlined in Section 2.3.2 on page 13, the following result is obtained.

$$SS_{res} = \sum y^2 - a \sum y - b \sum xy$$

(9.10)

---

[5]This is an unrealistically high proportion to attribute to medical care access but we wanted to impress you with a strong model as your first introduction to SLR.

Substituting previously calculated values into Equation 9.10 produces

$$SS_{res} = 575 - (.8744)(83) - (.7059)(700) = 8.2948$$

which, except for rounding,[6] is the same result obtained previously.
SS_{reg} may be computed directly as follows.

$$\boxed{SS_{reg} = b^2 SS_x} \tag{9.11}$$

Substituting previously calculated values gives

$$SS_{reg} = \left(.7059^2\right)(215.6) = 107.4324$$

which again, except for rounding, is the result previously calculated.

9.2.4 Further Comments on the Coefficients of Determination and Nondetermination

When the correlation between x and y is one or negative one, the regression model will achieve perfect prediction of y. This means that for each subject $\hat{y} = y$ so that $SS_{res} = \sum(y - \hat{y})^2$ becomes $\sum(y - y)^2 = 0$ which means that $1 - \widehat{R}^2 = \frac{SS_{res}}{SS_y}$ is zero. This also implies that $SS_{reg} = \sum(\hat{y} - \bar{y})$ becomes $\sum(y - \bar{y})^2 = SS_y$ so that $\widehat{R}^2 = \frac{SS_{reg}}{SS_y} = \frac{SS_y}{SS_y}$ is one. So when $r = 1$ or -1, all of the variation in y is accounted for by x so that there is no unexplained variation in y.

By contrast, when the correlation between y and x is zero, b is, by Equation 9.4, also zero so that $a = \bar{y} - b\bar{x}$ becomes simply $\bar{y}$. The regression model is then $\hat{y} = \bar{y}$. Thus, when x has no linear relationship to y, the model will predict $\bar{y}$ for every subject regardless of the subject's x score. In this case, $SS_{res} = \sum(y - \hat{y})^2$ becomes $\sum(y - \bar{y})^2 = SS_y$ so that $1 - \widehat{R}^2$ becomes $\frac{SS_y}{SS_y} = 1.0$. At the same time, SS_{reg} will be $\sum(\bar{y} - \bar{y})^2$ which is zero. So when there is no linear relationship between x and y, x will explain none of the variation in y so that all this variation is unaccounted for.

The symbols $\widehat{R}^2$ and $1 - \widehat{R}^2$ for the coefficients of determination and nondetermination respectively are not arbitrary. If we let r_{yx} represent the Pearson product-moment correlation between y and x and $r_{y\hat{y}}$ represent the correlation between y and $\hat{y}$, then in general, the symbol $\widehat{R}$ represents $r_{y\hat{y}}$. It follows that $\widehat{R}^2 = r_{y\hat{y}}^2$ and that $1 - \widehat{R}^2 = 1 - r_{y\hat{y}}^2$. In the present case this is easy to verify since for simple linear regression (but not multiple regression which we discuss below), $r_{y\hat{y}} = r_{yx}$ which we previously calculated to be .9635. We note that for the data in Table 8.2, $\widehat{R}^2 = r_{y\hat{y}}^2 = r_{yx}^2 = .9635^2 = .928$ which is the same result obtained with Equation 9.9. The coefficient of nondetermination for this data would be $1 - .9635^2 = .072$ which is the result obtained by Equation 9.8.

---

[6]We previously calculated this value to be 8.2898.

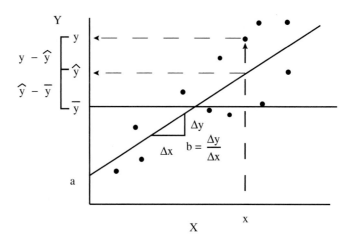

FIGURE 9.1: Depiction of the relationship between a simple linear regression model, predicted values of y and $y - \hat{y}$ and $\hat{y} - \bar{y}$.

In addition to the expressions in Equations 9.8 and 9.9 then, we can characterize the coefficients of nondetermination and determination as

$$1 - \widehat{R}^2 = 1 - r_{y\hat{y}}^2 \tag{9.12}$$

and

$$\widehat{R}^2 = r_{y\hat{y}}^2 \tag{9.13}$$

The relationship between the SLR model, the predicted values and the quantities $y - \hat{y}$ and $\hat{y} - \bar{y}$ are shown in Figure 9.1. You will recognize $\hat{y} = a + bx$ as the equation for a line. This **least squares regression line** is depicted in the figure along with the bivariate plot of some arbitrary data set. As shown in the figure, the a and b terms in the model represent the y intercept and slope of the line respectively. Geometrically speaking, this line relates each value of x to a predicted value of y as is shown for one of the x values. The line is constructed so as to minimize the squared vertical distances of the observations from the line. That is, $\sum(y - \hat{y})^2$ is less than would be achieved with any other line.

9.2.5 Inference Regarding b and $\widehat{R}^2$

Just as the sample mean and other statistics with which you are familiar can be tested for significance, so too can a, b, and $\widehat{R}^2$. A test of a is usually of little interest and will not be covered here. Tests of b and $\widehat{R}^2$ produce the same result in SLR but not in MLR. We will cover both tests here so that you will be familiar with them when you begin your study of MLR models.

A test of the null hypothesis

$$H_0 : \beta = 0$$

where β is the parameter counterpart of b can be conducted by means of Equation 9.14.[7]

$$t = \frac{b}{\sqrt{\frac{MS_{res}}{SS_x}}} \tag{9.14}$$

MS_{res} is the **mean square residual** which is defined as SS_{res} divided by its associated degrees of freedom which are $n - 2$. SS_x is the x sum of squares as discussed on page 320. The resultant t statistic is referenced to a t distribution with $n - 2$ degrees of freedom.

The lower and upper limits of a confidence interval for the estimation of β can be obtained by

$$L = b - t\sqrt{\frac{MS_{res}}{SS_x}} \tag{9.15}$$

$$U = b + t\sqrt{\frac{MS_{res}}{SS_x}} \tag{9.16}$$

where t has $n - 2$ degrees of freedom.

A test of the null hypothesis

$$H_0 : R^2 = 0$$

where R^2 is the parameter counterpart of $\widehat{R}^2$ can be conducted by means of Equation 9.17.

$$F = \frac{\widehat{R}^2}{\frac{1 - \widehat{R}^2}{n - 2}} \tag{9.17}$$

The resultant F statistic is referenced to an F distribution with numerator and denominator degrees of freedom of one and $n - 2$ respectively.

In the case of SLR, these two tests will always produce the same result in terms of the reject or fail to reject decision. This is reasonable since if there is no linear relationship between x and y, b will be zero as will the proportion of variance accounted for by x. In MLR models these t and F tests will not generally produce the same result because they will address different questions and hence test different hypotheses.

EXAMPLE 9.3

Use the results obtained in connection with the information contained in Table 9.1 to test the hypotheses $H_0 : \beta = 0$ and $\widehat{R}^2 = 0$. Form and interpret a two-sided 95% confidence interval for the estimation of β.

Solution SS_{res} was previously found to be 8.2898. Since the sample size (n) was 15, the associated degrees of freedom are $15 - 2 = 13$ so that

$$MS_{res} = \frac{SS_{res}}{n - 2} = \frac{8.2898}{13} = .6377$$

---

[7]Don't confuse this use of β with the same symbol when used to denote the probability of a Type II error.

SS_x was previously calculated to be 215.60. Using theses results in Equation 9.14 along with $b = .7059$ yields

$$t = \frac{b}{\sqrt{\frac{MS_{res}}{SS_x}}} = \frac{.7059}{\sqrt{\frac{.6377}{215.60}}} = 12.980$$

Appendix B shows that, for a two-tailed test conducted at $\alpha = .05$, critical t is ± 2.160. Because obtained t of 12.980 is greater than the critical value $+2.160$, the null hypothesis is rejected.

By Equations 9.15 and 9.16, lower and upper limits for a 95% confidence interval are

$$L = b - t\sqrt{\frac{MS_{res}}{SS_x}} = .7059 - 2.160\sqrt{\frac{.6377}{215.60}} = .588$$

and

$$U = b + t\sqrt{\frac{MS_{res}}{SS_x}} = .7059 + 2.160\sqrt{\frac{.6377}{215.60}} = .823$$

This means that we can be 95% confident that in the population, for each one unit change in access score, the change in predicted wellness score is between .588 and .823 Notice that the result of the significance test is confirmed since zero is not in this interval.

The coefficient of determination can be tested via Equation 9.17 which yields

$$F = \frac{\widehat{R}^2}{\frac{1 - \widehat{R}^2}{n - 2}} = \frac{.928}{\frac{1 - .928}{13}} = 167.56$$

Except for rounding, this is the square of the obtained t value calculated above. Appendix C shows that, for one and 13 degrees of freedom, critical F for $\alpha = .05$ is 4.67. Because obtained F of 167.56 exceeds this value, the null hypothesis is rejected. As a result, we can be confident that the population coefficient of determination is greater than zero. ∎

9.2.6   A Logical Inconsistency

At the beginning of this chapter we indicated that regression models can be used for predicting the value of one variable (y) from that of another variable (x). We also said that these models are used to assess the proportion of variation in one variable (y) accounted for or explained by some other variable (x). We demonstrated how this second role is carried out by means of the coefficients of determination and nondetermination. But to this point, there is a logical difficulty associated with the prediction role.

You must have both x and y available in order to construct the model which is to be used for predictions. But if you know y, why would you not just look at it rather than predict it? The answer is that studies designed to produce predictions take place in two phases. In the first, both x and y are collected for each subject and then used to calculate the model (i.e., a and b). Then, the model is used for predictions in situations where no y value is available.

For example, suppose a hospital administrator is concerned because a significant portion of new nurse hires leave the hospital's employ after a short time. The administrator would like to conduct a study to determine whether length of employment can be predicted at time of employment application.

To this end, all applicants for nursing positions are administered the Maslach Burnout Inventory [35] (discussed on page 94) because the administrator suspects that burnout is at least one component of the problem. Records are then kept as to the length of time the nurses remain on the hospital staff. When sufficient data have been collected, the model is constructed with length of employment (y) being predicted from the burnout scale score (x). If the model appears to be an adequate predictor, future applicants for nursing positions can be administered the burnout scale which can be subsequently used to predict length of employment.[8]

9.3 MULTIPLE LINEAR REGRESSION

The multiple linear regression model is one of the most powerful and flexible statistical tools available to researchers. For this reason, entire books and courses are often devoted to its study. The coverage here will necessarily be restricted to some basic concepts related to this model. Likewise, computations will be restricted to the two predictor case since models with more predictors are computationally difficult without the aid of computers.

9.3.1 The Model

The multiple linear regression model can be expressed as

$$\hat{y} = a + b_1x_1 + b_2x_2 + \cdots + b_px_p \tag{9.18}$$

where $\hat{y}$ is the predicted value of y, a, b_1, $b_2 \ldots b_p$ are constants and x_1, $x_2 \ldots x_p$ are the variables from which predictions are to be made. The subscript p is used to indicate the number of predictors. The number of predictors is limited only by the amount of data available for construction of the model.

In this and the following sections, we will show you how, in the case of two predictors, to construct the model and how to perform *full model* and *partial F tests* for models of any size. We will also provide comments regarding the utility of the MLR model.

9.3.2 Calculation of the Model

We will use the data in Table 9.2 on the next page to demonstrate the calculations involved in formulating a two predictor model.[9]

---

[8]Rarely would a single variable be an adequate predictor of such a multifaceted phenomenon as time of employment. This implies that MLR would likely be used for a study of this sort.

[9]In general, calculations for MLR models are carried out via matrix manipulations rather than the arithmetic methods demonstrated here. However, such calculations are quite complex and far beyond the scope of this chapter.

TABLE 9.2: Data used for construction of a two predictor MLR model.

Subject Number	y	x_1	x_2	y^2	x_1^2	x_2^2	yx_1	yx_2	x_1x_2
1	9	2	5	81	4	25	18	45	10
2	15	8	9	225	64	81	120	135	72
3	4	1	3	16	1	9	4	12	3
4	8	5	5	64	25	25	40	40	25
5	11	1	7	121	1	49	11	77	7
6	7	7	5	49	49	25	49	35	35
7	23	10	10	529	100	100	230	230	100
8	4	9	10	16	81	100	36	40	90
9	5	2	1	25	4	1	10	5	2
10	13	7	5	169	49	25	91	65	35
11	15	9	7	225	81	49	135	105	63
12	22	9	10	484	81	100	198	220	90
13	11	5	1	121	25	1	55	11	5
14	4	8	3	16	64	9	32	12	24
15	17	8	10	289	64	100	136	170	80
$\sum$	168	91	91	2430	693	699	1165	1202	641

We begin by calculating some intermediate values that will be used in the equations that follow.[10]

$$SS_y = \sum y^2 - \frac{\left(\sum y\right)^2}{n} = 2430 - \frac{(168)^2}{15} = 548.4000$$

$$SS_{x_1} = \sum x_1^2 - \frac{\left(\sum x_1\right)^2}{n} = 693 - \frac{(91)^2}{15} = 140.9333$$

$$SS_{x_2} = \sum x_2^2 - \frac{\left(\sum x_2\right)^2}{n} = 699 - \frac{(91)^2}{15} = 146.9333$$

$$SS_{yx_1} = \sum yx_1 - \frac{\left(\sum y\right)\left(\sum x_1\right)}{n} = 1165 - \frac{(168)\,(91)}{15} = 145.8000$$

$$SS_{yx_2} = \sum yx_2 - \frac{\left(\sum y\right)\left(\sum x_2\right)}{n} = 1202 - \frac{(168)\,(91)}{15} = 182.8000$$

$$SS_{x_1x_2} = \sum x_1x_2 - \frac{\left(\sum x_1\right)\left(\sum x_2\right)}{n} = 641 - \frac{(91)\,(91)}{15} = 88.9333$$

---

[10]Better make yourself comfortable, this could get tedious.

The equations for a, b_1 and b_2 are then

$$\boxed{a = \bar{y} - b_1\bar{x}_1 - b_2\bar{x}_2} \tag{9.19}$$

$$\boxed{b_1 = \frac{(SS_{x_2})(SS_{yx_1}) - (SS_{x_1x_2})(SS_{yx_2})}{(SS_{x_1})(SS_{x_2}) - (SS_{x_1x_2})^2}} \tag{9.20}$$

$$\boxed{b_2 = \frac{(SS_{x_1})(SS_{yx_2}) - (SS_{x_1x_2})(SS_{yx_1})}{(SS_{x_1})(SS_{x_2}) - (SS_{x_1x_2})^2}} \tag{9.21}$$

Substituting the results obtained above into Equations 9.20 and 9.21 yields

$$b_1 = \frac{(146.9333)(145.8000) - (88.9333)(182.8000)}{(140.9333)(146.9333) - (88.9333)^2} = .4036$$

and

$$b_2 = \frac{(140.9333)(182.8000) - (88.9333)(145.8000)}{(140.9333)(146.9333) - (88.9333)^2} = .9998$$

Noting that

$$\bar{y} = \frac{\sum y}{n} = \frac{168}{15} = 11.2000$$

$$\bar{x}_1 = \frac{\sum x_1}{n} = \frac{91}{15} = 6.0667$$

and

$$\bar{x}_2 = \frac{\sum x_2}{n} = \frac{91}{15} = 6.0667$$

so that the computation for a is by Equation 9.19

$$a = \bar{y} - b_1\bar{x}_1 - b_2\bar{x}_2 = 11.2000 - (.4036)(6.0667) - (.9998)(6.0667) = 2.6860$$

The two predictor model is then

$$\hat{y} = a + b_1x_1 + b_2x_2 = 2.6860 + .4036x_1 + .9998x_2$$

The value of .4036 for b_1 means that for an increase of one unit in x_1 with no change in x_2, predicted $\hat{y}$ will increase by .4036. The value of .9998 for b_2 is interpreted in a similar manner with x_1 being held constant.

We could use this model to compute all values of $\hat{y}$ so that Equations 9.6 and 9.7 can be used in conjunction with SS_y to find $\widehat{R}^2$ and $1 - \widehat{R}^2$ but SS_{reg} can be computed more directly as

$$\boxed{SS_{reg} = b_1 SS_{yx_1} + b_2 SS_{yx_2} + \cdots + b_p SS_{yx_p}} \tag{9.22}$$

where p represents the number of predictors in the model. In the present two predictor case, SS_{reg} is calculated as

$$SS_{reg} = b_1 SS_{yx_1} + b_2 SS_{yx_2} = (.4036)(145.8000) + (.9998)(182.8000) = 241.6083$$

From Equation 9.9 on page 324

$$\widehat{R}^2 = \frac{SS_{reg}}{SS_y} = \frac{241.6083}{548.4000} = .4406$$

This means that approximately 44 percent of the variation in y is accounted for by the combination of x_1 and x_2. This also implies that approximately $1 - \widehat{R}^2 = 1 - .4406 = .5594$ of the variation in y is unexplained.

9.3.3 Tests of Significance for $\widehat{R}^2$ and bs

A test of the null hypothesis

$$H_0 : R^2_{y.1,\cdots,p} = 0$$

or equivalently

$$H_0 : \beta_1 = \beta_2 = \cdots = \beta_p = 0$$

for models with any number of predictors can be carried out by means of the following F test.

$$F = \frac{\frac{\widehat{R}^2}{p}}{\frac{1-\widehat{R}^2}{N-p-1}} \qquad (9.23)$$

In this equation p is the number of predictor variables in the model and N is the number of observations (e.g., subjects). The obtained F statistic has numerator and denominator degrees of freedom of p and $N - p - 1$ respectively.

A test of the model constructed from the data in Table 9.2 on page 330 would be carried out as follows.

$$F = \frac{\frac{\widehat{R}^2}{p}}{\frac{1-\widehat{R}^2}{N-p-1}} = \frac{\frac{.4406}{2}}{\frac{1-.4406}{12}} = 4.73$$

From Appendix C it can be seen that for numerator and denominator degrees of freedom of two and 12 respectively and $\alpha = .05$, critical F is 3.89. Because obtained F of 4.73 is greater than this value, the null hypothesis is rejected. This means that the researcher can be assured that the model does account for some of the variation in the y variable.

EXAMPLE 9.4

A researcher is interested in attempting to explain the variation in blood pressures that occur among adult men. To this end the researcher gathers data on (1) average daily exercise measured in hours (x_1), (2) age (x_2), and (3) weight (x_3) for 74 subjects. The outcome variable (y) is systolic blood pressure. The (fictitious) model produced by these data has an associated $\widehat{R}^2$ of .2106 and is as follows.

$$\hat{y} = 33.5522 + .1710x_1 + .1033x_2 + .4471x_3$$

(a) What would the predicted systolic blood pressure be for a 21-year-old man who exercised one hour per day on average and weighed 165 pounds? (b) What would the predicted value be for a man of the same age who also exercised one hour per day on average but weighed 185 pounds? (c) What does the value of $b_3 = .4471$ mean? (d) Test the model $\widehat{R}^2$ for significance. (e) Interpret the results of this test.

Solution (a) The predicted systolic blood pressure for a 21 year old man who exercised one hour per day on average and weighed 165 pounds would be

$$\hat{y} = 33.5522 + (.1710)(1) + (.1033)(21) + (.4471)(165) = 109.6640$$

(b) The predicted value for a man of the same age who also exercised one hour per day on average but weighed 20 pounds more would be

$$109.6640 + (20)(.4471) = 118.6060$$

(c) The value of $b_3 = .4471$ means that, when other variables remain unchanged, for each pound of weight increase (decrease), predicted systolic blood pressure will increase (decrease) by .4471. This is the basis of the calculation given in (b).

(d) A test of the null hypothesis

$$H_0 : R^2 = 0$$

or equivalently

$$H_0 : \beta_1 = \beta_2 = \beta_3 = 0$$

can be conducted via Equation 9.23 as

$$F = \frac{\frac{\widehat{R}^2}{p}}{\frac{1-\widehat{R}^2}{N-p-1}} = \frac{\frac{.2106}{3}}{\frac{1-.2106}{74-3-1}} = 6.22$$

Appendix C shows that for a test at level $\alpha = .05$ with numerator and denominator degrees of freedom of three and 70 respectively, critical F is 2.74. Because obtained F of 6.22 is greater than this value the null hypothesis is rejected.

(e) The significant test assures the researcher that some portion of the variation in systolic blood pressures is accounted for by the combination of exercise, age and weight. An estimate of this proportion is provided by $\widehat{R}^2 = .2106$. ■

9.3.4 The Partial F Test

To this point we have treated MLR models as static entities. That is, we have spoken of one, two and three predictor models and have indicated that models may have any number of predictors so long as the researcher has sufficient data to support their construction.[11] But how is the decision made as to how many predictors should be in a given model? The answer to this question can be quite complex and depends on, among other things, the purpose for which the model is being used and the hypotheses in which the researcher has an interest.

Regardless of the specifics underlying the question as to the number of predictors to be included in the model, the answer often revolves around the question as to whether adding

---

[11]It is mathematically necessary that $p < N$. From a practical point of view, N should be several times greater than p.

more variables to an existing model will enhance R^2. For example, a researcher who is interested in identifying variables that explain the variation in systolic blood pressure may wish to determine whether adding subject's weight to a model that already contains his average daily exercise and age will add significantly to the model $\widehat{R}^2$. Said differently, can a patient's weight explain a significant portion of the variation in systolic blood pressure beyond that which is explained by exercise and age? Before looking at the methods used to answer such questions, it will be helpful to introduce the following notation.

Let $x_1, x_2, \ldots, x_p$ be a set of variables available for use in a MLR model. Let $\widehat{R}^2_{y.1}$ represent the value of $\widehat{R}^2$ obtained when only x_1 is used as a predictor. Likewise, $\widehat{R}^2_{y.2}$ would represent the value of $\widehat{R}^2$ obtained when only x_2 is used as a predictor. We use the notation $\widehat{R}^2_{y.1,2}$ to represent the value of $\widehat{R}^2$ obtained when both x_1 and x_2 are used as predictors. The same notation is used for any subset of predictors. For example, $\widehat{R}^2_{y.1,3,4}$ would represent the value of $\widehat{R}^2$ obtained when x_1, x_3, and x_4 are the variables used as predictors. We also use the notation $\widehat{R}^2_{y.L}$ and $\widehat{R}^2_{y.S}$ to represent the $\widehat{R}^2$s of two models whose variables are unspecified but in which the variables that make up $\widehat{R}^2_{y.S}$ are a lesser subset of those that make up $\widehat{R}^2_{y.L}$.[12] For example, $\widehat{R}^2_{y.L}$ might be $\widehat{R}^2_{y.1,3,4}$ and $\widehat{R}^2_{y.S}$ be $\widehat{R}^2_{y.1,4}$. Notice that the variables in $\widehat{R}^2_{y.S}$ are a lesser subset of those in $\widehat{R}^2_{y.L}$.

EXAMPLE 9.5

Given $\widehat{R}^2_{y.L} = \widehat{R}^2_{y.1,3,5,7,9}$, indicate which of the following coefficients of determination could constitute $\widehat{R}^2_{y.S}$. Specify why each of the alternatives is included or excluded as a possibility.

(a) $\widehat{R}^2_{y.1,3,5,7,9}$

(b) $\widehat{R}^2_{y.1,7,9}$

(c) $\widehat{R}^2_{y.1,4,7,9}$

(a) $\widehat{R}^2_{y.1,3,5,7}$

Solution

(a) is excluded because it is not a *lesser* subset of $\widehat{R}^2_{y.L}$. That is, it has the same number of variables as does $\widehat{R}^2_{y.L}$.

(b) is included because it has fewer variables than does $\widehat{R}^2_{y.L}$ and all of its variables are in $\widehat{R}^2_{y.L}$.

(c) is excluded because it has a variable (x_4) that is not in $\widehat{R}^2_{y.L}$.

---

[12]L represents the Larger of the two models and S the Smaller.

(d) is included because it has fewer variables than does $\widehat{R}^2_{y.L}$ and all of its variables are in $\widehat{R}^2_{y.L}$. ∎

An important question arises as to whether, or under what conditions,

$$\widehat{R}^2_{y.1} + \widehat{R}^2_{y.2} + \cdots + \widehat{R}^2_{y.p} = \widehat{R}^2_{y.1,2,\ldots,p}$$

The answer is that this equality will be true only when all x variables are uncorrelated. This result has many implications for the interpretation of results obtained from MLR models.

A test of the null hypothesis

$$H_0 : R^2_{y.L} - R^2_{y.S} = 0$$

can be conducted by means of the following F test.

$$F = \frac{\dfrac{\widehat{R}^2_{y.L} - \widehat{R}^2_{y.S}}{p_L - p_S}}{\dfrac{1 - \widehat{R}^2_{y.L}}{N - p_L - 1}} \tag{9.24}$$

In this equation p_L and p_S refer to the number of variables in the larger and smaller models respectively and N is the total number of observations (e.g., subjects) in the analysis. The resultant F statistic has numerator and denominator degrees of freedom of $p_L - p_S$ and $N - p_L - 1$ respectively. F tests that compare models in the manner shown here are referred to as **partial F tests**.

EXAMPLE 9.6

On page 332 we used the data from Table 9.2 to calculate $\widehat{R}^2_{y.12} = .4406$. In addition, we have calculated[13] $\widehat{R}^2_{y.1} = .2750$ and $\widehat{R}^2_{y.2} = .4147$. Use these results to answer the following questions.

(a) Does x_1 alone account for a significant proportion of the variation in y?

(b) Does x_2 alone account for a significant proportion of the variation in y?

(c) Do x_1 and x_2 when used together in a model account for a significant proportion of the variation in y?

(d) Does a model that uses both x_1 and x_2 as predictors account for a significantly greater proportion of the variation in y than does a model that uses only one of the two variables?

---

[13]Calculation of these results is not shown but you should feel free to use the methods of Section 9.2.3 to verify their validity.

Solution   The answers to questions (a), (b), and (c) can be obtained through application of Equation 9.23 on page 332. For question (a) we wish to test the hypothesis

$$H_0 : R_{y.1}^2 = 0$$

Application of Equation 9.23 yields

$$F = \frac{\frac{\widehat{R}^2}{p}}{\frac{1-\widehat{R}^2}{N-p-1}} = \frac{\frac{.2750}{1}}{\frac{1-.2750}{15-1-1}} = 4.93$$

Reference to Appendix C shows that for $\alpha = .05$ and numerator and denominator degrees of freedom of $p = 1$ and $N - p - 1 = 15 - 1 - 1 = 13$ respectively, critical F is 4.67. Because obtained F of 4.93 is greater than this value, the null hypothesis is rejected. We conclude, therefore, that x_1 alone accounts for a significant proportion of the variation in y.

For question (b) we wish to test the hypothesis

$$H_0 : R_{y.2}^2 = 0$$

Application of Equation 9.23 yields

$$F = \frac{\frac{\widehat{R}^2}{p}}{\frac{1-\widehat{R}^2}{N-p-1}} = \frac{\frac{.4147}{1}}{\frac{1-.4147}{15-1-1}} = 9.21$$

Again, reference to Appendix C shows that for $\alpha = .05$ and numerator and denominator degrees of freedom of $p = 1$ and $N - p - 1 = 15 - 1 - 1 = 13$ respectively, critical F is 4.67. Because obtained F of 9.21 is greater than this value, the null hypothesis is rejected. We conclude, therefore, that x_2 alone accounts for a significant proportion of the variation in y.

For question (c) we wish to test the hypothesis

$$H_0 : R_{y.12}^2 = 0$$

Application of Equation 9.23 yields

$$F = \frac{\frac{\widehat{R}^2}{p}}{\frac{1-\widehat{R}^2}{N-p-1}} = \frac{\frac{.4406}{2}}{\frac{1-.4406}{15-2-1}} = 4.73$$

Reference to Appendix C shows that for $\alpha = .05$ and numerator and denominator degrees of freedom of $p = 2$ and $N - p - 1 = 15 - 2 - 1 = 12$ respectively, critical F is 3.89. Because obtained F of 4.73 is greater than this value, the null hypothesis is rejected. We conclude, therefore, that x_1 and x_2 used together in the model account for a significant proportion of the variation in y.

For question (d) we will use Equation 9.24 on the previous page to test the hypotheses

$$H_0 : R_{y.12}^2 - R_{y.1}^2 = 0$$

and

$$H_0 : R^2_{y.12} - R^2_{y.2} = 0$$

Notice that the first hypothesis addresses the question "Does a model that uses both x_1 and x_2 account for a significantly greater proportion of the variation in y than does a model that uses only x_1? The second hypothesis addresses the question "Does a model that uses both x_1 and x_2 account for a significantly greater proportion of the variation in y than does a model that uses only x_2?

For a test of the first hypothesis we compute

$$F = \frac{\frac{\widehat{R}^2_{y.L} - \widehat{R}^2_{y.S}}{p_L - p_S}}{\frac{1 - \widehat{R}^2_{y.L}}{N - p_L - 1}} = \frac{\frac{.4406 - .2750}{2 - 1}}{\frac{1 - .4406}{15 - 2 - 1}} = 3.55$$

Using numerator and denominator degrees of freedom of one and 12 respectively, critical F for a test at the .05 level is found in Appendix C to be 4.75. Because obtained F of 3.55 is less than this value, we fail to reject the null hypothesis. We are unable to demonstrate, therefore, that the two predictor model accounts for a greater proportion of the variation in y than does a model that contains only x_1.

For a test of the second hypothesis we compute

$$F = \frac{\frac{\widehat{R}^2_{y.L} - \widehat{R}^2_{y.S}}{p_L - p_S}}{\frac{1 - \widehat{R}^2_{y.L}}{N - p_L - 1}} = \frac{\frac{.4406 - .4147}{2 - 1}}{\frac{1 - .4406}{15 - 2 - 1}} = .56$$

Using numerator and denominator degrees of freedom of one and 12 respectively, critical F for a test at the .05 level is found in Appendix C to be 4.75. Because obtained F of .56 is less than this value, we fail to reject the null hypothesis. We are unable to demonstrate, therefore, that the two predictor model accounts for a greater proportion of the variation in y than does a model that contains only x_2.

While we were unable to demonstrate that the two predictor model is a better predictor of y than is either of the one predictor models, we would likely choose x_2 if we decided to use only one predictor. Indeed, the difference between $\widehat{R}^2_{y.12}$ and $\widehat{R}^2_{y.2}$ appears small. You should note, however, that choice of variables for the model is not solely a function of statistical tests. Other factors such as cost, difficulty of acquiring data and other practical matters must be considered. ∎

EXAMPLE 9.7

In Example 9.4 on page 332 we presented an example in which a researcher is interested in attempting to explain the variation in blood pressures that occur among adult men. To this end the researcher gathers data on (1) average daily exercise measured in hours (x_1), (2) age (x_2), and (3) weight (x_3) for 74 subjects. The outcome variable (y) is systolic blood pressure. The (fictitious) model produced by these data has an associated $\widehat{R}^2$ of .2106. Does exercise (x_1) significantly increase $\widehat{R}^2$ when this variable is added to a model that already contains age and weight if $\widehat{R}^2_{y.23} = .1905$?

Solution   We wish to test the hypothesis

$$H_0 : R^2_{y.123} - R^2_{y.23} = 0$$

Notice that this compares a model that uses age, weight *and* exercise to a model that contains only age and weight. The partial F test for a test of this hypothesis is

$$F = \frac{\frac{\widehat{R}^2_{y.L} - \widehat{R}^2_{y.S}}{p_L - p_S}}{\frac{1 - \widehat{R}^2_{y.L}}{N - p_L - 1}} = \frac{\frac{.2106 - .1905}{3 - 2}}{\frac{1 - .2106}{74 - 3 - 1}} = 1.78$$

The critical value for an F statistic with $3 - 2 = 1$ and $74 - 3 - 1 = 70$ degrees of freedom is 3.98 so that the null hypothesis is not rejected. This means that we were unable to show that adding the exercise variable to a model that contains age and weight will increase the proportion of blood pressure variation accounted for by the model.   ■

9.4   ASSUMPTIONS

The assumptions underlying SLR and MLR can be summarized as (1) normality, (2) homogeneity of variance, and (3) independence. We will consider each of these in the context of SLR and then elaborate to MLR.

On page 321 we used the data from Table 8.2 on page 297 to construct a SLR model for the prediction of wellness scores from access to medical care scores. We subsequently performed statistical tests related to this model. In order to understand the assumptions underlying these tests, consider a series of populations made up of wellness scores. The first population is made up entirely of wellness scores of persons with the same access score, say a score of 1. The second population is made up entirely of wellness scores of persons with access scores of 2, the third of wellness scores of persons with access scores of 3 and so on. The normality and homogeneity of variance assumptions relate to these populations. More specifically, it is assumed that all of these populations are normally distributed and have the same variance.

If we consider a two predictor MLR model then we consider a series of populations in which each subject has the same x_1 score and the same x_2 score. We don't mean that x_1 and x_2 are the same value, rather that all subjects in a particular population have the same value of x_1 (e.g., 5) and the same value of x_2 (e.g., 20). This concept is carried out to any number of predictors. As before it is assumed that all populations are normally distributed with the same variance. Notice that the x values themselves need not meet these assumptions.

The independence assumption means that the response (y) of each subject is unrelated to the response of any other subject. Though not necessarily true under a specific set of conditions, it is nevertheless the case that SLR and MLR models tend to be most robust to violations of the normality assumption and least robust to violations of the independence assumption.

9.5   SOME ADDITIONAL COMMENTS REGARDING THE UTILITY OF MLR

Earlier in this chapter we stated that the " $\cdots$ multiple linear regression model is one of the most powerful and flexible statistical tools available to researchers." You have seen some of this as you viewed its ability to predict and explain variation. But this hardly imparts the full picture.

The MLR model is an expression of what is termed the General Linear Model. This means that with judicious choice of coded[14] independent and dependent variables, this model can be used to conduct all of the following common statistical tests.

1. Independent samples t test.

2. One-way ANOVA.

3. Factorial ANOVA.

4. Analysis of Covariance.

5. Discriminant Analysis (two group).

Also, various tests for interactions can be performed via this model. In addition, tests for curvilinear relationships such as that depicted in Figure 8.6 on page 308 as well as other forms can be detected.[15]

KEY WORDS AND PHRASES

After reading this chapter you should be able to demonstrate familiarity with the following words and phrases.

$1 - \widehat{R}^2$ 323	$\widehat{R}^2$ 324
coefficient of determination 324	coefficient of nondetermination 323
full model F test 329	least squares model 323
least squares regression line 326	mean square residual 327
multiple linear regression 329	partial F test 333
regression sum of squares 322	residual 320
residual sum of squares 322	simple linear regression 320

---

[14] A simple coding scheme might assign 1's to female and 0's to male subjects.

[15] Needless to say, you should enroll in a MLR course as soon as you finish this one.

EXERCISES

9.1 Respond to each of the following.

(a) Write two different expressions for the coefficient of determination (R^2).

(b) Write two different expressions for the coefficient of nondetermination ($1 - R^2$).

(c) For a SLR model, what values do SS_{res} and SS_{reg} take when $r_{yx} = 1.0$?

(d) For a SLR model, what values do SS_{res} and SS_{reg} take when $r_{yx} = 0.0$?

9.2 Indicate which of the following statements are true and which false.

(a) In general, $\widehat{R}^2_{y.1,2,3} = \widehat{R}^2_{y.1} + \widehat{R}^2_{y.2} + \widehat{R}^2_{y.3}$.

(b) A test of the hypothesis

$$H_0 : R^2_{y.1,2,3}$$

is equivalent to a test of the hypothesis

$$H_0 : \beta_1 = \beta_2 = \beta_3 = 1.$$

(c) A good prediction model will produce large (relatively speaking) residuals.

(d) Large (relatively speaking) values of $\hat{y} - \bar{y}$ are generally associated with good prediction models.

9.3 A test to determine whether adding variables x_1 and x_3 to a model that contains variables x_2 and x_4 increases R^2 can be carried out by testing which of the following hypotheses?

(a) $R^2_{y.1,2,3,4} - R^2_{y.2,4} = 0$

(b) $R^2_{y.1,2,3,4} - R^2_{y.1,3} = 0$

(c) $R^2_{y.1,3} = 0$

(d) $R^2_{y.2,4} = 0$

9.4 A researcher interested in the relationship between body mass index (BMI) and total serum cholesterol wishes to fit a SLR model in which total serum cholesterol is predicted from BMI to the following data. Use this data to respond to each of the following.

Total Cholesterol	BMI
165	25.9
155	20.1
141	22.2
228	30.7
190	28.0
155	29.4
132	20.2
170	20.7
188	26.3
150	18.2

(a) Construct a SLR model.

(b) Find the residuals for the first three subjects.

(c) Find the coefficients of determination and nondetermination.

(d) Test the hypothesis $H_0 : R^2 = 0$.

(e) Test the hypothesis $H_0 : \beta = 0$.

(f) Form a two-sided 95% CI for the estimation of β. Does this CI agree with the result of your hypothesis tests? Explain.

9.5 Answer the following questions by performing the indicated test of significance based on the following information. Assume $N = 40$ in all cases.

(1) $\widehat{R}^2_{y.1,2,3,4} = .68$

(2) $\widehat{R}^2_{y.2,3,4} = .43$

(3) $\widehat{R}^2_{y.4} = .03$

(4) $\widehat{R}^2_{y.1,3} = .27$

(5) $\widehat{R}^2_{y.2,4} = .14$

(6) $\widehat{R}^2_{y.1} = .10$

(a) Do all four variables used together account for any of the variation in Y?

(b) Does adding x_1 to a model that already contains x_2, x_3, and x_4 account for any additional variation in Y?

(c) Do x_2 and x_4 when used together account for any of the variation in Y?

(d) Does adding variables x_2 and x_4 to a model that contains variables x_1 and x_3 increase the proportion of variation accounted for in y?

D. The following questions refer to Case Study D (page 471)

9.6 The researchers conclude from their study that "These correlations suggest quantitation of PBV may offer an objective, easily acquired surrogate predictor of neuropsychological impairment and clinically apparent cognitive/motor dysfunction among HIV-infected persons."

(a) Construct a SLR model to predict NPZ-8 scores from PBV. Use only data from the 15 infected subjects for the calculation.

(b) Interpret the b term in the model in terms of NPZ-8 and PBV.

(c) Find the coefficients of determination and nondetermination associated with this model.

(d) Use the model to predict the NPZ-8 scores of the first five subjects.

(e) From your observation of the coefficient of determination and the predicted values for the first five subjects, would you agree with the authors conclusion?

9.7 Construct a MLR model to predict NPZ-8 scores from PBV and CD4 values. Use only the data from the 15 infected subjects.

(a) Find the coefficient of determination for the two predictor model.

(b) Does adding CD4 to the model significantly increase R^2 at the .05 level?

(c) Use the two predictor model to predict NPZ-8 scores from PBV and CD4 values for the first five subjects. Does the two predictor model appear to predict better than did the one predictor model?

(d) Do PBV and CD4 together account for a statistically significant proportion of the variation in NPZ-8 scores?

M. The following questions refer to Case Study M (page 475)

9.8 Construct a SLR model to predict "after" carbon dioxide scores from "after" oxygen values.

(a) Interpret b.

(b) Estimate β with a 95% confidence interval. Interpret this interval.

(c) Do you believe that a two-tailed test of the null hypothesis $R^2 = 0$ conducted at $\alpha = .05$ would be significant? Give the reason for your answer.

(d) Compute R^2 for the model.

(e) Do you think the correlation between the two variables is positive or negative? Why?

Methods Based on the Permutation Principle

10.1 INTRODUCTION

Most introductory statistics texts have a chapter with a title similar to, "Nonparametric[1] Methods" or "Distribution-Free Tests." Typically, this chapter presents (usually briefly) certain well known rank-based nonparametric[2] tests whose sampling distributions are derived via the permutation method. These tests are special cases of a broader class of permutation-based procedures that employ original observations rather than ranks. With few exceptions, the broader class is not discussed so that only the special cases are reviewed. Additionally, the permutation principle that underlies both types of tests is not explained.

We believe that this approach leaves the student with an unduly narrow view of the topic. As a result, in this text we abandon the more traditional approach (as we have done several times previously) and focus on the broader class of tests with the more familiar rank-based methods being presented as special cases. We have chosen this path for three reasons. First, with the advent of fast algorithms and powerful computers, the broader class of tests are seeing more use in applied research. While these methods have been well understood for many years, the absence of powerful computing platforms has made their application difficult. Second, students without strong mathematics backgrounds can gain deeper insights into these tests than is possible with other forms and third, the flexibility of these tests add a powerful dimension to the researcher's bag of statistical tools.

An understanding of permutation-based methods hinges on an understanding of the mathematical concepts of permutations and combinations. For this reason, we will begin with these concepts. We will then discuss a series of permutation based procedures and demonstrate how their familiar rank-based special case counterparts are obtained. In a final section we will discuss some of the characteristics of these methods and compare them to tests with which you are already familiar.

---

[1] Some authors prefer to hyphenate the term as non-parametric while others do not.
[2] This and other terms will be defined later in the chapter.

343

10.2 SOME PRELIMINARIES

The sampling distributions of many permutation tests rely on permutations and combinations of integers or other numbers. You have probably studied these methods in the past but we will review them here. We will then use them in the applications that follow.

10.2.1 Permutations

Consider an arbitrary left to right ordering of the integers 1, 2, and 3. One such ordering would be 3, 2, 1 while another would be 1, 3, 2. Each of these arrangements of the numbers is termed a **permutation** of the integers 1, 2, and 3. We are interested in determining the number of such arrangements that can be obtained from these numbers. With such a small number of numbers, we could list each such ordering and simply count the number thus obtained. This is done for the permutations listed below.

1	1	2	3
2	1	3	2
3	2	1	3
4	2	3	1
5	3	1	2
6	3	2	1

Because there are no other arrangements of the data that will not duplicate one of those given above, we can be sure that we have listed all possible permutations of the integers 1, 2, and 3. We further note that there are six such possible arrangements.

Closer inspection of the six arrangements reveals that we had three choices for filling the first position in each permutation. That is, we could choose the numbers 1, 2, or 3 for the first position. Once we had filled the first position we had only two remaining choices for the second position. Thus, in permutations 1 and 2 where we filled the first position with the integer 1, we could choose the integers 2 or 3 for the second position. Once the first two positions were filled, there was only one integer left to place in the third position. Thus, with three ways to fill the first position, and for each of these, two ways to fill the second position and for each of these only one way to fill the third position, the total number of permutations was $3 \cdot 2 \cdot 1 = 6$.

As a second example, let us consider the number of permutations that can be realized from the numbers 1, 2, 3, and 4. We list all such arrangements on the facing page. The first thing we notice is that the addition of one digit has dramatically increased the number of permutations from six to 24. We also see that we now have four choices for the first position. After filling the first position we are left with three choices for the second position,

TABLE 10.1: All possible permutations of the integers 1, 2, 3, and 4.

1	1	2	3	4		13	3	1	2	4
2	1	2	4	3		14	3	1	4	2
3	1	3	2	4		15	3	2	1	4
4	1	3	4	2		16	3	2	4	1
5	1	4	2	3		17	3	4	1	2
6	1	4	3	2		18	3	4	2	1
7	2	1	3	4		19	4	1	2	3
8	2	1	4	3		20	4	1	3	2
9	2	3	1	4		21	4	2	1	3
10	2	3	4	1		22	4	2	3	1
11	2	4	1	3		23	4	3	1	2
12	2	4	3	1		24	4	3	2	1

two for the third and but a single value for the fourth. This means that we have $4 \cdot 3 \cdot 2 \cdot 1 = 24$ permutations of the four digits.

This leads to a general conclusion. If we have n things (numbers, dogs, friends, etc.) we can arrange them into $n!$ different permutations. The notation $n!$ is read "n factorial" and means $n(n-1)(n-2)\cdots(1)$. If we let P_n represent the number of ways in which n objects can be permuted then we can write the following.

$$\boxed{P_n = n!} \tag{10.1}$$

EXAMPLE 10.1

How many distinct permutations of the numbers 3.1, 0.0, 13.0, 99.7, and .6 are possible?

Solution   Because there are five numbers, Equation 10.1 shows that the number of distinct permutations is $n! = 5! = 5 \cdot 4 \cdot 3 \cdot 2 \cdot 1 = 120$. ∎

10.2.2   Combinations

Consider now the problem of determining the number of ways in which we can divide n objects, numbers, etc., into two groups. To begin, let us suppose that we wish to divide the numbers 1 through 4 into two groups with two numbers being placed in each group. Order within the two groups is to be ignored. One possibility is to place 1 and 2 in the first group and 3 and 4 in the second. Such a partitioning is called a **combination**. Notice that placing the numbers 2 and 1 in group one and 4 and 3 in the second does not constitute a new combination since the two groups contain the same numbers as before—we have simply listed them in a different order. A second combination could be achieved by placing 1 and 3 in the first group and 2 and 4 in the second. But how many ways can this task be accomplished? As we did earlier with permutations, we could write down all possible combinations and simply count the number thus obtained. As may be seen from this listing, there are six possible combinations.

	Group One		Group Two	
1	1	2	3	4
2	1	3	2	4
3	1	4	2	3
4	2	3	1	4
5	2	4	1	3
6	3	4	1	2

As we did with permutations, it would be helpful if we could find a mathematical expression that would provide us with this number directly rather than having to list all possibilities. Before broaching this problem directly, it will be helpful to re-examine the permutations on the previous page.

EXAMPLE 10.2

Because the 4! *permutations* on the preceding page arrange the numbers in all possible configurations, it must be true that all possible *combinations* are represented therein. How many *combinations* can you find in the 4! = 24 *permutations* listed on the previous page? Assume that the first two positions for each permutation represent group one and the last two group two. List the permutations that represent each combination.

Solution   Each combination with its associated permutations is listed below.

Combination	Permutations
one	1, 2, 7, 8
two	3, 4, 13, 14
three	5, 6, 19, 20
four	9, 10, 15, 16
five	11, 12, 21, 22
six	17, 18, 23, 24

We can see from this listing that there are four permutations in each combination. This is not surprising because for each combination there are two numbers in the first group that can be permuted in two ways and for each of these permutations the two numbers in the second group can be permuted in two ways. For example, in the first combination the first permutation of the numbers in group one is 1 and 2. For this permutation in the first group there are two permutations in the second group—namely 3, 4 and 4, 3. For the second permutation in the first group—2, 1 there are also two permutations in the second group. Since there are two ways to permute the numbers in the first group and two ways to permute the numbers in the second group the total number of permutations for each combination is $2 \cdot 2 = 4$.

In general, if we let n_1 represent the number of objects in the first group and n_2 the number in the second, then the number of permutations in the first group for a given combination will be $n_1!$ and the number in the second group will be $n_2!$ so that each combination will have $n_1!n_2!$ permutations contained therein. ∎

EXAMPLE 10.3

If seven objects are to be divided into two groups with four in the first and three in the second, how many permutations will be in each of the possible combinations?

Solution   Because $n_1 = 4$ and $n_2 = 3$, there will be $4!3! = 24 \cdot 6 = 144$ permutations in each combination.

If we let $C_{n_2}^{n_1}$ represent the number of combinations that can be achieved by placing n_1 objects in the first group and n_2 in the second, then we can reason as follows. For any set of $n!$ permutations there will be $C_{n_2}^{n_1}$ combinations each of which will contain $n_1!n_2!$ permutations so that $n! = C_{n_2}^{n_1}n_1!n_2!$ where $n = n_1 + n_2$. Solving for $C_{n_2}^{n_1}$ gives us the number of ways we can divide n objects into two groups.

$$C_{n_2}^{n_1} = \frac{n!}{n_1!n_2!}$$

(10.2)

∎

EXAMPLE 10.4

How many ways can seven people be divided into two groups if four people are to be placed in the first group and three in the second?

Solution   Equation 10.2 gives

$$C_3^4 = \frac{n!}{n_1!n_2!} = \frac{7!}{4!3!}$$

Some computational effort can be saved by noting that $7! = 7 \cdot 6 \cdot 5 \cdot 4!$ so that

$$C_3^4 = \frac{7 \cdot 6 \cdot 5 \cdot \cancel{4!}}{\cancel{4!}\,3!} = \frac{210}{6} = 35$$

∎

EXAMPLE 10.5

How many ways can the integers 1 through 8 be divided into two groups if four integers are to be placed in each group?

Solution   By Equation 10.2

$$C_4^4 = \frac{n!}{n_1!n_2!} = \frac{8!}{4!4!} = \frac{8 \cdot 7 \cdot 6 \cdot 5 \cdot \cancel{4!}}{4!\cancel{4!}} = \frac{1680}{24} = 70$$ ∎

10.3   APPLICATIONS

We will now consider several tests whose sampling distributions are based on the permutation principle.[3] These tests are usually characterized as being "distribution-free" or "nonparametric." Nonparametric refers to the fact that the sampling distributions of these tests do not depend upon the specification of any population parameters.[4] For example, the sampling distribution of the one mean Z test requires that σ (as well as other parameters) be specified. Likewise, the sampling distribution of the independent samples t test requires that the populations be normally distributed which in turn implies the values of certain parameters such as skew and kurtosis.[5] As a result of this requirement, these tests are characterized as being **parametric**. By contrast, the Wilcoxon rank-sum test, which you will encounter shortly, is often characterized as the nonparametric counterpart of the independent samples t test. The Wilcoxon test does not require that any of the population parameters be specified so that population skew may be zero or any other (finite) value. This test is characterized, therefore, as **nonparametric**.

Further, because parameters need not be specified in order to obtain the sampling distributions for these tests, the shape (e.g., normal) of the sampled population need not be specified. For this reason these tests are often referred to as **distribution-free** tests.[6] So, while the independent samples t test has population normality as one of its underlying assumptions, and is therefore neither nonparametric nor distribution-free, the Wilcoxon rank-sum test does not and therefore is both nonparametric and distribution-free. While there is a distinction to be made between the terms nonparametric and distribution-free, most tests that possess one of the two characteristics also possess the other so that the terms are often used interchangeably.[7]

We wish to also point out that while this chapter is restricted to hypothesis testing, the methods outlined here can also be used to form confidence intervals. However, constructing confidence intervals via the permutation method can be even more computationally difficult than is the hypothesis test so that the topic is best treated with the aid of computers. It is also the problem of computational complexity that necessitates the use of small sample data sets used to illustrate each of the following tests. This is not a particular hardship since we are primarily interested in showing you the principles underlying these tests.

---

[3] As you are now aware, combinations are subsets of permutations so that tests based on combinations can also be considered as being permutation tests.

[4] There are a few exceptions. For example, the nonparametric Wilcoxon rank-sum test assumes that sampling is from a symmetric population which in turn implies a skew parameter of zero.

[5] Skew for the normal distribution is zero while kurtosis is three.

[6] Indeed, though it is beyond the scope of this book, these tests can be justified without the concept of a population.

[7] Statisticians sometimes quibble over this issue but it is not important for our purposes.

We now turn attention to several permutation tests. In the following, two tests will be presented for each statistical concept (e.g., correlation) addressed. The first will use original observations while the second will employ ranks.

10.3.1   Correlation

On page 309 you learned to use Equation 8.4 to test the null hypothesis $H_0 : \rho = 0$ where ρ is the population Pearson product-moment correlation coefficient. One of the assumptions underlying that test is that the sample came from a bivariate normal population. In this section you will learn to test hypotheses concerning the population correlation without having to make this restrictive assumption. Two tests will be demonstrated. The first, known as Pitman's test for correlation, uses original observations while the second, known as the Hoteling and Pabst test, uses ranks in place of the original observations. The underlying logic of these tests is emphasized as well as the relationship between the two.

Emphasis in this and later sections will be placed on understanding underlying principles rather than efficiency. For all practical purposes, tests based on original observations require computers for their implementation which in turn use various algorithms to enhance efficiency. We will not deal with such methods here.

Pitman's Test for Correlation.

Rationale

Suppose that a researcher obtains two measurements on each of four subjects with the result being the data shown below.

Subj.	x	y
1	4	7
2	0	0
3	3	3
4	1	4

By Equation 8.2 on page 296 the Pearson product-moment correlation between x and y is .822. The researcher would like to test the null hypothesis $H_0 : \rho = 0$ but is fairly certain that the population from which the data were sampled is skewed and therefore not normally distributed. As a result, the researcher feels that a test based on Equation 8.4 on page 309 cannot be depended upon to produce valid results. A test that does not require the normality assumption to ensure its validity can be conducted by evoking the following logic.

If the null hypothesis is true—i.e., if there is no correlation between x and y—then the pairings observed in the above data are arbitrary. That is, each x value might just as likely have been paired with any of the other y values as to have been paired with the ones observed in the data set. By chance, the pairings were as observed but could have just as likely been any other pairing under a true null hypothesis.

By contrast, if there is a positive or negative relationship between x and y, pairings will not be random. Rather, for a positive relationship, high x values will tend to be associated with high y values and low x values with low y values. An inverse relationship will exist for a negative association.

Lower Half of Sampling Distribution

$-.949 -.822 -.759 -.759 -.569 -.569 -.569 -.316 -.253 -.253 -.190 -.190$

Upper Half of Sampling Distribution

$.190 .190 .253 .253 .316 .569 .569 .569 .759 .759 .822 .949$

Obtained r

FIGURE 10.1: Permutation sampling distribution of *r* for a particular data set.

It follows that *under a true null hypothesis*, the *y* values could take any of $n! = 4! = 24$ arrangements as shown on the facing page. This means that any of the 24 correlation coefficients obtained from the 24 arrangements of *y* would be equally likely *if the null hypothesis is true*.

The permutation sampling distribution of *r* is constructed by arranging the 24 values of *r* obtained from the 24 permutations of *y* in order from smallest to largest as is shown in Figure 10.1. If the null hypothesis is true, we would expect *obtained r* (i.e., the one obtained from the data collected in the study) to be near zero since a true null hypothesis means that the correlation in the population is zero. On the other hand, if $\rho > 0$ we would expect *r* to take some value in the extreme right tail of the distribution with the opposite being true when $\rho < 0$. In the present case, obtained *r* is .822. What is the probability of obtaining such an extreme value of *r* when the null hypothesis is true? It is simply the proportion of *r*'s that take a value as extreme (i.e., far from zero), or more so, as that obtained from the data. In the present case, only two of the 24 possible values of *r* are as far, or further, from zero as that obtained from the data—namely, .822 and .949. We can now say that of the 24 possible values of *r* that might have been obtained under a true null hypothesis, only 2 are as extreme as that observed from the data. The probability of achieving such an extreme value of *r* under a true null hypothesis is then $2/24 = .083$. This is in fact the *p*-value for a test with alternative $\rho > 0$.

In general, the *p*-value for a test with one-tailed alternative $\rho > 0$ is the proportion of values in the permutation sampling distribution that are greater than or equal to the one observed from the data. When the one-tailed alternative is of the form $\rho < 0$, the *p*-value for this one-tailed alternative is the proportion of values in the permutation distribution that are less than or equal to the one observed from the data. The two-tailed *p* value for alternative $\rho \neq 0$ is calculated by finding the proportion of observations in the upper (lower) tail of the distribution that are greater (less) than or equal to obtained *r* and multiplying this value by 2. The proportion greater than or equal to obtained *r* is multiplied by two when *r* is greater than zero while the proportion less than or equal to obtained *r* is multiplied by two when obtained *r* is less than zero.

TABLE 10.2: All possible sets of data pairs and correlation coefficients for a particular data set.

	1			2			3			4			5
4	7		4	0		4	3		4	7		4	0
0	0		0	7		0	7		0	3		0	3
3	3		3	3		3	0		3	0		3	7
1	4		1	4		1	4		1	4		1	4

$r = .822$ $r = -.949$ $r = -.759$ $r = .253$ $r = -.190$

	6			7			8			9			10
4	3		4	4		4	0		4	7		4	4
0	0		0	0		0	4		0	4		0	7
3	7		3	7		3	7		3	0		3	0
1	4		1	3		1	3		1	3		1	3

$r = .569$ $r = .759$ $r = -.253$ $r = .190$ $r = -.569$

	11			12			13			14			15
4	0		4	7		4	7		4	3		4	4
0	7		0	0		0	3		0	7		0	7
3	4		3	4		3	4		3	4		3	3
1	3		1	3		1	0		1	0		1	0

$r = -.822$ $r = .949$ $r = .759$ $r = -.253$ $r = -.190$

	16			17			18			19			20
4	7		4	3		4	4		4	4		4	3
0	4		0	4		0	3		0	3		0	4
3	3		3	7		3	7		3	0		3	0
1	0		1	0		1	0		1	7		1	7

$r = .569$ $r = .316$ $r = .569$ $r = -.316$ $r = -.569$

	21			22			23			24
4	0		4	4		4	3		4	0
0	4		0	0		0	0		0	3
3	3		3	3		3	4		3	4
1	7		1	7		1	7		1	7

$r = -.759$ $r = .253$ $r = .190$ $r = -.569$

Conducting the Test

The steps for conducting Pitman's Test For Correlation are as follows.

1. Compute r for the data obtained from the study. Call this value obtained r.

2. Having fixed one of the two variables (e.g., x), form all possible sets of pairings of x and y by forming all possible $n!$ permutations of the unfixed (e.g., y) variable as shown for an example data set on page 351.

3. Calculate r for each set of data pairings as shown on page 351.

4. Arrange the values of r obtained from step 3 from smallest to largest. This is the permutation sampling distribution of r.

5. For a one-tailed test with alternative $H_A : \rho > 0$, calculate the proportion of values in the permutation distribution that are greater than or equal to obtained r. This is the p-value for the one-tailed test.

6. For a one-tailed test with alternative $H_A : \rho < 0$, calculate the proportion of values in the permutation distribution that are less than or equal to obtained r. This is the p-value for the one-tailed test.

7. For the two-tailed test with alternative $H_A : \rho \neq 0$:

   (a) calculate the proportion of values in the permutation distribution that are greater than or equal to obtained r if obtained r is greater than zero. Multiply this value by two to find the two-tailed p-value.

   (b) calculate the proportion of values in the permutation distribution that are less than or equal to obtained r if obtained r is less than zero. Multiply this value by two to find the two-tailed p-value.

EXAMPLE 10.6

Use the permutation sampling distribution of r shown in Figure 10.1 to find the p-value for the following test

$$H_0 : \rho = 0$$
$$H_A : \rho < 0$$

assuming obtained r is $-.949$.

Solution   Because the alternative is of the form $\rho < 0$, we are to perform a one-tailed test with the p-value being the proportion of values in the sampling distribution that are less than or equal to the one obtained from the data. In this case only the permutation value of r equals $-.949$ meets the criterion of being less than or equal to obtained r of $-.949$ so that $p = \frac{1}{24} = .042$. If α were .05, we would conclude that $\rho < 0$.   ∎

EXAMPLE 10.7

Use the permutation sampling distribution of r shown in Figure 10.1 to find the p-value for the following test

$$H_0 : \rho = 0$$
$$H_A : \rho \neq 0$$

assuming obtained r is .759.

Solution   Because obtained r is greater than zero, we find the proportion of permutation values that are greater than or equal to obtained r. We note that two values of .759, one value of .822, and one of .949 are greater than or equal to obtained r so that the desired p-value is $(2)(4/24) = .333$.   ∎

EXAMPLE 10.8

Given five xy data pairs, what is the smallest p-value that could be obtained from a permutation test of $H_0 : \rho = 0$? Would it be possible to conduct a two-tailed test at $\alpha = .01$. What about a one-tailed test?

Solution   Because there are $n! = 5! = 120$ possible sets of xy pairings each of which will contribute a value to the permutation distribution, the smallest p-value will be realized when obtained r takes the value of the largest or smallest of the permutation r's. Assuming no tied values at the extremes of the permutation distribution, the smallest possible one-tailed p-value would be $\frac{1}{120} = .0083$ and the smallest for a two-tailed test would be $(2)\left(\frac{1}{120}\right) = .0167$. It would not be possible, therefore, to conduct a two-tailed test at $\alpha = .01$ because a significant finding could never be achieved at this sample size. It would be possible, however, to conduct a one-tailed test at $\alpha = .01$.   ∎

Assumptions

The principle assumption underlying Pitman's test for correlation relates to independence. Not only must it be true that x and y are uncorrelated under a true null hypothesis, it must also be true that there is independence among all values of x as well as all values of y. For example, assume that visual acuities have been taken on the left (x_1) and right (y_1) eyes of a patient. At another point in time, the patient is tested once again producing the measures x_2 and y_2. Thus, in a set of n pairs of measurements to be correlated, x_1 and x_2 as well as y_1 and y_2 are from the same person. Suppose further that the subject has poor eye sight. Under these conditions, the values of (y_1) and (y_2) are dependent so that the value of y_2 will depend upon the value of y_1. In this circumstance all $n!$ permutations of y are not all equally likely since for fixed values of x_1 and x_2, y_2 must take some value near y_1 rather than any of the other y values. Pitman's test cannot be depended upon to be robust against violations of this sort.

Hoteling and Pabst Test for Rank Correlation.   The Hoteling and Pabst test (H & P) is similar to Pitman's test for correlation except that the original data are converted to ranks before obtained r and the permutation distribution of r are calculated. It can be shown that, in the absence of duplicate values for x or y, calculating the Pearson product-moment correlation coefficient on ranks as shown below produces the same result as does a well known nonparametric correlation statistic called **Spearman's rho**. For this reason, the test described in this section which employs the Pearson product-moment correlation coefficient is also a test of Spearman's rho.

Rationale

In order to conduct Pitman's test for eight pairs of observations, we would have to find $8! = 40, 320$ sets of data pairs and compute the correlation coefficient for each. The number of data set pairs rises rapidly as we consider larger n. It might be worth the trouble of generating this permutation distribution for a single test but we cannot help but be discouraged by the fact that for each new data set, the task must be repeated. But consider what would happen if we replaced each observation in the original data set by its rank with the smallest x value receiving a rank of 1, the second smallest a rank of 2, and so on with the largest value receiving a rank of n. We would then rank the y variable in the same manner.

We could then compute the correlation coefficient on the ranks rather than the original scores. This would produce a different result but would still provide a measure of the relationship between x and y. The attraction of this strategy is that the permutations would have to be carried out only once for any given sample size. When a new set of data with the same n were to be tested, the same ranks would appear in the analysis as before even though the original data had changed. It follows that the permutation distribution would be the same regardless of the values of the original data since the correlations are being conducted on their ranks. It remains only to find the permutation distribution for each sample size (n).

Let us now conduct a H & P test on the same data used previously for Pitman's test. These data are repeated here.

Subj.	x	y
1	4	7
2	0	0
3	3	3
4	1	4

The H & P test requires that we replace the original data with their respective ranks. The result is shown below. We use the notation R_x and R_y to denote the ranks of x and y.

Subj.	R_x	R_y
1	4	4
2	1	1
3	3	2
4	2	3

Obtained r for the ranks is .800. The 24 sets of permutations for the ranks with their associated values of r are shown on the facing page. The resulting permutation sampling distribution is shown in Figure 10.2 on page 356. As may be seen from this distribution, there are four values of the permutation sampling distribution, i.e., three values of .80 and one of 1.00, that are greater than or equal to the obtained r value of .80. Thus, the p-value for a one-tailed test would be $4/24 = .167$ and for a two-tailed test would be $(2)(4/24) = .333$.

TABLE 10.3: All possible sets of rank pairs and correlation coefficients for $n = 4$ data pairs.

1	
4	4
1	1
3	2
2	3

$r = .80$

2	
4	1
1	4
3	2
2	3

$r = -1.00$

3	
4	2
1	4
3	1
2	3

$r = -.80$

4	
4	4
1	2
3	1
2	3

$r = .40$

5	
4	1
1	2
3	4
2	3

$r = -.20$

6	
4	2
1	1
3	4
2	3

$r = .40$

7	
4	3
1	1
3	4
2	2

$r = .80$

8	
4	1
1	3
3	4
2	2

$r = -.40$

9	
4	4
1	3
3	1
2	2

$r = .20$

10	
4	3
1	4
3	1
2	2

$r = -.40$

11	
4	1
1	4
3	3
2	2

$r = -.80$

12	
4	4
1	1
3	3
2	2

$r = 1.00$

13	
4	4
1	2
3	3
2	1

$r = .80$

14	
4	2
1	4
3	3
2	1

$r = -.40$

15	
4	3
1	4
3	2
2	1

$r = -.20$

16	
4	4
1	3
3	2
2	1

$r = .40$

17	
4	2
1	3
3	4
2	1

$r = .00$

18	
4	3
1	2
3	4
2	1

$r = .60$

19	
4	3
1	2
3	1
2	4

$r = .00$

20	
4	2
1	3
3	1
2	4

$r = -.60$

21	
4	1
1	3
3	2
2	4

$r = -.80$

22	
4	3
1	1
3	2
2	4

$r = .40$

23	
4	2
1	1
3	3
2	4

$r = .20$

24	
4	1
1	2
3	3
2	4

$r = -.40$

Lower Half of Sampling Distribution

$-1.00 -.80 -.80 -.80 -.60 -.40 -.40 -.40 -.40 -.20 -.20 .00$

Upper Half of Sampling Distribution

$.00 .20 .20 .40 .40 .40 .40 .60 .80 .80 .80 1.00$

Obtained r

FIGURE 10.2: Permutation sampling distribution of r calculated on ranks where $n = 4$.

Unlike Pitman's test which uses original observations, the H & P test on ranks does not test $H_0 : \rho = 0$ but rather tests the less specific hypothesis that x and y are independent. The two-tailed alternative claims that the two variables are not independent while the one-tailed versions assert a positive or negative relationship between the variables. It should also be understood that the correlation on ranks is, unlike the correlation on original scores, not an expression of the degree of linearity between x and y, but rather an assessment of the monotonic relationship between x and y. A **monotonic relationship** is one in which increases or decreases in one variable are accompanied by increases or decreases in the other variable but not necessarily in a straight line fashion. An example would be if x took values 1, 2, and 3 while y took values 1, 4, and 11.

Let us now conduct the H & P test with a new data set as shown here.

Subj.	x	y
1	942	13
2	101	14
3	313	18
4	800	10

The H & P test requires that we replace the original data with their respective ranks. The result is shown below.

Subj.	R_x	R_y
1	4	2
2	1	3
3	2	4
4	3	1

Obtained r for these ranks is $-.60$. Reference to the permutation sampling distribution in Figure 10.2 shows that five permutation values are less than or equal to $-.60$ so that the lower tail p-value is $5/24 = .208$ while that for a two-tailed test is $(2)(5/24) = .417$. It is important to understand that had we chosen to perform Pitman's test we would have had to generate a new permutation sampling distribution for the data. Because we replaced the original data with ranks and had already generated the permutation distribution for ranks 1 through 4 for each variable, we already had the appropriate distribution.

Replacing original data with ranks also allows us to construct tables of critical values for the H & P test. In order to do so, we need only generate the permutation sampling distribution for sets of ranks for various sample sizes and locate the permutation values such that α or $\alpha/2$ of the permutation values are greater than or equal to the identified value. For example, examination of Figure 10.2 on the preceding page shows that a correlation of 1.0 has an associated (one-tailed) p-value of $1/24 = .042$ while .80 has an associated value of $4/24 = .167$ so that there is no critical value for a one-tailed test conducted at $\alpha = .005$, .010, or .025. Likewise, there is no critical value for a two-tailed test at level .010, .020, or .050. However, the value of 1.0 can be used for a one-tailed test at level .05 or a two-tailed test at .10. The level of significance would not be exactly .05 or .10 but would not exceed those values. The same critical value (1.0) could be used for one- and two-tailed tests at levels .10 and .20 respectively.

The point to be made here is that critical values from permutation based tests may not provide the exact level of significance desired. As a result, critical values are chosen so that the Type I error rate will be less than or equal to the stated level. As sample size increases this problem virtually disappears.

Appendix F gives upper tail critical values for $n = 4$ to 90 for the rank correlation coefficient. For upper tail tests you need only compare the obtained rank correlation coefficient to the appropriate critical value. If obtained r is greater than or equal to critical r the null hypothesis is rejected. For lower tail tests, obtained r is compared to the negative of the tabled value. If obtained r is less than or equal to this value, the null hypothesis is rejected. For two-tailed tests the null hypothesis is rejected if obtained r is greater than or equal to the table value or if obtained r is less than or equal to the negative of the table value.

For n greater than 90, the rank correlation can be substituted into Equation 8.4 (page 309) with the result being referenced to a t table with $n - 2$ degrees of freedom. The result is a very good approximation to the permutation value when $n > 90$.

Conducting the Test

The steps for conducting the H & P Test For Rank Correlation are as follows.

1. Replace original observations with their respective ranks with the lowest x value receiving a rank of 1, the second lowest a rank of two etc. until the largest value is replaced with rank n. The y variable is replaced by ranks in the same manner.

2. Compute r for the ranks obtained in step 1. Call this value obtained r.

3. For $4 \leq n \leq 90$, reference obtained r to Appendix F and

   (a) for an upper tail test reject the null hypothesis if obtained r is greater than or equal to the table value.

   (b) for a lower tail test reject the null hypothesis if obtained r is less than or equal to the negative of the table value.

(c) for a two-tailed test reject the null hypothesis if obtained r is greater than or equal to the table value *or* if obtained r is less than or equal to the negative of the table value.

4. For $n > 90$, use Equation 8.4 on page 309 with the result then being referenced to table B with $n - 2$ degrees of freedom.

EXAMPLE 10.9

Use the data provided to calculate the rank correlation coefficient. Use the resulting value of r to perform a two-tailed H & P test at $\alpha = .05$ of the null hypothesis that x and y are independent.

Subj.	x	y
1	1314	20
2	880	16
3	414	10
4	1774	18
5	101	11
6	902	15
7	544	12
8	722	13
9	377	9
10	1200	17

Solution   We begin by replacing the original observations with their ranks as shown here.

Subj.	R_x	R_y
1	9	10
2	6	7
3	3	2
4	10	9
5	1	3
6	7	6
7	4	4
8	5	5
9	2	1
10	8	8

For computational purposes, it will be convenient to arrange the ranks as shown in Table 10.4.

Using the above sums with Equation 8.2 (suitably modified for ranks) gives

$$r = \frac{\sum R_x R_y - \frac{(\sum R_x)(\sum R_y)}{n}}{\sqrt{\left[\sum R_x^2 - \frac{(\sum R_x)^2}{n}\right]\left[\sum R_y^2 - \frac{(\sum R_y)^2}{n}\right]}}$$

TABLE 10.4: Arrangement of ranks for Example 10.9.

Subj.	R_x	R_y	$R_x R_y$	R_x^2	R_y^2
1	9	10	90	81	100
2	6	7	42	36	49
3	3	2	6	9	4
4	10	9	90	100	81
5	1	3	3	1	9
6	7	6	42	49	36
7	4	4	16	16	16
8	5	5	25	25	25
9	2	1	2	4	1
10	8	8	64	64	64
$\sum$	55	55	380	385	385

$$= \frac{380 - \frac{(55)(55)}{10}}{\sqrt{\left[385 - \frac{(55)^2}{10}\right]\left[385 - \frac{(55)^2}{10}\right]}}$$

$$= \frac{77.5}{\sqrt{[82.5][82.5]}}$$

$$= .939.$$

Reference to Appendix F shows that for $n = 10$, critical r for a two-tailed test conducted at $\alpha = .05$ is $\pm.648$. Because obtained r of .939 exceeds .648, the null hypothesis is rejected. We may conclude, therefore, that x and y are not independent and that a positive relationship exists between the two variables. ∎

EXAMPLE 10.10

A researcher believes that a negative relationship exists between a measure of bone density and average daily consumption of caffeine in women 70 years of age and older. To test this hypothesis the two assessments are made on a group of 80 elderly women. Because the researcher suspects that the distribution of caffeine consumption is skewed, he chooses a rank correlation coefficient to evaluate the relationship between the two variables. The computed rank correlation is $-.122$. Use the H & P test to test the null hypothesis that there is no relationship between bone density and caffeine consumption in elderly women against the alternative that a negative relationship exists between the two variables. Conduct the test at level .05. Interpret the results.

Solution   Reference to Appendix F shows that for $n = 80$, critical r for a one-tailed test at $\alpha = .05$ whose alternative postulates a negative relationship is $-.185$. Because obtained r of $-.122$ is greater than this value, the null hypothesis is not rejected. As a result, the researcher was not able to demonstrate a negative relationship between the two variables. ∎

EXAMPLE 10.11

Given $n = 120$ and a rank correlation of .208, conduct a two-tailed test of the null hypothesis of no relationship between x and y at $\alpha = .10$.

Solution By Equation 8.4 on page 309

$$t = \frac{r}{\sqrt{\frac{1-r^2}{n-2}}} = \frac{.208}{\sqrt{\frac{1-.208^2}{120-2}}} = 2.310$$

Reference to Appendix B shows that critical t values for a two-tailed test with $n - 2 = 120 - 2 = 118$ degrees of freedom are ±1.658. Because obtained t of 2.310 exceeds 1.658, the null hypothesis is rejected leading to the conclusion that a positive relationship exists between the two variables. ∎

Assumptions

There are two principle assumptions underlying Hoteling and Pabst test for rank correlation. The first is that all x values are independent as are all y values. (See assumptions underlying Pitman's test for correlation on page 353 for further details.) The second assumption requires that there be no duplicate values among the x observations nor among the y observations. Values of x that duplicate values of y do not violate this assumption and are of no consequence. Consider the following data set.

Subj.	x	y
1	9	12
2	13	16
3	7	10
4	4	18
5	7	13

Notice that there are two x values of seven. This is the assumption violation. The fact that there is an x and y with value 13 makes no difference. The problem arises when we attempt to replace the original observations with ranks. When substituting ranks for the x variable the 4 would be replaced with rank 1 but what can we do with the two scores of 7? There are a variety of strategies but the most common assigns the average of the ranks that would have been assigned had there been no tie. Normally, the second lowest score would receive a rank of 2 while the third lowest would receive rank 3. The **average rank** strategy assigns the average of rank 2 and 3 or 2.5 to the two scores of 7. By this method the ranks would appear as follows.

Subj.	R_x	R_y
1	4	2
2	5	4
3	2.5	1
4	2	5
5	2.5	3

Had there been three ties rather than two, we would replace each with the average of the three ranks that would have been assigned to the three values had there been no ties.

But our problem is not solved. The critical values for $n = 5$ in Appendix F were obtained for ranks 1 through 5, not ranks 1, 2.5, 2.5, 4, and 5. The following are strategies that may be used when tied observations are encountered.[8]

1. Retain original observations and apply Pitman's test.[9]

2. Use the average rank method described above and generate the permutation sampling distribution of these ranks. (See footnote 9.)

3. Use the average rank method described above and apply Equation 8.4 on page 309 to the rank correlation. The result is referenced to a t distribution with $n - 2$ degrees of freedom. This method will usually produce good results if n is sufficiently large (e.g., greater than 30) and the proportion of tied observations is small (e.g., 20% or less).

10.3.2   Paired Samples Tests

In Chapter 5 on page 162 you learned to conduct a paired samples t test by means of Equation 5.1 to test the null hypothesis $H_0 : \mu_d = 0$ where μ_d is the mean of a difference score population. One of the assumptions underlying that test is that the sample difference scores came from a normally distributed population. In this section you will learn to conduct paired samples tests that do not require this restrictive assumption. Two tests will be demonstrated. The first uses original observations while the second, known as Wilcoxon's signed-ranks test, uses ranks in place of the original observations. The underlying logic of these tests is emphasized as well as the relationship between the two.

The Permutation Paired Samples t Test.

Rationale

Suppose that the following results are obtained from a study in which pre-treatment and post-treatment observations are obtained for four individuals as shown below.

Subject	Pre-treatment	Post-treatment	(Difference) d
1	95	99	4
2	111	120	9
3	97	102	5
4	132	130	-2

By Equation 5.1 on page 162 the paired samples t statistic for this data is 1.760. If the researcher wishes to test the null hypothesis $H_0 : \mu_d = 0$ but wants to avoid the assumption of difference score population normality, a permutation test based on the following logic may be performed.

---

[8]Other strategies are outlined in nonparametric texts. See for example [7].

[9]This can be done by means of commercially available software.

The first subject had a pre-treatment score of 95 and a post-treatment score of 99. But if the treatment nor any other factor had any impact on the scores, then the order of these scores for the first subject is a matter of chance. That is, in the absence of a treatment effect, this subject was just as likely to receive the 99 as a pre-treatment score and the 95 as a post-treatment score. The same is true for the other subjects as well. Thus, the ordering of scores for each subject *under a true null hypothesis of no treatment effect* are just a matter of chance.

From this argument, it can be seen that by reversing the scores of the subjects in all possible ways and computing a paired samples t statistic for each such configuration, we can generate all possible t statistics that might have been realized from the study data if the null hypothesis were true. But reversing a pair of scores simply changes the sign of the difference score so that positive differences become negative and visa versa. This means that in order to compute all possible t statistics we need only form all possible positive/negative patterns of difference scores and compute the statistic on each such set. Notice that the first difference score may be expressed in two ways—i.e., positive or negative. For each of these the second difference may also be expressed in two ways so that the number of ways of characterizing the first two difference scores is $2 \cdot 2 = 4$. Carrying this logic to all n difference scores we see that the number of ways of representing n difference scores is 2^n which for the present problem is $2^4 = 16$. All 16 configurations are shown in Table 10.5.

Having formed all 16 possible sign change sets of difference scores and computing a paired samples t statistic for each, we can arrange the 16 t statistics into a permutation sampling distribution as shown in Figure 10.3 on page 363. If the null hypothesis is true, we would expect *obtained t* (i.e., the one obtained from the data collected in the study) to be near zero since a true null hypothesis means that the mean of the difference score population is zero. On the other hand, if $\mu_d > 0$ we would expect t to take some value in the extreme right tail of the distribution with the opposite being true when $\mu_d < 0$. In the present case, obtained t for the data derived from the study is 1.760. What is the probability of obtaining such an extreme value of t when the null hypothesis is true? It is simply the proportion of t's in the sampling distribution that take a value as extreme (i.e., far from zero), or more so, as that obtained from the data. In the present case, only two of the 16 possible values of t are as far, or further, from zero as that obtained from the data—namely, 1.760 and 3.397. We can now say that of the 16 possible values of t that might have been obtained under a true null hypothesis, only 2 are as extreme as that observed from the data. The probability of achieving such an extreme value of t under a true null hypothesis is then $2/16 = .125$. This is in fact the p-value for a test with alternative $\mu_d > 0$. The two-tailed p-value would be

$$(2)\left(\frac{2}{16}\right) = .250.$$

In general, the p-value for a test with one-tailed alternative $\mu_d > 0$ is the proportion of values in the permutation sampling distribution that are greater than or equal to the one observed from the data. When the one-tailed alternative is of the form $\mu_d < 0$, the p-value for this one-tailed alternative is the proportion of values in the permutation distribution that are less than or equal to the one observed from the data. The two-tailed p-value for alternative $\mu_d \neq 0$ is calculated by finding the proportion of observations in the upper (lower) tail of the distribution that are greater (less) than or equal to obtained t and multiplying this

TABLE 10.5: All possible sets of algebraic sign changes and paired samples t statistics for a particular set of difference scores.

1	2	3	4	5	6
−4	4	−4	4	−4	4
9	−9	−9	9	9	−9
5	5	5	−5	−5	−5
−2	−2	−2	−2	−2	−2
$t = .661$	$t = −.155$	$t = −.862$	$t = .480$	$t = −.155$	$t = −1.095$

7	8	9	10	11	12
−4	4	−4	4	−4	4
−9	9	9	−9	−9	9
−5	5	5	5	5	−5
−2	2	2	2	2	2
$t = −3.397$	$t = 3.397$	$t = 1.095$	$t = .155$	$t = −.480$	$t = .862$

13	14	15	16
−4	4	−4	4
9	−9	−9	9
−5	−5	−5	5
2	2	2	−2
$t = .155$	$t = −.661$	$t = −1.760$	$t = 1.760$

Lower Half of Sampling Distribution

−3.397 −1.760 −1.095 −.862 −.661 −.480 −.155 −.155

Upper Half of Sampling Distribution

.155 .155 .480 .661 .862 1.095 1.760 3.397

↑

Obtained t

FIGURE 10.3: Permutation sampling distribution of paired samples t statistics for a particular data set.

value by 2. The proportion greater than or equal to obtained t is multiplied by two when t is greater than zero while the proportion less than or equal to obtained t is multiplied by two when obtained t is less than zero.

Conducting the Test

The steps for conducting the permutation paired samples t test are as follows.

1. Compute t for the data obtained from the study. Call this value obtained t.

2. Form all possible algebraic sign change sets of difference scores as shown for an example data set on the previous page.

3. Calculate the paired samples t statistic for each set of difference scores as shown on page 363.

4. Arrange the values of t obtained from step 3 from smallest to largest. This is the permutation sampling distribution of t.

5. For a one-tailed test with alternative $H_A : \mu_d > 0$, calculate the proportion of values in the permutation distribution that are greater than or equal to obtained t. This is the p-value for the one-tailed test.

6. For a one-tailed test with alternative $H_A : \mu_d < 0$, calculate the proportion of values in the permutation distribution that are less than or equal to obtained t. This is the p-value for the one-tailed test.

7. For the two-tailed test with alternative $H_A : \mu_d \neq 0$:

   (a) calculate the proportion of values in the permutation distribution that are greater than or equal to obtained t if obtained t is greater than zero. Multiply this value by two to find the two-tailed p-value.

   (b) calculate the proportion of values in the permutation distribution that are less than or equal to obtained t if obtained t is less than zero. Multiply this value by two to find the two-tailed p-value.

EXAMPLE 10.12

Use the permutation sampling distribution of the paired samples t statistic shown in Figure 10.3 to find the p-value for the following test

$$H_0 : \mu_d = 0$$
$$H_A : \mu_d < 0$$

assuming obtained t is $-.862$.

Solution   Because the alternative is of the form $\mu_d < 0$, we are to perform a one-tailed test with the p-value being the proportion of values in the sampling distribution that are less than or equal to the one obtained from the data. In this case the values of t that satisfy the criterion of being less than or equal to obtained t of $-.862$ are -3.397, -1.760, -1.095, and $-.862$ so that $p = \frac{4}{16} = .250$. If α were .05, we would fail to reject the null hypothesis. ∎

EXAMPLE 10.13

Use the permutation sampling distribution of t shown in Figure 10.3 to find the p-value for the following test

$$H_0 : \mu_d = 0$$
$$H_A : \mu_d \neq 0$$

assuming obtained t is 3.397.

Solution   Because obtained t is greater than zero, we find the proportion of permutation values that are greater than or equal to obtained t. We note that only the permutation value of 3.397 meets this criterion so that the desired p-value is $(2)(1/16) = .125$.   ∎

EXAMPLE 10.14

Given $n = 10$, how many values of t would make up the permutation sampling distribution for the paired samples t test?

Solution
$$2^n = 2^{10} = 1024$$ ∎

Assumptions

The principle assumption underlying the permutation paired samples t test is that all difference scores are independent. A violation of this assumption might occur, for example, if in a pre- and post-treatment study, the same subject was treated at two points in time with her/his pre- and post-treatment scores being entered into the analysis twice. Thus, this subject would have two difference scores in the data set. If the two difference scores were related in some fashion, a violation would occur.

As a second example, a violation might occur if difference scores for siblings were included in an analysis. If the study dealt with weight loss and if there were a genetic component to weight loss, then the amount of weight lost by brothers might be related. This test cannot be counted upon to be robust in the face of a violation of the independence assumption.

Wilcoxon's Signed-Ranks Test.   This test is similar to the permutation paired samples t test except that difference scores are converted to ranks before obtained t is calculated and the permutation sampling distribution generated.

Rationale

In order to conduct the permutation paired samples t test for 10 data pairs (i.e., difference scores), we would have to find $2^{10} = 1024$ sets of difference scores and compute a t statistic for each set. For 20 data pairs the number of difference score sets and t statistics would be $2^{20} = 1,048,576$. This daunting task, along with other concerns, prompted Frank Wilcoxon [49] to propose a paired samples permutation test based on ranks. One advantage of such a strategy is that once the permutation distribution for the test statistic is generated for a given sample size (n), the distribution can be used for any sample of the same size regardless of the values of the original data. This follows from the fact that the original data is replaced by ranks for which the distribution has already been constructed.

The conversion of original data to ranks is accomplished as follows. First, all difference scores of zero are removed from the analysis and the sample size (n) suitably reduced. Then, the difference scores are replaced by the ranks of their absolute values with the difference score with smallest absolute value being replaced by one, the second smallest by two and so on until the largest difference score (ignoring algebraic sign) is replaced by n. The algebraic sign (i.e., plus or minus) of each original observation is then assigned to the rank for that observation. For example, suppose the following set of difference scores is to be converted to ranks.

Difference

3
−9
1
−4
−2
5
8
−7
0

The conversion would be as shown in the following table.

(1)	(2)	(3)	(4)
3	3	3	3
−9	9	8	−8
1	1	1	1
−4	4	4	−4
−2	2	2	−2
5	5	5	5
8	8	7	7
−7	7	6	−6

Column (1) of the above table shows original difference scores with the zero difference removed, (2) shows the absolute values of the difference scores, (3) the ranks of the absolute valued difference scores, and (4) the ranks with signs of the original scores attached. Notice that the difference score of zero was dropped before the ranking process was carried out so that n is now 8.

Using the data employed for the permutation paired samples t test (repeated here for your convenience) the ranking would be as follows.

Subject	Pre-treatment	Post-treatment	(Difference) d
1	95	99	4
2	111	120	9
3	97	102	5
4	132	130	−2

Ignoring algebraic sign, the rank of one would be assigned to 2, two to 4, three to 5, and four to 9. Re-attaching the algebraic sign produces the following set of signed ranks.

TABLE 10.6: All possible sets of algebraic sign changes and paired samples t statistics for ranked difference scores when $n = 4$.

1	2	3	4	5	6
-2	2	-2	2	-2	2
4	-4	-4	4	4	-4
3	3	3	-3	-3	-3
-1	-1	-1	-1	-1	-1
$t = .679$	$t = .000$	$t = -.679$	$t = .322$	$t = -.322$	$t = -1.134$

7	8	9	10	11	12
-2	2	-2	2	-2	2
-4	4	4	-4	-4	4
-3	3	3	3	3	-3
-1	1	1	1	1	1
$t = -3.873$	$t = 3.873$	$t = 1.134$	$t = .322$	$t = -.322$	$t = .679$

13	14	15	16
-2	2	-2	2
4	-4	-4	4
-3	-3	-3	3
1	1	1	-1
$t = .000$	$t = -.679$	$t = -1.852$	$t = 1.852$

Subject	(Difference) d	Signed Ranks
1	4	2
2	9	4
3	5	3
4	-2	-1

Applying Equation 5.1 on page 162 to the ranks yields a paired samples t statistic of 1.852.

In order to generate the permutation sampling distribution for this rank test, we must generate the integers 1 through 4 and attach positive and negative values in all possible patterns. This is done in Table 10.6. The t values from each set are formed into a permutation sampling distribution as shown in Figure 10.4 on the next page.

Technically speaking, the hypothesis tested by this statistic maintains that the difference score population is symmetric about zero while the alternative maintains that this is not

Lower Half of Sampling Distribution

−3.873 −1.852 −1.134 −.679 −.679 −.322 −.322 .000

Upper Half of Sampling Distribution

.000 .322 .322 .679 .679 1.134 1.852 3.873

Obtained t

FIGURE 10.4: Permutation sampling distribution of rank based paired samples t statistics for $n = 4$.

the case. If we assume that the difference score population is symmetric[10], then the null hypothesis tested is the same as that tested by the permutation paired samples t test, i.e., $H_0 : \mu_d = 0$. Thus, to test this null hypothesis against the alternative $H_A : \mu_d > 0$, we note from Figure 10.4 that there are two values of the permutation distribution that are greater than, or equal to, obtained t of 1.852. These values are 1.852 and 3.873 so that the p-value for the one-tailed test is $\frac{2}{16} = .125$.

Replacing original data with ranks allows us to construct tables of critical values for this test. In order to do so, we need only generate the permutation sampling distribution for sets of signed ranks for various sample sizes and locate the permutation values such that α or $\alpha/2$ of the permutation values are greater than or equal to the identified value. For example, examination of Figure 10.4 shows that a t value of 3.873 has an associated (one-tailed) p-value of $1/16 = .0625$ while a value of 1.852 has an associated value of $2/16 = .125$ so that there is no critical value for a one-tailed test conducted at $\alpha = .005, .010, .025,$ or $.050$. Likewise, there is no critical value for a two-tailed test at level .010, .020, .050, or .100. However, the critical value of 3.873 can be used for a one-tailed test at level .100 or a two-tailed test at .200. The level of significance would not be exactly .100 or .200 but would not exceed those values. This problem diminishes as sample size increases.

Appendix G gives critical values for $n = 4$ to 90 for the rank paired samples t statistic or Wilcoxon's signed-ranks test. For upper tail tests you need only compare the obtained rank t statistic to the appropriate critical value. If obtained t is greater than or equal to critical t the null hypothesis is rejected. For lower tail tests, obtained t is compared to the negative of the tabled value. If obtained t is less than or equal to this value, the null hypothesis is rejected. The null hypothesis for a two-tailed test is rejected if obtained t is greater than or equal to the table value or is less than or equal to the negative of the table value. Notice that critical values in this table are referenced to the sample size (n) rather than to degrees of freedom as is done with conventional t tables.

---

[10]A topic to be discussed later.

For n greater than 90, the obtained rank t statistic can be referenced to a t table with $n-1$ degrees of freedom. The result is a very good approximation to the permutation value when $n > 90$.

Conducting the Test

The steps for conducting Wilcoxon's signed-ranks test are as follows.

1. Calculate difference scores by subtracting one observation in each data pair from the other with the observation chosen for subtraction being consistent across all data pairs. Remove any difference scores of zero.

2. Replace difference scores obtained in step 1 with the ranks of their absolute values with the smallest receiving rank 1 the second smallest rank 2 and so on and the largest rank n. Attach the sign of the difference scores to the ranks thus obtained.

3. Use Equation 5.1 on page 162 to calculate the paired samples t statistic for the ranks obtained in step 2. Call this value obtained t.

4. For $4 \leq n \leq 90$, reference obtained t to Appendix G and

   (a) for an upper tail test reject the null hypothesis if obtained t is greater than or equal to the table value.

   (b) for a lower tail test reject the null hypothesis if obtained t is less than or equal to the negative of the table value.

   (c) for a two-tailed test reject the null hypothesis if obtained t is greater than or equal to the table value *or* if obtained t is less than or equal to the negative of the table value.

5. For $n > 90$, reference obtained t to Appendix B with $n-1$ degrees of freedom.

EXAMPLE 10.15

Conduct a two-tailed Wilcoxon's signed-ranks test at $\alpha = .05$ on the (fictitious) blood pressure data in Table 10.7 on the next page.

Solution   The data are arranged for analysis as follows.

Difference	Absolute Difference	Rank	Signed Rank
−4	4	4	−4
−14	14	8	−8
−1	1	1	−1
2	2	2	2
3	3	3	3
−9	9	7	−7
6	6	6	6
−17	17	9	−9
−30	30	10	−10
5	5	5	5

TABLE 10.7: Pre- and post-treatment systolic blood pressure measurements.

Pre-Treatment	Post-Treatment
151	147
174	160
150	150
171	170
144	146
139	142
159	150
140	146
137	120
179	149
146	151

The first column shows difference scores—i.e., the post-treatment score of each subject minus the pre-treatment score. Notice that the difference score of zero has been deleted. The second column gives the absolute values of the difference scores while the third shows the ranks of the absolute values of the difference scores. The final column gives the signed rank for each subject which is obtained by attaching the sign of the difference score to the rank shown in the third column.

Obtained t is the paired samples t statistic calculated on the signed ranks. To facilitate this calculation we note that the sum of the signed ranks and the sum of the squared signed ranks are respectively

$$\sum R_d = (-4) + (-8) + (-1) + (2) + (3) + (-7) + (6) + (-9) + (-10) + (5) = -23$$

and

$$\sum R_d^2 = (-4)^2 + (-8)^2 + (-1)^2 + (2)^2 + (3)^2 + (-7)^2 + (6)^2 + (-9)^2 + (-10)^2 + (5)^2 = 385.$$

The paired samples t statistic (Equation 5.1 on page 162) with R_d substituted for d to indicate that the analysis is performed on signed ranks of difference scores rather than difference scores is as follows.

$$t = \frac{\bar{R}_d}{\frac{s_{R_d}}{\sqrt{n}}}$$

The sample standard deviation of the signed ranks (using Equation 2.16 on page 37 with R_d substituted for x) is

$$s_{R_d} = \sqrt{\frac{\sum R_d^2 - \frac{(\sum R_d)^2}{n}}{n-1}} = \sqrt{\frac{385 - \frac{(-23)^2}{10}}{10-1}} = 6.075$$

TABLE 10.8: Reaction times to simulated driving emergencies related to cell phone use.

Without Cell Phone	With Cell Phone
.318	.322
.301	.341
.384	.391
.290	.289
.411	.401
.371	.399
.371	.400
.333	.338

The mean signed rank is then

$$\bar{R}_d = \frac{\sum R_d}{n} = \frac{-23}{10} = -2.3$$

Substituting these values into the above equation for t yields

$$t = \frac{-2.3}{\frac{6.075}{\sqrt{10}}} = -1.197$$

Appendix G shows that for $n = 10$, the critical values for a two-tailed Wilcoxon's signed-ranks test conducted at $\alpha = .05$ are 2.424 and -2.424. Because obtained t of -1.197 is between these two values, the null hypothesis is not rejected. ■

EXAMPLE 10.16

Researchers are interested in determining the effect of cell phone use on reaction times to simulated driving emergencies. To this end, eight subjects are placed in driving simulators where they "drive" for a period of one hour. At a randomly chosen point in the one hour period the subject experiences a simulated emergency that calls for immediate braking in order to avoid a collision. Each subject undergoes the driving simulation twice—once while talking on a cell phone and once without the cell phone. The order of the cell phone and no cell phone simulations is randomized. Reaction time is defined as elapsed time, measured in fractions of a second, between introduction of the emergency and application of the brake. These times are provided in Table 10.8.

Use these reaction times to perform a one-tailed Wilcoxon's signed-ranks test with the alternative indicating an increase in reaction time under cell phone use. Conduct the test at $\alpha = .05$.

Solution   Difference scores—i.e., reaction times while using cell phones minus reaction times without cell phones, absolute values of difference scores, ranks of absolute values of difference scores, and signed-ranks of absolute values of difference scores are shown below.

	Absolute		Signed
Difference	Difference	Rank	Rank
.004	.004	2	2
.040	.040	8	8
.007	.007	4	4
−.001	.001	1	−1
−.010	.010	5	−5
.028	.028	6	6
.029	.029	7	7
.005	.005	3	3

Obtained t is the paired samples t statistic calculated on the signed ranks. To facilitate this calculation we note that the sum of the signed ranks and the sum of the squared signed ranks are respectively

$$\sum R_d = (2) + (8) + (4) + (-1) + (-5) + (6) + (7) + (3) = 24$$

and

$$\sum R_d^2 = (2)^2 + (8)^2 + (4)^2 + (-1)^2 + (-5)^2 + (6)^2 + (7)^2 + (3) = 204$$

Using these results,

$$s_{R_d} = \sqrt{\frac{\sum R_d^2 - \frac{(\sum R_d)^2}{n}}{n-1}} = \sqrt{\frac{204 - \frac{(24)^2}{8}}{8-1}} = 4.342$$

and

$$\bar{R}_d = \frac{\sum R_d}{n} = \frac{24}{8} = 3.0$$

Obtained t is then

$$t = \frac{\bar{R}_d}{\frac{s_{R_d}}{\sqrt{n}}} = \frac{3.0}{\frac{4.342}{\sqrt{8}}} = 1.954$$

Appendix G shows that for $n = 8$, the critical value for a one-tailed Wilcoxon's signed-ranks test conducted at $\alpha = .05$ is 2.225. Because obtained t of 1.954 is less than this value, the null hypothesis is not rejected. We were unable, therefore, to demonstrate that cell phone use increases reaction time. ■

EXAMPLE 10.17

Given $n = 120$ and obtained t of 2.626, conduct a two-tailed Wilcoxon signed-ranks test of the null Hypothesis $H_0 : \mu_d = 0$ at $\alpha = .10$.

Solution Because $n > 90$, obtained t is referenced to a t table with $n - 1$ degrees of freedom. Reference to Appendix B with $120 - 1 = 119$ degrees of freedom for $\alpha = .10$ gives two-tailed critical values of -1.658 and 1.658. Because obtained t of 2.626 is greater than critical t of 1.658, the null hypothesis is rejected.

Assumptions
There are two principle assumptions underlying Wilcoxon's signed-ranks test. The first is that all values of d and hence all signed ranks are independent. (See assumptions underlying the paired samples permutation test on page 365 for further details.) The second assumption requires that there be no duplicate values among the absolute values of d. Consider the following data set.

Difference	Absolute Difference	Rank	Signed Rank
2	2	2.5	2.5
4	4	6.0	6.0
7	7	8.0	8.0
1	1	1.0	1.0
4	4	6.0	6.0
−4	4	6.0	−6.0
−2	2	2.5	−2.5
−8	8	9.0	−9.0
3	3	4.0	4.0
9	9	10.0	10.0

Notice that for the absolute differences, there are two values of two and three of four. This is the assumption violation. The problem arises when we attempt to replace these absolute values with ranks. When substituting ranks for the $|d|$, the one would be replaced with rank 1 but what can we do with the two values of two? There are a number of different strategies but the most common assigns the average of the ranks that would have been assigned had there been no ties. Normally, the second lowest score would receive a rank of 2 while the third lowest would receive rank 3. The **average rank** strategy assigns the average of rank 2 and 3 or 2.5 to the two scores of two. The fourth smallest absolute value is three which receives rank 4. Again we face tied values of four since there are three such values. Had there been no ties the 5th, 6th, and 7th lowest values would receive ranks 5, 6, and 7 so the tied observations will receive the average of ranks 5, 6, and 7 or 6.0. The absolute values of seven, eight, and nine are untied and so receive ranks 8.0, 9.0, and 10.0.

But our problem is not solved. The critical values for $n = 10$ in Appendix G were obtained for ranks 1 through 10, not ranks 1, 2.5, 2.5, 4, 6, 6, 6, 7, 8, 9, and 10. The following are strategies that may be used when tied observations are encountered.[11]

1. Retain original observations and apply the permutation test as described previously (see page 361).[12]

2. Use the average rank method described above and generate the permutation sampling distribution of the t statistic computed on these ranks. (See footnote 12.)

3. Use the average rank method described above and reference obtained t to a t table using $n - 1$ degrees of freedom. This method will usually produce good results if n

[11] Other strategies are outlined in nonparametric texts. See for example [7].
[12] This can be done by means of commercially available software.

is sufficiently large (e.g., greater than 30) and the proportion of tied observations is small (e.g., 20% or less).

Earlier we said that, technically speaking, the hypothesis tested by the signed-ranks statistic maintains that the difference score population is symmetric about zero while the alternative maintains that this is not the case. If we assume that the difference score population is symmetric, then the null hypothesis tested is the same as that tested by the permutation paired samples t test, i.e., $H_0 : \mu_d = 0$. But how realistic is the assumption that the difference score population is symmetric?

When pre and post populations are identical as would usually be the case in the absence of a treatment effect, the difference score population will be symmetric *regardless of the shape of the pre and post populations*. Thus, the pre and post populations may be radically skewed but so long as they have the same shape the difference score population will be symmetric. It follows that the null hypothesis tested will be $H_0 : \mu_d = 0$. In situations where there is reason to believe that pre and post populations do not have the same shape, this will likely not be true.   ■

10.3.3   Two Independent Samples

In Chapter 6 on page 219 you learned to conduct an independent samples t test by means of Equation 6.1 to test the null hypothesis $H_0 : \mu_1 = \mu_2$ where μ_1 and μ_2 are the means of the populations from which the two samples were drawn. One of the assumptions underlying that test is that the samples came from normally distributed populations. In this section you will learn to conduct independent samples tests that do not require this restrictive assumption. Two tests will be demonstrated. The first uses original observations while the second, known as Wilcoxon's rank-sum test,[13] uses ranks in place of the original observations. The underlying logic of these tests is emphasized as well as the relationship between the two.

The Permutation Independent Samples t Test.

Rationale

Suppose that the following outcomes are obtained from a study in which six individuals have been randomly assigned to two groups.

group one	group two
9	0
12	4
13	6

By Equation 6.1 on page 219 the independent samples t statistic for this data is 3.748. If the researcher wishes to test the null hypothesis $H_0 : \mu_1 = \mu_2$ but wants to avoid the assumption of population normality, a permutation test based on the following logic may be performed.

Because the six subjects were randomly assigned to the two groups, the score of any particular subject is just as likely to appear in one of the two groups as in the other *if the null*

---

[13] Also known as the Mann-Whitney U test.

Lower Half of Sampling Distribution

−3.748 −1.765 −1.200 −.983 −.791 −.615 −.452 −.452 −.147 .000

Upper Half of Sampling Distribution

.000 .147 .452 .452 .615 .791 .983 1.200 1.765 3.748

↑

Obtained t

FIGURE 10.5: Permutation sampling distribution of independent samples t statistics for a particular data set.

hypothesis of no treatment effect is true. On the other hand, in the presence of a treatment effect that tends to increase or decrease the scores of the members of one group, the scores of that group will tend to be higher or lower than those of the other group.

Thus, when the null hypothesis of no treatment effect is true, subject scores are randomly distributed between the two groups with the result that neither group tends to have significantly higher or lower scores than does the other. It follows that a t statistic computed on these data would generally take a value near zero. When a treatment effect is present so that the scores of one group tend to be higher than those of the other, the t statistic will tend away from zero.

As previously stated, in the absence of a treatment effect, the score of any given subject is just as likely to appear in one group as the other. Thus, if the null hypothesis is true, the observed data arrangement is just one of the possible arrangements implied by the random assignment process. But how many ways could the scores have been assigned to the two groups?

From the discussion in Section 10.2.2 and Equation 10.2 on page 347 we see that the six observations could be assigned to the two groups in

$$C_{n_2}^{n_1} = \frac{n!}{n_1!n_2!} = \frac{6!}{3!3!} = 20$$

different ways. All 20 combinations are shown in Table 10.9 on the next page.

Having formed all 20 possible arrangements of the data into two groups and computing an independent samples t statistic for each, we can arrange the 20 t statistics into a permutation sampling distribution as shown in Figure 10.5.

If the null hypothesis is true, we would expect *obtained t* (i.e., the one obtained from the data collected in the study) to be near zero since a true null hypothesis implies that the difference between the means of the two populations from which the samples were drawn is zero. A second interpretation is that any difference between the sample means is due to the random assignment process and not to any treatment effect. On the other hand, if $\mu_1 > \mu_2$

TABLE 10.9: All possible combinations of six observations and associated t statistics for a particular set of scores.

group one	group two
0	9
4	12
6	13

$t = -3.748$

group one	group two
0	6
4	12
9	13

$t = -1.765$

group one	group two
0	6
4	9
12	13

$t = -.983$

group one	group two
0	6
4	9
13	12

$t = -.791$

group one	group two
0	4
6	12
9	13

$t = -1.200$

group one	group two
0	4
6	9
12	13

$t = -.615$

group one	group two
0	4
6	9
13	12

$t = -.452$

group one	group two
0	4
9	6
12	13

$t = -.147$

group one	group two
0	4
9	6
13	12

$t = .000$

group one	group two
0	4
12	6
13	9

$t = -.452$

group one	group two
4	0
6	12
9	13

$t = -.452$

group one	group two
4	0
6	9
12	13

$t = .000$

group one	group two
4	0
6	9
13	12

$t = .147$

group one	group two
4	0
9	6
12	13

$t = .452$

group one	group two
4	0
9	6
13	12

$t = .615$

group one	group two
4	0
12	6
13	9

$t = 1.200$

group one	group two
6	0
9	4
12	13

$t = .791$

group one	group two
6	0
9	4
13	12

$t = .983$

group one	group two
6	0
12	4
13	9

$t = 1.765$

group one	group two
9	0
12	4
13	6

$t = 3.748$

we would expect t to take some value in the extreme right tail of the distribution since the scores in sample one would tend to be larger than those in the second sample. The opposite would be true when $\mu_1 < \mu_2$. In the present case, obtained t for the data derived from the study is 3.748. What is the probability of obtaining such an extreme value of t when the null hypothesis is true? It is simply the proportion of t's in the sampling distribution that take a value as extreme (i.e., far from zero), or more so, as that obtained from the data. In the present case, only one of the 20 possible values of t is as far, or further, from zero as that obtained from the data—namely, 3.748. We can now say that of the 20 possible values of t that might have been obtained under a true null hypothesis, only one is as extreme as that observed from the data. The probability of achieving such an extreme value of t under a true null hypothesis is then $1/20 = .050$. This is in fact the p-value for a test with alternative $\mu_1 > \mu_2$. The two-tailed p-value would be $(2)\left(\frac{1}{20}\right) = .100$.

In general, the p-value for a test with one-tailed alternative $\mu_1 > \mu_2$ is the proportion of values in the permutation sampling distribution that are greater than or equal to the one observed from the data. When the one-tailed alternative is of the form $\mu_1 < \mu_2$, the p-value for this one-tailed alternative is the proportion of values in the permutation distribution that are less than or equal to the one observed from the data. The two-tailed p value for alternative $\mu_1 \neq \mu_2$ is calculated by finding the proportion of observations in the upper (lower) tail of the distribution that are greater (less) than or equal to obtained t and multiplying this value by 2. The proportion greater than or equal to obtained t is multiplied by two when t is greater than zero while the proportion less than or equal to obtained t is multiplied by two when obtained t is less than zero.

Conducting the Test

The steps for conducting the permutation independent samples t test are as follows.

1. Compute t for the data obtained from the study. Call this value obtained t.

2. Form all possible two group data combinations as shown for an example data set on the preceding page.

3. Calculate the independent samples t statistic for each combination as shown on page 376.

4. Arrange the values of t obtained from step 3 from smallest to largest. This is the permutation sampling distribution of t.

5. For a one-tailed test with alternative $H_A : \mu_1 > \mu_2$, calculate the proportion of values in the permutation distribution that are greater than or equal to obtained t. This is the p-value for the one-tailed test.

6. For a one-tailed test with alternative $H_A : \mu_1 < \mu_2$, calculate the proportion of values in the permutation distribution that are less than or equal to obtained t. This is the p-value for the one-tailed test.

7. For the two-tailed test with alternative $H_A : \mu_1 \neq \mu_2$:

(a) calculate the proportion of values in the permutation distribution that are greater than or equal to obtained t if obtained t is greater than zero. Multiply this value by two to find the two-tailed p-value.

(b) calculate the proportion of values in the permutation distribution that are less than or equal to obtained t if obtained t is less than zero. Multiply this value by two to find the two-tailed p-value.

EXAMPLE 10.18

Use the permutation sampling distribution of the independent samples t statistic shown in Figure 10.5 to find the p-value for the following test

$$H_0 : \mu_1 = \mu_2$$
$$H_A : \mu_1 < \mu_2$$

assuming obtained t is -1.200.

Solution Because the alternative is of the form $\mu_1 < \mu_2$, we are to perform a one-tailed test with the p-value being the proportion of values in the sampling distribution that are less than or equal to the one obtained from the data. In this case the values of t that satisfy the criterion of being less than or equal to obtained t of -1.200 are -3.748, -1.765, and -1.200 so that $p = \frac{3}{20} = .150$. If α were .05, we would fail to reject the null hypothesis. ■

EXAMPLE 10.19

Given $n_1 = n_2 = 5$, how many values of t would make up the permutation sampling distribution for the independent samples t test?

Solution

$$C_5^5 = \frac{n!}{n_1! n_2!} = \frac{10!}{5! 5!} = 252$$
■

Assumptions

The principle assumption underlying the permutation independent samples t test is that all scores are independent. A violation of this assumption might occur, for example, if in a clinical trial, the same subject were assessed at two points in time with the two resultant outcome scores being entered into the analysis. Thus, this subject would have two scores in the data set. If the two scores were related in some fashion, a violation would occur.

As a second example, a violation might occur if scores for siblings were included in an analysis. If the study dealt with weight loss and if there were a genetic component to weight loss, then the amount of weight lost by brothers might be related. This test cannot be counted upon to be robust in the face of a violation of the independence assumption.

Wilcoxon Rank-Sum Test. This test[14] is similar to the permutation independent samples t test except that scores are converted to ranks before obtained t is calculated and the permutation sampling distribution generated.

---

[14] Also known as the Mann-Whitney U test.

Rationale

In order to conduct the permutation independent samples t test for eight observations in each of two groups, we would have to find $C_8^8 = 12,870$ data combinations and compute an independent samples t statistic for each. The number of calculations increases rapidly with increases in sample size.

This daunting task, along with other considerations, prompted Frank Wilcoxon [49] to propose an independent samples permutation test based on ranks. One advantage of such a strategy is that once the permutation distribution for the test statistic is generated for given sample sizes (n_1 and n_2), the distribution can be used for any samples of the same size regardless of the values of the original data. This follows from the fact that the original data are replaced by ranks for which the distribution has already been constructed.

The conversion of original data to ranks is accomplished as follows. The smallest score in the data set is assigned rank 1, the second smallest rank 2 and so on with the largest observation being assigned rank $n_1 + n_2$. That is, if there are three observations in each sample, the smallest value is assigned rank 1 with the largest being assigned rank 6. Notice that ranks are assigned without regard to group membership. For example, suppose we wish to convert the data used previously in conjunction with the permutation independent samples t test to ranks. These data are repeated here for your convenience.

Treat-ment	Control
9	0
12	4
13	6

After replacing original observations with ranks the result is as follows.

Treat-ment	Control
4	1
5	2
6	3

Applying Equation 6.1 on page 219 to the ranks yields an independent samples t statistic of 3.674.

In order to generate the permutation sampling distribution for this rank test, we must generate the integers 1 through 6 and form all possible combinations in which 3 ranks are assigned to each of the two groups. A t statistic must then be computed for each such combination. This is done in Table 10.10 on the next page. The t values from each combination are formed into a permutation sampling distribution as shown in Figure 10.6 on page 381.

Strictly speaking, the hypothesis tested by this statistic maintains that the two samples were drawn from *identical* populations while the alternative maintains that this is not the case. Thus, a treatment that increases (decreases) variance, mean, median, skew, or any other aspect of the treated group would cause the null hypothesis to be false. However, this test is not equally powerful against all of these conditions. It is particularly powerful for

TABLE 10.10: All possible combinations of ranks one through six and associated t statistics.

group one	group two
1	4
2	5
3	6

$t = -3.674$

group one	group two
1	3
2	5
4	6

$t = -1.871$

group one	group two
1	3
2	4
5	6

$t = -1.118$

group one	group two
1	3
2	4
6	5

$t = -.612$

group one	group two
1	2
3	5
4	6

$t = -1.118$

group one	group two
1	2
3	4
5	6

$t = -.612$

group one	group two
1	2
3	4
6	5

$t = -.196$

group one	group two
1	2
4	3
5	6

$t = -.196$

group one	group two
1	2
4	3
6	5

$t = .196$

group one	group two
1	2
5	3
6	4

$t = .612$

group one	group two
2	1
3	5
4	6

$t = -.612$

group one	group two
2	1
3	4
5	6

$t = -.196$

group one	group two
2	1
3	4
6	5

$t = .196$

group one	group two
2	1
4	3
5	6

$t = .196$

group one	group two
2	1
4	3
6	5

$t = .612$

group one	group two
2	1
5	3
6	4

$t = 1.118$

group one	group two
3	1
4	2
5	6

$t = .612$

group one	group two
3	1
4	2
6	5

$t = 1.118$

group one	group two
3	1
5	2
6	4

$t = 1.871$

group one	group two
4	1
5	2
6	3

$t = 3.674$

Lower Half of Sampling Distribution

−3.674 −1.871 −1.118 −1.118 −.612 −.612 −.612 −.196 −.196 −.196

Upper Half of Sampling Distribution

.196 .196 .196 .612 .612 .612 1.118 1.118 1.871 3.674

$\uparrow$

Obtained t

FIGURE 10.6: Permutation sampling distribution of rank based independent samples t statistics for $n_1 = n_2 = 3$.

shift alternatives as described on page 275. James V. Bradley [4] states the following in this regard.

> Those who are naive in practical experimentation may be disheartened at the vagueness of the alternative hypothesis, prefering a definite statement of unequal means to a statement of nonidentity which is likely to include inequality of means. However, while it is theoretically possible for treatment-populations to be nonidentical while having exactly equal means, and while it is absurdly easy to invent such populations out of one's head, it is fantastically difficult to create them in the laboratory. That is, in most areas of research, it is virtually impossible by changing the manipulated variable (i.e., treatment or condition) to induce a change in *any* one aspect (i.e., mean, median, mode, variance, interquartile range, tenth percentile, shape etc.) or combinations of aspects of the population distribution of the measured variable without producing *some* change in *every* aspect. Nor is the sought situation any easier to find in nature. Thus, as a practical matter in most areas of research, a statistical test's verdict of nonidentical populations is tantamount to a verdict of unequal means, unequal variances etc.

By this argument, a significant rank t (or Wilcoxon rank-sum) test can be interpreted as indicating a difference between population means. We must bear in mind, however, that the null hypothesis encompasses a broader statement than one that focuses solely on a difference between means.

If we wish to test the null hypothesis of identical populations (or no treatment effect of any sort) for the sample data presented above in which obtained t was 3.674, we note that only one value in the permutation sampling is greater than or equal to obtained t (namely 3.674) so that the p-value for a one-tailed test is $\frac{1}{20} = .05$. The p-value for a two-tailed alternative would be $(2)\left(\frac{1}{20}\right) = .100$. If we assume this finding significant, by the argument presented above, we can interpret this result as indicating a treatment effect that (among other things) results in a higher mean response for one group than the other.

EXAMPLE 10.20

Use the data given here to conduct a one-tailed Wilcoxon rank-sum test at $\alpha = .20$ to show that the mean response for the treatment group is less than that for the control group. Interpret the result.

Treat-ment	Control
.9	.0
1.2	2.4
.3	6.0

Solution Replacing these observations with ranks yields

Treat-ment	Control
3	1
4	5
2	6

Applying Equation 6.1 on page 219 to the ranks yields an independent samples t statistic of $-.612$. Reference to Figure 10.6 on the previous page shows that seven values of the permutation sampling distribution are less than or equal to obtained t so that the one-tailed p-value is $\frac{7}{20} = .35$ which results in failure to reject the null hypothesis. In practical terms, this means that we were unable to demonstrate a treatment impact. ■

Replacing original data with ranks allows us to construct tables of critical values for this test. In order to do so, we need only generate the permutation sampling distribution for sets of ranks for various sample sizes and locate the permutation values such that α or $\alpha/2$ of the permutation values are greater than or equal to the identified value. For example, examination of Figure 10.6 on the preceding page shows that a t value of 3.674 has an associated (one-tailed) p-value of $1/20 = .050$ while a value of 1.871 has an associated value of $2/20 = .100$ so that there is no critical value for a one-tailed test conducted at $\alpha = .005, .010,$ or $.025$. Likewise, there is no critical value for a two-tailed test at level $.010, .020,$ or $.050$. However, the critical value of 3.674 can be used for a one-tailed test at level $.050$ or a two-tailed test at $.100$. As sample sizes increase and the permutation sampling distribution is composed of more distinct values of t, tests can be conducted at more levels of α. Because the permutation sampling distribution is symmetric, lower tail critical values are obtained as negatives of the table values.

Appendix H gives critical values for $n_1 = n_2 = 3$ to 60 for the rank independent samples t statistic or Wilcoxon's rank-sum test. This table provides critical values only for situations in which sample sizes for the two groups are equal.[15] For upper tail tests you need only

---

[15]Extensive tables for unequal sample sizes are currently under construction.

compare the obtained rank t statistic to the appropriate critical value. If obtained t is greater than or equal to critical t, the null hypothesis is rejected. For lower tail tests, obtained t is compared to the negative of the tabled value. If obtained t is less than or equal to this value, the null hypothesis is rejected. The null hypothesis for a two-tailed test is rejected if obtained t is greater than or equal to the table value or is less than or equal to the negative of the table value. Notice that critical values in this table are referenced to the sample size (n_1 or n_2) rather than to degrees of freedom as is done with conventional t tables.

For $n_1 = n_2$ greater than 60, the obtained rank t statistic can be referenced to a t table with $n_1 + n_2 - 2$ degrees of freedom. The result is a very good approximation to the permutation value when $n_1 = n_2 > 60$. This approximation is also generally acceptable for unequal sample sizes provided sample sizes are sufficiently large—e.g., n_1 and n_2 are both greater than 15. The approximation is very good when both sample sizes are greater than 60.

Conducting the Test

The steps for conducting Wilcoxon's rank-sum test are as follows.

1. Replace original scores with their respective ranks with the lowest score receiving rank 1, the second lowest rank 2, and so on with the highest score receiving rank $n_1 + n_2$. Ranking is carried out without regard to group.

2. Use Equation 6.1 on page 219 to calculate the independent samples t statistic for the ranks obtained in step 1. Call this value obtained t.

3. For equal sample sizes where $3 \leq n_1 = n_2 \leq 60$, reference obtained t to Appendix H and

   (a) for an upper tail test reject the null hypothesis if obtained t is greater than or equal to the table value.

   (b) for a lower tail test reject the null hypothesis if obtained t is less than or equal to the negative of the table value.

   (c) for a two-tailed test reject the null hypothesis if obtained t is greater than or equal to the table value *or* if obtained t is less than or equal to the negative of the table value.

4. For $n_1 = n_2 > 60$, reference obtained t to Appendix B with $n_1 + n_2 - 2$ degrees of freedom.

5. For unequal sample sizes, obtained t can be referenced to tables constructed for unequal sample sizes (available from the authors) or, if n_1 and n_2 are both greater than 15, reference may be to a t table with $n_1 + n_2 - 2$ degrees of freedom.

EXAMPLE 10.21

Conduct a two-tailed Wilcoxon's rank-sum test at $\alpha = .05$ on the (fictitious) blood pressure data in Table 10.11 on the following page.

Solution   We begin by replacing the blood pressure measures with their respective ranks and forming the result for analysis as shown in Table 10.12.

TABLE 10.11: Blood pressures of hypertensive patients after treatment via drug and placebo therapies.

Drug	Placebo
129	146
131	130
128	127
117	160
144	165
119	133
138	148
142	150
110	171
140	141

TABLE 10.12: Ranks of blood pressures arranged for analysis via the independent samples t test.

Drug		Placebo	
R_1	R_1^2	R_2	R_2^2
6	36	15	225
8	64	7	49
5	25	4	16
2	4	18	324
14	196	19	361
3	9	9	81
10	100	16	256
13	169	17	289
1	1	20	400
11	121	12	144
$\sum$ 73	725	137	2145

Using the sums from Table 10.12 we calculate the independent samples t statistic as demonstrated on page 219 as follows.[16]

The mean rank for the first group is

$$\bar{R}_1 = \frac{\sum R_1}{n_1} = \frac{73}{10} = 7.3$$

---

[16]We use the symbol R rather than X in the calculations as a reminder that we are analyzing ranks rather than original observations.

while that for the second is

$$\bar{R}_2 = \frac{\sum R_2}{n_2} = \frac{137}{10} = 13.7$$

The pooled variance estimate (by Equation 6.2 on page 220) is

$$s_{P_R}^2 = \frac{\left(\sum R_1^2 - \frac{(\sum R_1)^2}{n_1}\right) + \left(\sum R_2^2 - \frac{(\sum R_2)^2}{n_2}\right)}{n_1 + n_2 - 2}$$

$$= \frac{\left(725 - \frac{(73)^2}{10}\right) + \left(2145 - \frac{(137)^2}{10}\right)}{10 + 10 - 2}$$

$$= 25.567$$

By Equation 6.1 on page 219, obtained t, with appropriate substitutions of R for x, is then

$$t = \frac{\bar{R}_1 - \bar{R}_2}{\sqrt{s_{P_R}^2 \left(\frac{1}{n_1} + \frac{1}{n_2}\right)}}$$

$$= \frac{7.3 - 13.7}{\sqrt{25.567 \left(\frac{1}{10} + \frac{1}{10}\right)}}$$

$$= -2.830$$

Reference to Appendix H shows that for $n_1 = n_2 = 10$, critical t for a two-tailed Wilcoxon's rank-sum test conducted at $\alpha = .05$ is -2.248 and 2.248. Because obtained t of -2.830 is less than critical t of -2.248, the null hypothesis is rejected. Technically, this means that the samples did not come from the same population. By earlier arguments, we can conclude that drug treated patients produced lower blood pressures than did placebo treated patients. ∎

EXAMPLE 10.22

Researchers are interested in determining the effect of cell phone use on reaction times to simulated driving emergencies. To this end, 16 subjects are randomly assigned to one of two groups. Both groups are placed in driving simulators where they "drive" for a period of one hour. At a randomly chosen point in the one-hour period each subject experiences a simulated emergency that calls for immediate braking in order to avoid a collision. Throughout the simulations, the subjects in the first group are required to carry on a cell phone conversation with one of the researchers while subjects in the second group simply "drive" without the cell phone encumbrance. Reaction time is defined as elapsed time, measured in fractions of a second, between introduction of the simulated emergency and application of the brake. These times for the two groups are provided in Table 10.13.

Use these reaction times to perform a one-tailed Wilcoxon's rank-sum test with the alternative indicating an increase in reaction time under cell phone use. Conduct the test at $\alpha = .05$.

TABLE 10.13: Reaction times to a simulated emergency of cell phone and non cell phone users.

| | No |
Phone	Phone
.322	.297
.295	.298
.399	.211
.215	.199
.444	.370
.377	.299
.300	.236
.331	.290

Solution We begin by replacing the reaction time measures with their respective ranks and forming the result for analysis as shown in Table 10.14.

Using the sums from Table 10.14 we calculate the independent samples t statistic as demonstrated on page 219 as follows.[17]

The mean rank for the cell phone use group is

$$\bar{R}_1 = \frac{\sum R_1}{n_1} = \frac{87}{8} = 10.875$$

while that for the no cell phone group is

$$\bar{R}_2 = \frac{\sum R_2}{n_2} = \frac{49}{8} = 6.125$$

The pooled variance estimate (by Equation 6.2 on page 220) is

$$s^2_{P_R} = \frac{\left(\sum R_1^2 - \frac{(\sum R_1)^2}{n_1}\right) + \left(\sum R_2^2 - \frac{(\sum R_2)^2}{n_2}\right)}{n_1 + n_2 - 2}$$

$$= \frac{\left(1087 - \frac{(87)^2}{8}\right) + \left(409 - \frac{(49)^2}{8}\right)}{8 + 8 - 2}$$

$$= 17.839$$

---

[17]We use the symbol R rather than X in the calculations as a reminder that we are analyzing ranks rather than original observations.

TABLE 10.14: Ranks of reaction times arranged for analysis via the independent samples t test.

Phone		No Phone	
R_1	R_1^2	R_2	R_2^2
11	121	7	49
6	36	8	64
15	225	2	4
3	9	1	1
16	256	13	169
14	196	9	81
10	100	4	16
12	144	5	25
$\sum$ 87	1087	49	409

By Equation 6.1 on page 219, obtained t, with appropriate substitutions of R for x, is then

$$t = \frac{\bar{R}_1 - \bar{R}_2}{\sqrt{s_{P_R}^2 \left(\frac{1}{n_1} + \frac{1}{n_2}\right)}}$$

$$= \frac{10.875 - 6.125}{\sqrt{17.839 \left(\frac{1}{8} + \frac{1}{8}\right)}}$$

$$= 2.249$$

Reference to Appendix H shows that for $n_1 = n_2 = 8$, critical t for a one-tailed Wilcoxon's rank-sum test conducted at $\alpha = .05$ is 1.944. Because obtained t of 2.249 is greater than critical t of 1.944, the null hypothesis is rejected. By earlier arguments, we can conclude that subjects using cell phones had greater reaction times than did subjects not using cell phones. ■

EXAMPLE 10.23

Given $n_1 = n_2 = 63$ and obtained t of -2.626, conduct a two-tailed Wilcoxon rank-sum test at $\alpha = .05$.

Solution   Because $n_1 = n_2 > 60$, obtained t is referenced to a t table with $n_1 + n_2 - 2$ degrees of freedom. Reference to Appendix B with $63 + 63 - 2 = 124$ degrees of freedom for $\alpha = .05$ gives two-tailed critical values of -1.979 and 1.979. Because obtained t of -2.626 is less than critical t of -1.979, the null hypothesis is rejected. ■

TABLE 10.15: Example of tied observations in a two group design.

Group One	Group Two
14	76
3	44
17	73
22	76
3	90
0	11
76	49
23	53

EXAMPLE 10.24

Given $n_1 = 18$ and $n_2 = 24$ and obtained t of 1.334, conduct a one-tailed Wilcoxon rank-sum test at $\alpha = .05$ to show that values from group one are greater than values from group two.

Solution   Because $n_1 \neq n_2$ and $n_1 > 15$ and $n_2 > 15$, we can perform an approximate test by referencing obtained t to a t distribution with $n_1 + n_2 - 2 = 18 + 24 - 2 = 40$ degrees of freedom. Appendix B shows that critical t for a one-tailed test with 40 degrees of freedom conducted at $\alpha = .05$ is 1.684. Because obtained t of 1.334 is less than this value, the null hypothesis is not rejected.   ∎

Assumptions

There are two principle assumptions underlying Wilcoxon's rank-sum test. The first is that all values in the two samples and hence all ranks are independent. (See assumptions underlying the independent samples permutation test on page 378 for further details.) The second assumption requires that there be no duplicate values among the data in the two samples. Consider the data in Table 10.15.

Notice that there are two values of three and three of 76. This is the assumption violation. The problem arises when we attempt to replace these original data values with ranks. When substituting ranks, the value zero would be replaced with rank 1 but what can we do with the two values of three? There are a number of different strategies but the most common assigns the average of the ranks that would have been assigned had there been no tie. Normally, the second lowest score would receive a rank of 2 while the third lowest would receive rank 3. The "average rank" strategy assigns the average of ranks 2 and 3 or 2.5 to the two scores of three. The fourth smallest value is 11 which receives rank 4. We continue with the ranking by assigning rank five to 14, rank six to 17, rank seven to 22, rank eight to 23, rank nine to 44, rank 10 to 49, rank 11 to 53, and rank 12 to 73. The next highest value is 76 of which there are three. Since the next three values would receive ranks 13, 14, and 15 had there been no ties, we assign the average of these ranks, namely 14, to the three values of 76. The score 90 receives rank 16. The completed set of ranks is shown in Table 10.16.

TABLE 10.16: Example of ranks assigned to tied observations in a two group design.

Group One	Group Two
5	14
2.5	9
6	12
7	14
2.5	16
1	4
14	10
8	11

The question now arises as to how this tied rank data might form the basis of an hypothesis test. After all, the critical values for $n_1 = n_2 = 8$ in Appendix H were obtained for ranks 1 through 16, not ranks 1, 2.5, 2.5, 4, and so on. The following are strategies that may be used when tied observations are encountered. [18]

1. Retain original observations and apply the permutation test described previously (see page 374).[19]

2. Use the average rank method described above and generate the permutation sampling distribution of these ranks. (See footnote 19.)

3. Use the average rank method described above and reference obtained t to a t table using $n_1 + n_2 - 2$ degrees of freedom. This method will usually produce good results if $n_1 + n_2$ is sufficiently large (e.g., greater than 30) and the proportion of tied observations is small (e.g., 20% or less).

10.3.4   Multiple Independent Samples

In Chapter 7 you learned to conduct a one-way ANOVA F test in order to test the null hypothesis $H_0 : \mu_1 = \mu_2 = \cdots = \mu_k$ where $\mu_1, \mu_2 \cdots \mu_k$ are the means of the populations from which the samples were drawn. One of the assumptions underlying this test is that the samples came from normally distributed populations. In this section you will learn to conduct multi sample ANOVA tests that do not require this restrictive assumption. Two tests will be demonstrated. The first uses original observations while the second, known as the Kruskal-Wallis test, uses ranks in place of the original observations. The underlying logic of these tests is emphasized as well as the relationship between the two.

---

[18]Other strategies are outlined in nonparametric texts. See for example [7].

[19]This can be done by means of commercially available software.

The Permutation one-way ANOVA F Test.

Rationale

Suppose that the following outcomes are obtained from a study in which six individuals have been randomly assigned to three treatment groups.

Group One	Group Two	Group Three
4	−1	7
9	0	12

By Equation 7.2 on page 264 the one-way ANOVA F statistic for this data is 6.196. If the researcher wishes to test the null hypothesis $H_0 : \mu_1 = \mu_2 = \mu_3$ but wants to avoid the assumption of population normality, a permutation test based on the following logic may be performed.

Because the six subjects were randomly assigned to the three groups, the score of any particular subject is just as likely to appear in one of the three groups as in either of the others *if the null hypothesis of no differential treatment effect is true*. On the other hand, in the presence of a treatment effect that tends to increase or decrease scores of the members of certain groups, the scores of those groups will tend to be higher or lower than those of other groups.

Thus, when the null hypothesis of no treatment effect is true, subject scores are randomly distributed between the three groups with the result that no group tends to have significantly higher or lower scores than does the other. It follows that an F statistic computed on these data would generally take a value near one. When a treatment effect is present the F statistic will tend to be larger than one.[20]

As previously stated, in the absence of a treatment effect, the score of any given subject is just as likely to appear in one group as in either of the other two. Thus, if the null hypothesis is true, the observed data arrangement is just one of the possible arrangements implied by the random assignment process. But how many ways could the scores have been assigned to the three groups?

By extension of the logic in Section 10.2.2 and Equation 10.2 on page 347 we can deduce that the six observations could be assigned to the three groups in

$$\frac{n!}{n_1! n_2! n_3!} = \frac{6!}{2!2!2!} = 90$$

ways.

In general, n subjects can be assigned to k groups in

$$\boxed{\frac{n!}{n_1! n_2! \cdots n_k!}} \tag{10.3}$$

ways. As may be seen, Equation 10.2 on page 347 is just a special case of the more general form provided by expression 10.3. We will not attempt to show all 90 data arrangements for the data considered here but provide six examples of such arrangements in Table 10.17.

---

[20]See the discussion in Section 7.2.5 on page 272.

TABLE 10.17: Six of 90 possible arrangements of six observations to three groups with associated F statistics for a particular set of scores.

group one	group two	group three
4	−1	7
9	0	12

$F = 6.196$

group one	group two	group three
−1	4	9
0	7	12

$F = 19.158$

group one	group two	group three
−1	0	9
4	7	12

$F = 3.229$

group one	group two	group three
0	−1	9
4	7	12

$F = 2.910$

group one	group two	group three
0	−1	9
7	4	12

$F = 3.229$

group one	group two	group three
−1	0	9
7	4	12

$F = 2.910$

Upper 12 Observations in Sampling Distribution

6.196 6.196 6.196 6.196 6.196 6.196 19.158 19.158 19.158 19.158 19.158 19.158

↑
Obtained F

FIGURE 10.7: Upper 12 observations in the permutation sampling distribution of F statistics for a particular data set.

Having formed all 90 possible arrangements of the data into three groups and computing an F statistic for each, we can arrange the 90 F statistics into a permutation sampling distribution as shown in Figure 10.7. Note that we do not show all 90 permutation F statistics here but, because of space considerations, show only the upper 12 statistics in the distribution.[21]

If the null hypothesis is true, we would expect *obtained F* (i.e., the one obtained from the data collected in the study) to be near one since a true null hypothesis implies that the numerator and denominator of the F statistic both estimate the variance of the populations from which the samples were drawn.[22] A second interpretation is that any variation between the sample means is due to the random assignment process and not to any treatment effect.

---

[21] You will recall from Chapter 7 that the one-way ANOVA F test is inherently one tailed.

[22] See Chapter 7.

On the other hand, if H_0 is not true, we would expect F to take some value in the extreme right tail of the distribution since the between sample variation (i.e., MS_b) would be greater than would be expected if this variation were due to the random assignment process alone. In the present case, obtained F for the data derived from the study is 6.196. What is the probability of obtaining such an extreme value of F when the null hypothesis is true? It is simply the proportion of F's in the sampling distribution that take a value as large as, or larger than, that obtained from the data. In the present case, 12 of the 90 possible values of F are equal to or exceed that obtained from the data—namely, 6.196. We can now say that of the 90 possible values of F that might have been obtained under a true null hypothesis, only 12 are as extreme as that observed from the data. The probability of achieving such an extreme value of F under a true null hypothesis is then $12/90 = .133$. This is the p-value for a test of the null hypothesis $\mu_1 = \mu_2 = \mu_3$.

Conducting the Test

The steps for conducting the permutation one-way ANOVA F test are as follows.

1. Compute F for the data obtained from the study. Call this value obtained F.

2. Form all possible arrangements of the data into the k groups.

3. Calculate the one-way ANOVA F statistic for each such arrangement.

4. Arrange the values of F obtained from step 3 from smallest to largest. This is the permutation sampling distribution of F.

5. Calculate the proportion of values in the permutation distribution that are greater than or equal to obtained F. This is the p-value for the test.

EXAMPLE 10.25

Use the permutation sampling distribution of the one-way ANOVA F statistic shown in Figure 10.7 to find the p-value for the following test

$$H_0 : \mu_1 = \mu_2 = \mu_3$$

assuming obtained F is 19.158 and $n_1 = n_2 = n_3 = 2$.

Solution   The p-value for the test is the proportion of F statistics in the permutation sampling distribution that are equal to or exceed obtained F of 19.158. In this case the values of F that satisfy the criterion of being greater than or equal to obtained F are the six values of 19.158 so that $p = \frac{6}{90} = .067$. If α were .100, we would reject the null hypothesis. *Given the data in this study,* could we ever reject H_0 at $\alpha = .050$? (Hint: NO! Because the smallest possible p-value for these data is .067.)   ■

EXAMPLE 10.26

How many ways can 12 subjects be assigned to four groups if three subjects are to be assigned to each group? Suppose two subjects are to be assigned to the first group, four to the second, five to the third and one to the fourth group?

Solution   For the case where three subjects are to be assigned to each of the four groups we obtain by expression 10.3

$$\frac{n!}{n_1!n_2!\cdots n_k!} = \frac{12!}{3!3!3!3!} = \frac{479,001,600}{1296} = 369,600$$

For $n_1 = 2$, $n_2 = 4$, $n_3 = 5$, and $n_4 = 1$,

$$\frac{n!}{n_1!n_2!\cdots n_k!} = \frac{12!}{2!4!5!1!} = \frac{479,001,600}{5760} = 83,160 \qquad \blacksquare$$

Assumptions

The underlying assumptions for the permutation one-way ANOVA F test are the same as those underlying the permutation independent samples t test discussed on page 378.

Kruskal-Wallis Test.   This test is similar to the permutation one-way ANOVA F test except that scores are converted to ranks before obtained F is calculated and the permutation sampling distribution generated.

Rationale

In order to conduct the permutation one-way ANOVA F test with three observations in each of four groups, we previously found that we would have to form 369,600 data arrangements with an F statistic being calculated for each such arrangement. The number of calculations increases rapidly with increases in sample size and number of groups.

This daunting task, along with other considerations, prompted W. H. Kruskal and W. A. Wallis [28] to propose an independent multi-sample permutation test based on ranks. One advantage of such a strategy is that once the permutation distribution for the test statistic is generated for given sample sizes (n_1, n_2, etc.), the distribution can be used for any samples of the same sizes regardless of the values of the original data. This follows from the fact that the original data is replaced by ranks for which the distribution has already been constructed.

The conversion of original data to ranks is accomplished as follows. The smallest score in the data set is assigned rank 1, the second smallest rank 2 etc. with the largest observation being assigned rank $n_1 + n_2 + \cdots + n_k$ where n_k is the number of observations in the kth group. That is, if there are four observations in each of three samples, the smallest value is assigned rank 1 with the largest being assigned rank $4 + 4 + 4 = 12$. Notice that ranks are assigned without regard to group membership. For example, suppose we wish to convert the data used previously in conjunction with the permutation one-way ANOVA F test to ranks. These data are repeated here for your convenience.

Group One	Group Two	Group Three
4	−1	7
9	0	12

The original observations would be replaced by ranks as follows.

Group One	Group Two	Group Three
3	1	4
5	2	6

TABLE 10.18: Six of 90 possible arrangements of ranks one through six to three groups with associated F statistics.

group one	group two	group three
3	1	4
5	2	6

$F = 4.333$

group one	group two	group three
1	3	5
2	4	6

$F = 16.000$

group one	group two	group three
1	2	5
3	4	6

$F = 4.333$

group one	group two	group three
2	1	5
3	4	6

$F = 3.273$

group one	group two	group three
2	1	5
4	3	6

$F = 4.333$

group one	group two	group three
1	2	5
4	3	6

$F = 3.273$

Applying Equation 7.2 on page 264 to the ranks yields a one-way ANOVA F statistic of 4.333.

In order to generate the permutation sampling distribution for this rank test, we must generate the integers 1 through 6 and form all possible arrangements in which 2 ranks are assigned to each of the three groups. An F statistic must then be computed for each such arrangement. It was shown on page 390 that 90 such arrangements of the ranks are possible. For demonstration purposes, six of the 90 possible arrangements are shown in Table 10.18.

Having formed all 90 possible arrangements of the ranks into three groups and computing an F statistic for each, we can arrange the 90 F statistics into a permutation sampling distribution as shown in Figure 10.8 on the facing page. Note that we do not show all 90 permutation F statistics here but, due to space considerations, show only the upper 18 statistics in the distribution.[23]

If the null hypothesis is true, we would expect *obtained* F (i.e., the one obtained from the ranks of the data collected in the study) to be near one.[24] On the other hand, if H_0 is not true, we would expect F to take some value in the extreme right tail of the distribution since the between sample variation (i.e., MS_b) would be greater than would be expected if this variation were due to the random assignment process alone. In the present case, obtained F

---

[23] You will recall from Chapter 7 that the one-way ANOVA F test is inherently one-tailed.
[24] See Chapter 7.

Upper 18 Observations in Sampling Distribution

4.333 4.333 4.333 4.333 4.333 4.333 4.333 4.333 4.333 4.333 4.333 4.333

↑
Obtained F

Upper 18 Observations in Sampling Distribution (continued)

16.000 16.000 16.000 16.000 16.000 16.000

FIGURE 10.8: Upper 18 observations in the permutation sampling distribution of rank F statistics obtained from all possible arrangements of ranks one through six into three groups composed of two ranks each.

for the ranks of the data derived from the study is 4.333. What is the probability of obtaining such an extreme value of F when the null hypothesis is true? It is simply the proportion of F statistics in the sampling distribution that take a value as large as, or larger than, that obtained from the data. In the present case, 12 of the 90 possible values of F are equal to obtained F of 4.333 while another six exceed this value. We can now say that of the 90 possible values of F that might have been obtained under a true null hypothesis, only $12 + 6 = 18$ are as extreme as that observed from the data. The probability of achieving such an extreme value of F under a true null hypothesis is then $18/90 = .20$. This is the p-value for a test of the null hypothesis of identical populations or equal treatment effects.[25]

Replacing original data with ranks allows us to construct tables of critical values for this test. In order to do so, we need only generate the permutation sampling distribution for sets of ranks for various numbers of groups and sample sizes and locate the permutation values such that α of the permutation values are greater than or equal to the identified value. For example, examination of Figure 10.8 shows that an F value of 16.0 has an associated p-value of $6/90 = .067$. Since 16.0 is the most extreme value in the distribution, it is obvious that for three groups with two observations in each, tests may be conducted at $\alpha = .10$ but not at levels .05, .025, .01, or .005. As sample sizes increase and the permutation sampling distribution is composed of more distinct values of F, tests can be conducted at more levels of α.

Appendix I gives critical values for the Kruskal-Wallis test when the number of observations per group is three to 30 and the number of groups is two to 10. Notice that in order to use this table the number of observations must be the same for each group.[26]

EXAMPLE 10.27

Use the data provided here to conduct a Kruskal-Wallis test at $\alpha = .05$.

Group One	Group Two	Group Three	Group Four
3	4	16	0
1	2	12	7
5	9	8	10

---

[25]See the discussion of the null hypothesis tested by the rank-sum procedure on page 381.

[26]Critical values for unequal sample sizes are currently under construction.

Solution Replacing original observations with ranks produces the following table.

Group One	Group Two	Group Three	Group Four
4	5	12	1
2	3	11	7
6	9	8	10

If we let R_1 represent the ranks of the first group, R_1^2 the squared ranks of the first group and the subscripts 2, 3, and 4 for the remaining groups then we compute

$$\sum R_1 = 4 + 2 + 6 = 12$$
$$\sum R_1^2 = 4^2 + 2^2 + 6^2 = 56$$
$$\sum R_2 = 5 + 3 + 9 = 17$$
$$\sum R_2^2 = 5^2 + 3^2 + 9^2 = 115$$
$$\sum R_3 = 12 + 11 + 8 = 31$$
$$\sum R_3^2 = 12^2 + 11^2 + 8^2 = 329$$
$$\sum R_4 = 1 + 7 + 10 = 18$$
$$\sum R_4^2 = 1^2 + 7^2 + 10^2 = 150$$

The sums of squares for the four groups are as follows.

$$SS_1 = \sum R_1^2 - \frac{\left(\sum R_1\right)^2}{n_1} = 56 - \frac{(12)^2}{3} = 8.000$$

$$SS_2 = \sum R_2^2 - \frac{\left(\sum R_2\right)^2}{n_2} = 115 - \frac{(17)^2}{3} = 18.667$$

$$SS_3 = \sum R_3^2 - \frac{\left(\sum R_3\right)^2}{n_2} = 329 - \frac{(31)^2}{3} = 8.667$$

$$SS_4 = \sum R_4^2 - \frac{\left(\sum R_4\right)^2}{n_4} = 150 - \frac{(18)^2}{3} = 42.000$$

By Equation 7.4 on page 265 the sum of squares within is

$$SS_w = SS_1 + SS_2 + SS_3 + SS_4 = 8.000 + 18.667 + 8.667 + 42.000 = 77.334$$

By Equation 7.3 on page 265 the mean square within is

$$MS_w = \frac{SS_w}{N - k} = \frac{77.334}{12 - 4} = 9.667$$

where SS_w is the sum of squares within, N is the total number of observations, and k is the number of groups.

By Equation 7.7 on page 267 with substitution of R for x to indicate that the calculation is for ranks rather than original observations, the sum of squares between is

$$SS_b = n \left[\sum_{j=1}^{k} \bar{R}_j^2 - \frac{\left(\sum_{j=1}^{k} \bar{R}_j \right)^2}{k} \right]$$

where n is the number of observations in *each* group, and $\bar{R}_j$ are the rank group means.

From earlier calculations we obtain

$$\bar{R}_1 = \frac{\sum R_1}{n_1} = \frac{12}{3} = 4.000$$

$$\bar{R}_2 = \frac{\sum R_2}{n_2} = \frac{17}{3} = 5.667$$

$$\bar{R}_3 = \frac{\sum R_3}{n_3} = \frac{31}{3} = 10.333$$

$$\bar{R}_4 = \frac{\sum R_4}{n_4} = \frac{18}{3} = 6.000$$

Then

$$\sum \bar{R} = 4.000 + 5.667 + 10.333 + 6.000 = 26.000$$

and

$$\sum \bar{R}^2 = 4.000^2 + 5.667^2 + 10.333^2 + 6.000^2 = 190.886$$

Making the proper substitutions in Equation 7.7 yields

$$SS_b = n \left[\sum_{j=1}^{k} \bar{R}_j^2 - \frac{\left(\sum_{j=1}^{k} \bar{R}_j \right)^2}{k} \right] = 3 \left[190.886 - \frac{(26.000)^2}{4} \right] = 65.658$$

By Equation 7.6 on page 267 the mean square between is

$$MS_b = \frac{SS_b}{k-1} = \frac{65.658}{4-1} = 21.886$$

where SS_b is the sum of squares between and k is the number of groups. Finally, by Equation 7.2 on page 264 obtained F is

$$F = \frac{MS_b}{MS_w} = \frac{21.886}{9.667} = 2.264$$

With $\alpha = .05$, four groups and three observations per group, Appendix I gives critical F as 4.483. Because obtained F of 2.264 is less than this value, we are unable to reject the null hypothesis. Thus, we are unable to demonstrate a treatment effect in this study. ∎

Conducting the Test

The steps for conducting the Kruskal-Wallis rank test are as follows.

1. Replace original scores with their respective ranks with the lowest score receiving rank 1, the second lowest rank 2, and so on with the highest score receiving rank $n_1 + n_2 + \cdots + n_k$. Ranking is carried out without regard to group.

2. Use Equation 7.2 on page 264 to calculate the one-way ANOVA F statistic for the ranks obtained in step 1. Call this value obtained F.

3. For equal sample sizes where $3 \leq n \leq 30$ and $2 \leq k \leq 10$, reference obtained F to Appendix I and reject H_0 if obtained F is greater than or equal to critical F.

4. For equal sample sizes with $n > 30$, reference obtained F to an F table such as Appendix C with numerator and denominator degrees of freedom of $k - 1$ and $N - k$ respectively. Here k represents the number of groups and N the total number of observations in the k groups combined.

5. For unequal sample sizes, obtained F can be referenced to tables constructed for unequal sample sizes (available from the authors) or, if all groups have sample sizes greater than 15, reference may be to an F table with $k - 1$ and $N - k$ degrees of freedom.

EXAMPLE 10.28

In a quality control study, three insurance plans are compared as to length of hospital stay (measured in days) for patients admitted under the three plans. When the data are collected, it is found that 119 patients were admitted under Plan I, 142 under Plan II, and 92 under plan III. Because length of hospital stay data are typically skewed, researchers decide that a Kruskal-Wallis test would be more appropriate than the normality assuming one-way ANOVA F test. To this end, length of hospital stay for the 353 patients are converted to ranks with an F statistic being computed on the resultant ranks. Obtained F for the ranks was 3.22. Test the significance of this finding at $\alpha = .05$. What is your conclusion concerning length of hospital stay for the three plans?

Solution   While Appendix I does not provide a critical value for this test, the large sample sizes make the F distribution an appropriate approximation. Using numerator degrees of freedom of $k - 1 = 3 - 1 = 2$ and denominator degrees of freedom $N - k = 353 - 3 = 350$, Appendix C provides a critical value of 3.02 for $\alpha = .05$. Because obtained F of 3.22 exceeds this value, we can reject the null hypothesis of identical populations. Based on the argument made by James Bradley outlined on page 381, we may also conclude that average length of stay differs for some of the three insurance plans. ∎

Assumptions

The assumptions underlying the Kruskal-Wallis test are the same as those underlying Wilcoxon's rank-sum test discussed on page 388. See also the discussions on page 388 and 389 regarding tied observations. Note that, in the presence of ties, while the rank-sum test can be referenced to a t distribution, the Kruskal-Wallis test would be referenced to an F distribution. In the case of two groups, the Wilcoxon rank-sum and Kruskal-Wallis tests provide the same result.

TABLE 10.19: Contingency table showing positive and negative results from two treatment groups.

	+	−	
Treatment One	7	3	10
Treatment Two	1	9	10
	8	12	

10.3.5 Contingency Tables

To this point we have focused on continuous outcomes such as blood pressures or reaction times that are amenable to analysis by means of correlations or tests based on t or F statistics. But the permutation principle is equally applicable to the analysis of contingency tables. While the permutation principle may be applied to contingency tables of any size, we will restrict attention to the two by two table since these tables can be analyzed without use of specialized computer software. We will begin by discussing the rationale underlying the testing of two by two tables as it relates to Fisher's exact test, then present the method of calculation.

Rationale.   Suppose that in the course of a clinical trial, 20 patients are randomly assigned to one of two treatments with 10 patients being assigned to each. The outcome for each patient is assessed as either positive (+) or negative (−). The results of the trial are shown in Table 10.19.

As may be seen, 7 or $\frac{7}{10} = .7$ of the patients receiving treatment one obtained a positive outcome while 1 or $\frac{1}{10} = .1$ of the patients receiving treatment two realized a positive outcome. The question arises as to what accounts for this difference in positive outcomes between the two groups. One possibility is that the treatment afforded the patients in group one was more effective than that provided patients in group two thereby producing more positive results. But a second possibility exists. It might be that the random assignment process put seven of the patients achieving a positive outcome in group one while, by random happenstance, placing only one positive outcome patient in group two. Notice that the first explanation implies a difference in the effectiveness of the treatments provided the two groups while the second denies any such difference. But how might a decision be made as to which of the explanations is to be believed?

A permutation based decision method would generate all of the outcome configurations possible under a true null hypothesis of no treatment difference and then use this distribution to access the probability of the result achieved in the study. In the present case, the random assignment process might have placed none, one, two, three, four, five, six, seven, or all eight of the + patients in group one. Figure 10.20 on the next page shows all such table arrangements. Notice that specifying the number of + patients in group one automatically specifies the values in the other three cells of the table. For example, Table 2 in Figure 10.20 shows the data arrangement when one positive result is observed in group one. In this case the number of negative results in group one *must* be nine because a total of 10 patients were assigned to group one. Likewise, because there was a total of eight positive outcomes in the

TABLE 10.20: All possible fixed margin tables formed from a particular data set.

	(1) +	(1) −	
one	0	10	10
two	8	2	10
	8	12	

	(2) +	(2) −	
one	1	9	10
two	7	3	10
	8	12	

	(3) +	(3) −	
one	2	8	10
two	6	4	10
	8	12	

	(4) +	(4) −	
one	3	7	10
two	5	5	10
	8	12	

	(5) +	(5) −	
one	4	6	10
two	4	6	10
	8	12	

	(6) +	(6) −	
one	5	5	10
two	3	7	10
	8	12	

	(7) +	(7) −	
one	6	4	10
two	2	8	10
	8	12	

	(8) +	(8) −	
one	7	3	10
two	1	9	10
	8	12	

	(9) +	(9) −	
one	8	2	10
two	0	10	10
	8	12	

study, group two must have seven positive outcomes when group one has one such outcome. The same logic mandates that three negative outcomes will be in group two. Thus, when generating a permutation distribution we need only specify the number of positive outcomes in group one.[27]

Based on your knowledge of permutation methods gained to this point, it would seem that a valid permutation testing procedure could be constructed by listing all values of the number of positive outcomes realized in group one under all possible randomizations to the two groups. To see this intuitively, suppose that we were to repeatedly randomly assign 10 subjects to each group. For each randomization we record the number of positive outcomes in group one. We continue this process until all possible data arrangements in the four cells have been obtained. The numbers of positive outcomes in group one could then be listed from smallest to largest to form the permutation sampling distribution. A test could then be conducted by counting the number of outcomes in the permutation distribution that are greater than (less than) or equal to the number of positive outcomes in group one realized in the study. This is in fact the (computer aided) method most often employed for the permutation testing of contingency tables.[28]

In the case of two by two tables, there is a more direct approach to the generation of the permutation sampling distribution. This approach, with the accompanying test of significance, is known as Fisher's exact test.

---

[27]We could have used any of the other three cells had we so chosen.

[28]Actually, in recent years sophisticated algorithms have been developed that make the listing of all outcomes unnecessary.

By Fisher's method, we can determine how many zeros, ones, twos, and so on will appear in the permutation sampling distribution thereby relieving us of the burden of generating all possible randomizations in order to find these values. As an example, let us see how many twos will appear in the distribution.

In order to make this determination we introduce the notation $+_1, +_2, \ldots, +_8$ to identify the eight patients achieving a positive outcome. Two positive results will be observed in group one if $+_1$ and $+_2$ are randomized to group one or if $+_1$ and $+_3$ are randomized to group one or if any of the possible combinations involving two positive results being assigned to group one is realized. But how many ways can the eight positive outcome subjects be divided between the two groups so that two such subjects are assigned to group one? From your study of Section 10.2.2 on page 345 you will recognize this as a combinations problem. By Equation 10.2 on page 347

$$C_{n_2}^{n_1} = \frac{n!}{n_1! n_2!} = C_6^2 = \frac{8!}{2! 6!} = 28$$

Thus, there are 28 ways in which random assignment of the eight positive outcome patients can be carried out so that two such patients are assigned to group one. But the negative outcome patients must also be randomized. For *each* of the 28 possible assignments of the positive outcome patients there will be

$$C_{n_2}^{n_1} = \frac{n!}{n_1! n_2!} = C_4^8 = \frac{12!}{8! 4!} = 495$$

assignments of the negative outcome patients.

This means that there are

$$\left(C_6^2\right)\left(C_4^8\right) = (28)(495) = 13,860$$

random assignments that will produce two positive outcomes in group one. This process can be carried out for all nine tables in Figure 10.20. This would provide the number of zeros, ones, twos, and so on that could result from the random assignment procedure. The results could then be combined to form the desired permutation sampling distribution.

Fisher's Exact Test.   Table 10.21 on the following page represents the results of a randomization of subjects to one of two groups. In the randomization, g subjects were assigned to group one while h subjects were assigned to group two. Notice that specifying the frequency of any cell automatically determines the frequencies of the other three cells.

It will also be useful to note that the maximum and minimum values of a for the specified margins e, f, g, and h will be respectively

$$\boxed{a_{min} = g - min\,(f, g)} \tag{10.4}$$

and

$$\boxed{a_{max} = min\,(e, g)} \tag{10.5}$$

TABLE 10.21: Contingency table showing the results of a random assignment of subjects to one of two groups.

	$+$	$-$	
Treatment One	a	b	$g = a + b$
Treatment Two	c	d	$h = c + d$
	$e = a + c$	$f = b + d$	

We wish to find the probability that the randomization will produce each of the possible a values. That is, the values from a_{min} to a_{max}.[29] Note that this probability would simply be the proportion of all possible randomizations that produce the specified value a.

In order to find the proportion of randomizations that produce a specified value of a, we must first find the number of such randomizations. This number can then be divided by the total number of randomizations to produce the desired probability.

We begin by noting that the fixed number of positive outcome patients is e. We must find the number of randomizations that produce exactly a positive outcomes in group one. There are two components to this problem. First we must ask, "How many ways can we divide e subjects into two groups with a subjects being placed in one group and c in the other?" From your study of Section 10.2.2 on page 345 you know this value is C_c^a. But we must also consider the assignments of the f negative outcome patients. By the same reasoning, this will be C_d^b. We can now say that for *each* of the C_c^a possible assignments of the positive outcome subjects there will be C_d^b assignments of the negative outcome patients. Thus, the number of possible randomizations that will produce the frequency a is $\left(C_c^a\right)\left(C_d^b\right)$. We can obtain the *probability* of achieving the frequency a by dividing $\left(C_c^a\right)\left(C_d^b\right)$ by the total number of possible randomizations. This number will be C_h^g. Notice that this number reflects *all* possible randomizations without fixing a to a specific value.

To summarize, we can say that the probability associated with any value of a is

$$P(a) = \frac{\left(C_c^a\right)\left(C_d^b\right)}{C_h^g}$$

Using the notation of Equation 10.2 on page 347 and letting n represent the total number of subjects (i.e., $e + f$ or $g + h$) we can write this expression as

$$P(a) = \frac{\left(\frac{e!}{a!c!}\right)\left(\frac{f!}{b!d!}\right)}{\frac{n!}{g!h!}}$$

which simplifies to

$$P(a) = \frac{e!f!g!h!}{a!b!c!d!n!} \qquad (10.6)$$

[29] The notation $max(x, y)$ and $min(x, y)$ indicate, respecitvely, that the maximum and minimum of x and y are to be employed.

TABLE 10.22: Probabilities of a.

a	$P(a)$
0	.00036
1	.00953
2	.07502
3	.24006
4	.35008
5	.24006
6	.07502
7	.00953
8	.00036

which is known as the **hypergeometric function**.

EXAMPLE 10.29

Given the marginal values in Table 10.19 on page 399, find the maximum and minimum values for the numbers of positive outcome patients assigned to group one as a result of the randomization process. Use Equation 10.6 on the facing page to find probabilities associated with each such value.

Solution   By Equations 10.4 and 10.5, the lowest and highest possible number of positive outcomes in group one are respectively,

$$a_{min} = g - min(f, g) = 10 - min(12, 10) = 10 - 10 = 0$$

and

$$a_{max} = min(e, g) = min(8, 10) = 8$$

Table frequencies for positive outcome values in group one of zero through eight are shown in Figure 10.20 on page 400. Using the frequencies in Table 1 of Figure 10.20 in Equation 10.6 gives

$$P(0) = \frac{8!12!10!10!}{0!10!8!2!20!} = .00036$$

The probability for Table 2 would be

$$P(1) = \frac{8!12!10!10!}{1!9!7!3!20!} = .00953$$

We provide the remaining probabilities in Table 10.22 but leave their calculations to you.[30]
We can perform a test of the null hypothesis

$$H_0 : \pi_1 = \pi_2$$

---

[30]Don't forget that you can cancel terms as you did in Section 4.2.5 on page 81 in order to save computational effort.

where π_1 and π_2 are the population proportions of positive outcomes for groups one and two by referencing the obtained value of a (7 in this example) to the distribution in 10.22. For the one-tailed alternative

$$H_A : \pi_1 > \pi_2$$

the probability of obtaining a value of a greater than or equal to 7 is $P(7) + P(8) = .00953 + .00036 = .00989$ which is the p-value for the test. We can, therefore, reject the null hypothesis at $\alpha = .01$ and thereby conclude that the treatment afforded group one was more effective at producing positive outcomes than was the treatment provided group two. ■

Conducting the Test

The steps for conducting the Fisher's exact test are as follows.

1. Use Equations 10.4 and 10.5 to find the minimum and maximum values for a.

2. Use the results from step 1 to construct all possible randomization tables having the same marginal values as those obtained from the original data set.

3. Use Equation 10.6 on page 402 to find the probabilities for each table enumerated in step 2.

4. Order the values of a along with their associated probabilities from a_{min} to a_{max}.

5. For one-tailed tests the p-value is found by

    (a) summing the probabilities for all values of a that are less than or equal to obtained a if the alternative specifies $H_A : \pi_1 < \pi_2$, or

    (b) summing the probabilities for all values of a that are greater than or equal to obtained a if the alternative specifies $H_A : \pi_1 > \pi_2$.

6. For two-tailed tests the p-value is found by[31]

    (a) noting the sum of the probabilities for all values of a that are less than or equal to obtained a as well as the sum for all values of a that are greater than or equal to obtained a. Call the lesser of these values P_1.

    (b) If P_1 is calculated from the lower tail, begin summing the probabilities in the upper tail beginning with a_{max}. Continue summing so long as the sum is less than or equal to P_1. Call this sum P_2.

    (c) If P_1 is calculated from the upper tail, begin summing the probabilities in the lower tail beginning with a_{min}. Continue summing so long as the sum is less than or equal to P_1. Call this sum P_2.

    (d) The two-tailed p-value is $P_1 + P_2$.

---

[31] Various methods for calculating the two-tailed p-value for Fisher's exact test have been proposed. The one presented here appears to be among those most commonly used.

TABLE 10.23: Contingency table relating smoking to disease.

	D	$\overline{D}$	
S	4	2	6
$\overline{S}$	1	8	9
	5	10	

EXAMPLE 10.30

Suppose that a study is conducted to determine whether the proportion of smokers who suffer from a particular disease differs from the proportion of non-smokers who manifest the same disease. Use the data in Table 10.23 and Fisher's exact test[32] to test the null hypothesis

$$H_0 : \pi_1 = \pi_2$$

against the alternative

$$H_A : \pi_1 > \pi_2$$

Then test the null hypothesis against the alternative

$$H_A : \pi_1 \neq \pi_2$$

Conduct both tests at the .05 level of significance.

Solution   By Equations 10.4 and 10.5 on page 401, the minimum and maximum numbers of smokers with disease appearing in the randomization tables will be

$$a_{min} = g - min\,(f, g) = 6 - min\,(10, 6) = 6 - 6 = 0$$

and

$$a_{max} = min\,(e, g) = min\,(5, 6) = 5$$

Figure 10.24 on the following page shows all possible table configurations associated with Table 10.23. Using these table values with Equation 10.6 on page 402 produces the probabilities in Table 10.25 on the following page.

We calculate $P\,(0)$ and $P\,(5)$ here but leave the remaining calculations to you.

$$P\,(0) = \frac{e!\,f!\,g!\,h!}{a!\,b!\,c!\,d!\,n!} = \frac{5!\,10!\,6!\,9!}{0!\,6!\,5!\,4!\,15!} = .04196$$

$$P\,(5) = \frac{5!\,10!\,6!\,9!}{5!\,1!\,0!\,9!\,15!} = .00200$$

For the one-tailed alternative $H_A : \pi_1 > \pi_2$, we apply step 5b and find

$$P\,(4) + P\,(5) = .04496 + .00200 = .04696$$

---

[32]What other methods have you encountered that might be used to perform this test?

TABLE 10.24: All possible fixed margin tables formed from a particular data set.

(1)

	D	$\overline{D}$	
S	0	6	6
$\overline{S}$	5	4	9
	5	10	

(2)

	D	$\overline{D}$	
S	1	5	6
$\overline{S}$	4	5	9
	5	10	

(3)

	D	$\overline{D}$	
S	2	4	6
$\overline{S}$	3	6	9
	5	10	

(4)

	D	$\overline{D}$	
S	3	3	6
$\overline{S}$	2	7	9
	5	10	

(5)

	D	$\overline{D}$	
S	4	2	6
$\overline{S}$	1	8	9
	5	10	

(6)

	D	$\overline{D}$	
S	5	1	6
$\overline{S}$	0	9	9
	5	10	

TABLE 10.25: Probabilities of a.

a	$P(a)$
0	.04196
1	.25175
2	.41958
3	.23976
4	.04496
5	.00200

which is the test p-value. Because this value is less than $\alpha = .05$, the null hypothesis is rejected.

To find the two-tailed p-value, we apply step 6a by first finding the probability that a takes a value greater than or equal to obtained a of 4. This value is

$$P(4) + P(5) = .04496 + .00200 = .04696$$

The probability that a takes a value less than or equal to the obtained value of 4 can be found by finding $P(0) + P(1) + P(2) + P(3) + P(4)$ or more simply by noting that $P(0) + P(1) + P(2) + P(3) = 1 - .04696 = .95304$ so that the probability that a takes a value less than or equal to 4 is $.95304 + P(4) = .95304 + .04496 = .99800$.

Because .04696 is less than .99800, we determine $P_1 = .04696$.

Because P_1 was obtained from the upper tail of the permutation distribution, we apply step 6c and begin the summation with a_{min} which is zero in this case. We find that $P(0) = .04196$. We can add no other probabilities to this value because $P(0) + P(1) = .04196 + .25175 = .29371$ which is greater than $P_1 = .04696$. Therefore, $P_2 = .04196$ and the two-tailed p-value for this test is $P_1 + P_2 = .04696 + .04196 = .08892$ so that the two-tailed test is not significant at $\alpha = .05$. ∎

TABLE 10.26: Contingency table relating childhood abuse to eating disorders.

	Disorder	No Disorder	
Abused	7	4	11
Not Abused	3	3	6
	10	7	

EXAMPLE 10.31

A researcher wishes to determine whether childhood abuse is related to eating disorders developed later in life. To this end, the data shown in Table 10.26 are collected. Use these data to conduct Fisher's exact test at the .05 level of significance. Test the hypothesis

$$H_0 : \pi_1 = \pi_2$$

against the alternative

$$H_A : \pi_1 \neq \pi_2$$

Would you guess that this data was gathered from the general population or some specific group?

Solution   The minimum and maximum values of a are by Equations 10.4 and 10.5

$$a_{min} = g - min\,(f, g) = 11 - min\,(7, 11) = 11 - 7 = 4$$

and

$$a_{max} = min\,(e, g) = min\,(10, 11) = 10$$

Using these values we form the randomization tables in Table 10.27 on the next page.[33]

The probability of observing each of the tables in Table 10.27 under the randomization process is given in Table 10.28 on the following page. We show the computation of two of these probabilities but leave the remainder to you. Using the cell and marginal frequencies from Tables 1 and 4 in Figure 10.27, we compute by means of Equation 10.6

$$P\,(4) = \frac{e!\,f!\,g!\,h!}{a!\,b!\,c!\,d!\,n!} = \frac{10!\,7!\,11!\,6!}{4!\,7!\,6!\,0!\,17!} = .01697$$

$$P\,(7) = \frac{10!\,7!\,11!\,6!}{7!\,4!\,3!\,3!\,17!} = .33937$$

To find the two-tailed p-value, we apply step 6a on page 404 by first finding the probability that a takes a value less than or equal to obtained a of 7. This value is

$$P\,(4) + P\,(5) + P\,(6) + P\,(7) = .01697 + .14253 + .35633 + .33937 = .85520$$

---

[33]What happens if you use a value for a that is outside this range? Try setting $a = 3$.

TABLE 10.27: All possible fixed margin tables formed from a data set relating abuse to eating disorders.

(1)

	Disorder	No Disorder	
Abused	4	7	11
Not Abused	6	0	6
	10	7	

(2)

	Disorder	No Disorder	
Abused	5	6	11
Not Abused	5	1	6
	10	7	

(3)

	Disorder	No Disorder	
Abused	6	5	11
Not Abused	4	2	6
	10	7	

(4)

	Disorder	No Disorder	
Abused	7	4	11
Not Abused	3	3	6
	10	7	

(5)

	Disorder	No Disorder	
Abused	8	3	11
Not Abused	2	4	6
	10	7	

(6)

	Disorder	No Disorder	
Abused	9	2	11
Not Abused	1	5	6
	10	7	

(7)

	Disorder	No Disorder	
Abused	10	1	11
Not Abused	0	6	6
	10	7	

TABLE 10.28: Probabilities of a.

a	$P(a)$
4	.01697
5	.14253
6	.35633
7	.33937
8	.12726
9	.01697
10	.00057

The upper tail probability is

$$P(7) + P(8) + P(9) + P(10) = .33937 + .12726 + .01697 + .00057 = .48417$$

Because .48417 is less than .85520, we set $P_1 = .48417$.

Because P_1 was obtained as a sum from the upper tail, we find P_2 by beginning to sum the probabilities in the lower tail beginning with $a_{min} = 4$. This sum is

$$P(4) + P(5) = .01697 + .14253 = .15950$$

Notice that we cannot add the probability for $P(6)$ because this would result in a sum of .51583 which is larger than P_1. Therefore the two-tailed p-value is $P_1 + P_2 = .48417 + .15950 = .64367$ so that the null hypothesis is not rejected. We also notice that the data were likely not sampled from the general population because even the non-abused group had a high incidence of eating disorders—namely, $3/6 = .5$.   ■

Assumptions

The primary assumption underlying Fisher's exact test is that all observations are independent. We have discussed the assumption of independence in connection with several other tests so will not repeat it here.[34] Suffice it to say that this test cannot be depended upon for control of Type I errors when this assumption is violated.

10.4   FURTHER COMMENTS REGARDING PERMUTATION BASED METHODS

We have devoted approximately 100 pages to this topic but leave much unsaid. We cannot close this chapter without saying just a bit more regarding permutation based methods and our presentation thereof.

- The computational forms for the nonparametric statistics presented in this chapter are not those commonly used in nonparametric texts. For example, for rank based tests we have taught you to simply convert original scores to ranks and compute the standard parametric statistics with which you are already familiar on the resultant ranks. In standard treatments, simpler computational forms are given. But this requires the student to master new statistics with which she/he is not familiar. These forms are also often associated with tables of critical values that are non-intuitive[35] and confusing. In earlier times, computational ease was of prime importance which made the simpler forms highly desirable. With the advent of hand calculators and computers, ease of calculation has diminished in importance. Thus, we chose not to have you learn a new set of statistics but to simply apply those with which you are already familiar. We emphasize that both computational methods provide the same result.

- Not all rank based nonparametric tests can be expressed as some parametric counterpart conducted on rank transformations of original data. However, as demonstrated in this chapter, this is true of most of the rank based tests presented in introductory texts. We venture to add that this is true of most commonly employed rank based nonparametric tests.

---

[34]See for example the discussion of independence in connection with the chi-square test presented on page 282.

[35]For example, some tables require that lower tail critical values be used for upper tail tests.

- A long held, and oft repeated, myth maintains that rank-based nonparametric tests are inherently less powerful (or efficient) than are their parametric counterparts. For example, Kuzma and Bohnenblust [30] state in regards to nonparametric tests "They are less efficient (i.e., they require a larger sample size to reject a false hypothesis) than comparable parametric tests." These same authors conclude, "In using nonparametric methods, you should be careful to view them as complementary statistical methods rather than attractive alternatives."

  The first statement is untrue. As a result, we do not support the conclusion rendered in the second statement. The truth of the matter is that, *nonparametric tests are necessarily less powerful than a parametric counterpart only when the assumptions of the latter are perfectly met.*[36] In point of fact, some nonparametric tests, e.g., Wilcoxon's signed-ranks and rank-sum tests, may have *very large* power advantages over their parametric counterparts, i.e., the paired and independent samples t tests, in commonly encountered situations. [1, 2] In the practice of data analysis, parametric methods may hold power advantages in some situations while nonparametric methods maintain advantages in others. In truth, neither set of tests hold absolute domination over the other in this regard.

- At the time of their development, most permutation methods that utilized original scores, were largely of theoretical rather than practical importance. This was because overwhelming computational difficulties made their use not only impractical, but for other than very small samples, impossible. With the advent of modern computers and newly devised computational algorithms, these tests have now entered the realm of practical statistical methods. This is witnessed to by the fact that many new commercial software packages have become available that implement these methods. We will not attempt a comprehensive list but note that among the best known and widely used is the StatXact package from Cytel Software Corporation [9]. Among the most comprehensive is SC (Statistical Calculator) from Mole Software [10]. A fairly simple and flexible programming language designed specifically for permutation and similar problems is provided by Resampling Stats software from Resampling Stats Inc. [41]. Additionally, many of the older packages that implemented parametric and rank-based nonparametric methods have added permutation-based methods for original scores and contingency tables to their cadre of procedures.

- In Section 2.2.2 on page 10, we indicated that ordinal level measurements have traditionally presented some difficulty insofar as inferential testing is concerned. Permutation methods overcome many of these problems.

- **Interaction** is an important statistical concept in which the effect of one variable depends upon its association with some other variable(s). A major shortcoming of rank-based permutation methods is their inability to meaningfully detect interactions. Permutation methods based on original data fair somewhat better in this regard but provide only approximate rather than exact results.

- We have touched on only a small number of permutation-based methods in this chapter. Many more are available. In fact, permutation methods may be used to develop ad hoc procedures for problems for which no parametric procedure is available.

---

[36]A situation rarely if ever encountered in practice.

KEY WORDS AND PHRASES

After reading this chapter you should be able to demonstrate familiarity with the following words and phrases.

average rank method 360
distribution-free 348
Fisher's exact test 401
hypergeometric function 403
Kruskal-Wallis test 393
nonparametric 348
permutation independent samples t test 374
permutation one-way ANOVA F test 390
permutation sampling distribution 350
Pitman's test for correlation 349
ranks 354
tied ranks 360
Wilcoxon's signed-ranks test 365

combinations 346
factorial 345
Hoteling and Pabst test 353
interaction 410
monotonic relationship 356
parametric 348
permutation method 348
permutation paired samples t test 361
permutations 344
rank-based tests 353
Spearman's rho 354
Wilcoxon's rank-sum test 378

EXERCISES

10.1 A series of five distinct medical tests are to be sequentially administered to a patient. In how many orders might the five tests be administered? Suppose there were six tests.

10.2 In a clinical trial patients are to be randomly assigned to treatment groups.

   (a) If 10 patients are assigned to a treatment and a control group with five patients being assigned to each, how many distinct groupings of patients are possible?

   (b) If 15 patients are assigned to three groups with five patients in each group, how many distinct groupings of patients are possible?

10.3 Given three pairs of observations,

   (a) how many values of r would appear in the sampling distribution of Pitman's test for correlation?

   (b) would it be possible to conduct a test at $\alpha = .05$? Why or why not?

10.4 A study is conducted to determine whether there is a relationship between age and prostate specific antigen (PSA) cancer screening test results. The data are shown below. Use these data to perform the Hoteling and Pabst test for rank correlation to test the null hypothesis of no relationship against the alternative of a positive relationship. Use $\alpha = .05$.

PSA	Age
9.1	78
1.2	24
2.3	60
6.0	77
3.9	48
1.0	33
6.1	75
2.7	51
5.9	88
9.9	79

10.5 Suppose that a rank correlation of .88 is obtained for $n = 124$ data pairs. Test this coefficient for significance by means of a two-tailed test at $\alpha = .10$.

10.6 Given six data pairs, how many data arrangements would be necessary in order to form the sampling distribution for the permutation-paired samples t test?

10.7 The data provided below represent the weights of 10 women recorded before using an oral contraceptive and at a later point in time after use of an oral contraceptive for six months. Use Wilcoxon's signed-ranks test to determine whether a significant difference exists between the weights recorded under the two conditions. Use $\alpha = .05$.

No Contraception	Contraception
128	130
145	145
103	108
139	135
125	124
160	163
144	153
122	129
134	128
161	169

10.8 Suppose a permutation-independent samples t test is to be performed on data collected from two groups consisting of five subjects each. How many t statistics will make up the sampling distribution? How many statistics would make up the sampling distribution if Wilcoxon's rank-sum test were used?

10.9 The data provided below represent waiting times (expressed in minutes) for patients brought into a public and private emergency care facility for treatment. Use Wilcoxon's rank-sum test to compare waiting times for the two facilities. Perform a two-tailed test at $\alpha = .05$. What is your conclusion.

Public	Private
128	34
56	18
13	49
94	66
155	177
22	4
19	8
3	1
99	15
41	17

10.10 Suppose that a permutation one-way ANOVA F test is to be used to analyze the data obtained from a clinical trial in which there were five subjects in each of three groups. How many F statistics will appear in the sampling distribution?

10.11 Students enrolling in an introduction to public health class are randomly assigned to one of three instructional sections. Traditional lecture methods are used as the medium of instruction in the first section, an Internet based mode of instruction is used with the second that includes instructor participation, while the third is taught via filmed lectures without instructor participation. Final course averages are given below for the 15 students participating in the study. Use the Kruskal-Wallis test, conducted at $\alpha = .05$, to determine whether the instructional methods differed in their impact on course averages.

Lecture	Internet	Film
87	90	100
77	76	99
56	44	89
92	58	88
80	97	70

10.12 Fisher's exact test is particularly useful in situations where expected cell frequencies are too small to permit valid use of the chi-square test. Use the table given below to

(a) calculate minimum and maximum values of a,

(b) list all randomization tables,

(c) calculate the probability associated with each table, and

(d) conduct tests of the null hypothesis
   $$H_0 : \pi_1 = \pi_2$$
   against the alternatives
   $$H_A : \pi_1 \neq \pi_2$$
   and
   $$H_A : \pi_1 < \pi_2$$
   at the .05 level of significance.

	+	−
Group One	6	4
Group Two	3	2

A. The following questions refer to Case Study A (page 469)

10.13 The authors point out that both parametric and nonparametric methods were used to analyze the ocular health data. When discussing the results of these analyses they state, "Differences, where found, between the parametric and non-parametric data are consistent with reduced power of non-parametric tests." Comment on this statement.

B. The following questions refer to Case Study B (page 470)

10.14 Suppose that the researchers conducting the analysis alluded to in 7.9 (on page 293) decide to perform the analysis via a nonparametric test. What tests might be used for this purpose? What null hypotheses would be associated with the tests you suggest?

D. The following questions refer to Case Study D (page 471)

10.15 The researchers used independent samples t tests to show that HIV-infected patients had significantly lower NPZ-8 and PBV scores than did the five healthy control subjects. Given the suspected non-normal population distribution of the NPZ-8 scores, some researchers might prefer to use a nonparametric procedure to make these determinations.

   (a) What nonparametric tests might be used for this purpose?
   (b) Use a two-tailed nonparametric test to compare the NPZ-8 and PBV data for HIV-infected and healthy subjects at the .05 level.
   (c) Were any assumptions of the nonparametric tests you used in 10.15b violated? Explain.
   (d) Did you use an exact or approximate test for the analyses in 10.15b? Why?
   (e) Explain how an exact version of the tests conducted in 10.15b might have been conducted.

10.16 The researchers report a $-.50$ Spearman rank order correlation between PBV and NPZ-8 for the 20 study participants.

   (a) Calculate the Spearman rank order correlation between PBV and NPZ-8 for the 20 study participants. Do you obtain the result reported by the researchers?
   (b) Calculate the Spearman rank order correlation between PBV and NPZ-8 for the 15 HIV-infected study participants. Did exclusion of the 5 healthy study participants make a difference?

   (c) Test the coefficients obtained in 10.16a and 10.16b for significance. Use two-tailed tests at $\alpha = .05$.
   (d) Given the goal of this study, which of the above correlation coefficients do you believe is most appropriate?

10.17 Use a nonparametric test to test the hypothesis that NPZ-8 scores for HIV-infected subjects who have positive ADC Stage assessments, HIV-infected subjects who have negative ADC Stage assessments, and subjects who are negative for HIV come from identical populations. Interpret the result. Compare this result to that obtained for 7.14 on page 294.

K. The following questions refer to Case Study K (page 474)

10.18 Use the school means in Table 1 to perform a nonparametric test to determine whether posttest school means differ significantly from the pretest means.

10.19 Does it seem likely that the assumption of population normality would be a major concern if these data were analyzed by means of a paired samples t test? Give the reasons for your answer.

N. The following questions refer to Case Study N (page 476)

10.20 Perform Fisher's exact test for each of the potential risk factors. (You can save yourself considerable effort by using some creative canceling as demonstrated on page 82.) Interpret the results.

O. The following questions refer to Case Study O (page 477)

10.21 Do you agree with point (b)? Explain.

10.22 Is the comment made in point (c) regarding nonparametric tests looking only at rank order strictly true? Explain.

10.23 Is the assertion made in point (f) true? Explain.

10.24 Does the example given in point (g) constitute a legitimate application of the Kruskal-Wallis test? Explain.

10.25 Is point (h) strictly true? Explain

APPENDIX A

Normal Curve Table

	Areas Under the Standard Normal Curve							
Z (1)	(2)	(3)	Z (1)	(2)	(3)	Z (1)	(2)	(3)
0.00	.0000	.5000	0.26	.1026	.3974	0.52	.1985	.3015
0.01	.0040	.4960	0.27	.1064	.3936	0.53	.2019	.2981
0.02	.0080	.4920	0.28	.1103	.3897	0.54	.2054	.2946
0.03	.0120	.4880	0.29	.1141	.3859	0.55	.2088	.2912
0.04	.0160	.4840	0.30	.1179	.3821	0.56	.2123	.2877
0.05	.0199	.4801	0.31	.1217	.3783	0.57	.2157	.2843
0.06	.0239	.4761	0.32	.1255	.3745	0.58	.2190	.2810
0.07	.0279	.4721	0.33	.1293	.3707	0.59	.2224	.2776
0.08	.0319	.4681	0.34	.1331	.3669	0.60	.2257	.2743
0.09	.0359	.4641	0.35	.1368	.3632	0.61	.2291	.2709
0.10	.0398	.4602	0.36	.1406	.3594	0.62	.2324	.2676
0.11	.0438	.4562	0.37	.1443	.3557	0.63	.2357	.2643
0.12	.0478	.4522	0.38	.1480	.3520	0.64	.2389	.2611
0.13	.0517	.4483	0.39	.1517	.3483	0.65	.2422	.2578
0.14	.0557	.4443	0.40	.1554	.3446	0.66	.2454	.2546
0.15	.0596	.4404	0.41	.1591	.3409	0.67	.2486	.2514
0.16	.0636	.4364	0.42	.1628	.3372	0.68	.2517	.2483
0.17	.0675	.4325	0.43	.1664	.3336	0.69	.2549	.2451
0.18	.0714	.4286	0.44	.1700	.3300	0.70	.2580	.2420
0.19	.0753	.4247	0.45	.1736	.3264	0.71	.2611	.2389
0.20	.0793	.4207	0.46	.1772	.3228	0.72	.2642	.2358
0.21	.0832	.4168	0.47	.1808	.3192	0.73	.2673	.2327
0.22	.0871	.4129	0.48	.1844	.3156	0.74	.2703	.2297
0.23	.0910	.4090	0.49	.1879	.3121	0.75	.2734	.2266
0.24	.0948	.4052	0.50	.1915	.3085	0.76	.2764	.2236
0.25	.0987	.4013	0.51	.1950	.3050	0.77	.2793	.2207

Areas Under the Standard Normal Curve

Z (1)	(2)	(3)	Z (1)	(2)	(3)	Z (1)	(2)	(3)
0.78	.2823	.2177	1.04	.3508	.1492	1.30	.4032	.0968
0.79	.2852	.2148	1.05	.3531	.1469	1.31	.4049	.0951
0.80	.2881	.2119	1.06	.3554	.1446	1.32	.4066	.0934
0.81	.2910	.2090	1.07	.3577	.1423	1.33	.4082	.0918
0.82	.2939	.2061	1.08	.3599	.1401	1.34	.4099	.0901
0.83	.2967	.2033	1.09	.3621	.1379	1.35	.4115	.0885
0.84	.2995	.2005	1.10	.3643	.1357	1.36	.4131	.0869
0.85	.3023	.1977	1.11	.3665	.1335	1.37	.4147	.0853
0.86	.3051	.1949	1.12	.3686	.1314	1.38	.4162	.0838
0.87	.3078	.1922	1.13	.3708	.1292	1.39	.4177	.0823
0.88	.3106	.1894	1.14	.3729	.1271	1.40	.4192	.0808
0.89	.3133	.1867	1.15	.3749	.1251	1.41	.4207	.0793
0.90	.3159	.1841	1.16	.3770	.1230	1.42	.4222	.0778
0.91	.3186	.1814	1.17	.3790	.1210	1.43	.4236	.0764
0.92	.3212	.1788	1.18	.3810	.1190	1.44	.4251	.0749
0.93	.3238	.1762	1.19	.3830	.1170	1.45	.4265	.0735
0.94	.3264	.1736	1.20	.3849	.1151	1.46	.4279	.0721
0.95	.3289	.1711	1.21	.3869	.1131	1.47	.4292	.0708
0.96	.3315	.1685	1.22	.3888	.1112	1.48	.4306	.0694
0.97	.3340	.1660	1.23	.3907	.1093	1.49	.4319	.0681
0.98	.3365	.1635	1.24	.3925	.1075	1.50	.4332	.0668
0.99	.3389	.1611	1.25	.3944	.1056	1.51	.4345	.0655
1.00	.3413	.1587	1.26	.3962	.1038	1.52	.4357	.0643
1.01	.3438	.1562	1.27	.3980	.1020	1.53	.4370	.0630
1.02	.3461	.1539	1.28	.3997	.1003	1.54	.4382	.0618
1.03	.3485	.1515	1.29	.4015	.0985	1.55	.4394	.0606

Areas Under the Standard Normal Curve

Z (1)	(2)	(3)	Z (1)	(2)	(3)	Z (1)	(2)	(3)
1.56	.4406	.0594	1.82	.4656	.0344	2.08	.4812	.0188
1.57	.4418	.0582	1.83	.4664	.0336	2.09	.4817	.0183
1.58	.4429	.0571	1.84	.4671	.0329	2.10	.4821	.0179
1.59	.4441	.0559	1.85	.4678	.0322	2.11	.4826	.0174
1.60	.4452	.0548	1.86	.4686	.0314	2.12	.4830	.0170
1.61	.4463	.0537	1.87	.4693	.0307	2.13	.4834	.0166
1.62	.4474	.0526	1.88	.4699	.0301	2.14	.4838	.0162
1.63	.4484	.0516	1.89	.4706	.0294	2.15	.4842	.0158
1.64	.4495	.0505	1.90	.4713	.0287	2.16	.4846	.0154
1.65	.4505	.0495	1.91	.4719	.0281	2.17	.4850	.0150
1.66	.4515	.0485	1.92	.4726	.0274	2.18	.4854	.0146
1.67	.4525	.0475	1.93	.4732	.0268	2.19	.4857	.0143
1.68	.4535	.0465	1.94	.4738	.0262	2.20	.4861	.0139
1.69	.4545	.0455	1.95	.4744	.0256	2.21	.4864	.0136
1.70	.4554	.0446	1.96	.4750	.0250	2.22	.4868	.0132
1.71	.4564	.0436	1.97	.4756	.0244	2.23	.4871	.0129
1.72	.4573	.0427	1.98	.4761	.0239	2.24	.4875	.0125
1.73	.4582	.0418	1.99	.4767	.0233	2.25	.4878	.0122
1.74	.4591	.0409	2.00	.4772	.0228	2.26	.4881	.0119
1.75	.4599	.0401	2.01	.4778	.0222	2.27	.4884	.0116
1.76	.4608	.0392	2.02	.4783	.0217	2.28	.4887	.0113
1.77	.4616	.0384	2.03	.4788	.0212	2.29	.4890	.0110
1.78	.4625	.0375	2.04	.4793	.0207	2.30	.4893	.0107
1.79	.4633	.0367	2.05	.4798	.0202	2.31	.4896	.0104
1.80	.4641	.0359	2.06	.4803	.0197	2.32	.4898	.0102
1.81	.4649	.0351	2.07	.4808	.0192	2.33	.4901	.0099

Areas Under the Standard Normal Curve

Z (1)	(2)	(3)	Z (1)	(2)	(3)	Z (1)	(2)	(3)
2.34	.4904	.0096	2.60	.4953	.0047	2.86	.4979	.0021
2.35	.4906	.0094	2.61	.4955	.0045	2.87	.4979	.0021
2.36	.4909	.0091	2.62	.4956	.0044	2.88	.4980	.0020
2.37	.4911	.0089	2.63	.4957	.0043	2.89	.4981	.0019
2.38	.4913	.0087	2.64	.4959	.0041	2.90	.4981	.0019
2.39	.4916	.0084	2.65	.4960	.0040	2.91	.4982	.0018
2.40	.4918	.0082	2.66	.4961	.0039	2.92	.4982	.0018
2.41	.4920	.0080	2.67	.4962	.0038	2.93	.4983	.0017
2.42	.4922	.0078	2.68	.4963	.0037	2.94	.4984	.0016
2.43	.4925	.0075	2.69	.4964	.0036	2.95	.4984	.0016
2.44	.4927	.0073	2.70	.4965	.0035	2.96	.4985	.0015
2.45	.4929	.0071	2.71	.4966	.0034	2.97	.4985	.0015
2.46	.4931	.0069	2.72	.4967	.0033	2.98	.4986	.0014
2.47	.4932	.0068	2.73	.4968	.0032	2.99	.4986	.0014
2.48	.4934	.0066	2.74	.4969	.0031	3.00	.4987	.0013
2.49	.4936	.0064	2.75	.4970	.0030	3.01	.4987	.0013
2.50	.4938	.0062	2.76	.4971	.0029	3.02	.4987	.0013
2.51	.4940	.0060	2.77	.4972	.0028	3.03	.4988	.0012
2.52	.4941	.0059	2.78	.4973	.0027	3.04	.4988	.0012
2.53	.4943	.0057	2.79	.4974	.0026	3.05	.4989	.0011
2.54	.4945	.0055	2.80	.4974	.0026	3.06	.4989	.0011
2.55	.4946	.0054	2.81	.4975	.0025	3.07	.4989	.0011
2.56	.4948	.0052	2.82	.4976	.0024	3.08	.4990	.0010
2.57	.4949	.0051	2.83	.4977	.0023	3.09	.4990	.0010
2.58	.4951	.0049	2.84	.4977	.0023	3.10	.4990	.0010
2.59	.4952	.0048	2.85	.4978	.0022	3.11	.4991	.0009

Areas Under the Standard Normal Curve

Z (1)	(2)	(3)	Z (1)	(2)	(3)
3.12	.4991	.0009	3.38	.4996	.0004
3.13	.4991	.0009	3.39	.4997	.0003
3.14	.4992	.0008	3.40	.4997	.0003
3.15	.4992	.0008	3.41	.4997	.0003
3.16	.4992	.0008	3.42	.4997	.0003
3.17	.4992	.0008	3.43	.4997	.0003
3.18	.4993	.0007	3.44	.4997	.0003
3.19	.4993	.0007	3.45	.4997	.0003
3.20	.4993	.0007	3.46	.4997	.0003
3.21	.4993	.0007	3.47	.4997	.0003
3.22	.4994	.0006	3.48	.4997	.0003
3.23	.4994	.0006	3.49	.4998	.0002
3.24	.4994	.0006	3.50	.4998	.0002
3.25	.4994	.0006			
3.26	.4994	.0006			
3.27	.4995	.0005			
3.28	.4995	.0005			
3.29	.4995	.0005			
3.30	.4995	.0005			
3.31	.4995	.0005			
3.32	.4995	.0005			
3.33	.4996	.0004			
3.34	.4996	.0004			
3.35	.4996	.0004			
3.36	.4996	.0004			
3.37	.4996	.0004			

APPENDIX B

Critical Values of Student's t Distribution

	Percentiles of Student's t Distribution				
CI 1 sided	.900	.950	.975	.990	.995
CI 2 sided	.800	.900	.950	.980	.990
HT 1 tailed	.100	.050	.025	.010	.005
HT 2 tailed	.200	.100	.050	.020	.010
Degrees of Freedom					
2	1.886	2.920	4.303	6.965	9.925
3	1.638	2.353	3.182	4.541	5.841
4	1.533	2.132	2.776	3.747	4.604
5	1.476	2.015	2.571	3.365	4.032
6	1.440	1.943	2.447	3.143	3.707
7	1.415	1.895	2.365	2.998	3.499
8	1.397	1.860	2.306	2.896	3.355
9	1.383	1.833	2.262	2.821	3.250
10	1.372	1.812	2.228	2.764	3.169
11	1.363	1.796	2.201	2.718	3.106
12	1.356	1.782	2.179	2.681	3.055
13	1.350	1.771	2.160	2.650	3.012
14	1.345	1.761	2.145	2.624	2.977
15	1.341	1.753	2.131	2.602	2.947
16	1.337	1.746	2.120	2.583	2.921
17	1.333	1.740	2.110	2.567	2.898
18	1.330	1.734	2.101	2.552	2.878
19	1.328	1.729	2.093	2.539	2.861
20	1.325	1.725	2.086	2.528	2.845
21	1.323	1.721	2.080	2.518	2.831
22	1.321	1.717	2.074	2.508	2.819
23	1.319	1.714	2.069	2.500	2.807
24	1.318	1.711	2.064	2.492	2.797
25	1.316	1.708	2.060	2.485	2.787

Percentiles of Student's t Distribution

CI 1 sided	.900	.950	.975	.990	.995
CI 2 sided	.800	.900	.950	.980	.990
HT 1 tailed	.100	.050	.025	.010	.005
HT 2 tailed	.200	.100	.050	.020	.010
Degrees of Freedom					
26	1.315	1.706	2.056	2.479	2.779
27	1.314	1.703	2.052	2.473	2.771
28	1.313	1.701	2.048	2.467	2.763
29	1.311	1.699	2.045	2.462	2.756
30	1.310	1.697	2.042	2.457	2.750
31	1.309	1.696	2.040	2.453	2.744
32	1.309	1.694	2.037	2.449	2.738
33	1.308	1.692	2.035	2.445	2.733
34	1.307	1.691	2.032	2.441	2.728
35	1.306	1.690	2.030	2.438	2.724
36	1.306	1.688	2.028	2.434	2.719
37	1.305	1.687	2.026	2.431	2.715
38	1.304	1.686	2.024	2.429	2.712
39	1.304	1.685	2.023	2.426	2.708
40	1.303	1.684	2.021	2.423	2.704
41	1.303	1.683	2.020	2.421	2.701
42	1.302	1.682	2.018	2.418	2.698
43	1.302	1.681	2.017	2.416	2.695
44	1.301	1.680	2.015	2.414	2.692
45	1.301	1.679	2.014	2.412	2.690
46	1.300	1.679	2.013	2.410	2.687
47	1.300	1.678	2.012	2.408	2.685
48	1.299	1.677	2.011	2.407	2.682
49	1.299	1.677	2.010	2.405	2.680
50	1.299	1.676	2.009	2.403	2.678

Percentiles of Student's t Distribution

CI 1 sided	.900	.950	.975	.990	.995
CI 2 sided	.800	.900	.950	.980	.990
HT 1 tailed	.100	.050	.025	.010	.005
HT 2 tailed	.200	.100	.050	.020	.010
Degrees of Freedom					
51	1.298	1.675	2.008	2.402	2.676
52	1.298	1.675	2.007	2.400	2.674
53	1.298	1.674	2.006	2.399	2.672
54	1.297	1.674	2.005	2.397	2.670
55	1.297	1.673	2.004	2.396	2.668
56	1.297	1.673	2.003	2.395	2.667
57	1.297	1.672	2.002	2.394	2.665
58	1.296	1.672	2.002	2.392	2.663
59	1.296	1.671	2.001	2.391	2.662
60	1.296	1.671	2.000	2.390	2.660
61	1.296	1.670	2.000	2.389	2.659
62	1.295	1.670	1.999	2.388	2.657
63	1.295	1.669	1.998	2.387	2.656
64	1.295	1.669	1.998	2.386	2.655
65	1.295	1.669	1.997	2.385	2.654
66	1.295	1.668	1.997	2.384	2.652
67	1.294	1.668	1.996	2.383	2.651
68	1.294	1.668	1.995	2.382	2.650
69	1.294	1.667	1.995	2.382	2.649
70	1.294	1.667	1.994	2.381	2.648
71	1.294	1.667	1.994	2.380	2.647
72	1.293	1.666	1.993	2.379	2.646
73	1.293	1.666	1.993	2.379	2.645
74	1.293	1.666	1.993	2.378	2.644
75	1.293	1.665	1.992	2.377	2.643

Percentiles of Student's t Distribution

CI 1 sided	.900	.950	.975	.990	.995
CI 2 sided	.800	.900	.950	.980	.990
HT 1 tailed	.100	.050	.025	.010	.005
HT 2 tailed	.200	.100	.050	.020	.010
Degrees of Freedom					
76	1.293	1.665	1.992	2.376	2.642
77	1.293	1.665	1.991	2.376	2.641
78	1.292	1.665	1.991	2.375	2.640
79	1.292	1.664	1.990	2.374	2.640
80	1.292	1.664	1.990	2.374	2.639
81	1.292	1.664	1.990	2.373	2.638
82	1.292	1.664	1.989	2.373	2.637
83	1.292	1.663	1.989	2.372	2.636
84	1.292	1.663	1.989	2.372	2.636
85	1.292	1.663	1.988	2.371	2.635
86	1.291	1.663	1.988	2.370	2.634
87	1.291	1.663	1.988	2.370	2.634
88	1.291	1.662	1.987	2.369	2.633
89	1.291	1.662	1.987	2.369	2.632
90	1.291	1.662	1.987	2.368	2.632
91	1.291	1.662	1.986	2.368	2.631
92	1.291	1.662	1.986	2.368	2.630
93	1.291	1.661	1.986	2.367	2.630
94	1.291	1.661	1.986	2.367	2.629
95	1.291	1.661	1.985	2.366	2.629
96	1.290	1.661	1.985	2.366	2.628
97	1.290	1.661	1.985	2.365	2.627
98	1.290	1.661	1.984	2.365	2.627
99	1.290	1.660	1.984	2.365	2.626
100	1.290	1.660	1.984	2.364	2.626

Percentiles of Student's t Distribution

CI 1 sided	.900	.950	.975	.990	.995
CI 2 sided	.800	.900	.950	.980	.990
HT 1 tailed	.100	.050	.025	.010	.005
HT 2 tailed	.200	.100	.050	.020	.010
Degrees of Freedom					
101	1.290	1.660	1.984	2.364	2.625
102	1.290	1.660	1.983	2.363	2.625
103	1.290	1.660	1.983	2.363	2.624
104	1.290	1.660	1.983	2.363	2.624
105	1.290	1.659	1.983	2.362	2.623
106	1.290	1.659	1.983	2.362	2.623
107	1.290	1.659	1.982	2.362	2.623
108	1.289	1.659	1.982	2.361	2.622
109	1.289	1.659	1.982	2.361	2.622
110	1.289	1.659	1.982	2.361	2.621
111	1.289	1.659	1.982	2.360	2.621
112	1.289	1.659	1.981	2.360	2.620
113	1.289	1.658	1.981	2.360	2.620
114	1.289	1.658	1.981	2.360	2.620
115	1.289	1.658	1.981	2.359	2.619
116	1.289	1.658	1.981	2.359	2.619
117	1.289	1.658	1.980	2.359	2.619
118	1.289	1.658	1.980	2.358	2.618
119	1.289	1.658	1.980	2.358	2.618
120	1.289	1.658	1.980	2.358	2.617
121	1.289	1.658	1.980	2.358	2.617
122	1.289	1.657	1.980	2.357	2.617
123	1.288	1.657	1.979	2.357	2.616
124	1.288	1.657	1.979	2.357	2.616
∞	1.282	1.645	1.960	2.326	2.576

APPENDIX C

Critical Values of the F Distribution

Critical Values of the F Distribution, $\alpha = .10$

| | One-Sided CI = .90 | | | | α | | $\alpha = .10$ | | |
	Two-Sided CI = .80								
Den. df	Numerator df								
	1	2	3	4	5	6	7	8	9
2	8.53	9.00	9.16	9.24	9.29	9.33	9.35	9.37	9.38
3	5.54	5.46	5.39	5.34	5.31	5.28	5.27	5.25	5.24
4	4.54	4.32	4.19	4.11	4.05	4.01	3.98	3.95	3.94
5	4.06	3.78	3.62	3.52	3.45	3.40	3.37	3.34	3.32
6	3.78	3.46	3.29	3.18	3.11	3.05	3.01	2.98	2.96
7	3.59	3.26	3.07	2.96	2.88	2.83	2.78	2.75	2.72
8	3.46	3.11	2.92	2.81	2.73	2.67	2.62	2.59	2.56
9	3.36	3.01	2.81	2.69	2.61	2.55	2.51	2.47	2.44
10	3.29	2.92	2.73	2.61	2.52	2.46	2.41	2.38	2.35
11	3.23	2.86	2.66	2.54	2.45	2.39	2.34	2.30	2.27
12	3.18	2.81	2.61	2.48	2.39	2.33	2.28	2.24	2.21
13	3.14	2.76	2.56	2.43	2.35	2.28	2.23	2.20	2.16
14	3.10	2.73	2.52	2.39	2.31	2.24	2.19	2.15	2.12
15	3.07	2.70	2.49	2.36	2.27	2.21	2.16	2.12	2.09
16	3.05	2.67	2.46	2.33	2.24	2.18	2.13	2.09	2.06
17	3.03	2.64	2.44	2.31	2.22	2.15	2.10	2.06	2.03
18	3.01	2.62	2.42	2.29	2.20	2.13	2.08	2.04	2.00
19	2.99	2.61	2.40	2.27	2.18	2.11	2.06	2.02	1.98
20	2.97	2.59	2.38	2.25	2.16	2.09	2.04	2.00	1.96
21	2.96	2.57	2.36	2.23	2.14	2.08	2.02	1.98	1.95
22	2.95	2.56	2.35	2.22	2.13	2.06	2.01	1.97	1.93
23	2.94	2.55	2.34	2.21	2.11	2.05	1.99	1.95	1.92
24	2.93	2.54	2.33	2.19	2.10	2.04	1.98	1.94	1.91
25	2.92	2.53	2.32	2.18	2.09	2.02	1.97	1.93	1.89
26	2.91	2.52	2.31	2.17	2.08	2.01	1.96	1.92	1.88
27	2.90	2.51	2.30	2.17	2.07	2.00	1.95	1.91	1.87
28	2.89	2.50	2.29	2.16	2.06	2.00	1.94	1.90	1.87
29	2.89	2.50	2.28	2.15	2.06	1.99	1.93	1.89	1.86
30	2.88	2.49	2.28	2.14	2.05	1.98	1.93	1.88	1.85

Critical Values of the F Distribution, $\alpha = .10$

Den.	Numerator df								
	One-Sided CI = .90					α	$\alpha = .10$		
	Two-Sided CI = .80								
df	1	2	3	4	5	6	7	8	9
31	2.87	2.48	2.27	2.14	2.04	1.97	1.92	1.88	1.84
32	2.87	2.48	2.26	2.13	2.04	1.97	1.91	1.87	1.83
33	2.86	2.47	2.26	2.12	2.03	1.96	1.91	1.86	1.83
34	2.86	2.47	2.25	2.12	2.02	1.96	1.90	1.86	1.82
35	2.85	2.46	2.25	2.11	2.02	1.95	1.90	1.85	1.82
36	2.85	2.46	2.24	2.11	2.01	1.94	1.89	1.85	1.81
37	2.85	2.45	2.24	2.10	2.01	1.94	1.89	1.84	1.81
38	2.84	2.45	2.23	2.10	2.01	1.94	1.88	1.84	1.80
39	2.84	2.44	2.23	2.09	2.00	1.93	1.88	1.83	1.80
40	2.84	2.44	2.23	2.09	2.00	1.93	1.87	1.83	1.79
41	2.83	2.44	2.22	2.09	1.99	1.92	1.87	1.82	1.79
42	2.83	2.43	2.22	2.08	1.99	1.92	1.86	1.82	1.78
43	2.83	2.43	2.22	2.08	1.99	1.92	1.86	1.82	1.78
44	2.82	2.43	2.21	2.08	1.98	1.91	1.86	1.81	1.78
45	2.82	2.42	2.21	2.07	1.98	1.91	1.85	1.81	1.77
46	2.82	2.42	2.21	2.07	1.98	1.91	1.85	1.81	1.77
47	2.82	2.42	2.20	2.07	1.97	1.90	1.85	1.80	1.77
48	2.81	2.42	2.20	2.07	1.97	1.90	1.85	1.80	1.77
49	2.81	2.41	2.20	2.06	1.97	1.90	1.84	1.80	1.76
50	2.81	2.41	2.20	2.06	1.97	1.90	1.84	1.80	1.76
55	2.80	2.40	2.19	2.05	1.95	1.88	1.83	1.78	1.75
60	2.79	2.39	2.18	2.04	1.95	1.87	1.82	1.77	1.74
65	2.78	2.39	2.17	2.03	1.94	1.87	1.81	1.77	1.73
70	2.78	2.38	2.16	2.03	1.93	1.86	1.80	1.76	1.72
80	2.77	2.37	2.15	2.02	1.92	1.8	1.79	1.75	1.71
100	2.76	2.36	2.14	2.00	1.91	1.83	1.78	1.73	1.69
150	2.74	2.34	2.12	1.98	1.89	1.81	1.76	1.71	1.67
350	2.72	2.32	2.10	1.96	1.86	1.79	1.73	1.69	1.65
∞	2.71	2.30	2.08	1.94	1.85	1.77	1.72	1.67	1.63

Critical Values of the F Distribution, $\alpha = .10$

Den.	Numerator df								
df	10	12	14	16	20	50	100	350	∞
2	9.39	9.41	9.42	9.43	9.44	9.47	9.48	9.49	9.49
3	5.23	5.22	5.20	5.20	5.18	5.15	5.14	5.14	5.13
4	3.92	3.90	3.88	3.86	3.84	3.80	3.78	3.77	3.76
5	3.30	3.27	3.25	3.23	3.21	3.15	3.13	3.11	3.10
6	2.94	2.90	2.88	2.86	2.84	2.77	2.75	2.73	2.72
7	2.70	2.67	2.64	2.62	2.59	2.52	2.50	2.48	2.47
8	2.54	2.50	2.48	2.45	2.42	2.35	2.32	2.30	2.29
9	2.42	2.38	2.35	2.33	2.30	2.22	2.19	2.17	2.16
10	2.32	2.28	2.26	2.23	2.20	2.12	2.09	2.06	2.06
11	2.25	2.21	2.18	2.16	2.12	2.04	2.01	1.98	1.97
12	2.19	2.15	2.12	2.09	2.06	1.97	1.94	1.91	1.90
13	2.14	2.10	2.07	2.04	2.01	1.92	1.88	1.86	1.85
14	2.10	2.05	2.02	2.00	1.96	1.87	1.83	1.81	1.80
15	2.06	2.02	1.99	1.96	1.92	1.83	1.79	1.77	1.76
16	2.03	1.99	1.95	1.93	1.89	1.79	1.76	1.73	1.72
17	2.00	1.96	1.93	1.90	1.86	1.76	1.73	1.70	1.69
18	1.98	1.93	1.90	1.87	1.84	1.74	1.70	1.67	1.66
19	1.96	1.91	1.88	1.85	1.81	1.71	1.67	1.64	1.63
20	1.94	1.89	1.86	1.83	1.79	1.69	1.65	1.62	1.61
21	1.92	1.87	1.84	1.81	1.78	1.67	1.63	1.60	1.59
22	1.90	1.86	1.83	1.80	1.76	1.65	1.61	1.58	1.57
23	1.89	1.84	1.81	1.78	1.74	1.64	1.59	1.56	1.55
24	1.88	1.83	1.80	1.77	1.73	1.62	1.58	1.55	1.53
25	1.87	1.82	1.79	1.76	1.72	1.61	1.56	1.53	1.52
26	1.86	1.81	1.77	1.75	1.71	1.59	1.55	1.52	1.50
27	1.85	1.80	1.76	1.74	1.70	1.58	1.54	1.50	1.49
28	1.84	1.79	1.75	1.73	1.69	1.57	1.53	1.49	1.48
29	1.83	1.78	1.75	1.72	1.68	1.56	1.52	1.48	1.47
30	1.82	1.77	1.74	1.71	1.67	1.55	1.51	1.47	1.46

One-Sided CI = .90
Two-Sided CI = .80

α $\alpha = .10$

Critical Values of the F Distribution, $\alpha = .10$

Den. df	One-Sided CI = .90 Two-Sided CI = .80	α	$\alpha = .10$						
	Numerator df								
df	10	12	14	16	20	50	100	350	∞
31	1.81	1.77	1.73	1.70	1.66	1.54	1.50	1.46	1.45
32	1.81	1.76	1.72	1.69	1.65	1.53	1.49	1.45	1.44
33	1.80	1.75	1.72	1.69	1.64	1.53	1.48	1.44	1.43
34	1.79	1.75	1.71	1.68	1.64	1.52	1.47	1.44	1.42
35	1.79	1.74	1.70	1.67	1.63	1.51	1.47	1.43	1.41
36	1.78	1.73	1.70	1.67	1.63	1.51	1.46	1.42	1.40
37	1.78	1.73	1.69	1.66	1.62	1.50	1.45	1.41	1.40
38	1.77	1.72	1.69	1.66	1.61	1.49	1.45	1.41	1.39
39	1.77	1.72	1.68	1.65	1.61	1.49	1.44	1.40	1.38
40	1.76	1.71	1.68	1.65	1.61	1.48	1.43	1.39	1.38
41	1.76	1.71	1.67	1.64	1.60	1.48	1.43	1.39	1.37
42	1.75	1.71	1.67	1.64	1.60	1.47	1.42	1.38	1.37
43	1.75	1.70	1.67	1.64	1.59	1.47	1.42	1.38	1.36
44	1.75	1.70	1.66	1.63	1.59	1.46	1.41	1.37	1.35
45	1.74	1.70	1.66	1.63	1.58	1.46	1.41	1.37	1.35
46	1.74	1.69	1.65	1.63	1.58	1.46	1.40	1.36	1.34
47	1.74	1.69	1.65	1.62	1.58	1.45	1.40	1.36	1.34
48	1.73	1.69	1.65	1.62	1.57	1.45	1.40	1.35	1.34
49	1.73	1.68	1.65	1.62	1.57	1.44	1.39	1.35	1.33
50	1.73	1.68	1.64	1.61	1.57	1.44	1.39	1.35	1.33
55	1.72	1.67	1.63	1.60	1.55	1.43	1.37	1.33	1.31
60	1.71	1.66	1.62	1.59	1.54	1.41	1.36	1.31	1.29
65	1.70	1.65	1.61	1.58	1.53	1.40	1.35	1.30	1.28
70	1.69	1.64	1.60	1.57	1.53	1.39	1.34	1.29	1.27
80	1.68	1.63	1.59	1.56	1.51	1.38	1.32	1.27	1.24
100	1.66	1.61	1.57	1.54	1.49	1.35	1.29	1.24	1.21
150	1.64	1.59	1.55	1.52	1.47	1.33	1.26	1.20	1.17
350	1.62	1.56	1.52	1.49	1.44	1.29	1.22	1.15	1.11
∞	1.60	1.55	1.50	1.47	1.42	1.26	1.18	1.10	1.00

Critical Values of the F Distribution, $\alpha = .05$

					Numerator df				
	One-Sided CI = .95 Two-Sided CI = .90				α		$\alpha = .05$		
Den. df	1	2	3	4	5	6	7	8	9
2	18.51	19.00	19.16	19.25	19.30	19.33	19.35	19.37	19.38
3	10.13	9.55	9.28	9.12	9.01	8.94	8.89	8.85	8.81
4	7.71	6.94	6.59	6.39	6.26	6.16	6.09	6.04	6.00
5	6.61	5.79	5.41	5.19	5.05	4.95	4.88	4.82	4.77
6	5.99	5.14	4.76	4.53	4.39	4.28	4.21	4.15	4.10
7	5.59	4.74	4.35	4.12	3.97	3.87	3.79	3.73	3.68
8	5.32	4.46	4.07	3.84	3.69	3.58	3.50	3.44	3.39
9	5.12	4.26	3.86	3.63	3.48	3.37	3.29	3.23	3.18
10	4.96	4.10	3.71	3.48	3.33	3.22	3.14	3.07	3.02
11	4.84	3.98	3.59	3.36	3.20	3.09	3.01	2.95	2.90
12	4.75	3.89	3.49	3.26	3.11	3.00	2.91	2.85	2.80
13	4.67	3.81	3.41	3.18	3.03	2.92	2.83	2.77	2.71
14	4.60	3.74	3.34	3.11	2.96	2.85	2.76	2.70	2.65
15	4.54	3.68	3.29	3.06	2.90	2.79	2.71	2.64	2.59
16	4.49	3.63	3.24	3.01	2.85	2.74	2.66	2.59	2.54
17	4.45	3.59	3.20	2.96	2.81	2.70	2.61	2.55	2.49
18	4.41	3.55	3.16	2.93	2.77	2.66	2.58	2.51	2.46
19	4.38	3.52	3.13	2.90	2.74	2.63	2.54	2.48	2.42
20	4.35	3.49	3.10	2.87	2.71	2.60	2.51	2.45	2.39
21	4.32	3.47	3.07	2.84	2.68	2.57	2.49	2.42	2.37
22	4.30	3.44	3.05	2.82	2.66	2.55	2.46	2.40	2.34
23	4.28	3.42	3.03	2.80	2.64	2.53	2.44	2.37	2.32
24	4.26	3.40	3.01	2.78	2.62	2.51	2.42	2.36	2.30
25	4.24	3.39	2.99	2.76	2.60	2.49	2.40	2.34	2.28
26	4.23	3.37	2.98	2.74	2.59	2.47	2.39	2.32	2.27
27	4.21	3.35	2.96	2.73	2.57	2.46	2.37	2.31	2.25
28	4.20	3.34	2.95	2.71	2.56	2.45	2.36	2.29	2.24
29	4.18	3.33	2.93	2.70	2.55	2.43	2.35	2.28	2.22
30	4.17	3.32	2.92	2.69	2.53	2.42	2.33	2.27	2.21

Critical Values of the F Distribution, $\alpha = .05$

Den.	Numerator df								
df	1	2	3	4	5	6	7	8	9
31	4.16	3.30	2.91	2.68	2.52	2.41	2.32	2.25	2.20
32	4.15	3.29	2.90	2.67	2.51	2.40	2.31	2.24	2.19
33	4.14	3.28	2.89	2.66	2.50	2.39	2.30	2.23	2.18
34	4.13	3.28	2.88	2.65	2.49	2.38	2.29	2.23	2.17
35	4.12	3.27	2.87	2.64	2.49	2.37	2.29	2.22	2.16
36	4.11	3.26	2.87	2.63	2.48	2.36	2.28	2.21	2.15
37	4.11	3.25	2.86	2.63	2.47	2.36	2.27	2.20	2.14
38	4.10	3.24	2.85	2.62	2.46	2.35	2.26	2.19	2.14
39	4.09	3.24	2.85	2.61	2.46	2.34	2.26	2.19	2.13
40	4.08	3.23	2.84	2.61	2.45	2.34	2.25	2.18	2.12
41	4.08	3.23	2.83	2.60	2.44	2.33	2.24	2.17	2.12
42	4.07	3.22	2.83	2.59	2.44	2.32	2.24	2.17	2.11
43	4.07	3.21	2.82	2.59	2.43	2.32	2.23	2.16	2.11
44	4.06	3.21	2.82	2.58	2.43	2.31	2.23	2.16	2.10
45	4.06	3.20	2.81	2.58	2.42	2.31	2.22	2.15	2.10
46	4.05	3.20	2.81	2.57	2.42	2.30	2.22	2.15	2.09
47	4.05	3.20	2.80	2.57	2.41	2.30	2.21	2.14	2.09
48	4.04	3.19	2.80	2.57	2.41	2.29	2.21	2.14	2.08
49	4.04	3.19	2.79	2.56	2.40	2.29	2.20	2.13	2.08
50	4.03	3.18	2.79	2.56	2.40	2.29	2.20	2.13	2.07
55	4.02	3.16	2.77	2.54	2.38	2.27	2.18	2.11	2.06
60	4.00	3.15	2.76	2.53	2.37	2.25	2.17	2.10	2.04
65	3.99	3.14	2.75	2.51	2.36	2.24	2.15	2.08	2.03
70	3.98	3.13	2.74	2.50	2.35	2.23	2.14	2.07	2.02
80	3.96	3.11	2.72	2.49	2.33	2.21	2.13	2.06	2.00
100	3.94	3.09	2.70	2.46	2.31	2.19	2.10	2.03	1.97
150	3.90	3.06	2.66	2.43	2.27	2.16	2.07	2.00	1.94
350	3.87	3.02	2.63	2.40	2.24	2.12	2.04	1.96	1.91
∞	3.84	3.00	2.60	2.37	2.21	2.10	2.01	1.94	1.88

One-Sided CI = .95
Two-Sided CI = .90

$\alpha = .05$

Critical Values of the F Distribution, $\alpha = .05$

	One-Sided CI = .95 Two-Sided CI = .90				α $\alpha = .05$				
Den.	Numerator df								
df	10	12	14	16	20	50	100	350	∞
2	19.40	19.41	19.42	19.43	19.45	19.48	19.49	19.49	19.50
3	8.79	8.74	8.71	8.69	8.66	8.58	8.55	8.53	8.53
4	5.96	5.91	5.87	5.84	5.80	5.70	5.66	5.64	5.63
5	4.74	4.68	4.64	4.60	4.56	4.44	4.41	4.38	4.36
6	4.06	4.00	3.96	3.92	3.87	3.75	3.71	3.68	3.67
7	3.64	3.57	3.53	3.49	3.44	3.32	3.27	3.24	3.23
8	3.35	3.28	3.24	3.20	3.15	3.02	2.97	2.94	2.93
9	3.14	3.07	3.03	2.99	2.94	2.80	2.76	2.72	2.71
10	2.98	2.91	2.86	2.83	2.77	2.64	2.59	2.55	2.54
11	2.85	2.79	2.74	2.70	2.65	2.51	2.46	2.42	2.40
12	2.75	2.69	2.64	2.60	2.54	2.40	2.35	2.31	2.30
13	2.67	2.60	2.55	2.51	2.46	2.31	2.26	2.22	2.21
14	2.60	2.53	2.48	2.44	2.39	2.24	2.19	2.15	2.13
15	2.54	2.48	2.42	2.38	2.33	2.18	2.12	2.08	2.07
16	2.49	2.42	2.37	2.33	2.28	2.12	2.07	2.03	2.01
17	2.45	2.38	2.33	2.29	2.23	2.08	2.02	1.98	1.96
18	2.41	2.34	2.29	2.25	2.19	2.04	1.98	1.93	1.92
19	2.38	2.31	2.26	2.21	2.16	2.00	1.94	1.90	1.88
20	2.35	2.28	2.22	2.18	2.12	1.97	1.91	1.86	1.84
21	2.32	2.25	2.20	2.16	2.10	1.94	1.88	1.83	1.81
22	2.30	2.23	2.17	2.13	2.07	1.91	1.85	1.80	1.78
23	2.27	2.20	2.15	2.11	2.05	1.88	1.82	1.78	1.76
24	2.25	2.18	2.13	2.09	2.03	1.86	1.80	1.75	1.73
25	2.24	2.16	2.11	2.07	2.01	1.84	1.78	1.73	1.71
26	2.22	2.15	2.09	2.05	1.99	1.82	1.76	1.71	1.69
27	2.20	2.13	2.08	2.04	1.97	1.81	1.74	1.69	1.67
28	2.19	2.12	2.06	2.02	1.96	1.79	1.73	1.68	1.65
29	2.18	2.10	2.05	2.01	1.94	1.77	1.71	1.66	1.64
30	2.16	2.09	2.04	1.99	1.93	1.76	1.70	1.64	1.62

Critical Values of the F Distribution, $\alpha = .05$

	One-Sided CI = .95					$\alpha = .05$			
	Two-Sided CI = .90								
Den.	Numerator df								
df	10	12	14	16	20	50	100	350	∞
31	2.15	2.08	2.03	1.98	1.92	1.75	1.68	1.63	1.61
32	2.14	2.07	2.01	1.97	1.91	1.74	1.67	1.62	1.59
33	2.13	2.06	2.00	1.96	1.90	1.72	1.66	1.60	1.58
34	2.12	2.05	1.99	1.95	1.89	1.71	1.65	1.59	1.57
35	2.11	2.04	1.99	1.94	1.88	1.70	1.63	1.58	1.56
36	2.11	2.03	1.98	1.93	1.87	1.69	1.62	1.57	1.55
37	2.10	2.02	1.97	1.93	1.86	1.68	1.62	1.56	1.54
38	2.09	2.02	1.96	1.92	1.85	1.68	1.61	1.55	1.53
39	2.08	2.01	1.95	1.91	1.85	1.67	1.60	1.54	1.52
40	2.08	2.00	1.95	1.90	1.84	1.66	1.59	1.53	1.51
41	2.07	2.00	1.94	1.90	1.83	1.65	1.58	1.52	1.50
42	2.06	1.99	1.94	1.89	1.83	1.65	1.57	1.52	1.49
43	2.06	1.99	1.93	1.89	1.82	1.64	1.57	1.51	1.48
44	2.05	1.98	1.92	1.88	1.81	1.63	1.56	1.50	1.48
45	2.05	1.97	1.92	1.87	1.81	1.63	1.55	1.50	1.47
46	2.04	1.97	1.91	1.87	1.80	1.62	1.55	1.49	1.46
47	2.04	1.96	1.91	1.86	1.80	1.61	1.54	1.48	1.46
48	2.03	1.96	1.90	1.86	1.79	1.61	1.54	1.48	1.45
49	2.03	1.96	1.90	1.85	1.79	1.60	1.53	1.47	1.44
50	2.03	1.95	1.89	1.85	1.78	1.60	1.52	1.46	1.44
55	2.01	1.93	1.88	1.83	1.76	1.58	1.50	1.44	1.41
60	1.99	1.92	1.86	1.82	1.75	1.56	1.48	1.42	1.39
65	1.98	1.90	1.85	1.80	1.73	1.54	1.46	1.40	1.37
70	1.97	1.89	1.84	1.79	1.72	1.53	1.45	1.38	1.35
80	1.95	1.88	1.82	1.77	1.70	1.51	1.43	1.36	1.32
100	1.93	1.85	1.79	1.75	1.68	1.48	1.39	1.32	1.28
150	1.89	1.82	1.76	1.71	1.64	1.44	1.34	1.26	1.22
350	1.86	1.78	1.72	1.67	1.60	1.39	1.29	1.19	1.14
∞	1.83	1.75	1.69	1.64	1.57	1.35	1.24	1.13	1.00

Critical Values of the F Distribution, $\alpha = .025$

One-Sided CI = .975 Two-Sided CI = .95				α	$\alpha = .025$			
Den.				**Numerator df**				
df	**1**	**2**	**3**	**4**	**5**	**6**	**7**	**8**	**9**

Den. df	1	2	3	4	5	6	7	8	9
2	38.51	39.00	39.17	39.25	39.30	39.33	39.36	39.37	39.39
3	17.44	16.04	15.44	15.10	14.88	14.73	14.62	14.54	14.47
4	12.22	10.65	9.98	9.60	9.36	9.20	9.07	8.98	8.90
5	10.01	8.43	7.76	7.39	7.15	6.98	6.85	6.76	6.68
6	8.81	7.26	6.60	6.23	5.99	5.82	5.70	5.60	5.52
7	8.07	6.54	5.89	5.52	5.29	5.12	4.99	4.90	4.82
8	7.57	6.06	5.42	5.05	4.82	4.65	4.53	4.43	4.36
9	7.21	5.71	5.08	4.72	4.48	4.32	4.20	4.10	4.03
10	6.94	5.46	4.83	4.47	4.24	4.07	3.95	3.85	3.78
11	6.72	5.26	4.63	4.28	4.04	3.88	3.76	3.66	3.59
12	6.55	5.10	4.47	4.12	3.89	3.73	3.61	3.51	3.44
13	6.41	4.97	4.35	4.00	3.77	3.60	3.48	3.39	3.31
14	6.30	4.86	4.24	3.89	3.66	3.50	3.38	3.29	3.21
15	6.20	4.77	4.15	3.80	3.58	3.41	3.29	3.20	3.12
16	6.12	4.69	4.08	3.73	3.50	3.34	3.22	3.12	3.05
17	6.04	4.62	4.01	3.66	3.44	3.28	3.16	3.06	2.98
18	5.98	4.56	3.95	3.61	3.38	3.22	3.10	3.01	2.93
19	5.92	4.51	3.90	3.56	3.33	3.17	3.05	2.96	2.88
20	5.87	4.46	3.86	3.51	3.29	3.13	3.01	2.91	2.84
21	5.83	4.42	3.82	3.48	3.25	3.09	2.97	2.87	2.80
22	5.79	4.38	3.78	3.44	3.22	3.05	2.93	2.84	2.76
23	5.75	4.35	3.75	3.41	3.18	3.02	2.90	2.81	2.73
24	5.72	4.32	3.72	3.38	3.15	2.99	2.87	2.78	2.70
25	5.69	4.29	3.69	3.35	3.13	2.97	2.85	2.75	2.68
26	5.66	4.27	3.67	3.33	3.10	2.94	2.82	2.73	2.65
27	5.63	4.24	3.65	3.31	3.08	2.92	2.80	2.71	2.63
28	5.61	4.22	3.63	3.29	3.06	2.90	2.78	2.69	2.61
29	5.59	4.20	3.61	3.27	3.04	2.88	2.76	2.67	2.59
30	5.57	4.18	3.59	3.25	3.03	2.87	2.75	2.65	2.57

Critical Values of the F Distribution, α = .025

	One-Sided CI = .975 Two-Sided CI = .95					α = .025			
Den. df	Numerator df								
	1	2	3	4	5	6	7	8	9
31	5.55	4.16	3.57	3.23	3.01	2.85	2.73	2.64	2.56
32	5.53	4.15	3.56	3.22	3.00	2.84	2.71	2.62	2.54
33	5.51	4.13	3.54	3.20	2.98	2.82	2.70	2.61	2.53
34	5.50	4.12	3.53	3.19	2.97	2.81	2.69	2.59	2.52
35	5.48	4.11	3.52	3.18	2.96	2.80	2.68	2.58	2.50
36	5.47	4.09	3.50	3.17	2.94	2.78	2.66	2.57	2.49
37	5.46	4.08	3.49	3.16	2.93	2.77	2.65	2.56	2.48
38	5.45	4.07	3.48	3.15	2.92	2.76	2.64	2.55	2.47
39	5.43	4.06	3.47	3.14	2.91	2.75	2.63	2.54	2.46
40	5.42	4.05	3.46	3.13	2.90	2.74	2.62	2.53	2.45
41	5.41	4.04	3.45	3.12	2.89	2.74	2.62	2.52	2.44
42	5.40	4.03	3.45	3.11	2.89	2.73	2.61	2.51	2.43
43	5.39	4.02	3.44	3.10	2.88	2.72	2.60	2.50	2.43
44	5.39	4.02	3.43	3.09	2.87	2.71	2.59	2.50	2.42
45	5.38	4.01	3.42	3.09	2.86	2.70	2.58	2.49	2.41
46	5.37	4.00	3.42	3.08	2.86	2.70	2.58	2.48	2.41
47	5.36	3.99	3.41	3.07	2.85	2.69	2.57	2.48	2.40
48	5.35	3.99	3.40	3.07	2.84	2.69	2.56	2.47	2.39
49	5.35	3.98	3.40	3.06	2.84	2.68	2.56	2.46	2.39
50	5.34	3.97	3.39	3.05	2.83	2.67	2.55	2.46	2.38
55	5.31	3.95	3.36	3.03	2.81	2.65	2.53	2.43	2.36
60	5.29	3.93	3.34	3.01	2.79	2.63	2.51	2.41	2.33
65	5.26	3.91	3.32	2.99	2.77	2.61	2.49	2.39	2.32
70	5.25	3.89	3.31	2.97	2.75	2.59	2.47	2.38	2.30
80	5.22	3.86	3.28	2.95	2.73	2.57	2.45	2.35	2.28
100	5.18	3.83	3.25	2.92	2.70	2.54	2.42	2.32	2.24
150	5.13	3.78	3.20	2.87	2.65	2.49	2.37	2.28	2.20
350	5.07	3.73	3.15	2.82	2.60	2.44	2.32	2.23	2.15
∞	5.02	3.69	3.12	2.79	2.57	2.41	2.29	2.19	2.11

Critical Values of the F Distribution, $\alpha = .025$

	One-Sided CI = .975 Two-Sided CI = .95								

Den. df	Numerator df								
	10	12	14	16	20	50	100	350	∞
2	39.40	39.41	39.43	39.44	39.45	39.48	39.49	39.50	39.50
3	14.42	14.34	14.28	14.23	14.17	14.01	13.96	13.92	13.90
4	8.84	8.75	8.68	8.63	8.56	8.38	8.32	8.28	8.26
5	6.62	6.52	6.46	6.40	6.33	6.14	6.08	6.03	6.02
6	5.46	5.37	5.30	5.24	5.17	4.98	4.92	4.87	4.85
7	4.76	4.67	4.60	4.54	4.47	4.28	4.21	4.16	4.14
8	4.30	4.20	4.13	4.08	4.00	3.81	3.74	3.69	3.67
9	3.96	3.87	3.80	3.74	3.67	3.47	3.40	3.35	3.33
10	3.72	3.62	3.55	3.50	3.42	3.22	3.15	3.10	3.08
11	3.53	3.43	3.36	3.30	3.23	3.03	2.96	2.90	2.88
12	3.37	3.28	3.21	3.15	3.07	2.87	2.80	2.75	2.72
13	3.25	3.15	3.08	3.03	2.95	2.74	2.67	2.62	2.60
14	3.15	3.05	2.98	2.92	2.84	2.64	2.56	2.51	2.49
15	3.06	2.96	2.89	2.84	2.76	2.55	2.47	2.42	2.40
16	2.99	2.89	2.82	2.76	2.68	2.47	2.40	2.34	2.32
17	2.92	2.82	2.75	2.70	2.62	2.41	2.33	2.27	2.25
18	2.87	2.77	2.70	2.64	2.56	2.35	2.27	2.21	2.19
19	2.82	2.72	2.65	2.59	2.51	2.30	2.22	2.16	2.13
20	2.77	2.68	2.60	2.55	2.46	2.25	2.17	2.11	2.09
21	2.73	2.64	2.56	2.51	2.42	2.21	2.13	2.07	2.04
22	2.70	2.60	2.53	2.47	2.39	2.17	2.09	2.03	2.00
23	2.67	2.57	2.50	2.44	2.36	2.14	2.06	1.99	1.97
24	2.64	2.54	2.47	2.41	2.33	2.11	2.02	1.96	1.94
25	2.61	2.51	2.44	2.38	2.30	2.08	2.00	1.93	1.91
26	2.59	2.49	2.42	2.36	2.28	2.05	1.97	1.90	1.88
27	2.57	2.47	2.39	2.34	2.25	2.03	1.94	1.88	1.85
28	2.55	2.45	2.37	2.32	2.23	2.01	1.92	1.86	1.83
29	2.53	2.43	2.36	2.30	2.21	1.99	1.90	1.84	1.81
30	2.51	2.41	2.34	2.28	2.20	1.97	1.88	1.81	1.79

Critical Values of the F Distribution, $\alpha = .025$

Den. df	\multicolumn{9}{c}{Numerator df}

One-Sided CI = .975
Two-Sided CI = .95

α $\alpha = .025$

Den. df	10	12	14	16	20	50	100	350	∞
31	2.50	2.40	2.32	2.26	2.18	1.95	1.86	1.80	1.77
32	2.48	2.38	2.31	2.25	2.16	1.93	1.85	1.78	1.75
33	2.47	2.37	2.29	2.23	2.15	1.92	1.83	1.76	1.73
34	2.45	2.35	2.28	2.22	2.13	1.90	1.82	1.75	1.72
35	2.44	2.34	2.27	2.21	2.12	1.89	1.80	1.73	1.70
36	2.43	2.33	2.25	2.20	2.11	1.88	1.79	1.72	1.69
37	2.42	2.32	2.24	2.18	2.10	1.87	1.77	1.70	1.67
38	2.41	2.31	2.23	2.17	2.09	1.85	1.76	1.69	1.66
39	2.40	2.30	2.22	2.16	2.08	1.84	1.75	1.68	1.65
40	2.39	2.29	2.21	2.15	2.07	1.83	1.74	1.67	1.64
41	2.38	2.28	2.20	2.15	2.06	1.82	1.73	1.66	1.63
42	2.37	2.27	2.20	2.14	2.05	1.81	1.72	1.65	1.62
43	2.36	2.26	2.19	2.13	2.04	1.80	1.71	1.64	1.61
44	2.36	2.26	2.18	2.12	2.03	1.80	1.70	1.63	1.60
45	2.35	2.25	2.17	2.11	2.03	1.79	1.69	1.62	1.59
46	2.34	2.24	2.17	2.11	2.02	1.78	1.69	1.61	1.58
47	2.33	2.23	2.16	2.10	2.01	1.77	1.68	1.60	1.57
48	2.33	2.23	2.15	2.09	2.01	1.77	1.67	1.59	1.56
49	2.32	2.22	2.15	2.09	2.00	1.76	1.66	1.59	1.55
50	2.32	2.22	2.14	2.08	1.99	1.75	1.66	1.58	1.55
55	2.29	2.19	2.11	2.05	1.97	1.72	1.62	1.55	1.51
60	2.27	2.17	2.09	2.03	1.94	1.70	1.60	1.52	1.48
65	2.25	2.15	2.07	2.01	1.93	1.68	1.58	1.49	1.46
70	2.24	2.14	2.06	2.00	1.91	1.66	1.56	1.47	1.44
80	2.21	2.11	2.03	1.97	1.88	1.63	1.53	1.44	1.40
100	2.18	2.08	2.00	1.94	1.85	1.59	1.48	1.39	1.35
150	2.13	2.03	1.95	1.89	1.80	1.54	1.42	1.32	1.27
350	2.09	1.98	1.90	1.84	1.75	1.48	1.35	1.23	1.17
∞	2.05	1.94	1.87	1.80	1.71	1.43	1.30	1.15	1.00

Critical Values of the F Distribution, $\alpha = .01$

| | One-Sided CI = .99 | | | | α | | $\alpha = .01$ | | |
| | Two-Sided CI = .98 | | | | | | | | |

Den.	Numerator df								
df	1	2	3	4	5	6	7	8	9
2	98.50	99.00	99.17	99.25	99.30	99.33	99.36	99.37	99.39
3	34.12	30.82	29.46	28.71	28.24	27.91	27.67	27.49	27.35
4	21.20	18.00	16.69	15.98	15.52	15.21	14.98	14.80	14.66
5	16.26	13.27	12.06	11.39	10.97	10.67	10.46	10.29	10.16
6	13.75	10.92	9.78	9.15	8.75	8.47	8.26	8.10	7.98
7	12.25	9.55	8.45	7.85	7.46	7.19	6.99	6.84	6.72
8	11.26	8.65	7.59	7.01	6.63	6.37	6.18	6.03	5.91
9	10.56	8.02	6.99	6.42	6.06	5.80	5.61	5.47	5.35
10	10.04	7.56	6.55	5.99	5.64	5.39	5.20	5.06	4.94
11	9.65	7.21	6.22	5.67	5.32	5.07	4.89	4.74	4.63
12	9.33	6.93	5.95	5.41	5.06	4.82	4.64	4.50	4.39
13	9.07	6.70	5.74	5.21	4.86	4.62	4.44	4.30	4.19
14	8.86	6.51	5.56	5.04	4.69	4.46	4.28	4.14	4.03
15	8.68	6.36	5.42	4.89	4.56	4.32	4.14	4.00	3.89
16	8.53	6.23	5.29	4.77	4.44	4.20	4.03	3.89	3.78
17	8.40	6.11	5.19	4.67	4.34	4.10	3.93	3.79	3.68
18	8.29	6.01	5.09	4.58	4.25	4.01	3.84	3.71	3.60
19	8.18	5.93	5.01	4.50	4.17	3.94	3.77	3.63	3.52
20	8.10	5.85	4.94	4.43	4.10	3.87	3.70	3.56	3.46
21	8.02	5.78	4.87	4.37	4.04	3.81	3.64	3.51	3.40
22	7.95	5.72	4.82	4.31	3.99	3.76	3.59	3.45	3.35
23	7.88	5.66	4.76	4.26	3.94	3.71	3.54	3.41	3.30
24	7.82	5.61	4.72	4.22	3.90	3.67	3.50	3.36	3.26
25	7.77	5.57	4.68	4.18	3.85	3.63	3.46	3.32	3.22
26	7.72	5.53	4.64	4.14	3.82	3.59	3.42	3.29	3.18
27	7.68	5.49	4.60	4.11	3.78	3.56	3.39	3.26	3.15
28	7.64	5.45	4.57	4.07	3.75	3.53	3.36	3.23	3.12
29	7.60	5.42	4.54	4.04	3.73	3.50	3.33	3.20	3.09
30	7.56	5.39	4.51	4.02	3.70	3.47	3.30	3.17	3.07

Critical Values of the F Distribution, $\alpha = .01$

Den. df	Numerator df								
One-Sided CI = .99 Two-Sided CI = .98						$\alpha = .01$			
	1	2	3	4	5	6	7	8	9
31	7.53	5.36	4.48	3.99	3.67	3.45	3.28	3.15	3.04
32	7.50	5.34	4.46	3.97	3.65	3.43	3.26	3.13	3.02
33	7.47	5.31	4.44	3.95	3.63	3.41	3.24	3.11	3.00
34	7.44	5.29	4.42	3.93	3.61	3.39	3.22	3.09	2.98
35	7.42	5.27	4.40	3.91	3.59	3.37	3.20	3.07	2.96
36	7.40	5.25	4.38	3.89	3.57	3.35	3.18	3.05	2.95
37	7.37	5.23	4.36	3.87	3.56	3.33	3.17	3.04	2.93
38	7.35	5.21	4.34	3.86	3.54	3.32	3.15	3.02	2.92
39	7.33	5.19	4.33	3.84	3.53	3.30	3.14	3.01	2.90
40	7.31	5.18	4.31	3.83	3.51	3.29	3.12	2.99	2.89
41	7.30	5.16	4.30	3.81	3.50	3.28	3.11	2.98	2.87
42	7.28	5.15	4.29	3.80	3.49	3.27	3.10	2.97	2.86
43	7.26	5.14	4.27	3.79	3.48	3.25	3.09	2.96	2.85
44	7.25	5.12	4.26	3.78	3.47	3.24	3.08	2.95	2.84
45	7.23	5.11	4.25	3.77	3.45	3.23	3.07	2.94	2.83
46	7.22	5.10	4.24	3.76	3.44	3.22	3.06	2.93	2.82
47	7.21	5.09	4.23	3.75	3.43	3.21	3.05	2.92	2.81
48	7.19	5.08	4.22	3.74	3.43	3.20	3.04	2.91	2.80
49	7.18	5.07	4.21	3.73	3.42	3.19	3.03	2.90	2.79
50	7.17	5.06	4.20	3.72	3.41	3.19	3.02	2.89	2.78
55	7.12	5.01	4.16	3.68	3.37	3.15	2.98	2.85	2.75
60	7.08	4.98	4.13	3.65	3.34	3.12	2.95	2.82	2.72
65	7.04	4.95	4.10	3.62	3.31	3.09	2.93	2.80	2.69
70	7.01	4.92	4.07	3.60	3.29	3.07	2.91	2.78	2.67
80	6.96	4.88	4.04	3.56	3.26	3.04	2.87	2.74	2.64
100	6.90	4.82	3.98	3.51	3.21	2.99	2.82	2.69	2.59
150	6.81	4.75	3.91	3.45	3.14	2.92	2.76	2.63	2.53
350	6.71	4.67	3.84	3.37	3.07	2.85	2.69	2.56	2.46
∞	6.63	4.61	3.78	3.32	3.02	2.80	2.64	2.51	2.41

Critical Values of the F Distribution, $\alpha = .01$

| One-Sided CI = .99 | | | | | | | | |
| Two-Sided CI = .98 | | | | | | | | |

$\alpha = .01$

Den. df	Numerator df								
	10	12	14	16	20	50	100	350	∞
2	99.40	99.42	99.43	99.44	99.45	99.48	99.49	99.50	99.50
3	27.23	27.05	26.92	26.83	26.69	26.35	26.24	26.16	26.13
4	14.55	14.37	14.25	14.15	14.02	13.69	13.58	13.50	13.46
5	10.05	9.89	9.77	9.68	9.55	9.24	9.13	9.05	9.02
6	7.87	7.72	7.60	7.52	7.40	7.09	6.99	6.91	6.88
7	6.62	6.47	6.36	6.28	6.16	5.86	5.75	5.68	5.65
8	5.81	5.67	5.56	5.48	5.36	5.07	4.96	4.89	4.86
9	5.26	5.11	5.01	4.92	4.81	4.52	4.41	4.34	4.31
10	4.85	4.71	4.60	4.52	4.41	4.12	4.01	3.94	3.91
11	4.54	4.40	4.29	4.21	4.10	3.81	3.71	3.63	3.60
12	4.30	4.16	4.05	3.97	3.86	3.57	3.47	3.39	3.36
13	4.10	3.96	3.86	3.78	3.66	3.38	3.27	3.20	3.17
14	3.94	3.80	3.70	3.62	3.51	3.22	3.11	3.04	3.00
15	3.80	3.67	3.56	3.49	3.37	3.08	2.98	2.90	2.87
16	3.69	3.55	3.45	3.37	3.26	2.97	2.86	2.78	2.75
17	3.59	3.46	3.35	3.27	3.16	2.87	2.76	2.69	2.65
18	3.51	3.37	3.27	3.19	3.08	2.78	2.68	2.60	2.57
19	3.43	3.30	3.19	3.12	3.00	2.71	2.60	2.52	2.49
20	3.37	3.23	3.13	3.05	2.94	2.64	2.54	2.45	2.42
21	3.31	3.17	3.07	2.99	2.88	2.58	2.48	2.39	2.36
22	3.26	3.12	3.02	2.94	2.83	2.53	2.42	2.34	2.31
23	3.21	3.07	2.97	2.89	2.78	2.48	2.37	2.29	2.26
24	3.17	3.03	2.93	2.85	2.74	2.44	2.33	2.25	2.21
25	3.13	2.99	2.89	2.81	2.70	2.40	2.29	2.20	2.17
26	3.09	2.96	2.86	2.78	2.66	2.36	2.25	2.17	2.13
27	3.06	2.93	2.82	2.75	2.63	2.33	2.22	2.13	2.10
28	3.03	2.90	2.79	2.72	2.60	2.30	2.19	2.10	2.06
29	3.00	2.87	2.77	2.69	2.57	2.27	2.16	2.07	2.03
30	2.98	2.84	2.74	2.66	2.55	2.25	2.13	2.04	2.01

Critical Values of the F Distribution, $\alpha = .01$

	One-Sided CI = .99 Two-Sided CI = .98				α		$\alpha = .01$		
Den.	**Numerator df**								
df	**10**	**12**	**14**	**16**	**20**	**50**	**100**	**350**	**∞**
31	2.96	2.82	2.72	2.64	2.52	2.22	2.11	2.02	1.98
32	2.93	2.80	2.70	2.62	2.50	2.20	2.08	1.99	1.96
33	2.91	2.78	2.68	2.60	2.48	2.18	2.06	1.97	1.93
34	2.89	2.76	2.66	2.58	2.46	2.16	2.04	1.95	1.91
35	2.88	2.74	2.64	2.56	2.44	2.14	2.02	1.93	1.89
36	2.86	2.72	2.62	2.54	2.43	2.12	2.00	1.91	1.87
37	2.84	2.71	2.61	2.53	2.41	2.10	1.98	1.89	1.85
38	2.83	2.69	2.59	2.51	2.40	2.09	1.97	1.88	1.84
39	2.81	2.68	2.58	2.50	2.38	2.07	1.95	1.86	1.82
40	2.80	2.66	2.56	2.48	2.37	2.06	1.94	1.84	1.80
41	2.79	2.65	2.55	2.47	2.36	2.04	1.92	1.83	1.79
42	2.78	2.64	2.54	2.46	2.34	2.03	1.91	1.82	1.78
43	2.76	2.63	2.53	2.45	2.33	2.02	1.90	1.80	1.76
44	2.75	2.62	2.52	2.44	2.32	2.01	1.89	1.79	1.75
45	2.74	2.61	2.51	2.43	2.31	2.00	1.88	1.78	1.74
46	2.73	2.60	2.50	2.42	2.30	1.99	1.86	1.77	1.73
47	2.72	2.59	2.49	2.41	2.29	1.98	1.85	1.76	1.71
48	2.71	2.58	2.48	2.40	2.28	1.97	1.84	1.75	1.70
49	2.71	2.57	2.47	2.39	2.27	1.96	1.83	1.74	1.69
50	2.70	2.56	2.46	2.38	2.27	1.95	1.82	1.73	1.68
55	2.66	2.53	2.42	2.34	2.23	1.91	1.78	1.68	1.64
60	2.63	2.50	2.39	2.31	2.20	1.88	1.75	1.65	1.60
65	2.61	2.47	2.37	2.29	2.17	1.85	1.72	1.62	1.57
70	2.59	2.45	2.35	2.27	2.15	1.83	1.70	1.59	1.54
80	2.55	2.42	2.31	2.23	2.12	1.79	1.65	1.54	1.49
100	2.50	2.37	2.27	2.19	2.07	1.74	1.60	1.48	1.43
150	2.44	2.31	2.20	2.12	2.00	1.66	1.52	1.39	1.33
350	2.37	2.24	2.13	2.05	1.93	1.58	1.43	1.28	1.20
∞	2.32	2.18	2.08	2.00	1.88	1.52	1.36	1.18	1.00

Critical Values of the F Distribution, $\alpha = .005$

| | One-Sided CI = .995 | | | | | α | $\alpha = .005$ | | |
| | Two-Sided CI = .99 | | | | | | | | |

Den.	Numerator df								
df	1	2	3	4	5	6	7	8	9
2	198.50	199.00	199.17	199.25	199.30	199.33	199.36	199.37	199.39
3	55.55	49.80	47.47	46.19	45.39	44.84	44.43	44.13	43.88
4	31.33	26.28	24.26	23.15	22.46	21.97	21.62	21.35	21.14
5	22.78	18.31	16.53	15.56	14.94	14.51	14.20	13.96	13.77
6	18.64	14.54	12.92	12.03	11.46	11.07	10.79	10.57	10.39
7	16.24	12.40	10.88	10.05	9.52	9.16	8.89	8.68	8.51
8	14.69	11.04	9.60	8.81	8.30	7.95	7.69	7.50	7.34
9	13.61	10.11	8.72	7.96	7.47	7.13	6.88	6.69	6.54
10	12.83	9.43	8.08	7.34	6.87	6.54	6.30	6.12	5.97
11	12.23	8.91	7.60	6.88	6.42	6.10	5.86	5.68	5.54
12	11.75	8.51	7.23	6.52	6.07	5.76	5.52	5.35	5.20
13	11.37	8.19	6.93	6.23	5.79	5.48	5.25	5.08	4.94
14	11.06	7.92	6.68	6.00	5.56	5.26	5.03	4.86	4.72
15	10.80	7.70	6.48	5.80	5.37	5.07	4.85	4.67	4.54
16	10.58	7.51	6.30	5.64	5.21	4.91	4.69	4.52	4.38
17	10.38	7.35	6.16	5.50	5.07	4.78	4.56	4.39	4.25
18	10.22	7.21	6.03	5.37	4.96	4.66	4.44	4.28	4.14
19	10.07	7.09	5.92	5.27	4.85	4.56	4.34	4.18	4.04
20	9.94	6.99	5.82	5.17	4.76	4.47	4.26	4.09	3.96
21	9.83	6.89	5.73	5.09	4.68	4.39	4.18	4.01	3.88
22	9.73	6.81	5.65	5.02	4.61	4.32	4.11	3.94	3.81
23	9.63	6.73	5.58	4.95	4.54	4.26	4.05	3.88	3.75
24	9.55	6.66	5.52	4.89	4.49	4.20	3.99	3.83	3.69
25	9.48	6.60	5.46	4.84	4.43	4.15	3.94	3.78	3.64
26	9.41	6.54	5.41	4.79	4.38	4.10	3.89	3.73	3.60
27	9.34	6.49	5.36	4.74	4.34	4.06	3.85	3.69	3.56
28	9.28	6.44	5.32	4.70	4.30	4.02	3.81	3.65	3.52
29	9.23	6.40	5.28	4.66	4.26	3.98	3.77	3.61	3.48
30	9.18	6.35	5.24	4.62	4.23	3.95	3.74	3.58	3.45

Critical Values of the F Distribution, $\alpha = .005$

	One-Sided CI = .995 Two-Sided CI = .99				α		$\alpha = .005$		
Den.	Numerator df								
df	1	2	3	4	5	6	7	8	9
31	9.13	6.32	5.20	4.59	4.20	3.92	3.71	3.55	3.42
32	9.09	6.28	5.17	4.56	4.17	3.89	3.68	3.52	3.39
33	9.05	6.25	5.14	4.53	4.14	3.86	3.66	3.49	3.37
34	9.01	6.22	5.11	4.50	4.11	3.84	3.63	3.47	3.34
35	8.98	6.19	5.09	4.48	4.09	3.81	3.61	3.45	3.32
36	8.94	6.16	5.06	4.46	4.06	3.79	3.58	3.42	3.30
37	8.91	6.13	5.04	4.43	4.04	3.77	3.56	3.40	3.28
38	8.88	6.11	5.02	4.41	4.02	3.75	3.54	3.39	3.26
39	8.85	6.09	5.00	4.39	4.00	3.73	3.53	3.37	3.24
40	8.83	6.07	4.98	4.37	3.99	3.71	3.51	3.35	3.22
41	8.80	6.05	4.96	4.36	3.97	3.70	3.49	3.33	3.21
42	8.78	6.03	4.94	4.34	3.95	3.68	3.48	3.32	3.19
43	8.76	6.01	4.92	4.32	3.94	3.67	3.46	3.30	3.18
44	8.74	5.99	4.91	4.31	3.92	3.65	3.45	3.29	3.16
45	8.71	5.97	4.89	4.29	3.91	3.64	3.43	3.28	3.15
46	8.70	5.96	4.88	4.28	3.90	3.62	3.42	3.26	3.14
47	8.68	5.94	4.86	4.27	3.88	3.61	3.41	3.25	3.12
48	8.66	5.93	4.85	4.25	3.87	3.60	3.40	3.24	3.11
49	8.64	5.91	4.84	4.24	3.86	3.59	3.39	3.23	3.10
50	8.63	5.90	4.83	4.23	3.85	3.58	3.38	3.22	3.09
55	8.55	5.84	4.77	4.18	3.80	3.53	3.33	3.17	3.05
60	8.49	5.79	4.73	4.14	3.76	3.49	3.29	3.13	3.01
65	8.44	5.75	4.69	4.11	3.73	3.46	3.26	3.10	2.98
70	8.40	5.72	4.66	4.08	3.70	3.43	3.23	3.08	2.95
80	8.33	5.67	4.61	4.03	3.65	3.39	3.19	3.03	2.91
100	8.24	5.59	4.54	3.96	3.59	3.33	3.13	2.97	2.85
150	8.12	5.49	4.45	3.88	3.51	3.25	3.05	2.89	2.77
350	7.98	5.38	4.35	3.78	3.42	3.16	2.96	2.81	2.68
∞	7.88	5.30	4.28	3.72	3.35	3.09	2.90	2.74	2.62

Critical Values of the F Distribution, $\alpha = .005$

	One-Sided CI = .995 Two-Sided CI = .99								
Den.	Numerator df								
df	10	12	14	16	20	50	100	350	∞
2	199.40	199.42	199.43	199.44	199.45	199.48	199.49	199.50	199.50
3	43.69	43.39	43.17	43.01	42.78	42.21	42.02	41.88	41.83
4	20.97	20.70	20.51	20.37	20.17	19.67	19.50	19.37	19.32
5	13.62	13.38	13.21	13.09	12.90	12.45	12.30	12.19	12.14
6	10.25	10.03	9.88	9.76	9.59	9.17	9.03	8.92	8.88
7	8.38	8.18	8.03	7.91	7.75	7.35	7.22	7.12	7.08
8	7.21	7.01	6.87	6.76	6.61	6.22	6.09	5.99	5.95
9	6.42	6.23	6.09	5.98	5.83	5.45	5.32	5.23	5.19
10	5.85	5.66	5.53	5.42	5.27	4.90	4.77	4.68	4.64
11	5.42	5.24	5.10	5.00	4.86	4.49	4.36	4.26	4.23
12	5.09	4.91	4.77	4.67	4.53	4.17	4.04	3.94	3.90
13	4.82	4.64	4.51	4.41	4.27	3.91	3.78	3.69	3.65
14	4.60	4.43	4.30	4.20	4.06	3.70	3.57	3.47	3.44
15	4.42	4.25	4.12	4.02	3.88	3.52	3.39	3.30	3.26
16	4.27	4.10	3.97	3.87	3.73	3.37	3.25	3.15	3.11
17	4.14	3.97	3.84	3.75	3.61	3.25	3.12	3.02	2.98
18	4.03	3.86	3.73	3.64	3.50	3.14	3.01	2.91	2.87
19	3.93	3.76	3.64	3.54	3.40	3.04	2.91	2.82	2.78
20	3.85	3.68	3.55	3.46	3.32	2.96	2.83	2.73	2.69
21	3.77	3.60	3.48	3.38	3.24	2.88	2.75	2.65	2.61
22	3.70	3.54	3.41	3.31	3.18	2.82	2.69	2.59	2.55
23	3.64	3.47	3.35	3.25	3.12	2.76	2.62	2.52	2.48
24	3.59	3.42	3.30	3.20	3.06	2.70	2.57	2.47	2.43
25	3.54	3.37	3.25	3.15	3.01	2.65	2.52	2.42	2.38
26	3.49	3.33	3.20	3.11	2.97	2.61	2.47	2.37	2.33
27	3.45	3.28	3.16	3.07	2.93	2.57	2.43	2.33	2.29
28	3.41	3.25	3.12	3.03	2.89	2.53	2.39	2.29	2.25
29	3.38	3.21	3.09	2.99	2.86	2.49	2.36	2.25	2.21
30	3.34	3.18	3.06	2.96	2.82	2.46	2.32	2.22	2.18

Critical Values of the F Distribution, $\alpha = .005$

	One-Sided CI = .995 Two-Sided CI = .99					α	$\alpha = .005$		
Den.	Numerator df								
df	10	12	14	16	20	50	100	350	∞
31	3.31	3.15	3.03	2.93	2.79	2.43	2.29	2.19	2.14
32	3.29	3.12	3.00	2.90	2.77	2.40	2.26	2.16	2.11
33	3.26	3.09	2.97	2.88	2.74	2.37	2.24	2.13	2.09
34	3.24	3.07	2.95	2.85	2.72	2.35	2.21	2.11	2.06
35	3.21	3.05	2.93	2.83	2.69	2.33	2.19	2.08	2.04
36	3.19	3.03	2.90	2.81	2.67	2.30	2.17	2.06	2.01
37	3.17	3.01	2.88	2.79	2.65	2.28	2.14	2.04	1.99
38	3.15	2.99	2.87	2.77	2.63	2.27	2.12	2.02	1.97
39	3.13	2.97	2.85	2.75	2.62	2.25	2.11	2.00	1.95
40	3.12	2.95	2.83	2.74	2.60	2.23	2.09	1.98	1.93
41	3.10	2.94	2.82	2.72	2.58	2.21	2.07	1.96	1.91
42	3.09	2.92	2.80	2.71	2.57	2.20	2.06	1.94	1.90
43	3.07	2.91	2.79	2.69	2.55	2.18	2.04	1.93	1.88
44	3.06	2.89	2.77	2.68	2.54	2.17	2.03	1.91	1.87
45	3.04	2.88	2.76	2.66	2.53	2.16	2.01	1.90	1.85
46	3.03	2.87	2.75	2.65	2.51	2.14	2.00	1.89	1.84
47	3.02	2.86	2.74	2.64	2.50	2.13	1.99	1.87	1.82
48	3.01	2.85	2.72	2.63	2.49	2.12	1.97	1.86	1.81
49	3.00	2.83	2.71	2.62	2.48	2.11	1.96	1.85	1.80
50	2.99	2.82	2.70	2.61	2.47	2.10	1.95	1.84	1.79
55	2.94	2.78	2.66	2.56	2.42	2.05	1.90	1.78	1.73
60	2.90	2.74	2.62	2.53	2.39	2.01	1.86	1.74	1.69
65	2.87	2.71	2.59	2.49	2.36	1.98	1.83	1.70	1.65
70	2.85	2.68	2.56	2.47	2.33	1.95	1.80	1.67	1.62
80	2.80	2.64	2.52	2.43	2.29	1.90	1.75	1.62	1.56
100	2.74	2.58	2.46	2.37	2.23	1.84	1.68	1.55	1.49
150	2.67	2.51	2.38	2.29	2.15	1.76	1.59	1.44	1.37
350	2.58	2.42	2.30	2.20	2.06	1.66	1.48	1.32	1.23
∞	2.52	2.36	2.24	2.14	2.00	1.59	1.40	1.21	1.00

APPENDIX D

Critical Values of the Chi-Square Distribution

Critical Values of the Chi-Square Distribution

df	.200	.100	.050	.025	.010	.005	.001
1	1.642	2.706	3.841	5.024	6.635	7.879	10.828
2	3.219	4.605	5.991	7.378	9.210	10.597	13.816
3	4.642	6.251	7.815	9.348	11.345	12.838	16.266
4	5.989	7.779	9.488	11.143	13.277	14.860	18.467
5	7.289	9.236	11.070	12.833	15.086	16.750	20.515
6	8.558	10.645	12.592	14.449	16.812	18.548	22.458
7	9.803	12.017	14.067	16.013	18.475	20.278	24.322
8	11.030	13.362	15.507	17.535	20.090	21.955	26.124
9	12.242	14.684	16.919	19.023	21.666	23.589	27.877
10	13.442	15.987	18.307	20.483	23.209	25.188	29.588
11	14.631	17.275	19.675	21.920	24.725	26.757	31.264
12	15.812	18.549	21.026	23.337	26.217	28.300	32.909
13	16.985	19.812	22.362	24.736	27.688	29.819	34.528
14	18.151	21.064	23.685	26.119	29.141	31.319	36.123
15	19.311	22.307	24.996	27.488	30.578	32.801	37.697
16	20.465	23.542	26.296	28.845	32.000	34.267	39.252
17	21.615	24.769	27.587	30.191	33.409	35.718	40.790
18	22.760	25.989	28.869	31.526	34.805	37.156	42.312
19	23.900	27.204	30.144	32.852	36.191	38.582	43.820
20	25.038	28.412	31.410	34.170	37.566	39.997	45.315
21	26.171	29.615	32.671	35.479	38.932	41.401	46.797
22	27.301	30.813	33.924	36.781	40.289	42.796	48.268
23	28.429	32.007	35.172	38.076	41.638	44.181	49.728
24	29.553	33.196	36.415	39.364	42.980	45.559	51.179
25	30.675	34.382	37.652	40.646	44.314	46.928	52.620
26	31.795	35.563	38.885	41.923	45.642	48.290	54.052
27	32.912	36.741	40.113	43.195	46.963	49.645	55.476
28	34.027	37.916	41.337	44.461	48.278	50.993	56.892
29	35.139	39.087	42.557	45.722	49.588	52.336	58.301
30	36.250	40.256	43.773	46.979	50.892	53.672	59.703

Critical Values of the Chi-Square Distribution

df	.200	.100	.050	.025	.010	.005	.001
31	37.359	41.422	44.985	48.232	52.191	55.003	61.098
32	38.466	42.585	46.194	49.480	53.486	56.328	62.487
33	39.572	43.745	47.400	50.725	54.776	57.648	63.870
34	40.676	44.903	48.602	51.966	56.061	58.964	65.247
35	41.778	46.059	49.802	53.203	57.342	60.275	66.619
36	42.879	47.212	50.998	54.437	58.619	61.581	67.985
37	43.978	48.363	52.192	55.668	59.893	62.883	69.346
38	45.076	49.513	53.384	56.896	61.162	64.181	70.703
39	46.173	50.660	54.572	58.120	62.428	65.476	72.055
40	47.269	51.805	55.758	59.342	63.691	66.766	73.402
41	48.363	52.949	56.942	60.561	64.950	68.053	74.745
42	49.456	54.090	58.124	61.777	66.206	69.336	76.084
43	50.548	55.230	59.304	62.990	67.459	70.616	77.419
44	51.639	56.369	60.481	64.201	68.710	71.893	78.750
45	52.729	57.505	61.656	65.410	69.957	73.166	80.077
46	53.818	58.641	62.830	66.617	71.201	74.437	81.400
47	54.906	59.774	64.001	67.821	72.443	75.704	82.720
48	55.993	60.907	65.171	69.023	73.683	76.969	84.037
49	57.079	62.038	66.339	70.222	74.919	78.231	85.351
50	58.164	63.167	67.505	71.420	76.154	79.490	86.661
55	63.577	68.796	73.311	77.380	82.292	85.749	93.168
60	68.972	74.397	79.082	83.298	88.379	91.952	99.607
65	74.351	79.973	84.821	89.177	94.422	98.105	105.988
70	79.715	85.527	90.531	95.023	100.425	104.215	112.317
75	85.066	91.061	96.217	100.839	106.393	110.286	118.599
80	90.405	96.578	101.879	106.629	112.329	116.321	124.839
85	95.734	102.079	107.522	112.393	118.236	122.325	131.041
90	101.054	107.565	113.145	118.136	124.116	128.299	137.208
95	106.364	113.038	118.752	123.858	129.973	134.247	143.344
100	111.667	118.498	124.342	129.561	135.807	140.169	149.449

Critical Values of q for Tukey's HSD Test

Critical Values of q for Tukey's HSD Test, $\alpha = .025$ (1 tailed), .05 (2 tailed)

df	\multicolumn{9}{c}{Number of Means}
	2	3	4	5	6	7	8	9	10
2	6.085	8.331	9.798	10.881	11.734	12.435	13.027	13.539	13.988
3	4.501	5.910	6.825	7.502	8.037	8.478	8.852	9.177	9.462
4	3.926	5.040	5.757	6.287	6.706	7.053	7.347	7.602	7.826
5	3.635	4.602	5.218	5.673	6.033	6.330	6.582	6.801	6.995
6	3.460	4.339	4.896	5.305	5.628	5.895	6.122	6.319	6.493
7	3.344	4.165	4.681	5.060	5.359	5.606	5.815	5.997	6.158
8	3.261	4.041	4.529	4.886	5.167	5.399	5.596	5.767	5.918
9	3.199	3.948	4.415	4.755	5.024	5.244	5.432	5.595	5.738
10	3.151	3.877	4.327	4.654	4.912	5.124	5.304	5.460	5.598
11	3.113	3.820	4.256	4.574	4.823	5.028	5.202	5.353	5.486
12	3.081	3.773	4.199	4.508	4.750	4.950	5.119	5.265	5.395
13	3.055	3.734	4.151	4.453	4.690	4.884	5.049	5.192	5.318
14	3.033	3.701	4.111	4.407	4.639	4.829	4.990	5.130	5.253
15	3.014	3.673	4.076	4.367	4.595	4.782	4.940	5.077	5.198
16	2.998	3.649	4.046	4.333	4.557	4.741	4.896	5.031	5.150
17	2.984	3.628	4.020	4.303	4.524	4.705	4.858	4.991	5.108
18	2.971	3.609	3.997	4.276	4.494	4.673	4.824	4.955	5.071
19	2.960	3.593	3.977	4.253	4.468	4.645	4.794	4.924	5.037
20	2.950	3.578	3.958	4.232	4.445	4.620	4.768	4.895	5.008
21	2.941	3.565	3.942	4.213	4.424	4.597	4.743	4.870	4.981
22	2.933	3.553	3.927	4.196	4.405	4.577	4.722	4.847	4.957
23	2.926	3.542	3.914	4.180	4.388	4.558	4.702	4.826	4.935
24	2.919	3.532	3.901	4.166	4.373	4.541	4.684	4.807	4.915
25	2.913	3.523	3.890	4.153	4.358	4.526	4.667	4.789	4.897
26	2.907	3.514	3.880	4.141	4.345	4.511	4.652	4.773	4.880
27	2.902	3.506	3.870	4.130	4.333	4.498	4.638	4.758	4.864
28	2.897	3.499	3.861	4.120	4.322	4.486	4.625	4.745	4.850
29	2.892	3.493	3.853	4.111	4.311	4.475	4.613	4.732	4.837
30	2.888	3.486	3.845	4.102	4.301	4.464	4.601	4.720	4.824

Critical Values of q for Tukey's HSD Test, $\alpha = .025$ (1 tailed), .05 (2 tailed)

df	\multicolumn{9}{c}{Number of Means}
	2	3	4	5	6	7	8	9	10
31	2.884	3.481	3.838	4.094	4.292	4.454	4.591	4.709	4.812
32	2.881	3.475	3.832	4.086	4.284	4.445	4.581	4.698	4.802
33	2.877	3.470	3.825	4.079	4.276	4.436	4.572	4.689	4.791
34	2.874	3.465	3.820	4.072	4.268	4.428	4.563	4.680	4.782
35	2.871	3.461	3.814	4.066	4.261	4.421	4.555	4.671	4.773
36	2.868	3.457	3.809	4.060	4.255	4.414	4.547	4.663	4.764
37	2.865	3.453	3.804	4.054	4.249	4.407	4.540	4.655	4.756
38	2.863	3.449	3.799	4.049	4.243	4.400	4.533	4.648	4.749
39	2.861	3.445	3.795	4.044	4.237	4.394	4.527	4.641	4.741
40	2.858	3.442	3.791	4.040	4.232	4.389	4.521	4.635	4.735
41	2.857	3.439	3.788	4.035	4.227	4.383	4.515	4.629	4.728
42	2.855	3.436	3.784	4.031	4.222	4.378	4.509	4.623	4.722
43	2.853	3.434	3.780	4.027	4.218	4.373	4.504	4.617	4.716
44	2.851	3.431	3.777	4.023	4.213	4.368	4.499	4.612	4.711
45	2.849	3.428	3.773	4.019	4.209	4.364	4.494	4.607	4.705
46	2.847	3.426	3.770	4.015	4.205	4.360	4.490	4.602	4.700
47	2.846	3.423	3.767	4.012	4.201	4.355	4.485	4.597	4.695
48	2.844	3.421	3.764	4.009	4.198	4.351	4.481	4.593	4.691
49	2.843	3.419	3.762	4.006	4.194	4.348	4.477	4.588	4.686
50	2.841	3.416	3.759	4.003	4.191	4.344	4.473	4.584	4.682
55	2.835	3.407	3.747	3.989	4.176	4.328	4.456	4.566	4.663
60	2.829	3.399	3.738	3.978	4.163	4.314	4.441	4.551	4.647
65	2.825	3.392	3.729	3.969	4.153	4.303	4.429	4.538	4.633
70	2.821	3.387	3.722	3.961	4.144	4.293	4.419	4.527	4.622
75	2.818	3.382	3.716	3.954	4.136	4.285	4.410	4.517	4.612
80	2.814	3.377	3.711	3.948	4.130	4.278	4.402	4.509	4.603
100	2.806	3.365	3.695	3.929	4.110	4.256	4.379	4.484	4.577
120	2.800	3.356	3.685	3.917	4.096	4.241	4.363	4.468	4.560
240	2.786	3.335	3.659	3.887	4.063	4.205	4.324	4.427	4.517
∞	2.772	3.314	3.633	3.858	4.030	4.170	4.286	4.387	4.474

Critical Values of q for Tukey's HSD Test, $\alpha = .005$ (1 tailed), .01 (2 tailed)

df	Number of Means								
	2	3	4	5	6	7	8	9	10
2	14.036	19.019	22.294	24.717	26.629	28.201	29.530	30.679	31.689
3	8.260	10.619	12.170	13.324	14.241	14.998	15.641	16.199	16.691
4	6.511	8.120	9.173	9.958	10.583	11.101	11.542	11.925	12.264
5	5.702	6.976	7.804	8.421	8.913	9.321	9.669	9.971	10.239
6	5.243	6.331	7.033	7.556	7.972	8.318	8.612	8.869	9.097
7	4.949	5.919	6.542	7.005	7.373	7.678	7.939	8.166	8.367
8	4.745	5.635	6.204	6.625	6.959	7.237	7.474	7.680	7.863
9	4.596	5.428	5.957	6.347	6.657	6.915	7.134	7.325	7.494
10	4.482	5.270	5.769	6.136	6.428	6.669	6.875	7.054	7.213
11	4.392	5.146	5.621	5.970	6.247	6.476	6.671	6.841	6.992
12	4.320	5.046	5.502	5.836	6.101	6.320	6.507	6.670	6.814
13	4.260	4.964	5.404	5.726	5.981	6.192	6.372	6.528	6.666
14	4.210	4.895	5.322	5.634	5.881	6.085	6.258	6.409	6.543
15	4.167	4.836	5.252	5.556	5.796	5.994	6.162	6.309	6.438
16	4.131	4.786	5.192	5.489	5.722	5.915	6.079	6.222	6.348
17	4.099	4.742	5.140	5.430	5.659	5.847	6.007	6.147	6.270
18	4.071	4.703	5.094	5.379	5.603	5.787	5.944	6.081	6.201
19	4.046	4.669	5.054	5.334	5.553	5.735	5.889	6.022	6.141
20	4.024	4.639	5.018	5.293	5.510	5.688	5.839	5.970	6.086
21	4.004	4.612	4.986	5.257	5.470	5.646	5.794	5.924	6.038
22	3.986	4.588	4.957	5.225	5.435	5.608	5.754	5.882	5.994
23	3.970	4.566	4.931	5.195	5.403	5.573	5.718	5.844	5.955
24	3.955	4.546	4.907	5.168	5.373	5.542	5.685	5.809	5.919
25	3.942	4.527	4.885	5.144	5.347	5.513	5.655	5.778	5.886
26	3.930	4.510	4.865	5.121	5.322	5.487	5.627	5.749	5.856
27	3.918	4.495	4.847	5.101	5.300	5.463	5.602	5.722	5.828
28	3.908	4.481	4.830	5.082	5.279	5.441	5.578	5.697	5.802
29	3.898	4.467	4.814	5.064	5.260	5.420	5.556	5.674	5.778
30	3.889	4.455	4.799	5.048	5.242	5.401	5.536	5.653	5.756

Critical Values of q for Tukey's HSD Test, $\alpha = .005$ (1 tailed), .01 (2 tailed)

df	\multicolumn{9}{c}{Number of Means}								
	2	3	4	5	6	7	8	9	10
31	3.881	4.443	4.786	5.032	5.225	5.383	5.517	5.633	5.736
32	3.873	4.433	4.773	5.018	5.210	5.367	5.500	5.615	5.716
33	3.865	4.423	4.761	5.005	5.195	5.351	5.483	5.598	5.698
34	3.859	4.413	4.750	4.992	5.181	5.336	5.468	5.581	5.682
35	3.852	4.404	4.739	4.980	5.169	5.323	5.453	5.566	5.666
36	3.846	4.396	4.729	4.969	5.156	5.310	5.439	5.552	5.651
37	3.840	4.388	4.720	4.959	5.145	5.298	5.427	5.538	5.637
38	3.835	4.381	4.711	4.949	5.134	5.286	5.414	5.526	5.623
39	3.830	4.374	4.703	4.940	5.124	5.275	5.403	5.513	5.611
40	3.828	4.370	4.700	4.936	5.119	5.269	5.396	5.505	6.602
41	3.826	4.366	4.692	4.928	5.111	5.259	5.385	5.494	5.591
42	3.822	4.360	4.685	4.920	5.102	5.250	5.375	5.484	5.580
43	3.817	4.354	4.679	4.912	5.094	5.241	5.366	5.474	5.570
44	3.813	4.349	4.672	4.905	5.085	5.232	5.357	5.465	5.560
45	3.809	4.343	4.666	4.898	5.078	5.224	5.349	5.456	5.551
46	3.806	4.338	4.660	4.891	5.071	5.217	5.340	5.447	5.542
47	3.802	4.334	4.654	4.885	5.064	5.209	5.333	5.439	5.533
48	3.799	4.329	4.649	4.879	5.057	5.202	5.325	5.431	5.525
49	3.796	4.325	4.644	4.873	5.051	5.195	5.318	5.424	5.517
50	3.793	4.321	4.639	4.867	5.045	5.189	5.311	5.416	5.510
55	3.779	4.302	4.617	4.843	5.018	5.160	5.281	5.385	5.477
60	3.761	4.287	4.598	4.822	4.996	5.136	5.255	5.358	5.449
65	3.758	4.274	4.583	4.805	4.977	5.116	5.234	5.336	5.427
70	3.750	4.263	4.570	4.790	4.961	5.099	5.216	5.318	5.407
75	3.743	4.253	4.559	4.777	4.947	5.085	5.201	5.301	5.390
80	3.737	4.245	4.549	4.766	4.935	5.073	5.187	5.287	5.375
100	3.718	4.220	4.520	4.733	4.900	5.035	5.147	5.245	5.331
120	3.702	4.200	4.497	4.709	4.872	5.005	5.118	5.214	5.299
240	3.672	4.160	4.450	4.655	4.814	4.943	5.052	5.145	5.227
	3.643	4.120	4.403	4.603	4.757	4.882	4.987	5.078	5.157

Critical Values of the Rank Correlation Coefficient

Critical Values for the Rank Correlation Coefficient

One Tailed	.100	.050	.025	.010	.005
Two Tailed	.200	.100	.050	.020	.010
n					
4	1.000	1.000	—	—	—
5	.800	.900	1.000	1.000	—
6	.657	.829	.886	.943	1.000
7	.607	.714	.786	.893	.929
8	.524	.643	.738	.833	.881
9	.483	.600	.700	.783	.833
10	.455	.564	.648	.745	.794
11	.427	.536	.618	.709	.755
12	.406	.503	.587	.678	.727
13	.385	.484	.560	.648	.703
14	.367	.464	.538	.622	.675
15	.354	.446	.521	.604	.654
16	.341	.429	.503	.585	.635
17	.328	.414	.488	.566	.615
18	.319	.401	.472	.550	.600
19	.309	.391	.460	.535	.584
20	.299	.380	.448	.522	.570
21	.291	.370	.435	.509	.556
22	.284	.361	.425	.497	.544
23	.278	.353	.416	.486	.532
24	.271	.344	.406	.477	.522
25	.265	.338	.398	.466	.511
26	.259	.330	.389	.456	.501
27	.255	.324	.383	.448	.491
28	.250	.318	.375	.440	.483
29	.245	.312	.369	.433	.475
30	.240	.307	.362	.425	.467

Critical Values for the Rank Correlation Coefficient

One Tailed	.100	.050	.025	.010	.005
Two Tailed	.200	.100	.050	.020	.010
n					
31	.236	.302	.356	.419	.460
32	.232	.296	.351	.412	.453
33	.229	.291	.345	.405	.445
34	.225	.287	.340	.400	.439
35	.222	.283	.335	.394	.433
36	.219	.279	.330	.388	.427
37	.216	.275	.326	.384	.422
38	.213	.271	.321	.378	.416
39	.210	.267	.317	.373	.410
40	.207	.264	.313	.368	.405
41	.204	.261	.309	.363	.400
42	.202	.257	.305	.360	.396
43	.199	.254	.301	.355	.391
44	.197	.251	.298	.351	.386
45	.194	.248	.294	.347	.382
46	.192	.246	.291	.343	.378
47	.190	.243	.288	.340	.374
48	.188	.240	.285	.336	.371
49	.186	.238	.282	.333	.366
50	.184	.235	.279	.330	.363
51	.182	.233	.276	.326	.359
52	.181	.231	.274	.323	.356
53	.179	.229	.271	.320	.353
54	.177	.226	.268	.317	.349
55	.175	.224	.266	.314	.346
56	.174	.222	.264	.311	.343
57	.172	.220	.261	.308	.340
58	.171	.218	.259	.306	.337
59	.169	.216	.257	.303	.335
60	.168	.214	.254	.301	.332

Critical Values for the Rank Correlation Coefficient

One Tailed	.100	.050	.025	.010	.005
Two Tailed	.200	.100	.050	.020	.010
n					
61	.166	.212	.252	.298	.328
62	.165	.211	.250	.296	.326
63	.164	.209	.248	.293	.323
64	.162	.207	.247	.291	.321
65	.161	.206	.244	.289	.318
66	.160	.204	.243	.287	.316
67	.158	.203	.241	.284	.314
68	.157	.201	.239	.282	.311
69	.156	.200	.237	.280	.309
70	.155	.198	.235	.278	.307
71	.154	.197	.234	.276	.305
72	.153	.195	.232	.274	.303
73	.151	.194	.231	.272	.301
74	.151	.193	.229	.271	.298
75	.150	.191	.227	.269	.297
76	.149	.190	.226	.267	.295
77	.147	.189	.224	.265	.292
78	.147	.188	.223	.264	.291
79	.146	.186	.222	.262	.289
80	.145	.185	.220	.260	.288
81	.144	.184	.218	.258	.285
82	.143	.183	.217	.257	.284
83	.142	.182	.216	.255	.282
84	.141	.181	.215	.254	.281
85	.140	.180	.213	.252	.279
86	.139	.179	.212	.251	.277
87	.139	.178	.211	.249	.275
88	.138	.177	.210	.248	.274
89	.137	.175	.209	.247	.273
90	.136	.174	.207	.245	.271

Critical Values for Wilcoxon's Signed-Ranks Test (as a t statistic)

Critical Values for Wilcoxon's Signed-Ranks Test

One Tailed	.100	.050	.025	.010	.005
Two Tailed	.200	.100	.050	.020	.010
n					
4	3.873	—	—	—	—
5	1.773	4.243	—	—	—
6	1.872	2.371	4.583	—	—
7	1.721	2.420	2.925	4.899	—
8	1.508	2.225	2.934	4.123	5.196
9	1.605	1.975	2.704	3.414	4.545
10	1.450	2.049	2.424	3.160	3.862
11	1.502	2.009	2.477	3.074	3.592
12	1.481	1.906	2.416	3.071	3.488
13	1.414	1.869	2.296	2.958	3.468
14	1.394	1.875	2.245	2.919	3.339
15	1.407	1.833	2.238	2.815	3.286
16	1.382	1.827	2.261	2.770	3.170
17	1.385	1.785	2.236	2.763	3.114
18	1.359	1.774	2.178	2.709	3.096
19	1.357	1.785	2.152	2.690	3.034
20	1.373	1.767	2.150	2.637	3.006
21	1.365	1.769	2.167	2.612	3.003
22	1.338	1.748	2.156	2.610	2.966
23	1.363	1.746	2.122	2.580	2.953
24	1.336	1.758	2.108	2.570	2.912
25	1.354	1.750	2.110	2.577	2.893
26	1.327	1.726	2.092	2.560	2.891
27	1.341	1.744	2.089	2.559	2.867
28	1.339	1.719	2.098	2.538	2.858
29	1.325	1.733	2.090	2.532	2.831
30	1.322	1.731	2.069	2.510	2.848

Critical Values for Wilcoxon's Signed-Ranks Test

One Tailed	.100	.050	.025	.010	.005
Two Tailed	.200	.100	.050	.020	.010
n					
31	.236	.302	.356	.419	.460
32	.232	.296	.351	.412	.453
33	.229	.291	.345	.405	.445
34	.225	.287	.340	.400	.439
35	.222	.283	.335	.394	.433
36	.219	.279	.330	.388	.427
37	.216	.275	.326	.384	.422
38	.213	.271	.321	.378	.416
39	.210	.267	.317	.373	.410
40	.207	.264	.313	.368	.405
41	.204	.261	.309	.363	.400
42	.202	.257	.305	.360	.396
43	.199	.254	.301	.355	.391
44	.197	.251	.298	.351	.386
45	.194	.248	.294	.347	.382
46	.192	.246	.291	.343	.378
47	.190	.243	.288	.340	.374
48	.188	.240	.285	.336	.371
49	.186	.238	.282	.333	.366
50	.184	.235	.279	.330	.363
51	.182	.233	.276	.326	.359
52	.181	.231	.274	.323	.356
53	.179	.229	.271	.320	.353
54	.177	.226	.268	.317	.349
55	.175	.224	.266	.314	.346
56	.174	.222	.264	.311	.343
57	.172	.220	.261	.308	.340
58	.171	.218	.259	.306	.337
59	.169	.216	.257	.303	.335
60	.168	.214	.254	.301	.332

Critical Values for Wilcoxon's Signed-Ranks Test

One Tailed	.100	.050	.025	.010	.005
Two Tailed	.200	.100	.050	.020	.010
n					
61	1.304	1.680	2.013	2.406	2.688
62	1.301	1.675	2.007	2.405	2.680
63	1.301	1.673	2.012	2.408	2.683
64	1.298	1.675	2.005	2.406	2.682
65	1.297	1.673	2.009	2.407	2.677
66	1.300	1.674	2.009	2.398	2.675
67	1.300	1.678	2.005	2.398	2.676
68	1.296	1.671	2.004	2.402	2.673
69	1.301	1.675	2.007	2.402	2.674
70	1.297	1.674	2.005	2.398	2.670
71	1.295	1.671	2.001	2.397	2.669
72	1.297	1.670	1.999	2.392	2.672
73	1.295	1.672	2.000	2.397	2.670
74	1.296	1.671	2.003	2.392	2.665
75	1.294	1.673	1.998	2.390	2.663
76	1.295	1.671	2.001	2.390	2.664
77	1.293	1.667	1.995	2.388	2.661
78	1.293	1.670	1.997	2.387	2.661
79	1.296	1.670	1.996	2.390	2.657
80	1.296	1.668	1.998	2.389	2.656
81	1.293	1.668	1.997	2.390	2.658
82	1.293	1.666	1.998	2.384	2.656
83	1.295	1.666	1.996	2.385	2.657
84	1.294	1.668	1.997	2.388	2.660
85	1.291	1.667	1.995	2.383	2.650
86	1.295	1.668	1.995	2.381	2.653
87	1.292	1.667	1.993	2.385	2.652
88	1.295	1.664	1.993	2.383	2.649
89	1.296	1.667	1.994	2.382	2.648
90	1.294	1.664	1.994	2.379	2.649

APPENDIX H

Critical Values for Wilcoxon's Rank-Sum Test (as a t statistic)

Critical Values for Wilcoxon's Rank-Sum Test					
One Tailed	.100	.050	.025	.010	.005
Two Tailed	.200	.100	.050	.020	.010
$n_1 = n_2$					
3	3.674	1.871	—	—	—
4	1.594	2.898	4.382	—	—
5	1.732	2.077	3.031	3.781	5.000
6	1.526	1.982	2.550	3.321	3.847
7	1.546	1.887	2.497	3.008	3.662
8	1.409	1.944	2.249	2.985	3.450
9	1.408	1.726	2.330	2.757	3.260
10	1.394	1.845	2.248	2.705	3.097
11	1.374	1.836	2.173	2.642	3.070
12	1.352	1.747	2.179	2.662	2.929
13	1.384	1.792	2.171	2.585	2.966
14	1.353	1.712	2.096	2.576	2.906
15	1.368	1.737	2.131	2.558	2.846
16	1.336	1.752	2.105	2.533	2.840
17	1.342	1.681	2.078	2.504	2.824
18	1.344	1.723	2.049	2.513	2.802
19	1.343	1.723	2.055	2.477	2.776
20	1.339	1.719	2.055	2.474	2.783
21	1.333	1.712	2.050	2.467	2.750
22	1.326	1.703	2.043	2.456	2.746
23	1.318	1.693	2.034	2.469	2.738
24	1.309	1.704	2.045	2.451	2.726
25	1.320	1.690	2.031	2.432	2.736
26	1.308	1.696	2.037	2.434	2.718
27	1.315	1.699	2.039	2.432	2.720
28	1.320	1.682	2.021	2.428	2.718
29	1.307	1.683	2.020	2.422	2.714
30	1.309	1.681	2.017	2.414	2.689

Critical Values for Wilcoxon's Rank-Sum Test

One Tailed	.100	.050	.025	.010	.005
Two Tailed	.200	.100	.050	.020	.010
$n_1 = n_2$					
31	1.310	1.679	2.012	2.421	2.698
32	1.310	1.689	2.022	2.410	2.688
33	1.309	1.684	2.014	2.412	2.692
34	1.294	1.678	2.020	2.413	2.679
35	1.305	1.672	2.010	2.412	2.679
36	1.301	1.677	2.013	2.409	2.677
37	1.298	1.680	2.002	2.405	2.673
38	1.305	1.682	2.002	2.400	2.669
39	1.300	1.673	1.991	2.394	2.663
40	1.305	1.674	2.000	2.398	2.667
41	1.299	1.674	1.997	2.390	2.659
42	1.303	1.673	2.004	2.391	2.660
43	1.296	1.671	1.999	2.391	2.660
44	1.299	1.669	1.994	2.391	2.658
45	1.300	1.667	1.998	2.389	2.647
46	1.293	1.672	1.992	2.378	2.644
47	1.294	1.668	2.001	2.383	2.649
48	1.287	1.672	1.994	2.379	2.652
49	1.294	1.667	1.995	2.382	2.647
50	1.294	1.669	1.994	2.377	2.648
51	1.286	1.664	1.993	2.379	2.642
52	1.291	1.665	1.991	2.380	2.641
53	1.296	1.666	1.989	2.373	2.641
54	1.294	1.667	1.993	2.372	2.639
55	1.292	1.666	1.990	2.371	2.637
56	1.296	1.660	1.987	2.376	2.640
57	1.293	1.665	1.983	2.374	2.637
58	1.296	1.664	1.991	2.371	2.633
59	1.293	1.662	1.986	2.368	2.629
60	1.289	1.666	1.987	2.370	2.635

Critical Values for the Kruskal-Wallis Test (expressed as *F* statistic)

Critical Values for the Kruskal-Wallis Test, $\alpha = .10$

n Per Group	Number of Groups								
	2	3	4	5	6	7	8	9	10
3	3.501	4.105	3.170	2.724	2.472	2.289	2.164	2.069	1.989
4	8.400	3.300	2.733	2.438	2.247	2.109	2.008	1.930	1.865
5	4.313	2.898	2.520	2.290	2.131	2.018	1.930	1.860	1.804
6	3.929	2.731	2.413	2.209	2.069	1.966	1.885	1.822	1.769
7	3.562	2.650	2.355	2.162	2.029	1.932	1.856	1.795	1.745
8	3.778	2.621	2.315	2.130	2.003	1.908	1.835	1.778	1.729
9	3.380	2.562	2.280	2.104	1.981	1.891	1.821	1.763	1.716
10	3.406	2.534	2.257	2.085	1.966	1.877	1.809	1.752	1.707
11	3.082	2.506	2.238	2.071	1.954	1.866	1.798	1.745	1.699
12	3.054	2.486	2.223	2.059	1.945	1.858	1.790	1.738	1.692
13	3.213	2.469	2.212	2.048	1.935	1.851	1.784	1.731	1.687
14	3.117	2.452	2.201	2.041	1.928	1.845	1.780	1.726	1.683
15	3.018	2.443	2.192	2.034	1.923	1.839	1.775	1.722	1.679
16	3.068	2.437	2.185	2.028	1.918	1.835	1.765	1.719	1.676
17	2.825	2.421	2.177	2.023	1.913	1.832	1.764	1.716	1.673
18	2.970	2.418	2.172	2.017	1.908	1.827	1.761	1.713	1.671
19	2.857	2.409	2.167	2.014	1.906	1.825	1.758	1.711	1.668
20	2.954	2.404	2.163	2.011	1.902	1.823	1.759	1.709	1.666
21	2.837	2.401	2.159	2.008	1.898	1.820	1.757	1.706	1.665
22	2.900	2.394	2.155	2.003	1.896	1.817	1.756	1.705	1.663
23	2.865	2.389	2.152	2.001	1.894	1.815	1.754	1.704	1.662
24	2.827	2.387	2.148	1.998	1.892	1.813	1.752	1.702	1.661
25	2.856	2.383	2.146	1.995	1.891	1.812	1.750	1.701	1.659
26	2.876	2.377	2.141	1.994	1.888	1.811	1.748	1.699	1.658
27	2.886	2.376	2.141	1.992	1.887	1.808	1.747	1.698	1.657
28	2.830	2.374	2.140	1.990	1.886	1.807	1.747	1.697	1.656
29	2.831	2.370	2.137	1.989	1.885	1.806	1.746	1.697	1.655
30	2.827	2.367	2.134	1.987	1.883	1.805	1.744	1.694	1.654

Critical Values for the Kruskal-Wallis Test, α=.05

n Per Group	Number of Groups								
	2	3	4	5	6	7	8	9	10
3	—	7.000	4.483	3.640	3.226	2.941	2.718	2.561	2.431
4	19.200	4.826	3.727	3.187	2.854	2.628	2.463	2.339	2.238
5	6.224	4.072	3.377	2.950	2.673	2.482	2.342	2.230	2.142
6	6.500	3.885	3.183	2.812	2.574	2.402	2.271	2.171	2.089
7	5.185	3.646	3.073	2.731	2.508	2.348	2.226	2.130	2.052
8	5.059	3.545	2.998	2.678	2.464	2.309	2.194	2.103	2.028
9	5.428	3.480	2.945	2.634	2.430	2.284	2.171	2.080	2.008
10	5.055	3.415	2.904	2.604	2.404	2.260	2.151	2.065	1.993
11	4.722	3.370	2.870	2.580	2.384	2.243	2.136	2.053	1.981
12	4.428	3.329	2.844	2.560	2.369	2.230	2.123	2.042	1.972
13	4.713	3.295	2.821	2.543	2.354	2.219	2.113	2.032	1.964
14	4.394	3.275	2.804	2.529	2.343	2.210	2.106	2.024	1.958
15	4.327	3.257	2.790	2.517	2.335	2.201	2.099	2.018	1.951
16	4.432	3.231	2.778	2.508	2.326	2.194	2.085	2.013	1.948
17	4.317	3.213	2.764	2.499	2.318	2.188	2.083	2.008	1.943
18	4.354	3.199	2.756	2.491	2.313	2.182	2.078	2.004	1.939
19	4.222	3.189	2.747	2.484	2.308	2.178	2.073	2.000	1.936
20	4.221	3.179	2.739	2.479	2.302	2.174	2.075	1.998	1.933
21	4.204	3.172	2.730	2.473	2.296	2.170	2.072	1.994	1.930
22	4.175	3.162	2.725	2.467	2.293	2.165	2.070	1.991	1.927
23	4.135	3.151	2.722	2.463	2.288	2.162	2.066	1.989	1.926
24	4.089	3.146	2.714	2.458	2.286	2.159	2.063	1.986	1.924
25	4.126	3.141	2.709	2.455	2.283	2.158	2.061	1.985	1.921
26	4.148	3.133	2.703	2.451	2.279	2.156	2.059	1.983	1.920
27	4.158	3.127	2.700	2.450	2.276	2.153	2.056	1.981	1.918
28	4.084	3.125	2.700	2.446	2.275	2.150	2.056	1.980	1.917
29	4.079	3.116	2.695	2.442	2.274	2.149	2.054	1.979	1.916
30	4.067	3.112	2.691	2.440	2.270	2.147	2.052	1.976	1.913

Critical Values for the Kruskal-Wallis Test, $\alpha = .025$

n Per Group	Number of Groups								
	2	3	4	5	6	7	8	9	10
3	—	8.739	6.133	4.792	4.084	3.642	3.315	3.081	2.895
4	—	6.789	4.860	3.995	3.501	3.171	2.936	2.758	2.615
5	9.188	5.570	4.304	3.644	3.236	2.959	2.758	2.602	2.479
6	11.029	5.110	4.016	3.443	3.092	2.843	2.656	2.516	2.404
7	9.046	4.750	3.842	3.321	2.994	2.763	2.592	2.459	2.353
8	7.735	4.589	3.717	3.239	2.930	2.708	2.547	2.421	2.318
9	6.078	4.447	3.636	3.173	2.880	2.671	2.514	2.389	2.290
10	6.680	4.370	3.568	3.127	2.842	2.637	2.487	2.368	2.270
11	6.469	4.292	3.523	3.092	2.813	2.614	2.466	2.349	2.254
12	6.221	4.226	3.480	3.063	2.790	2.595	2.447	2.335	2.241
13	6.315	4.164	3.444	3.037	2.769	2.579	2.434	2.322	2.230
14	6.009	4.126	3.420	3.017	2.752	2.566	2.424	2.311	2.221
15	5.727	4.095	3.397	2.998	2.740	2.555	2.413	2.303	2.212
16	5.926	4.061	3.379	2.986	2.728	2.544	2.393	2.294	2.207
17	5.628	4.034	3.356	2.972	2.716	2.535	2.389	2.289	2.201
18	5.728	4.012	3.344	2.960	2.708	2.527	2.382	2.283	2.195
19	5.609	3.997	3.333	2.950	2.699	2.522	2.376	2.278	2.191
20	5.641	3.980	3.320	2.939	2.692	2.515	2.380	2.274	2.186
21	5.643	3.963	3.306	2.935	2.684	2.508	2.375	2.269	2.183
22	5.624	3.954	3.300	2.927	2.680	2.503	2.371	2.266	2.180
23	5.588	3.935	3.292	2.918	2.673	2.499	2.366	2.262	2.177
24	5.539	3.928	3.281	2.913	2.671	2.494	2.363	2.258	2.174
25	5.482	3.917	3.275	2.908	2.665	2.492	2.359	2.256	2.171
26	5.516	3.902	3.267	2.903	2.661	2.489	2.358	2.253	2.170
27	5.440	3.897	3.261	2.898	2.658	2.486	2.354	2.251	2.166
28	5.448	3.890	3.259	2.893	2.654	2.482	2.352	2.248	2.165
29	5.443	3.881	3.253	2.889	2.652	2.479	2.350	2.248	2.163
30	5.351	3.874	3.247	2.885	2.648	2.477	2.347	2.244	2.160

Critical Values for the Kruskal-Wallis Test, $\alpha = .01$

n Per Group	Number of Groups								
	2	3	4	5	6	7	8	9	10
3	—	12.882	8.773	6.630	5.422	4.685	4.189	3.838	3.558
4	—	10.293	6.502	5.188	4.431	3.940	3.593	3.334	3.131
5	25.000	7.953	5.681	4.644	4.019	3.615	3.322	3.103	2.932
6	14.798	7.025	5.213	4.326	3.802	3.437	3.175	2.975	2.819
7	10.969	6.489	4.931	4.133	3.653	3.321	3.081	2.893	2.746
8	11.905	6.115	4.730	4.006	3.556	3.241	3.013	2.836	2.697
9	10.629	5.894	4.601	3.906	3.483	3.186	2.964	2.792	2.657
10	9.593	5.717	4.489	3.837	3.428	3.137	2.927	2.762	2.627
11	9.427	5.584	4.418	3.780	3.383	3.103	2.894	2.734	2.606
12	8.580	5.477	4.349	3.736	3.345	3.072	2.867	2.714	2.586
13	8.799	5.391	4.292	3.698	3.316	3.051	2.851	2.695	2.571
14	8.055	5.322	4.255	3.667	3.290	3.030	2.833	2.679	2.557
15	8.100	5.261	4.217	3.639	3.273	3.015	2.820	2.668	2.547
16	8.065	5.219	4.190	3.621	3.255	3.000	2.789	2.658	2.537
17	7.977	5.178	4.156	3.595	3.237	2.984	2.785	2.648	2.530
18	7.854	5.136	4.134	3.579	3.225	2.973	2.773	2.639	2.521
19	7.709	5.099	4.116	3.566	3.210	2.966	2.765	2.633	2.516
20	7.746	5.075	4.096	3.552	3.203	2.955	2.771	2.627	2.509
21	7.742	5.058	4.077	3.545	3.189	2.948	2.762	2.620	2.506
22	7.542	5.033	4.068	3.532	3.182	2.939	2.759	2.616	2.499
23	7.496	5.005	4.052	3.520	3.174	2.934	2.753	2.610	2.495
24	7.432	4.991	4.037	3.511	3.168	2.928	2.749	2.607	2.492
25	7.484	4.976	4.027	3.500	3.162	2.923	2.743	2.602	2.488
26	7.387	4.954	4.016	3.497	3.159	2.920	2.741	2.600	2.486
27	7.397	4.944	4.008	3.487	3.148	2.913	2.736	2.595	2.481
28	7.388	4.929	4.002	3.481	3.147	2.908	2.731	2.589	2.480
29	7.263	4.916	3.993	3.473	3.145	2.905	2.728	2.589	2.476
30	7.233	4.905	3.983	3.469	3.137	2.903	2.724	2.585	2.471

Critical Values for the Kruskal-Wallis Test, $\alpha = .005$

n Per Group	Number of Groups								
	2	3	4	5	6	7	8	9	10
3	—	27.000	11.285	8.106	6.591	5.593	4.936	4.459	4.100
4	—	12.000	7.930	6.205	5.218	4.574	4.120	3.794	3.534
5	99.900	10.092	6.813	5.443	4.650	4.135	3.768	3.494	3.277
6	20.643	8.614	6.204	5.031	4.366	3.902	3.575	3.331	3.132
7	16.608	7.938	5.818	4.782	4.174	3.754	3.458	3.220	3.043
8	13.836	7.382	5.548	4.611	4.045	3.650	3.368	3.152	2.981
9	13.364	7.022	5.379	4.485	3.946	3.579	3.304	3.095	2.932
10	12.598	6.851	5.215	4.388	3.876	3.518	3.259	3.058	2.896
11	10.942	6.636	5.121	4.317	3.818	3.472	3.218	3.022	2.869
12	11.029	6.505	5.034	4.259	3.772	3.436	3.184	2.997	2.843
13	10.339	6.384	4.956	4.208	3.733	3.408	3.163	2.975	2.824
14	10.186	6.282	4.901	4.166	3.701	3.377	3.139	2.954	2.809
15	9.958	6.195	4.854	4.134	3.676	3.360	3.123	2.941	2.795
16	10.049	6.134	4.819	4.109	3.656	3.346	3.087	2.927	2.781
17	9.719	6.082	4.773	4.078	3.630	3.321	3.078	2.915	2.771
18	9.673	6.016	4.744	4.052	3.617	3.310	3.067	2.905	2.764
19	9.324	5.975	4.721	4.037	3.598	3.296	3.055	2.898	2.757
20	9.443	5.939	4.688	4.018	3.588	3.287	3.063	2.889	2.746
21	9.289	5.919	4.669	4.007	3.572	3.276	3.053	2.879	2.743
22	9.120	5.875	4.655	3.988	3.558	3.265	3.047	2.875	2.734
23	9.116	5.835	4.638	3.976	3.548	3.258	3.038	2.867	2.730
24	9.081	5.820	4.618	3.967	3.543	3.251	3.035	2.862	2.726
25	9.021	5.807	4.604	3.954	3.537	3.245	3.027	2.857	2.722
26	8.944	5.770	4.594	3.944	3.534	3.239	3.023	2.856	2.719
27	8.982	5.764	4.580	3.929	3.517	3.232	3.016	2.849	2.710
28	8.750	5.738	4.569	3.925	3.518	3.227	3.012	2.840	2.711
29	8.868	5.718	4.557	3.912	3.511	3.221	3.009	2.842	2.706
30	8.739	5.704	4.543	3.909	3.505	3.218	3.002	2.838	2.701

APPENDIX J

Case Studies[1]

A: Contact Lens Wear

Coles et al. [6] addressed the question of whether new soft contact lenses treated with a conditioning solution would provide a higher degree of patient comfort than would similar lenses without such treatment. To this end,

> Sixty-one experienced contact lens wearers with no unusual ocular characteristics were recruited from the subject database of Breman Consultants and by advertisement in local newspapers. To be eligible to participate in the study, subjects were at least 18 years of age and had been successfully wearing soft contact lenses for at least one month prior to the study. Myopes up to -12.00 D with less then 0.75 D of anisometropia were included.

Data for 59 of the subjects, 24 males and 35 females, was considered appropriate for analysis. The average age of these subjects was 31.5 with standard deviation of 10.2.

Patients were categorized as belonging to one of two groups, those who upon entry into the study answered "yes" to the question "Are you able to wear your lenses for as long as you want?" and those who answered negatively. Patients in both categories had a conditioned lens placed in a randomly chosen eye with an unconditioned lens being placed in the remaining eye. The order of insertion i.e. conditioned lens first or unconditioned lens first was also randomized. Patients were masked as to which of the lenses was conditioned.

Lenses were inserted during a morning visit (a.m.) to the research facility. After insertion subjects were asked which lens felt more comfortable. Approximately eight hours later subjects returned for an afternoon visit (p.m.) at which time they were asked again to choose the more comfortable lens. Patients were not allowed to indicate "no difference" so that forced choice designations were required. The results for this portion of the study are given below.

In the table, no signifies no to duration question, yes signifies yes to duration question, treated lens signifies chose lens treated with conditioning solution, untreated lens signifies chose lens with no conditioning solution, a.m. signifies assessment during morning visit, and p.m. signifies assessment during afternoon visit.

In addition to the above, a number of ocular health variables were assessed for each eye on each visit. These measurements constitute a continuous scale.

---

[1]Descriptions of studies in this section are abbreviated and otherwise simplified for reasons of space and nonrelevance to the material in this book. The reader should refer to the original publication for details.

TABLE J.1: Counts of lens preferences.

	No		Yes	
	Treated Lens	Untreated Lens	Treated Lens	Untreated Lens
a.m.	10	2	23	24
p.m.	7	5	26	21

TABLE J.2: Means and standard deviations of WOMAC A pain scores at baseline and 12 weeks.

	Standard		Weak		Placebo	
Time	$\bar{x}_1$	s_1	$\bar{x}_2$	s_2	$\bar{x}_3$	s_3
Baseline	10.7	2.1	11.0	2.0	10.9	2.1
12 Weeks	7.8	3.9	8.8	3.2	9.3	3.2

B: Magnetic Bracelets for Relieving Pain of Osteoarthritis

Harlow et al. [21] conducted a clinical trial to study the efficacy of magnetic bracelets for relieving pain associated with osteoarthritis of the knee or hip. To this end, subjects were randomly assigned to one of three groups. The first group was provided standard strength magnetic bracelets (standard treatment), the second a weakened magnetic bracelet (weak treatment), and the third an unmagnetized bracelet (dummy or placebo). Subjects were masked as to the type of bracelet received.

The primary outcome was a (continuous) score on the Western Ontario and McMaster Universities osteoarthritis lower limb pain scale (WOMAC A). Higher scores on this scale are indicative of more pain than are lower scores. Pain assessment was made at baseline and at the end of 12 weeks of wear.

Participants in the study ranged in age from 45 to 80 years and were recruited from five rural general medical practices in Mid Devon, England. All subjects were diagnosed as having osteoarthritis of the knee or hip.

Means and standard deviations of the ages of the randomly formed groups were as follows. Standard treatment group, $\bar{x} = 66.6$, $s = 8.4$, weak treatment group, $\bar{x} = 66.8$, $s = 8.3$, dummy or placebo group, $\bar{x} = 66.3$, $s = 9.1$.

Means and standard deviations of WOMAC A pain scores at baseline and 12 weeks are given in Table J.2. Sample sizes for the three groups are $n_1 = 65$, $n_2 = 64$, and $n_3 = 64$.

In order to evaluate the effectiveness of the masking of subjects as to the type of bracelet they were wearing, at the end of the trial each subject was asked to indicate whether they believed they were wearing a real or dummy bracelet. The results are reported in Table J.3.[2]

A correct answer for members of the Standard group would be "Real" while for the Placebo group the correct answer would be "Dummy." Because the magnet placed in the

---

[2]Ten subjects were lost to followup and do not appear in this table.

TABLE J.3: Indications of subjects as to whether they believed they were wearing real or dummy bracelets.

Belief	Standard	Weak	Placebo
Real	35	24	10
Dummy	5	12	30
Don't Know	22	23	22

TABLE J.4: Number of deaths in invasive and non-invasive treatment groups for myocardial infarction at various points in time.

	Invasive		Non-Invasive	
Time	Dead	Alive	Dead	Alive
discharge	21	441	6	452
one month	23	439	9	449
one year	58	404	36	422
23 months	80	382	59	399

bracelets of the Weak group were believed to be too weak for any therapeutic value, a correct answer for members of this group would also be "Dummy."

C: Invasive Versus Non-Invasive Treatment of Myocardial Infarction

Boden et al. [3] randomly assigned 920 patients who had recently suffered a myocardial infarction to treatment by the standard invasive method ($n = 462$) or a non-invasive method ($n = 458$). One of the primary outcomes was whether the patient was still alive at specified periods of time after treatment. Table J.4 shows the number of patients dead and alive at time of hospital discharge, one month after treatment, one year after treatment, and 23 months after treatment.

D: Correlation Between Brain Volume and Neurocognitive Performance in HIV-Infected Patients

Patel et al. [37] conducted a study whose intent was to determine whether neuropsychological function in HIV-infected persons is correlated with loss of brain volume. Neuropsychological performance was assessed by application of a neuropsychological functioning (NPZ-8) test battery while brain volume was measured as percentage of brain parenchymal volume obtained from Magnetic Resonance imaging (MRI). The researchers hypothesized that a relationship exists between these two measures even before overt clinical dysfunction is evident.

Table J.5 shows measures of brain parenchymal volume (PBV), neuropsychological function (NPZ-8), HIV status (positive or negative), whether or not the patient has been diagnosed as manifesting AIDS dementia complex (ADC) Stage and immune system status (CD4) for the subjects involved in the study. Lower PBV values indicate less brain volume

TABLE J.5: Measures of brain volume, neuropsychological function and immune system status in 15 HIV positive and 5 HIV negative subjects.

Subject	PBV	NPZ-8	HIV Status	ADC Stage	CD4
1	.791	12	+	+	16
2	.782	5	+	+	324
3	.646	3	+	+	256
4	.740	5	+	+	563
5	.804	2	+	+	321
6	.858	6	+	+	190
7	.729	8	+	+	818
8	.803	0	+	−	355
9	.831	3	+	−	465
10	.826	0	+	−	519
11	.786	0	+	−	87
12	.882	7	+	−	108
13	.889	3	+	−	190
14	.917	0	+	−	573
15	.885	0	+	−	1032
16	.886	0	−		
17	.833	0	−		
18	.851	1	−		
19	.897	0	−		
20	.901	0	−		

while lower NPZ-8 scores indicate better neuropsychological function. A positive ADC value indicates that a diagnosis of AIDS dementia complex has been made. For AIDS patients, higher CD4 counts are associated with better immune system function.

E: Screening Test for Acute Pancreatitis

Acute pancreatitis can be difficult to diagnose in a timely manner. To assist in this endeavor, Kemppainen et al. [25] developed a rapid dipstick screening test for pancreatitis based on the immunochromatographic measurement of urinary trypsinogen-2.

In order to ascertain the properties of the new method, the test was administered to 500 consecutive patients complaining of acute abdominal pain who came to the emergency departments of two hospitals. Each patient was subsequently evaluated via standard methods to determine whether he/she was suffering from acute pancreatitis.

Of the 53 patients diagnosed as suffering from acute pancreatitis, 50 tested positive on the new test while 3 tested negative. Of the 447 patients diagnosed as not having acute pancreatitis, 21 tested positive while 426 tested negative.

F: Reinfection as a Cause of Recurrent Tuberculosis

It has long been assumed that recurrent tuberculosis is usually caused by reactivation of endogenous infection rather than by a new, exogenous infection. In order to test the validity of this assumption, van Rie et al. [48] performed DNA analysis on pairs of isolates of *Mycobacterium tuberculosis* from 16 patients who had a relapse of pulmonary tuberculosis after curative treatment of postprimary tuberculosis. The patients lived in areas of South Africa where tuberculosis is endemic.

DNA analysis showed that for 12 of the 16 patients, reinfection was related to a different causative agent rather than to recurrence of the original causative agent. The researchers concluded that "Exogenous reinfection appears to be a major cause of postprimary tuberculosis after a previous cure in an area with a high incidence of this disease."

G: Multiple Sclerosis in Twins

The relative role of environmental and genetic factors in the determination of susceptibility to multiple sclerosis has not been clearly determined. To assist in this task, Ebers et al. [11] surveyed the records of 5,463 patients attending 10 multiple sclerosis clinics across Canada. It was found that 70 of these patients had a twin (27 were monozygotic[3] and 43 were dizygotic[4]). Additionally, 4,582 patients were studied who did not have a twin but did have a sibling. It was found that seven of the monozygotic twins had a twin who also had the disease while one of the 43 dizygotic twins had a twin with the disease. Of the 4,582 nontwin patients, 87 had a sibling with the disease. (We will assume that the nontwin patients had but a single sibling.)

H: A Comparison of Two Treatments for Nephropathic Cystinosis

Smolin et al. [43] compared the effectiveness of two drugs, Cysteamine (MEA) and Phosphocysteamine (MEAP), for treating children with nephropathic cystinosis. MEA is currently used for treatment of this disease but MEAP has the advantage of tasting and smelling better and may be, therefore, more palatable to children. A primary outcome of the study was to determine whether a difference exists between the ability of the two drugs to lower leukocyte cystine content.

The study was conducted by administering the two drugs to six children. The authors report that "The percent decrease in leukocyte cystine content obtained with MEA administration (61.9%) was not significantly different from the decrease observed when MEAP was administered (65.3%)." They go on to conclude "MEA and MEAP appear to be equally effective in their cystine-depleting properties."

I: A Comparison of Two Treatments for Acute Myocardial Infarction

Rapid lysis (desolving) of coronary-artery thrombi (obstructions) is desirable in treating patients who experience an acute myocardial infarction. The Continuous Infusion versus Double-Bolus Administration of Alteplase (COBALT) Investigators [46] compared the

---

[3]From a single fertilized ovum.
[4]From two different ova.

death rates associated with two methods of lysis of the thrombi. The first method, acceler-ated infusion (AI) requires about 90 minutes while the double-bolus (D-B) method requires an even shorter period.

The study was conducted by randomly assigning 7,169 patients to the two treatments with the D-B group having 3,685 and the AI group having 3,684 patients assigned. The outcome of interest was death by any cause during the 30 day period after treatment. The purpose of the study was to determine whether D-B could be considered therapeutically equivalent to AI insofar as death after treatment is concerned. More precisely, the researchers would consider D-B at least as good as AI if the proportion of deaths associated with D-B treatment exceeds the proportion under AI treatment by less than .004. The proportion of deaths in the D-B group was .0798 while the proportion in the AI group was .0753.

J: Oral Versus Intravenous Treatment for Fever in Cancer Patients

Standard therapy for treatment of fever in cancer patients with granulocytopenia (low white blood cell counts) involves intravenously administered antimicrobial agents. If orally ad-ministered therapy could be shown to be as effective as intravenously administered therapy, marked improvements in quality of life and costs of therapy could be attained.

In order to determine whether such equivalence can be established, Kern et al. [26] con-ducted a study in which cancer patients suffering from fever and granulocytopenia were randomly assigned to oral and intravenous treatments. All patients were hospitalized and prescribed treatments administered. A treatment was deemed successful for a particular pa-tient if all the following were attained without a change in the treatment regimen, "... the temperature was normal for at least three consecutive days (or two days for patients with unexplained fever and rapid recovery from granulocytopenia), the symptoms and signs of infection at identifiable sites of infection had disappeared, the primary pathogen had been eradicated, and the primary documented infection had not recurred within one week after the end of treatment."

Results showed that 138 of 161 patients receiving oral therapy and 127 of 151 patients receiving intravenous therapy were deemed a success. The researchers decided that equiv-alence would be declared if the absolute difference between the proportions of successes attributed to the two treatment regimens was .10 or less.

K: Farm Safety Education

Farming is one of the most hazardous occupations in the U.S. Tractor overturns alone ac-count for approximately 132 deaths per year. Additionally, about 500 agricultural work-ers suffer disabling injures daily. Unfortunately, on average 104 adolescents aged 19 and younger are killed in farm accidents annually.

Liller et al. [32] conducted a large scale effort to educate young people regarding hazards and safety issues related to farming. As part of that effort, a curriculum focused on farm safety was developed. With the cooperation of the local school system, five schools located in primarily agricultural areas of the county were chosen for study. Over an extended period, the farm safety lessons were presented to all fifth graders in the five schools. The participat-ing fifth graders were administered a specially developed farm safety paper and pencil test before and after presentation of the lessons. All papers were graded and the average pre- and

TABLE J.6: Mean pre- and posttest scores on a farm safety test for five schools participating in a farm safety study.

School	Pretest Mean	Posttest Mean
1	5.93	8.68
2	5.65	7.83
3	6.84	8.09
4	5.68	7.78
5	6.05	8.03

TABLE J.7: Severity of injuries sustained by school employees classified by age.

Age	Minor Injury	Moderate Injury	Severe Injury
< 30	255	57	14
30 − 39	332	105	44
40 − 49	533	201	46
50+	561	205	66

posttest score *for each school* was recorded. These average scores are given in Table J.6.

L: School Nurse Impact Assessment

Perrin, Goad, and Williams [38] assessed the economic impact of employing on site nurses at high schools in a large Southeastern school system. The nurses were authorized to treat staff as well as students. As part of their evaluation, they examined the relationship between the age of employees suffering an on the job injury and the severity of the sustained injury. The data collected in this regard are provided in Table J.7.

Minor injures were defined as injures requiring no treatment, moderate injures were those requiring treatment but no hospitalization, and severe injures were those requiring hospitalization.

M: Treating Sleep Disorders in Mountain Climbers

A common complaint among mountain climbers is their inability to acquire sufficient sleep once they reach moderate to high altitudes. Sleeplessness can be treated with certain classes of drugs but questions have arisen as to whether such drugs can interfere with respiration—an important consideration at such altitudes.

Röggla, Moser and Röggla [40] conducted a study designed to assess the effect of temazepam on the oxygen and carbon dioxide levels of mountain climbers at 171 and 300 meters altitude. Oxygen and carbon dioxide levels were assessed on seven mountain climbers before and one hour after administration of the drug. Results obtained before and after taking

TABLE J.8: Oxygen and carbon dioxide levels of seven mountain climbers before and after taking a sleep medication.

Subject	Oxygen Before	Oxygen After	Carbon Dioxide Before	Carbon Dioxide After
1	9.3	8.6	4.3	4.4
2	8.6	8.4	4.4	4.7
3	8.9	8.2	4.4	4.7
4	9.1	8.3	4.0	4.3
5	8.5	8.1	4.1	4.4
6	9.1	8.1	4.4	4.7
7	9.4	8.4	4.0	4.8

TABLE J.9: Potential risk factors for sexual abuse by sexually abused adolescents.

Risk Factor	Abused Only (n=14)	Abused and Abuser (n=11)
Experiencing intrafamilial violence	5	10
Witnessing intrafamilial violance	5	9
Rejection by family	8	10
Rejection by peers	3	5

the medication at 300 meters are shown in Table J.8.

N: Risk Factors Related to Sexual Abuse by Sexually Abused Adolescent Boys

A large proportion of sexually abused adolescent boys become abusers in their own right while a sizable proportion do not. Skuse et al. [42] investigated the risk factors associated with abusive behavior on the part of abused adolescent boys. In their study they identified 25 sexually abused adolescent males between the ages of 11 years and 15 years and 11 months, 11 of whom had abused other boys and 14 of whom had not. Risk factors investigated included, experiencing intrafamilial violence, witnessing intrafamilial violence, rejection by family, poor identification with a father figure and others.

Partial results from their study are shown in Table J.9.

The first column in this table shows the potential risk factor while the second and third columns show the number of subjects in the two groups that experienced that particular factor.

O: How to Read a Paper

In an article titled "How to read a paper: Statistics for the non-statistician. I: Different types of data need different statistical tests," Greenhalgh [19] provides guidelines (or advice) to medical researchers on the use of statistical procedures. A number of points about this paper are noteworthy.[5]

1. The author provides examples of several measurement scales and notes that, "Numbers are often used to label the properties of things." As an example numbers might be assigned to represent height and weight. By contrast, numbers might also be used to label the property "city of origin" where 1 = London, 2 = Manchester, 3 = Birmingham, and so on. Finally, numbers might be assigned to represent "liking for x" where 1 = not at all, 2 = a bit, and 3 = a lot.

2. The author states, "In general, parametric tests are more powerful than non-parametric ones and so should be used if possible."

3. The author states, "Non-parametric tests look at the rank order of the values (which one is the smallest, which one comes next, and so on) and ignores the absolute differences between them. As you might imagine, statistical significance is more difficult to show with non-parametric tests⋯."

4. After discussing the normal curve, the author notes that some biological variables such as body weight show a "skew normal" distribution.

5. The author states, "Using tests based on the normal distribution to analyse non-normally distributed data, however, is definitely cheating."

6. The author describes the Kruskal-Wallis test as, "Effectively, a generalisation of the paired t or Wilcoxon matched pairs test where three or more sets of observations are made on a single sample"

7. An example situation in which the Kruskal-Wallis test might be used is given as "To determine whether plasma glucose level is higher one hour, two hours, or three hours after a meal."

8. The author indicates that Spearman's rank correlation, "Assesses the strength of the straight line association between two continuous variables."

9. While extensively discussing the normality assumption, the author does not address such issues as homoscedasticity and independence.

---

[5]The characterizations made here are intended to help you understand statistical issues and are not intended as an evaluation of this article. The usefulness of the comments made in this paper should be evaluated only after reading the article in its entirety.

TABLE J.10: Diabetic patients in Tayside, Scotland broken down by type of diabetes, gender, and age.

	Type I		Type II	
Age	Male	Female	Male	Female
0-14	29	42	0	0
15-24	70	65	0	0
25-44	252	157	22	16
45-65	90	74	163	152
> 65	10	18	195	242
$\sum$	451	356	380	410

P: Blood Glucose Monitoring

Evans et al. [12] conducted a study whose purpose was, in part, to investigate patterns of self monitoring of blood glucose concentration in type I and type II diabetic patients. In the course of that study the authors constructed a database representing the population of diabetic patients in Tayside, Scotland. Among other variables, the database showed numbers of diabetic patients broken down by type of diabetes (type I or type II), age, and gender. The patient categorizations are shown in Table J.10.

APPENDIX K

Answers to Exercises

Chapter One

1.1 (a) The organization and summarization of data and (b) the drawing of inferences about populations based on samples.

1.3 Data refers to the recordings of measurements made on characteristics.

A. The following answers refer to Case Study A.

1.5 The sample consists of the 24 male and 35 female participants in the study. This sample is best characterized as being from a popular population. We can characterize the counts of lens preferences in Table J.1 as a sample from a statistical population.

1.7 The popular population most likely consists of males and females who are at least 18 years of age who have no unusual ocular characteristics and have been successfully wearing soft contact lenses for at least one month. Myopes up to −12.00 D with less than 0.75 D of anisometropia are also present. The counts of lens preferences may be used to estimate lens preference in the statistical population.

1.9 No. The statistical population was not available for study so that no parameters can be obtained.

B. The following answers refer to Case Study B.

1.11 The sample consists of the $65 + 64 + 64 = 193$ subjects who participated in the study. Because this sample is characterized as consisting of people rather than characteristics, it is best thought of as being from a popular population. We also have three samples consisting of pain intensity values. These samples are from a statistical population.

1.13 The popular population likely consists of persons aged from about 45 to 80 who suffer from osteoarthritis of the hip or knee and who see general practice physicians in the Mid Devon area of England. The statistical population consists of pain whose intensity can be described as a value on the Western Ontario and McMaster Universities osteoarthritis lower limb pain scale. The mean and standard deviation of this statistical population are likely in the neighborhood of 67 and 8 or 9 respectively.

1.15 No parameters are provided because the statistical population was not available for study.

F. The following answers refer to Case Study F.

1.17 Probably not. These subjects lived in areas of South Africa where tuberculosis is endemic with high incidence. We cannot assume that the mechanism for reinfection found in an area of high incidence is the same as that found in an area of low incidence.

Chapter Two

2.1 (a) Ordinal. We can determine whether one nurse has more seniority than another but cannot tell how much more (or less).

(b) The sick days *variable* is continuous since days can be fractionalized to any level, e.g., 2.313 etc. However, the *data* given here could be considered discrete since only whole day measurements were recorded. The classification of data as continuous or discrete is sometimes a bit arbitrary.

(c)   i. 1, 8, 5

   ii. $2+9+1+0+5+4+6+7+8+8 = 50$

   iii. $6+3+7+8+9+2+8+9+6+8+5 = 71$

   iv. Here we are instructed to sum all scores. But we already summed the first 10 in 1(c)ii and the remainder in 1(c)iii so that the sum is $50+71 = 121$.

   v. Here we are instructed to sum *all* the *squared* values. That is, we must calculate $2^2 + 9^2 + \ldots + 8^2 + 5^2 = 853$.

(d) From the rules of summation on page 13 we know that

$$\sum_{i=1}^{n}(x_i + c) = \sum_{i=1}^{n} x_i + nc$$

so that for the problem at hand

$$\sum_{i=1}^{n}(x_i + 2) = \sum_{i=1}^{n} x_i + (21)(2) = \sum_{i=1}^{n} x_i + 42$$

(e)

Sick Days	Frequency	Cumulative Frequency	Relative Frequency	Cumulative Relative Frequency
9	3	21	.143	1.000
8	5	18	.238	.857
7	2	13	.095	.619
6	3	11	.143	.524
5	2	8	.095	.381
4	1	6	.048	.286
3	1	5	.048	.238
2	2	4	.095	.190
1	1	2	.048	.095
0	1	1	.048	.048

(f) The three graphs are shown in Figure K.1. Notice that we labeled the x axis with data values rather than upper and lower real limits of the data value intervals as we did with Figures 2.2, 2.3, and 2.4. This is common practice with ungrouped data but it would also be correct to use upper and lower real limit labels.

(g)   i. By Equation 2.1 we obtain

$$\bar{x} = \frac{\sum x}{n} = \frac{121}{21} = 5.762.$$

Placing the sick days in order from smallest to largest and applying Equation 2.3 yields

$$\text{Median (n odd)} = x_{\frac{n+1}{2}} = x_{\frac{21+1}{2}} = x_{11} = 6.$$

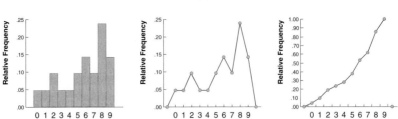

FIGURE K.1: Figures for Exercise 2.1f: (a) Relative Frequency (b) Relative Frequency (c) Cumulative Relative Frequency Polygon

From the discussion on page 31 and observation of the frequency distribution in exercise 1e we note that 8 is the most frequently occurring observation and is, therefore, the mode.

ii. By Equations 2.12 and 2.16

$$s^2 = \frac{\sum x^2 - \frac{(\sum x)^2}{n}}{n-1} = \frac{853 - \frac{(121)^2}{21}}{21-1} = 7.790$$

and

$$s = \sqrt{\frac{\sum x^2 - \frac{(\sum x)^2}{n}}{n-1}} = \sqrt{\frac{853 - \frac{(121)^2}{21}}{21-1}} = 2.791.$$

iii. By Equations 2.6 and 2.7

$$\text{Range (exclusive)} = x_L - x_S = 9 - 0 = 9$$

and

$$\text{Range (inclusive)} = URL_L - LRL_S = 9.5 - (-.5) = 10.$$

iv. The z scores given below were generated via Equation 2.23 using $\bar{x} = 5.762$ and $s = 2.791$. For example, nurse 1 had 2 sick days so that the associated z score is

$$z = \frac{x - \bar{x}}{s} = \frac{2.0 - 5.762}{2.791} = -1.348.$$

Nurse Number	z	Nurse Number	z	Nurse Number	z
1	-1.348	8	0.444	15	1.160
2	1.160	9	0.802	16	-1.348
3	-1.706	10	0.802	17	0.802
4	-2.064	11	0.085	18	1.160
5	-0.273	12	-0.990	19	0.085
6	-0.631	13	0.444	20	0.802
7	0.085	14	0.802	21	-0.273

v. By Equation 2.17 on page 38 and the frequency and cumulative frequency distributions constructed in exercise 1e, we obtain the following.

$$P_{15} = LRL + (w) \left[\frac{(pr)\,(n) - cf}{f} \right] = 1.5 + (1) \left[\frac{(.15)\,(21) - 2}{2} \right] = 2.075$$

$$P_{50} = 5.5 + (1) \left[\frac{(.5)\,(21) - 8}{3} \right] = 6.333$$

$$P_{80} = 7.5 + (1) \left[\frac{(.8)\,(21) - 13}{5} \right] = 8.26$$

Notice that $P_{50} = 6.33$, which constitutes one definition of the median, contrasts with the median value of 6 obtained by the cruder but more commonly used method employed in exercise 2.1(g)i.

vi. From the discussion on page 41 we know that the percentile rank will depend upon our definition of percentile rank as it relates to the three scores. Thus, we might use Equation 2.20, 2.21, or 2.22 for the calculation. We will use Equation 2.21 but use of either of the other two would also be correct.

$$PR_2 = 100 \left[\frac{(.5)\,(f) + cf}{n} \right] = 100 \left[\frac{(.5)\,(2) + 2}{21} \right] = 14.286 \text{ or } 14$$

$$PR_5 = 100 \left[\frac{(.5)\,(2) + 6}{21} \right] = 33.333 \text{ or } 33$$

$$PR_8 = 100 \left[\frac{(.5)\,(5) + 13}{21} \right] = 73.810 \text{ or } 74$$

vii. By Equations 2.25 and 2.26 on page 45

$$\text{Skew} = \frac{\sum z^3}{n} = \frac{-12.480}{21} = -.594$$

and

$$\text{Kurtosis} = \frac{\sum z^4}{n} = \frac{41.931}{21} = 1.997.$$

A. The following answers refer to Case Study A.

2.3 Yes, a.m./p.m., yes/no, treated/untreated. These are nominal level variables.

2.5 Because the mean age is of the sample of subjects used in the study, the proper symbol would be $\bar{x}$. We would use s to represent the sample (age) standard deviation.

2.7 By Equation 2.23

$$z = \frac{x - \bar{x}}{s}$$

so that the z scores for subjects aged 18 and 75 would be

$$z_{18} = \frac{18.0 - 31.5}{10.2} = -1.32$$

and

$$z_{75} = \frac{75.0 - 31.5}{10.2} = 4.26.$$

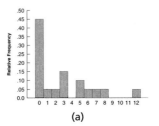 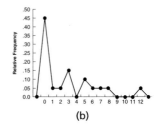

(a)   (b)

FIGURE K.2: Figures for Exercise 2.9: (a) NPZ-8 Relative Frequency Histogram (b) Frequency Polygon

D. The following answers refer to Case Study D.

2.9 See Figure K.2.

2.11 By Equation 2.1

$$\bar{x} = \frac{\sum x}{n} = \frac{16.537}{20} = .827.$$

By Equation 2.4

$$\text{Median (n even)} = \frac{x_{\frac{n}{2}} + x_{\frac{n}{2}+1}}{2} = \frac{x_{10} + x_{11}}{2} = \frac{.831 + .833}{2} = .832$$

or by the method demonstrated on page 30, we note that any point between .8315 and .8325 satisfies the definition of the median so that we take the midpoint of this interval or .832 as the median.

Because all scores occur with frequency one, there is no mode.

2.13 Noting that the first, second and third quartiles are, respectively, P_{25}, P_{50} and P_{75}, (see page 40) we calculate each of these values.

Because $(.25)(20) = 5$, P_{25} is the point on the scale below which 5 observations fall. Because zero observations fall below $-.5$ and nine fall below .5, the point below which 5 observations fall must be in the interval $-.5$ to .5 (i.e., the lower and upper real limits of the zero interval). Using this information with Equation 2.17 which states

$$P_p = LRL + (w)\left[\frac{(pr)(n) - cf}{f}\right]$$

yields

$$P_{25} = -.5 + (1)\left[\frac{(.25)(20) - 0}{9}\right] = .056.$$

Because $(.5)(20) = 10$ and 10 observations fall below the upper real limit of 1.5, $P_{50} = 1.5$.

Noting that $(.75)(20) = 15$ and that 14 observations fall below the upper real limit of 4.5 and 16 fall below 5.5, we employ Equation 2.17 as follows.

$$P_{75} = 4.5 + (1)\left[\frac{(.75)(20) - 14}{2}\right] = 5$$

2.15 Using the mean and standard deviation calculated in 2.10 and 2.12 and Equation 2.23 which states

$$z = \frac{x - \bar{x}}{s}$$

we calculate

$$z_0 = \frac{0 - 2.75}{3.432} = -.801$$

$$z_1 = \frac{1 - 2.75}{3.432} = -.510$$

$$z_2 = \frac{2 - 2.75}{3.432} = -.219$$

$$z_3 = \frac{3 - 2.75}{3.432} = .073$$

$$z_5 = \frac{5 - 2.75}{3.432} = .656$$

$$z_6 = \frac{6 - 2.75}{3.432} = .947$$

$$z_7 = \frac{7 - 2.75}{3.432} = 1.238$$

$$z_8 = \frac{8 - 2.75}{3.432} = 1.530$$

$$z_{12} = \frac{12 - 2.75}{3.432} = 2.695$$

Using Equation 2.25 and taking into account the frequency associated with each z value yields

$$\text{Skew} = \frac{\sum z^3}{n}$$

$$= \frac{(9)(-.801^3)+(-.510^3)+(-.219^3)+(3)(.073^3)+(2)(.656^3)+(.947^3)+(1.238^3)+(1.530^3)+(2.695^3)}{20}$$

$$= 1.085$$

This shows positive skew which was the conclusion reached by inspection in 2.14.

O. The following answers refer to Case Study O.

2.17 Height and weight are best characterized as ratio level measurements, city of origin as nominal and liking for x as ordinal.

Chapter Three

3.1   (a) By Equation 3.3 on page 56

$$P\left(\bar{I} \mid F\right) = \frac{P\left(\bar{I}F\right)}{P\left(F\right)} = \frac{.36}{.46} = .783.$$

(b) Reading from the table

$$P\left(F\right) = .46.$$

(c) By Equation 3.2 on page 56

$$P(M \cup I) = P(M) + P(I) - P(MI) = .54 + .32 - .22 = .64.$$

(d) Reading from the table

$$P(FI) = .10.$$

(e) By Equation 3.3 on page 56

$$P(F \mid I) = \frac{P(FI)}{P(I)} = \frac{.10}{.32} = .313.$$

(f) By Equation 3.2 on page 56

$$P(M \cup F) = P(M) + P(F) - P(MF) = .54 + .46 - .00 = 1.0.$$

(You have to be one or the other, right?)

3.3 From the information given we know that $P(F) = .50$, $P(T) = .20$ and $P(FT) = .10$ where F indicates female and T represents supporter of the tax increase. We note that $P(F)(T) = (.50)(.20) = .10$ which is equal to $P(FT)$ so that by 3.5 gender and support for the increase are independent.

We could just as well have applied 3.4 on page 56 by noting $P(T \mid F) = \frac{P(TF)}{P(F)} = \frac{.10}{.50} = .20$ which is equal to $P(T) = .20$. Would we have reached the same conclusion had we chosen to compare $P(F \mid T)$ to $P(F)$?

3.5 The probability table is as follows.

	B	$\overline{B}$	
A	.624	.176	.800
$\overline{A}$	.096	.104	.200
	.720	.280	

We begin by filling in the marginal values. We are given $P(\overline{A}) = .20$ so that $P(A) = 1 - P(\overline{A}) = 1 - .20 = .80$. Likewise, $P(B) = .72$ so that $P(\overline{B}) = 1 - .72 = .28$. We now turn attention to the joint probabilities. Because we know $P(\overline{B} \mid A) = \frac{P(A\overline{B})}{P(A)}$, we can multiply both sides by $P(A)$ to obtain

$$P(\overline{B} \mid A) P(A) = P(A\overline{B})$$

which is the joint probability we need. So we calculate $P(A\overline{B}) = (.22)(.80) = .176$. We can now use the marginals to fill in the other joint probabilities.

$$P(AB) = P(A) - P(A\overline{B}) = .800 - .176 = .624$$
$$P(\overline{A}B) = P(B) - P(AB) = .720 - .624 = .096$$
$$P(\overline{A}\,\overline{B}) = P(\overline{B}) - P(A\overline{B}) = .280 - .176 = .104$$

3.7 (a) By Equation 3.6 on page 58

$$\text{Sensitivity} = P(+ \mid D) = \frac{.065}{.070} = .929$$

(b) By Equation 3.7 on page 58

$$\text{Specificity} = P\left(- \mid \overline{D}\right) = \frac{.92}{.93} = .989$$

(c) By Equation 3.8 on page 58

$$PPV = P\left(D \mid +\right) = \frac{.065}{.075} = .867$$

(d) By Equation 3.9 on page 58

$$NPV = P\left(\overline{D} \mid -\right) = \frac{.920}{.925} = .995$$

3.9 Let U represent a person under 40, $\overline{U}$ a person 40 or over and S an inoculation supporter. Then from the information given we have $P\left(U\right) = .40$, $P\left(S \mid U\right) = .72$ and $P\left(S \mid \overline{U}\right) = .52$. We are asked to find $P\left(U \mid S\right)$.

We could proceed as in Exercise 3.5, but a more direct method is to apply Bayes rule as expressed by Equation 3.13 on page 61 which states

$$P\left(B \mid A\right) = \frac{P\left(A \mid B\right) P\left(B\right)}{P\left(A \mid B\right) P\left(B\right) + P\left(A \mid \overline{B}\right) P\left(\overline{B}\right)}$$

For the problem at hand this becomes

$$P\left(U \mid S\right) = \frac{P\left(S \mid U\right) P\left(U\right)}{P\left(S \mid U\right) P\left(U\right) + P\left(S \mid \overline{U}\right) P\left(\overline{U}\right)} = \frac{(.72)\,(.40)}{(.72)\,(.40) + (.52)\,(.60)} = .48$$

3.11 (a) We are to find the area between a point on the curve (98) and the mean (80) so we need only find the Z score for the point and read the associated area from Appendix A. So by Equation 2.24 on page 42

$$Z = \frac{x - \mu}{\sigma} = \frac{98 - 80}{12} = 1.5.$$

From column two of Appendix A we find that the area is .4332.

(b) This area is in the lower tail of the curve so we need only find the Z score for 74 and read the associated area from column three of the table.

$$Z = \frac{74 - 80}{12} = -.5.$$

From column three of Appendix A we find that the area is .3085.

(c) This is the area below a point that is above the mean of the distribution. We will find the area between 80 and 82 and add the result to .5 which is the area below the mean of the distribution.

$$Z = \frac{82 - 80}{12} = .17.$$

From column two we obtain .0675 so that the area below 82 is $.5000 + .0675 = .5675$.

(d) The area is between two points one of which is above, and the other below the distribution mean. We will find the area associated with each and sum the two to find the total area between the two points.

$Z = \frac{94-80}{12} = 1.17$ which has an associated area of .3790.

$Z = \frac{72-80}{12} = -.67$ which has an associated area of .2486.

The total area between the two points is therefore, $.3790 + .2486 = .6276$.

(e) This is the area between two points, both of which are below the distribution mean. The strategy is to find the areas associated with each of the two points and subtract the smaller area from the larger to find the area between the two points.

$Z = \frac{60-80}{12} = -1.67$ which has an associated area of .4525.

$Z = \frac{56-80}{12} = -2.00$ which has an associated area of .4772.

The area between the two points is then, $.4772 - .4525 = .0247$.

(f) This is the area above a point that is above the mean of the distribution. Thus, it is an area in the right hand tail of the distribution.

$Z = \frac{104-80}{12} = 2.00$ which has an associated area in column three of .0228.

(g) This is the area below a point that is below the mean of the distribution. Thus, it is an area in the left hand tail of the distribution.

$Z = \frac{54-80}{12} = -2.17$ which has an associated area in column three of .0150.

(h) This is the area between two points, both of which are above the distribution mean. The strategy is to find the areas associated with each of the two points and subtract the smaller area from the larger to find the area between the two points.

$Z = \frac{82-80}{12} = .17$ which has an associated area of .0675.

$Z = \frac{94-80}{12} = 1.17$ which has an associated area of .3790.

The area between the two points is then, $.3790 - .0675 = .3115$.

A. The following answers refer to Case Study A.

3.13 $P(M) = \frac{24}{59} = .41$.

3.15 (a) $P(T)$ can be read directly from the table margin and is .56.

(b) $P(Y)$ can be read directly from the table margin and is .80.

(c) $P(T \mid Y) = \frac{23}{47} = .49$.

(d) $P(T \mid N) = \frac{10}{12} = .83$.[1]

(e) For 3.15a: The probability of randomly choosing a subject who had chosen a treated lens.

For 3.15b: The probability of randomly choosing a subject who had answered yes to the duration question.

For 3.15c: The probability of randomly choosing *from among the subjects who answered yes to the duration question*, a subject who chose the treated lens.

For 3.15d: The probability of randomly choosing *from among the subjects who answered no to the duration question*, a subject who chose the treated lens.

---

[1]Notice that had we used the values .17 and .20 from the probability table we would have gotten a different result. That is because these values are rounded which introduces some error into the calculation.

(f) No. Following 3.5 on page 56 and noting that $P(T) P(N) = \left(\frac{33}{59}\right) \left(\frac{12}{59}\right) = .11$ is not equal to $P(TN) = \frac{10}{59} = .17$

3.17 By Equation 3.11

$$RR = \frac{P(U \mid N)}{P(U \mid Y)} = \frac{\frac{2}{12}}{\frac{24}{47}} = .33.$$

This indicates that persons who cannot wear the lenses for as long as they like are at less risk of choosing the untreated lens than are persons who can wear lenses for as long as they please. This may indicate that persons who cannot wear lenses for as long as they like might benefit more from treated lenses than do persons who can wear their lens for as long as they like.

3.19 The odds of choosing an untreated lens for persons who answered yes to the duration question is, by the discussion beginning on page 60,

$$\frac{P(U \mid Y)}{P(T \mid Y)} = \frac{\frac{24}{47}}{\frac{23}{47}} = 1.04.$$

For persons who answered no to the duration question the odds would be

$$\frac{P(U \mid N)}{P(T \mid N)} = \frac{\frac{2}{12}}{\frac{10}{12}} = .20.$$

E. The following answers refer to Case Study E.

3.21 We set up a table as follows.

	D	$\overline{D}$	
$+$	.100	.042	.142
$-$	.006	.852	.858
	.106	.894	

Then by Equation 3.6 on page 58

$$\text{Sensitivity} = P(+ \mid D) = \frac{.100}{.106} = .943.$$

By Equation 3.7 on page 58

$$\text{Specificity} = P\left(- \mid \overline{D}\right) = \frac{.852}{.894} = .953$$

By Equation 3.8 on page 58

$$PPV = P(D \mid +) = \frac{.100}{.142} = .704$$

By Equation 3.9 on page 58

$$NPV = P\left(\overline{D} \mid -\right) = \frac{.852}{.858} = .993$$

P. The following answers refer to Case Study P.

3.22 (a) The probability of selecting a patient who is female, has type I diabetes and is in the age range 25–44.

(b) This joint probability can be found by dividing the number of patients who have all three characteristics by the total number of patients or $\frac{157}{1,597} = .098$.

(c) This would include all patients who are female *or* have Type I diabetes *or* are in the age range 24–44. In short, this is everyone except men who have type II diabetes and are in an age range other than 24–44. Inspection of Table J.10 shows that there are $163 + 195$ such patients so that the probability is $1 - \frac{163+195}{1,597} = .776$.

(d) This is the probability of selecting a male from among patients who, are in the age range 15–24 *or* have type II diabetes.

(e) There are a total of $380 + 410$ patients who have type II diabetes and an additional $70 + 65$ patients in the age range 15–24 for a total of 925. Because there are $380 + 70 = 450$ males in this group, the probability of selecting a male from among the designated group would be $\frac{450}{925} = .486$.

Chapter Four

4.1 By Equation 4.2 on page 77

$$\sigma_{\bar{x}}^2 = \frac{\sigma^2}{n} = \frac{2500}{100} = 25$$

4.3 It decreases.

4.5 By Equation 4.3 on page 78

$$Z = \frac{\bar{x} - \mu}{\frac{\sigma}{\sqrt{n}}} = \frac{45 - 50}{\frac{10}{\sqrt{25}}} = -2.5.$$

Reference to column three of Appendix A shows that the desired probability is .0062.

4.7 Using the upper real limit of .35 for a proportion of .30 in Equation 4.6 on page 85 yields

$$Z = \frac{.35 - .40}{\sqrt{\frac{.40(1-.40)}{10}}} = -.32.$$

Reference to column three in Appendix A provides a probability estimate of .3745. This is quite close to the exact value which is a bit surprising given the small sample size.

4.9 Because σ is not given, a one mean t test will be appropriate for the hypothesis test.

The necessary sums for the calculation of s, $\bar{x}$, and t are

X	X²
3	9
3	9
2	4
1	1
0	0
6	36
5	25
4	16

Σ	24	100

Then by Equation 2.16 on page 37, Equation 2.1 on page 25, and Equation 4.8 on page 102

$$s = \sqrt{\frac{\sum x^2 - \frac{(\sum x)^2}{n}}{n-1}} = \sqrt{\frac{100 - \frac{(24)^2}{8}}{8-1}} = 2.0$$

$$\bar{x} = \frac{\sum x}{n} = \frac{24}{8} = 3.0$$

$$t = \frac{\bar{x} - \mu_0}{\frac{s}{\sqrt{n}}} = \frac{3.0 - 5.0}{\frac{2.0}{\sqrt{8.0}}} = -2.828.$$

Reference to Appendix B shows that the critical value for this one-tailed t test with $8 - 1 = 7$ degrees of freedom conducted at .05 is -1.895. Because critical t of -2.828 is less than this value, the null hypothesis is rejected.

4.11 The form of the stated hypotheses indicate that this is a two-tailed equivalence test. Z rather than t tests are used because σ is provided.

Calculating Z for Test One

$$Z_1 = \frac{\bar{x} - \mu_0}{\frac{\sigma}{\sqrt{n}}} = \frac{.52 - 1.0}{\frac{24}{\sqrt{144}}} = -.24$$

and for Test Two

$$Z_2 = \frac{.52 - (-1.0)}{\frac{24}{\sqrt{144}}} = .76.$$

The one-tailed critical values for the two tests are -1.65 and 1.65 respectively so that the equivalence null hypothesis is not rejected.

4.13 From the discussion of exact two-tailed tests for π beginning on page 112, we see that the critical value of $\hat{p}$ for this test is .30 for the lower tail critical region with no critical value existing in the upper region. This can be seen by noting that $P(0) + P(1) + P(2) + P(3) = .000006 + .000138 + 001447 + 009002 = .010593$ which is less than $\alpha/2 = .025$ while $P(0) + P(1) + P(2) + P(3) + P(4) = .000006 + .000138 + .001447 + .009002 + .036757 = .047350$ which is greater than $\alpha/2 = .025$. For the upper tail, $P(10) = .028248$ which is greater than $\alpha/2$ so that no value of $\hat{p}$ satisfies the criterion of having an associated probability less than $\alpha/2$.

Clearly, obtained $\hat{p}$ of .10 is in the lower tail critical region so that the null hypothesis is rejected. The two-tailed p-value is $2(.000006 + .000138) = .000288$ which is less than $\alpha = .05$ so that rejection is also indicated by this method. Calculations of probabilities for individual values of $\hat{p}$ are provided via Equation 4.5 on page 81 as follows.

$$P\,(0) = \frac{10!}{0!(10-0)!}.7^0\,(1-.7)^{10-0} = .000006$$

$$P\,(1) = \frac{10!}{1!(10-1)!}.7^1\,(1-.7)^{10-1} = .000138$$

$$P\,(2) = \frac{10!}{2!(10-2)!}.7^2\,(1-.7)^{10-2} = .001447$$

$$P\,(3) = \frac{10!}{3!(10-3)!}.7^3\,(1-.7)^{10-3} = .009002$$

$$P\,(4) = \frac{10!}{4!(10-4)!}.7^4\,(1-.7)^{10-4} = .036757$$

$$P\,(9) = \frac{10!}{9!(10-9)!}.7^9\,(1-.7)^{10-9} = .121061$$

$$P\,(10) = \frac{10!}{10!(10-10)!}.7^{10}\,(1-.7)^{10-10} = .028248$$

4.15 From the discussion of exact one-tailed tests for π beginning on page 108, we see that the p-value for this test is $P\,(0) + P\,(1)$ where by Equation 4.5 on page 81

$$P\,(0) = \frac{9!}{0!\,(9-0)!}.5^0\,(1-.5)^{9-0} = .001953$$

$$P\,(1) = \frac{9!}{1!\,(9-1)!}.5^1\,(1-.5)^{9-1} = .017578$$

so that the one-tailed p-value is $.001953 + .017578 = .019531$. Because this value is less than $\alpha = .05$, the null hypothesis is rejected.

Because $P\,(0) + P\,(1) = .001953 + .017578 = .019531$ which is less than $\alpha = .05$ and $P\,(0) + P\,(1) + P\,(2) = .019531 + .070313 = .089844$ is greater than $\alpha = .05$, critical $\hat{p}$ is $1/9 = .11$. Because obtained $\hat{p}$ of .11 falls in this critical region, the null hypothesis is rejected.

4.17 The level of significance of the test (α).

The sample size (n).

The form of the alternative distribution.

4.19 Again applying Equation 4.10 with the standard error of the mean reflecting $n = 100$ yields

$$Z_\beta = \frac{40-42}{2} + 1.65 = .65$$

which has associated power of .2578. Beta is then $1.0 - .2578 = .7422$.

4.21 By Equation 4.11 on page 136 with $\alpha/2$ substituted for alpha to reflect the two-tailed test, and the accompanying discussion, we calculate

$$n = \frac{\sigma^2\,(Z_\beta - Z_{\alpha/2})^2}{(\mu_0 - \mu)^2} = \frac{4\,(1.28 - (-1.96))^2}{(10-6)^2} = 2.6244$$

which we round to 3.[2] The small sample size reflects the fact that the mean of the alternative is

$$\frac{6-10}{\frac{2}{\sqrt{2.6244}}} = -3.24$$

standard errors below the mean of the null distribution.

---

[2]It is helpful to sketch the problem as we have done in Figure 4.26 on page 137.

4.23 Noting that $\hat{p} = 150/200 = .75$ and applying Equations 4.16 and 4.17 on page 148 yields

$$L = \hat{p} - Z\sqrt{\frac{\hat{p}\hat{q}}{n}} = .75 - 1.96\sqrt{\frac{(.75)(.25)}{200}} = .690$$

and

$$U = \hat{p} + Z\sqrt{\frac{\hat{p}\hat{q}}{n}} = .75 + 1.96\sqrt{\frac{(.75)(.25)}{200}} = .810.$$

4.25 From the discussion beginning on page 149 and Equations 4.20 and 4.21 on page 150 we calculate

$$df_{LN} = 2(n - S + 1) = 2(10 - 6 + 1) = 10$$
$$df_{LD} = 2S = 2 \times 6 = 12$$

Appendix C shows that for numerator and denominator degrees of freedom of 10 and 12 respectively, the appropriate F value for a one-sided 95% confidence interval is 2.75. Using this value in Equation 4.18 on page 149 gives

$$L = \frac{S}{S + (n - S + 1)F_L} = \frac{6}{6 + (10 - 6 + 1)2.75} = .304.$$

A. The following answers refer to Case Study A.

4.27 As noted on page 139, a sample statistic when used as an estimate of a parameter is termed a "point estimate."

In the article we are given

$$\bar{x} = 31.5$$
$$s = 10.2$$
$$n = 59$$

Then by Equations 4.14 and 4.15 on page 146

$$L = \bar{x} - t\frac{s}{\sqrt{n}} = 31.5 - 2.002\frac{10.2}{\sqrt{59}} = 28.8$$

and

$$U = \bar{x} + t\frac{s}{\sqrt{n}} = 31.5 + 2.002\frac{10.2}{\sqrt{59}} = 34.2.$$

Thus, we can be 95% confident that the population mean age is between 28.8 and 34.2.

B. The following answers refer to Case Study B.

4.29 In this case we will use an approximate method to form the required confidence intervals. To this end let the subscripts S, W, and P indicate, respectively, the Standard, Weak and Placebo groups. Then by Equations 4.16 and 4.17 which state

$$L = \hat{p} - Z\sqrt{\frac{\hat{p}\hat{q}}{n}}$$

and

$$U = \hat{p} + Z\sqrt{\frac{\hat{p}\hat{q}}{n}}$$

and noting that

$$\hat{p}_S = \frac{35}{62} = .565$$

$$\hat{p}_W = \frac{12}{59} = .203$$

$$\hat{p}_P = \frac{30}{62} = .484$$

so that

$$L_S = .565 - 1.96\sqrt{\frac{(.565)(.435)}{62}} = .442$$

$$U_S = .565 - 1 + 96\sqrt{\frac{(.565)(.435)}{62}} = .688$$

and

$$L_W = .203 - 1.96\sqrt{\frac{(.203)(.797)}{59}} = .100$$

$$U_W = .203 + 1.96\sqrt{\frac{(.203)(.797)}{59}} = .306$$

and

$$L_P = .484 - 1.96\sqrt{\frac{(.484)(.516)}{62}} = .360$$

$$U_P = .484 + 1.96\sqrt{\frac{(.484)(.516)}{62}} = .608.$$

It is likely that some subjects noticed their bracelets attracting metal objects while others noted that this was not the case.

D. The following answers refer to Case Study D.

4.31 No. In order to perform the Z test we would have to know σ which we do not. We could, of course, test this hypothesis with a one mean t test.

O. The following answers refer to Case Study O.

4.33 Not if the test is known to be robust to violation of the normality assumption under the circumstance for which it is being employed.

Chapter Five

5.1 The difference scores, their squares and the sums of each are as follows.

d	d^2
−14	196
−11	121
2	4
−2	4
−19	361
8	64
0	0
−18	324
$\sum$ −54	1074

Using these sums, the mean and standard deviation of the difference scores is

$$\bar{d} = \frac{\sum d}{n} = \frac{-54}{8} = -6.75$$

and

$$s_d = \sqrt{\frac{\sum d^2 - \frac{(\sum d)^2}{n}}{n-1}} = \sqrt{\frac{1074 - \frac{(-54)^2}{8}}{8-1}} = 10.068.$$

(a) By Equation 5.1 on page 162

$$t = \frac{\bar{d}}{\frac{s_d}{\sqrt{n}}} = \frac{-6.75}{\frac{10.068}{\sqrt{8}}} = -1.896.$$

Critical t for a two-tailed test with $n - 1 = 8 - 1 = 7$ degrees of freedom conducted at $\alpha = .05$ is, by Appendix B, ± 2.365 so that the null hypothesis is not rejected.

(b) By Equations 5.2 and 5.3 on page 172

$$L = \bar{d} - t\frac{s_d}{\sqrt{n}} = -6.75 - (2.365)\left(\frac{10.068}{\sqrt{8}}\right) = -15.168$$

and

$$U = \bar{d} + t\frac{s_d}{\sqrt{n}} = -6.75 + (2.365)\left(\frac{10.068}{\sqrt{8}}\right) = 1.668.$$

(c) The nonsignificant t test, or equivalently, the fact that the CI contained zero, indicates that we do not have sufficient evidence to claim a cholesterol changing effect for the diet.

5.3 (a) i. Using the discussion on page 179 and disregarding (S-S) and (U-U) as noninformative, we calculate

$$\hat{p} = \frac{5}{5+2} = .7143.$$

The binomial probabilities for $\pi = .5$ and $n = 7$ are by Equation 4.5 on page 81 as follows.

Proportion $\hat{p}$	Number of Successes y	Probability $P(y)$
.0000	0	.00781
.1429	1	.05469
.2857	2	.16406
.4286	3	.27344
.5714	4	.27344
.7143	5	.16406
.8571	6	.05469
1.0000	7	.00781

Employing the method outlined beginning on page 108 for finding the p-value for a two-tailed exact test, we obtain $p = 2(.00781 + .05469 + .16406) = .45312$. The null hypothesis is not rejected so that we cannot assert that the treatment is effective at improving patient's perception of their appearance.

ii. Using the discussion on page 149 and noting that $S = 5$ and $n = 7$, we find by Equations 4.20 and 4.21 that the numerator and denominator degrees of freedom for F_L are respectively,

$$df_{LN} = 2(n - S + 1) = 2(7 - 5 + 1) = 6$$

and

$$df_{LD} = 2S = 2 \cdot 5 = 10.$$

Then by Equation 4.18 on page 149, the lower limit of the CI is

$$L = \frac{S}{S + (n - S + 1) F_L} = \frac{5}{5 + (7 - 5 + 1) 4.07} = .291.$$

The numerator and denominator degrees of freedom for U are, by Equations 4.22 and 4.23

$$df_{UN} = 2(S + 1) = 2(5 + 1) = 12$$

and

$$df_{UD} = 2(n - S) = 2(7 - 5) = 4.$$

Then by Equation 4.19

$$U = \frac{(S + 1) F_U}{n - S + (S + 1) F_U} = \frac{(5 + 1) 8.75}{7 - 5 + (5 + 1) 8.75} = \frac{52.5}{54.5} = .963.$$

Are you surprised to find that .5 is in the interval? (We sincerely hope not.)

(b) i. Discarding S-S and U-U as noninformative, we calculate

$$\hat{p} = \frac{71}{71 + 20} = .780.$$

Then by Equation 5.4 on page 176

$$Z = \frac{\hat{p} - .5}{\frac{.5}{\sqrt{n}}} = \frac{.780 - .5}{\frac{.5}{\sqrt{91}}} = 5.342.$$

Reference to Appendix A shows that critical Z for a two-tailed test at $\alpha = .05$ is ± 1.96. The null hypothesis $H_0 : \pi = .5$ is rejected so that we may claim a beneficial effect for the treatment insofar as improving patient's perception of their appearance is concerned.

ii. From the discussion on page 186 and by Equations 4.16 and 4.17

$$L = \hat{p} - Z\sqrt{\frac{\hat{p}\hat{q}}{n}} = .780 - 1.96\sqrt{\frac{(.78)(.22)}{91}} = .695$$

and

$$U = \hat{p} + Z\sqrt{\frac{\hat{p}\hat{q}}{n}} = .780 + 1.96\sqrt{\frac{(.78)(.22)}{91}} = .865.$$

Are you surprised that .5 is not in the interval? (We sincerely hope not.)

5.5 (a) By Equation 5.12 on page 200

$$\widehat{OR} = \frac{b}{c} = \frac{24}{9} = 2.667$$

which means that, in this sample, the odds of being a motorcycle patrolman for officers with skin cancer are 2.667 times the odds for those who do not have skin cancer.

(b) By the first method, we begin by constructing a confidence interval for the estimation of π, then use the relationship between $\widehat{OR}$ and $\hat{p}$ to convert the interval into an estimation of OR. To this end, we first use the result from 5.5a and Equation 5.14 on page 205 to obtain

$$\hat{p} = \frac{\widehat{OR}}{1 + \widehat{OR}} = \frac{2.667}{1 + 2.667} = .727.$$

Then by Equations 5.15 and 5.16 on page 208, we compute

$$L = \hat{p} - Z\sqrt{\frac{\hat{p}(1 - \hat{p})}{n}} = .727 - 1.96\sqrt{\frac{.727(1 - .727)}{33}} = .575$$

and

$$U = \hat{p} + Z\sqrt{\frac{\hat{p}(1 - \hat{p})}{n}} = .727 + 1.96\sqrt{\frac{.727(1 - .727)}{33}} = .879.$$

We now employ Equation 5.17 on page 208 to convert these end points to odds ratios as follows.

$$L = \frac{\hat{p}}{1 - \hat{p}} = \frac{.575}{1 - .575} = 1.353$$

and

$$U = \frac{\hat{p}}{1 - \hat{p}} = \frac{.879}{1 - .879} = 7.264.$$

A second method produces the confidence interval directly by means of Equations 5.18 and 5.5.4 on page 210 as follows.

$$L = \exp\left(\ln\left(\widehat{OR}\right) - Z\sqrt{\frac{1}{b} + \frac{1}{c}}\right) = \exp\left(\ln(2.667) - 1.96\sqrt{\frac{1}{24} + \frac{1}{9}}\right) = 1.240$$

$$U = \exp\left(\ln\left(\widehat{OR}\right) + Z\sqrt{\frac{1}{b} + \frac{1}{c}}\right) = \exp\left(\ln(2.667) + 1.96\sqrt{\frac{1}{24} + \frac{1}{9}}\right) = 5.738.$$

As may be seen, the two approximations produce rather similar estimates for L but differ substantially when estimating U.

(c) From the discussion on page 210 and use of Equations 5.22, 5.23, 5.24, 5.25, 5.20, and 5.21 we obtain

$$df_{LN} = 2(c+1) = 2(9+1) = 20$$

and

$$df_{LD} = 2b = 2 \cdot 24 = 48$$

so that

$$L = \frac{b}{b+(c+1)F_L} = \frac{24}{24+(9+1)2.01} = .542.$$

The degrees of freedom for calculation of U are

$$df_{UN} = 2(b+1) = 2(24+1) = 50$$

and

$$df_{UD} = 2c = 2 \cdot 9 = 18$$

so that

$$U = \frac{(b+1)F_U}{c+(b+1)F_U} = \frac{(24+1)2.35}{9+(24+1)2.35} = .867.$$

Converting the end points of the interval to estimates of OR via Equation 5.17 yields

$$L = \frac{\hat{p}}{1-\hat{p}} = \frac{.542}{1-.542} = 1.183$$

and

$$U = \frac{\hat{p}}{1-\hat{p}} = \frac{.867}{1-.867} = 6.519.$$

Comparing the exact interval to the two approximations leads us to conclude that while both approximations were fairly accurate, we would always prefer the exact interval.

A. The following answers refer to Case Study A.

5.7 Since the data are paired and dichotomous, they likely used McNemar's test.[3]

If this were other than a textbook exercise, we would use an exact method for the test but because of the computations involved we will use an approximate method as provided by Equation 5.4. We first note that $\hat{p} = \frac{8}{16} = .50$ so that

$$Z = \frac{\hat{p}-.5}{\frac{.5}{\sqrt{n}}} = \frac{.5-.5}{\frac{.5}{\sqrt{16}}} = 0.$$

Because the critical values for a two-tailed test conducted at $\alpha = .05$ are ±1.96, we fail to reject the null hypothesis which agrees with the result reported by the authors. Thus, no change could be demonstrated.

---

[3]This is in fact the test they report using.

K. The following answers refer to Case Study K.

5.9 The assumption of independence might have been violated had individual blood pressure measurements been used for the analysis. Students in a particular school may be more similar to each other than to students in other schools because of socioeconomic or other factors much as siblings in a particular family may be more similar to each other than to children in a different family.

5.11 A confidence interval would have been more informative because it would have provided an estimate of *how much* mean change took place rather than simply asserting that such a change occurred.

Using the values for $\bar{d}$ and s_d calculated in 5.10 with Equations 5.2 and 5.3 gives

$$L = \bar{d} - t\frac{s_d}{\sqrt{n}} = 2.052 - 4.604\frac{.537}{\sqrt{5}} = .946$$

and

$$U = \bar{d} + t\frac{s_d}{\sqrt{n}} = 2.052 + 4.604\frac{.537}{\sqrt{5}} = 3.158.$$

Thus, we can be 99 percent confident that the mean school change was between .946 and 3.158 points.

M. The following answers refer to Case Study M.

5.13 Following the same steps given in 5.12, we obtain

$$\bar{d} = \frac{\sum d}{n} = \frac{2.4}{7} = .343$$

and

$$s_d = \sqrt{\frac{\sum d^2 - \frac{(\sum d)^2}{n}}{n-1}} = \sqrt{\frac{1.1 - \frac{(2.4)^2}{7}}{7-1}} = .215.$$

Then by 5.2 and 5.3

$$L = \bar{d} - t\frac{s_d}{\sqrt{n}} = .343 - 2.447\frac{.215}{\sqrt{7}} = .144$$

and

$$U = \bar{d} + t\frac{s_d}{\sqrt{n}} = .343 + 2.447\frac{.215}{\sqrt{7}} = .542.$$

Chapter Six

6.1 The data are arranged for analysis via an independent samples *t* test as shown in Table K.1. By Equation 6.2 on page 220

$$s_P^2 = \frac{\left(\sum x_1^2 - \frac{(\sum x_1)^2}{n_1}\right) + \left(\sum x_2^2 - \frac{(\sum x_2)^2}{n_2}\right)}{n_1 + n_2 - 2}$$

$$= \frac{\left(5913.29 - \frac{(239.3)^2}{10}\right) + \left(5957.57 - \frac{(240.7)^2}{10}\right)}{10 + 10 - 2}$$

$$= \frac{186.841 + 163.921}{18}$$

$$= 19.487.$$

TABLE K.1: Data arranged for analysis for Exercise 1.

	Group One		Group Two	
	X_1	X_1^2	X_2	X_2^2
	26.2	686.44	25.9	670.81
	24.5	600.25	20.1	404.01
	20.0	400.00	22.2	492.84
	30.2	912.04	29.7	882.09
	28.4	806.56	28.0	784.00
	18.6	345.96	29.4	864.36
	21.5	462.25	20.2	408.04
	21.7	470.89	20.7	428.49
	29.9	894.01	26.3	691.69
	18.3	334.89	18.2	331.24
$\sum$	239.3	5913.29	240.7	5957.57

Continuing,

$$\bar{x}_1 = \frac{\sum x_1}{n_1} = \frac{239.3}{10} = 23.93$$

and

$$\bar{x}_2 = \frac{\sum x_2}{n_2} = \frac{240.7}{10} = 24.07.$$

Then by Equation 6.1 on page 219

$$t = \frac{\bar{x}_1 - \bar{x}_2 - \delta_0}{\sqrt{s_P^2 \left(\frac{1}{n_1} + \frac{1}{n_2} \right)}} = \frac{23.93 - 24.07}{\sqrt{19.487 \left(\frac{1}{10} + \frac{1}{10} \right)}} = -0.071.$$

Appendix B shows the critical value for a one-tailed t test with 18 degrees of freedom conducted at $\alpha = .05$ to be 1.734. The null hypothesis is not rejected. Therefore, we did not find sufficient evidence to conclude that children living within one mile of a fast food restaurant have higher mean BMI than do children living further away.

6.3   Using some of the results obtained in Exercise 6.1 with Equations 6.3 and 6.4 on page 228 yields

$$L = (\bar{x}_1 - \bar{x}_2) - t \sqrt{s_P^2 \left(\frac{1}{n_1} + \frac{1}{n_2} \right)} = (23.93 - 24.07) - 2.101 \sqrt{19.487 \left(\frac{1}{10} + \frac{1}{10} \right)} = -4.288$$

and

$$U = (\bar{x}_1 - \bar{x}_2) + t \sqrt{s_P^2 \left(\frac{1}{n_1} + \frac{1}{n_2} \right)} = (23.93 - 24.07) + 2.101 \sqrt{19.487 \left(\frac{1}{10} + \frac{1}{10} \right)} = 4.008.$$

This gives us 95% confidence that $\mu_1 - \mu_2$ is between -4.288 and 4.008. From the researchers point of view, this indicates with 95% confidence that the difference between the average weights of the two groups is as specified by the CI. Notice that zero is in the interval. What does this mean insofar as an hypothesis test is concerned?

6.5 By Equation 6.5 on page 232

$$Z = \frac{\hat{p}_1 - \hat{p}_2 - \delta_0}{\sqrt{\frac{\hat{p}_1 \hat{q}_1}{n_1} + \frac{\hat{p}_2 \hat{q}_2}{n_2}}} = \frac{.312 - .288}{\sqrt{\frac{(.312)(.688)}{1777} + \frac{(.288)(.712)}{1821}}} = 1.57$$

Appendix A shows the critical values for a two-tailed Z test conducted at $\alpha = .05$ to be ± 1.96. Thus, the test is not significant so that we were unable to demonstrate a difference in the proportions of male and female smokers.

6.7 By Equations 6.6 and 6.7 on page 236,

$$L = (\hat{p}_1 - \hat{p}_2) - \left(Z \sqrt{\frac{\hat{p}_1 \hat{q}_1}{n_1 - 1} + \frac{\hat{p}_2 \hat{q}_2}{n_2 - 1} + \frac{1}{2}\left(\frac{1}{n_1} + \frac{1}{n_2}\right)} \right)$$

$$= (.312 - .288) - \left(1.96 \sqrt{\frac{(.312)(.688)}{1777 - 1} + \frac{(.288)(.712)}{1821 - 1} + \frac{1}{2}\left(\frac{1}{1777} + \frac{1}{1821}\right)} \right)$$

$$= -.007$$

and

$$U = (.312 - .288) + \left(1.96 \sqrt{\frac{(.312)(.688)}{1777 - 1} + \frac{(.288)(.712)}{1821 - 1} + \frac{1}{2}\left(\frac{1}{1777} + \frac{1}{1821}\right)} \right)$$

$$= .055$$

A two-tailed hypothesis test conducted at $\alpha = .05$ would not be significant as zero is in the interval. The interval estimates $\pi_1 - \pi_2$ or, in the context of the study, the difference between the proportions of male and female smokers.

6.9 (a) For purposes of analysis, it will be convenient to arrange the data as follows.

Cancer

		yes	no
Within	yes	590	9258
500 yards	no	577	12535

By Equation 6.9 on page 239

$$\widehat{RR} = \frac{a/(a+b)}{c/(c+d)} = \frac{590/(590 + 9258)}{577/(577 + 12535)} = 1.361.$$

We conduct Test One[4] by means of Equation 6.10 on page 240 as follows.

$$Z = \frac{\ln\left(\widehat{RR}\right) - \ln\left(RR_0\right)}{\sqrt{\frac{b/a}{a+b} + \frac{d/c}{c+d}}} = \frac{\ln\left(1.361\right) - \ln\left(1.100\right)}{\sqrt{\frac{9258/590}{590 + 9258} + \frac{12535/577}{577 + 12535}}} = 3.735.$$

---

[4]Actually, we can see by inspection that the test will be non-significant because the test statistic is not in the EI.

From Appendix A we find that the critical value for a one-tailed Z test conducted at $\alpha = .05$ is -1.65. Because obtained Z of 3.735 is greater than this value, the null hypothesis is not rejected.

Test Two is then

$$Z = \frac{\ln\left(\widehat{RR}\right) - \ln\left(RR_0\right)}{\sqrt{\frac{b/a}{a+b} + \frac{d/c}{c+d}}} = \frac{\ln(1.361) - \ln(.910)}{\sqrt{\frac{9258/590}{590+9258} + \frac{12535/577}{577+12535}}} = 7.061.$$

Because 7.061 is greater than critical Z of 1.65, Test Two is significant.

Because *both* Test One and Test Two must be significant in order for the two-tailed equivalence null hypothesis to be rejected, the equivalence test is not significant. Thus, safety was not established. The bottom line is that the previous studies did not show that the power lines were dangerous while the present study failed to show that they were safe.

(b) $H_{0E} : RR \le .91$ or $RR \ge 1.10$

$H_{AE} : .91 < RR < 1.1$

(c) $H_0 : RR = 1$

What would the conclusion be if this hypothesis were tested in this study?

(d) If $RR = 1$, we assume no relationship between the power lines and cancer. Thus, a test of the null hypothesis $RR = 1$ assumes safety and attempts to reject the safety hypothesis in favor of an alternative that denies safety.

The null hypothesis stated in Exercise 6.9b assumes danger ($RR \ge 1.1$ or $RR \le .91$) and attempts to reject the danger null hypothesis in favor of a safety alternative hypothesis.

6.11 (a) Using the table constructed for the answer to Exercise 6.8a with Equation 6.14 on page 249 yields

$$\widehat{OR} = \frac{ad}{bc} = \frac{(196)(237)}{(95)(191)} = 2.560$$

(b) By Equation 6.15 on page 249

$$Z = \frac{\ln\left(\widehat{OR}\right) - \ln\left(OR_0\right)}{\sqrt{\frac{1}{a} + \frac{1}{b} + \frac{1}{c} + \frac{1}{d}}} = \frac{\ln(2.560)}{\sqrt{\frac{1}{196} + \frac{1}{95} + \frac{1}{191} + \frac{1}{237}}} = 5.935.$$

Because critical Z is ± 1.96, the null hypothesis is rejected leading to the conclusion that the odds of developing MS for patients with lesions is greater than the odds for patients without such lesions.

6.13 By Equations 6.16 and 6.17 on page 254

$$L = \exp\left[\ln\left(\widehat{OR}\right) - Z\sqrt{\frac{1}{a} + \frac{1}{b} + \frac{1}{c} + \frac{1}{d}}\right]$$

$$= \exp\left[\ln(2.560) - 1.96\sqrt{\frac{1}{196} + \frac{1}{95} + \frac{1}{191} + \frac{1}{237}}\right]$$

$$= 1.877$$

end

$$U = \exp\left[\ln\left(\widehat{OR}\right) + Z\sqrt{\frac{1}{a} + \frac{1}{b} + \frac{1}{c} + \frac{1}{d}}\right]$$

$$= \exp\left[\ln\left(2.560\right) + 1.96\sqrt{\frac{1}{196} + \frac{1}{95} + \frac{1}{191} + \frac{1}{237}}\right]$$

$$= 3.492.$$

Thus, we can be 95% confident that the odds of developing MS for patients with lesions is between 1.877 and 3.492 times that for patients without such lesions.

This interval indicates that a two-tailed test of $H_0 : OR = 1$ would be rejected because 1 is not in the interval which is the result obtained in Exercise 6.11b.

A. The following answers refer to Case Study A.

6.15 The null hypothesis is rejected as evidenced by the fact that zero is not in the confidence interval.

B. The following answers refer to Case Study B.

6.17 No, because no treatments had been administered at that point and subjects had been randomly assigned to groups. Assuming random assignment had been properly conducted, a significant finding would likely be due to a Type I error.

6.19 Using the methods employed for question 6.18 we obtain

$$SS_S = (65 - 1)\left(3.9^2\right) = 973.44$$

and

$$SS_P = (64 - 1)\left(3.2^2\right) = 645.12.$$

We then obtain

$$s_P^2 = \frac{973.44 + 645.12}{65 + 64 - 2} = 12.745.$$

The limits are then

$$L = (7.8 - 9.3) - 1.979\sqrt{12.745\left(\frac{1}{65} + \frac{1}{64}\right)} = -2.744$$

and

$$U = (7.8 - 9.3) + 1.979\sqrt{12.745\left(\frac{1}{65} + \frac{1}{64}\right)} = -.256$$

This interval estimates the magnitude of the difference between the effects of magnetized and placebo bracelets as expressed by the difference between the mean pain scores produced by the two treatments. Because zero is not in the interval, we can claim a decreased level of pain as a result of the treatment.

6.21 (a) Noting that $\hat{p}_s = \frac{35}{62} = .565$ and $\hat{p}_w = \frac{12}{59} = .203$ and applying Equation 6.5 on page 232 gives

$$Z = \frac{\hat{p}_1 - \hat{p}_2 - \delta_0}{\sqrt{\frac{\hat{p}_1 \hat{q}_1}{n_1} + \frac{\hat{p}_2 \hat{q}_2}{n_2}}} = \frac{.565 - .203}{\sqrt{\frac{(.565)(.435)}{62} + \frac{(.203)(.797)}{59}}} = 4.420.$$

Because this value exceeds 1.96, we reject the null hypothesis and conclude that the proportion of subjects wearing the standard bracelet who were able to correctly identify the type bracelet they were wearing is greater than the proportion for those wearing the weak bracelet.

(b) $\hat{p}_p = \frac{30}{62} = .484$

$$Z = \frac{.565 - .484}{\sqrt{\frac{(.565)(.435)}{62} + \frac{(.484)(.516)}{62}}} = .906$$

The null hypothesis is not rejected so that we are unable to demonstrate a difference between the two proportions.

(c)

$$Z = \frac{.203 - .484}{\sqrt{\frac{(.203)(.797)}{59} + \frac{(.484)(.516)}{62}}} = -3.415$$

The null hypothesis is rejected.

6.23 We first note that the proportions of subjects in the Weak and Placebo groups who indicate that they are wearing a Dummy bracelet are $\frac{12}{59} = .203$ and $\frac{30}{62} = .484$, respectively. We further note that the difference between the two proportions is $.203 - .484 = -.281$, which is less than EI_L of $-.02$ so that Test Two will not be significant. We can, therefore, declare Test Two not significant without resorting to calculation of the test statistic so that the equivalence null hypothesis is not rejected. We cannot declare the two groups equivalent.

C. The following answers refer to Case Study C.

6.25 Confidence intervals because they are more informative than hypothesis tests.

6.27 Confidence intervals for RR may be formed by means of Equations 6.11 and 6.12 as follows.

$$L = \exp\left[\ln\left(\widehat{RR}\right) - Z\sqrt{\frac{b/a}{a+b} + \frac{d/c}{c+d}}\right]$$

and

$$U = \exp\left[\ln\left(\widehat{RR}\right) + Z\sqrt{\frac{b/a}{a+b} + \frac{d/c}{c+d}}\right]$$

Applying these equations to the data for the four time periods produces the following.

$$L = \exp\left[\ln(3.470) - 1.96\sqrt{\frac{441/21}{21+441} + \frac{452/6}{6+452}}\right] = 1.414$$

discharge

$$U = \exp\left[\ln(3.470) + 1.96\sqrt{\frac{441/21}{21+441} + \frac{452/6}{6+452}}\right] = 8.518$$

$$L = \exp\left[\ln(2.533) - 1.96\sqrt{\frac{439/23}{23 + 439} + \frac{449/9}{9 + 449}}\right] = 1.185$$

one month

$$U = \exp\left[\ln(2.533) + 1.96\sqrt{\frac{439/23}{23 + 439} + \frac{449/9}{9 + 449}}\right] = 5.415$$

$$L = \exp\left[\ln(1.597) - 1.96\sqrt{\frac{404/58}{58 + 404} + \frac{422/36}{36 + 422}}\right] = 1.076$$

one year

$$U = \exp\left[\ln(1.597) + 1.96\sqrt{\frac{404/58}{58 + 404} + \frac{422/36}{36 + 422}}\right] = 2.371$$

$$L = \exp\left[\ln(1.344) - 1.96\sqrt{\frac{382/80}{80 + 382} + \frac{399/59}{59 + 399}}\right] = .985$$

23 months

$$U = \exp\left[\ln(1.344) + 1.96\sqrt{\frac{382/80}{80 + 382} + \frac{399/59}{59 + 399}}\right] = 1.833$$

D. The following answers refer to Case Study D.

6.29 (a) Independent samples t test. It was used to compare two independent groups (HIV infected and noninfected subjects). Additionally, since there were 15 subjects in one group and 5 in the other, the degrees of freedom for the independent samples test would be $15 + 5 - 2 = 18$ which is the degrees of freedom reported for the test.

(b) First we note that an obtained t of 2.26 should have a smaller p-value than does t of 1.79. Yet, the authors report that the former is significant at the .05 level and the latter at the .01 level. Second, critical t for one- and two-tailed tests with 18 degrees of freedom conducted at $\alpha = .01$ are 2.552 and 2.878 respectively so that obtained t of 1.79 could not produce a significant result at the .01 level as reported by the authors.

(c) Calculating s_P^2 for the NPZ-8 data via Equation 6.2 gives

$$s_P^2 = \frac{\left(\sum x_1^2 - \frac{(\sum x_1)^2}{n_1}\right) + \left(\sum x_2^2 - \frac{(\sum x_2)^2}{n_2}\right)}{n_1 + n_2 - 2}$$

$$= \frac{\left(374 - \frac{(54)^2}{15}\right) + \left(1 - \frac{(1)^2}{5}\right)}{15 + 5 - 2}$$

$$= 10.022$$

then

$$t = \frac{\bar{x}_1 - \bar{x}_2 - \delta_0}{\sqrt{s_P^2\left(\frac{1}{n_1} + \frac{1}{n_2}\right)}} = \frac{3.6 - .2}{\sqrt{10.022\left(\frac{1}{15} + \frac{1}{5}\right)}} = 2.080$$

For the PBV data,

$$s_P^2 = \frac{\left(9.94394 - \frac{(12.169)^2}{15}\right) + \left(3.81950 - \frac{(4.368)^2}{5}\right)}{15 + 5 - 2} = .00418.$$

Then

$$t = \frac{.8113 - .8736}{\sqrt{.00418 \left(\frac{1}{15} + \frac{1}{5}\right)}} = -1.866.$$

Critical t for a two-tailed test at level .05 is ± 2.101 and at level .01 is ± 2.878. Thus, results for both tests are not significant. We also note that our obtained t values differ from those reported by the authors.

(d) From visual inspection, it appears that the variance of NPZ-8 scores in the HIV group is much greater than in the control group. This is confirmed by calculations which show that the former is approximately 12.829 and the latter .2. It appears that the homogeneity of variance assumption is violated. Further, because sample sizes are relatively small and decidedly unequal, we cannot depend on the robust properties of the independent samples t test to provide a valid result.

E. The following answers refer to Case Study E.

6.31 We begin by arranging the data as follows

Disease

		yes	no
Test	+	50	21
	−	3	426

Then by Equation 6.14

$$\widehat{OR} = \frac{ad}{bc} = \frac{(50)(426)}{(21)(3)} = 338.095.$$

By Equations 6.16 and 6.17

$$L = \exp\left[\ln\left(\widehat{OR}\right) - Z\sqrt{\frac{1}{a} + \frac{1}{b} + \frac{1}{c} + \frac{1}{d}}\right]$$

$$= \exp\left[\ln(338.095) - 1.96\sqrt{\frac{1}{50} + \frac{1}{21} + \frac{1}{3} + \frac{1}{426}}\right]$$

$$= 97.379$$

$$U = \exp\left[\ln\left(\widehat{OR}\right) + Z\sqrt{\frac{1}{a} + \frac{1}{b} + \frac{1}{c} + \frac{1}{d}}\right]$$

$$= \exp\left[\ln(338.095) + 1.96\sqrt{\frac{1}{50} + \frac{1}{21} + \frac{1}{3} + \frac{1}{426}}\right]$$

$$= 1173.848$$

H. The following answers refer to Case Study H.

6.33 The authors had no basis for the conclusion they reached. Essentially, they used failure to reject a null hypothesis as evidence that the null hypothesis is true. Their error was compounded by the fact that they were dealing with an extremely small sample giving rise to concerns as to whether the test had sufficient power for detecting any but the most gross differences.

6.35 No, there is nothing in the article to lead one to believe that any equivalence testing was carried out. Again, the authors appear to have simply used a conventional test of significance and did not know how to correctly interpret a nonsignificant result.

I. The following answers refer to Case Study I.

6.37 The following statement by the authors implies a one-tailed equivalence test. "More precisely, the researchers would consider D-B at least as good as AI if the proportion of deaths associated with D-B treatment exceeded the proportion under AI treatment by less than .004."

6.39 In reference to Test One, because $.0798 - .0753 = .0045$ is greater than .004, the Z value will be positive and, therefore, cannot be less than or equal to the critical value of -1.65 so that the equivalence null hypothesis will not be rejected.

6.41 No, for the same reason outlined in question 6.39

J. The following answers refer to Case Study J.

6.43 Taking $\hat{p}_1 = \frac{138}{161} = .857$ and $\hat{p}_2 = \frac{127}{151} = .841$, and conducting Test One via Equation 6.5 gives

$$Z_1 = \frac{\hat{p}_1 - \hat{p}_2 - \delta_0}{\sqrt{\frac{\hat{p}_1 \hat{q}_1}{n_1} + \frac{\hat{p}_2 \hat{q}_2}{n_2}}} = \frac{.857 - .841 - .10}{\sqrt{\frac{(.857)(.143)}{161} + \frac{(.841)(.159)}{151}}} = -2.070.$$

Test Two is then

$$Z_2 = \frac{.857 - .841 - (-.10)}{\sqrt{\frac{(.857)(.143)}{161} + \frac{(.841)(.159)}{151}}} = 2.859.$$

Critical values for Test One and Test Two are -1.65 and 1.65 respectively so that both null hypotheses are rejected so that the equivalence null hypothesis is rejected and equivalence established.

O. The following answers refer to Case Study O.

6.45 Tests such as the independent samples t test are generally less robust to violations of these assumptions than to the normality assumption. It is odd that so much effort would be devoted to one assumption while ignoring two such important assumptions.

Chapter Seven

7.1 For computational purposes, it will be convenient to arrange the data as follows. To simplify notation, we will subscript observations for the three groups by 1, 2, and 3 respectively.

Office		Hazardous		Non-Hazardous	
X_1	X_1^2	X_2	X_2^2	X_3	X_3^2
58	3364	88	7744	65	4225
64	4096	59	3481	70	4900
71	5041	74	5476	79	6241
66	4356	80	6400	66	4356
79	6241	81	6561	74	5476
74	5476	69	4761	79	6241
70	4900	90	8100	60	3600
$\sum$ 482	33474	541	42523	493	35039

(a) Using the sums from the above table, we calculate the sums of squares for the individual groups as follows.

$$SS_1 = \sum x_1^2 - \frac{\left(\sum x_1\right)^2}{n_1} = 33474 - \frac{(482)^2}{7} = 284.857$$

$$SS_2 = \sum x_2^2 - \frac{\left(\sum x_2\right)^2}{n_2} = 42523 - \frac{(541)^2}{7} = 711.429$$

$$SS_3 = \sum x_3^2 - \frac{\left(\sum x_3\right)^2}{n_3} = 35039 - \frac{(493)^2}{7} = 317.714$$

By Equation 7.4 on page 265

$$SS_w = SS_1 + SS_2 + \cdots + SS_k = 284.857 + 711.429 + 317.714 = 1314.000.$$

Then by Equation 7.3 on page 265

$$MS_w = \frac{SS_w}{N - k} = \frac{1314.0}{21 - 3} = 73.0.$$

(b) By Equation 7.8 on page 268

$$SS_b = \frac{\left(\sum\limits_{i=1}^{n_1} x_{i1}\right)^2}{n_1} + \frac{\left(\sum\limits_{i=1}^{n_2} x_{i2}\right)^2}{n_2} + \cdots + \frac{\left(\sum\limits_{i=1}^{n_k} x_{ik}\right)^2}{n_k} - \frac{\left(\sum\limits_{All} x_{..}\right)^2}{N}$$

$$= \frac{(482)^2}{7} + \frac{(541)^2}{7} + \frac{(493)^2}{7} - \frac{(1516)^2}{21}$$

$$= 281.238$$

And by Equation 7.6 on page 267

$$MS_b = \frac{SS_b}{k - 1} = \frac{281.238}{3 - 1} = 140.619.$$

(c) By Equation 7.2 on page 264 obtained F is

$$F = \frac{MS_b}{MS_w} = \frac{146.619}{73.0} = 1.93.$$

Appendix C shows that critical F for a test with 2 and 18 degrees of freedom conducted at $\alpha = .05$ is 3.55. Because obtained F of 1.93 does not exceed this value, the null hypothesis is not rejected. We conclude, therefore, that we failed to show any differences between mean heart rates of the three groups.

7.3 We will perform a 2 by 3 chi-square analysis since only three groups are involved.

	Facility One	Facility Two	Facility Three	
Participate	[46] (48.86)	[52] (48.36)	[36] (36.77)	134
Do Not Participate	[51] (48.14)	[44] (47.64)	[37] (36.23)	132
	97	96	73	$N = 266$

Expected values are then

$$f_{e11} = \frac{(134)\,(97)}{266} = 48.86$$

$$f_{e12} = \frac{(134)\,(96)}{266} = 48.36$$

$$f_{e13} = \frac{(134)\,(73)}{266} = 36.77$$

$$f_{e21} = \frac{(132)\,(97)}{266} = 48.14$$

$$f_{e22} = \frac{(132)\,(96)}{266} = 47.64$$

$$f_{e23} = \frac{(132)\,(73)}{266} = 36.23$$

Obtained chi-square is then

$$\chi^2 = \frac{(46.0 - 48.86)^2}{48.86} + \frac{(52.0 - 48.36)^2}{48.36} + \frac{(36.0 - 36.77)^2}{36.77} + \frac{(51.0 - 48.14)^2}{48.14}$$

$$+ \frac{(44.0 - 47.64)^2}{47.64} + \frac{(37.0 - 36.23)^2}{36.23}$$

$$= .167 + .274 + .016 + .170 + .278 + .016$$

$$= .921.$$

From Appendix D we see that for $k - 1 = 3 - 1 = 2$ degrees of freedom, critical χ^2 for a test at $\alpha = .05$ is 5.991. Because obtained χ^2 is less than critical χ^2, the null hypothesis is not rejected so that a claim of a differential effect for the three instructional methods cannot be supported.

7.5 NO, this statement is not correct. Familywise error is the probability that one or more of the *true* null hypotheses is *rejected* not the probability that one or more of the null hypotheses is false.

B. The following answers refer to Case Study B.

7.7 Given random assignment and the absence of any differential treatment afforded the groups, we would expect a significant result to be a Type I error. Thus, the probability of such an event would be α.

7.9 Using the reported standard deviations from Table 1 and the fact that the sum of squares (SS) for any given group can be expressed as $(n - 1)\,s^2$ we calculate

$$SS_s = (65 - 1)\,3.9^2 = 973.44$$

$$SS_w = (64 - 1)\,3.2^2 = 645.12$$

$$SS_p = (64 - 1)\,3.2^2 = 645.12$$

Then by Equations 7.4[5] and 7.3

$$SS_{\text{within}} = SS_s + SS_w + SS_p = 973.44 + 645.12 + 645.12 = 2263.68$$

$$MS_{\text{within}} = \frac{SS_{\text{within}}}{N - k} = \frac{2263.68}{193 - 3} = 11.914.$$

By noting that the sum of observations for any particular group can be expressed as $n\bar{x}$, we calculate the sums for the standard, weak and placebo groups as $(65)\,(7.8) = 507$, $(64)\,(8.8) = 563$, and $(64)\,(9.3) = 595$ respectively.[6] The sum for the three groups is then $507 + 563 + 595 = 1665$. Then by Equation 7.8

$$SS_b = \frac{\left(\sum_{i=1}^{n_1} x_{i1}\right)^2}{n_1} + \frac{\left(\sum_{i=1}^{n_2} x_{i2}\right)^2}{n_2} + \cdots + \frac{\left(\sum_{i=1}^{n_k} x_{ik}\right)^2}{n_k} - \frac{\left(\sum_{All} x_{..}\right)^2}{N}$$

$$= \frac{(507)^2}{65} + \frac{(563)^2}{64} + \frac{(595)^2}{64} - \frac{(1665)^2}{193}$$

$$= 75.021.$$

Then by Equation 7.6

$$MS_b = \frac{SS_b}{k - 1} = \frac{75.021}{2} = 37.511.$$

By Equation 7.2, obtained F is then

$$F = \frac{MS_b}{MS_{\text{within}}} = \frac{37.511}{11.914} = 3.148$$

Degrees of freedom for critical F are 2 and 190. Not finding this value in the F table, we conservatively use 2 and 150 degrees of freedom which gives critical F of 3.06. (Why do we characterize this value as conservative?) The null hypothesis is rejected leading to the conclusion that the three treatments were not equal in their effect on pain as measured by the WOMAC A scale.

7.11   The data are arranged for analysis as shown in the cell brackets.

	Standard	Weak	Placebo	
Correct	[35] (26.09)	[12] (24.83)	[30] (26.09)	77
Not Correct	[27] (35.91)	[47] (34.17)	[32] (35.91)	106
	62	59	62	$N = 183$

Expected values for the cells (in parentheses) are computed by Equation 7.11 as

$$f_e = \frac{(N_R)\,(N_C)}{N}.$$

---

[5]Note that we used SS_{within} to represent the sum of squares within because, for this problem, we used SS_w to represent the sum of squares for the weak bracelet group. For consistency, we will use the same subscript for the mean square within.

[6]The means reported in the paper were rounded so that rounding to the nearest integer is necessary to recover the sum of observations for some groups.

The expected values for the six cells are then

$$f_{e11} = \frac{(77)\,(62)}{183} = 26.09$$

$$f_{e12} = \frac{(77)\,(59)}{183} = 24.83$$

$$f_{e13} = \frac{(77)\,(62)}{183} = 26.09$$

$$f_{e21} = \frac{(106)\,(62)}{183} = 35.91$$

$$f_{e22} = \frac{(106)\,(59)}{183} = 34.17$$

$$f_{e23} = \frac{(106)\,(62)}{183} = 35.91$$

Then by Equation 7.10

$$\chi^2 = \sum_{\text{all cells}} \left[\frac{(f_o - f_e)^2}{f_e} \right].$$

So that

$$\chi^2 = \frac{(35\text{-}26.09)^2}{26.09} + \frac{(12\text{-}24.83)^2}{24.83} + \frac{(30\text{-}26.09)^2}{26.09} + \frac{(27\text{-}35.91)^2}{35.91}$$
$$+ \frac{(47\text{-}34.17)^2}{34.17} + \frac{(32\text{-}35.91)^2}{35.91}$$
$$= 3.04 + 6.63 + .59 + 2.21 + 4.82 + .43$$
$$= 17.72$$

From Appendix D we see that critical χ^2 for a test with 2 degrees of freedom conducted at the .05 level is 5.991. The null hypothesis is rejected leading to the conclusion that the three groups were not equally effective at determining the type bracelet they wore.

C. The following answers refer to Case Study C.

7.13 For purposes of analysis, it will be convenient to arrange the data for each time period as follows.

	Dead	Alive
Invasive	a	b
Non-Invasive	c	d

Then by Equations 6.9 and 6.10

$$\widehat{RR} = \frac{a/\,(a+b)}{c/\,(c+d)}$$

and

$$Z = \frac{\ln\left(\widehat{RR}\right) - \ln\left(RR_0\right)}{\sqrt{\frac{b/a}{a+b} + \frac{d/c}{c+d}}}.$$

Discharge

$$\widehat{RR} = \frac{21/\,(21 + 441)}{6/\,(6 + 452)} = 3.470$$

$$Z = \frac{\ln\,(3.470)}{\sqrt{\frac{441/21}{21+441} + \frac{452/6}{6+452}}} = 2.715 \qquad p = 2 \times .0033 = .0066$$

One Month

$$\widehat{RR} = \frac{23/\,(23 + 439)}{9/\,(9 + 449)} = 2.533$$

$$Z = \frac{\ln\,(2.533)}{\sqrt{\frac{439/23}{23+439} + \frac{449/9}{9+449}}} = 2.398 \qquad p = 2 \times .0082 = .0164$$

One Year

$$\widehat{RR} = \frac{58/\,(58 + 404)}{36/\,(36 + 422)} = 1.597$$

$$Z = \frac{\ln\,(1.597)}{\sqrt{\frac{404/58}{58+404} + \frac{422/36}{36+422}}} = 2.321 \qquad p = 2 \times .0102 = .0204$$

23 Months

$$\widehat{RR} = \frac{80/\,(80 + 382)}{59/\,(59 + 399)} = 1.344$$

$$Z = \frac{\ln\,(1.344)}{\sqrt{\frac{382/80}{80+382} + \frac{399/59}{59+399}}} = 1.866 \qquad p = 2 \times .0307 = .0614$$

Comparing the smallest p-value, .0066, to $\frac{.05}{4} = .0125$ results in a significant result. The second smallest p-value, .0164, compared to $\frac{.05}{3} = .0167$ is also significant as is the third smallest, .0204 when compared to $\frac{.05}{2} = .0250$. The remaining p-value, .0614, is compared to $\frac{.05}{1} = .0500$ and is not significant. Thus, we can declare significant differences for the discharge, one month, and one year date but cannot do so for the 23 month data. Additionally, we can make these declarations knowing that familywise error does not exceed .05.

G. The following answers refer to Case Study G.

7.15 The data are arranged for analysis as shown in the cell brackets.

	Monozygotic	Dizygotic	Nontwin	
MS	[7]	[1]	[87]	95
	(.55)	(.88)	(93.57)	
No MS	[20]	[42]	[4495]	4557
	(26.45)	(42.12)	(4488.43)	
	27	43	4582	$N = 4652$

Expected values for the cells (in parentheses) are computed by Equation 7.11 as

$$f_{e11} = \frac{(95)(27)}{4652} = .55$$

$$f_{e12} = \frac{(95)(43)}{4652} = .88$$

$$f_{e13} = \frac{(95)(4582)}{4652} = 93.57$$

$$f_{e21} = \frac{(4557)(27)}{4652} = 26.45$$

$$f_{e22} = \frac{(4557)(43)}{4652} = 42.12$$

$$f_{e23} = \frac{(4557)(4582)}{4652} = 4488.43$$

By Equation 7.10

$$\chi^2 = \sum_{\text{all cells}} \left[\frac{(f_o - f_e)^2}{f_e} \right].$$

So that

$$\chi^2 = \frac{(7-.55)^2}{.55} + \frac{(1-.88)^2}{.88} + \frac{(87-93.57)^2}{93.57} + \frac{(20-26.45)^2}{26.45}$$
$$+ \frac{(42-42.12)^2}{42.12} + \frac{(4495-4488.43)^2}{4488.43}$$
$$= 75.64 + .02 + .46 + 1.57 + .00 + .00$$
$$= 77.69$$

From Appendix D we see that critical χ^2 for a test with 2 degrees of freedom conducted at the .05 level is 5.991. The null hypothesis is rejected leading to the conclusion that the three groups did not have the same proportions of siblings with MS.

We note that two of the cells had expected values of .55 and .88. Expected values this small lead us to doubt the validity of this analysis.

Chapter Eight

8.1 Given variables x and y, define or explain the following terms.

(a) Indicates that high x values tend to be associated with high y values and low x values tend to be associated with low y values.

(b) Indicates that high x values tend to be associated with low y values and low x values tend to be associated with high y values.

(c) Indicates that there is no *linear* relationship between x and y.

(d) Indicates how strong the linear tendency is for x and y to be associated in a positive or negative fashion. Strength is at maximum value when $r = 1$ or $r = -1$ and is at minimum values when $r = 0$.

(e) The relationship between x and y can be characterized by a line as in Figures 8.1 and 8.2.

(f) Correlation of 1.0 indicates that x and y take the same value when expressed on a common scale (e.g., as z scores). Correlation of -1.0 indicates that when expressed on a common scale (e.g., as z scores), x and y have same absolute value but have opposite algebraic sign.

8.3 The data are arranged into the following table for analysis.

	Category One	Category Two	Category Three	
(+)	[19] (14.87)	[7] (8.40)	[13] (15.73)	39
(−)	[44] (47.27)	[29] (26.72)	[51] (50.01)	124
(I)	[6] (6.86)	[3] (3.88)	[9] (7.26)	18
	69	39	73	$N = 181$

By Equation 7.11

$$f_{e11} = \frac{(39)(69)}{181} = 14.87 \qquad f_{e12} = \frac{(39)(39)}{181} = 8.40 \qquad f_{e13} = \frac{(39)(73)}{181} = 15.73$$

$$f_{e21} = \frac{(124)(69)}{181} = 47.27 \qquad f_{e22} = \frac{(124)(39)}{181} = 26.72 \qquad f_{e23} = \frac{(124)(73)}{181} = 50.01$$

$$f_{e31} = \frac{(18)(69)}{181} = 6.86 \qquad f_{e32} = \frac{(18)(39)}{181} = 3.88 \qquad f_{e33} = \frac{(18)(73)}{181} = 7.26$$

Then by Equation 7.10 obtained chi-square is

$$\chi^2 = \frac{(19.0 - 14.87)^2}{14.87} + \frac{(7.0 - 8.40)^2}{8.40} + \frac{(13.0 - 15.73)^2}{15.73} + \frac{(44.0 - 47.27)^2}{47.27} + \frac{(29.0 - 26.72)^2}{26.72}$$

$$\frac{(51.0 - 50.01)^2}{50.01} + \frac{(6.0 - 6.86)^2}{6.86} + \frac{(3.0 - 3.88)^2}{3.88} + \frac{(9.0 - 7.26)^2}{7.26}$$

$$= 1.147 + .233 + .474 + .226 + .195 + .020 + .108 + .200 + .417$$

$$= 3.02.$$

From Appendix D we see that critical chi-square for a test with

$$(j - 1)(k - 1) = (3 - 1)(3 - 1) = 4$$

degrees of freedom conducted at $\alpha = .05$ is 9.488. The null hypothesis is not rejected so that we cannot claim a relationship between time of arrest and test result.

D. The following answers refer to Case Study D.

8.5 Yes, P-M represents the linear relationship between the two variables regardless of the shape of the bivariate population involved.

8.7 Yes, higher PBV assessments indicate greater brain volume while lower NPZ-8 scores indicate better neurophysiological function. Thus, there is a tendency for high PBV assessments to be associated with lower NPZ-8 scores.

8.9 No, they obtained a correlation for a mixture of HIV infected and non-HIV infected subjects which is contrary to their stated goal. Further, the correlation for HIV infected subjects alone is significantly lower than for the mixture.

L. The following answers refer to Case Study L.

8.11

	Minor Injury	Moderate Injury	Severe Injury	
< 30	[255] (226.54)	[57] (76.55)	[14] (22.91)	326
30 – 39	[332] (334.25)	[105] (112.94)	[44] (33.80)	481
40 – 49	[533] (542.03)	[201] (183.15)	[46] (54.82)	780
50+	[561] (578.17)	[205] (195.36)	[66] (58.47)	832
	1681	568	170	$N = 2419$

$$f_{e11} = \frac{(326)\,(1681)}{2419} = 226.54 \quad f_{e12} = \frac{(326)\,(568)}{2419} = 76.55 \quad f_{e13} = \frac{(326)\,(170)}{2419} = 22.91$$

$$f_{e21} = \frac{(481)\,(1681)}{2419} = 334.25 \quad f_{e22} = \frac{(481)\,(568)}{2419} = 112.94 \quad f_{e23} = \frac{(481)\,(170)}{2419} = 33.80$$

$$f_{e31} = \frac{(780)\,(1681)}{2419} = 542.03 \quad f_{e32} = \frac{(780)\,(568)}{2419} = 183.15 \quad f_{e33} = \frac{(780)\,(170)}{2419} = 54.82$$

$$f_{e41} = \frac{(832)\,(1681)}{2419} = 578.17 \quad f_{e42} = \frac{(832)\,(568)}{2419} = 195.36 \quad f_{e43} = \frac{(832)\,(170)}{2419} = 58.47$$

By Equation 7.10 obtained chi-square is

$$\chi^2 = \frac{(255 - 226.54)^2}{226.54} + \frac{(57 - 76.55)^2}{76.55} + \frac{(14 - 22.91)^2}{22.91}$$
$$+ \frac{(332 - 334.25)^2}{334.25} + \frac{(105 - 112.94)^2}{112.94} + \frac{(44 - 33.80)^2}{33.80}$$
$$+ \frac{(533 - 542.03)^2}{542.03} + \frac{(201 - 183.15)^2}{183.15} + \frac{(46 - 54.82)^2}{54.82}$$
$$+ \frac{(561 - 578.17)^2}{578.17} + \frac{(205 - 195.36)^2}{195.36} + \frac{(66 - 58.47)^2}{58.47}$$
$$= 3.58 + 4.99 + 3.47 + .02 + .56 + 3.08 + .15 + 1.74 + 1.42 + .51 + .48 + .97$$
$$= 20.97$$

Entering Appendix D with $(4 - 1)\,(3 - 1) = 6$ degrees of freedom yields a critical chi-square value of 12.592 for a test conducted at the .05 level. Because obtained chi-square of 20.97 exceeds this value, we reject the null hypothesis of independence and conclude that severity of injury is related to age of the injured subject.

M. The following answers refer to Case Study M.

8.13 Using Equation 8.2 we compute

$$r = \frac{522.25 - \frac{(62.9)(58.1)}{7}}{\sqrt{\left[565.89 - \frac{(62.9)^2}{7}\right]\left[482.43 - \frac{(58.1)^2}{7}\right]}}$$

$$= .485$$

By Equations 8.6 and 8.7

$$L = \frac{(1+F)r + (1-F)}{(1+F) + (1-F)r}$$

$$= \frac{(1+7.15)(.485) + (1-7.15)}{(1+7.15) + (1-7.15)(.485)}$$

$$= -.425$$

$$U = \frac{(1+F)r - (1-F)}{(1+F) - (1-F)r}$$

$$= \frac{(1+7.15)(.485) - (1-7.15)}{(1+7.15) - (1-7.15)(.485)}$$

$$= .907$$

F=7.15 was obtained by entering Appendix C for a two-sided confidence interval with numerator and denominator degrees of freedom of $7 - 2 = 5$.

Chapter Nine

9.1 (a) By Equation 9.9 on page 324 and Equation 9.13 on page 326

$$\widehat{R}^2 = \frac{SS_{reg}}{SS_y}$$

and

$$\widehat{R}^2 = r_{y\hat{y}}^2$$

(b) By Equation 9.8 on page 323 and Equation 9.12 on page 326

$$1 - \widehat{R}^2 = \frac{SS_{res}}{SS_y}$$

and

$$1 - \widehat{R}^2 = 1 - r_{y\hat{y}}^2$$

(c) From the discussion in Section 9.2.4 on page 325,

$$SS_{res} = 0 \text{ and } SS_{reg} = SS_y.$$

(d) From the discussion in Section 9.2.4 on page 325,

$$SS_{res} = SS_y \text{ and } SS_{reg} = 0.$$

9.3

$$R^2_{y.1,2,3,4} - R^2_{y.2,4}$$

A comparison is made between a model that contains all variables including x_1 and x_3 and a model that contains all variables *except* x_1 and x_3. Thus, if R^2 is greater for the larger model, it must be because of the presence of x_1 and x_3.

9.5 Answers are provided via Equation 9.23 on page 332

$$F = \frac{\frac{\widehat{R^2}}{p}}{\frac{1-\widehat{R^2}}{N-p-1}}$$

and Equation 9.24 on page 335

$$F = \frac{\frac{R^2_{y.L} - R^2_{y.S}}{p_L - p_S}}{\frac{1 - R^2_{y.L}}{N - p_L - 1}}$$

(a) By 9.23

$$F = \frac{\frac{\widehat{R}^2_{y.1,2,3,4}}{4}}{\frac{1 - \widehat{R}^2_{y.1,2,3,4}}{40-4-1}} = \frac{\frac{.68}{4}}{\frac{1-.68}{35}} = 18.59.$$

For four and 35 degrees of freedom, critical F is 2.64 so that the null hypothesis is rejected.
Answer: **yes**

(b) By 9.24

$$F = \frac{\frac{R^2_{y.1,2,3,4} - R^2_{y.2,3,4}}{4-3}}{\frac{1 - R^2_{y.1,2,3,4}}{40-4-1}} = \frac{\frac{.68-.43}{1}}{\frac{1-.68}{35}} = 27.34$$

For one and 35 degrees of freedom, critical F is 4.12 so that the null hypothesis is rejected.
Answer: **yes**

(c) By 9.23

$$F = \frac{\frac{\widehat{R}^2_{y.2,4}}{2}}{\frac{1 - \widehat{R}^2_{y.2,4}}{40-2-1}} = \frac{\frac{.14}{2}}{\frac{1-.14}{37}} = 3.01.$$

For two and 37 degrees of freedom, critical F is 3.25 so that the null hypothesis is not rejected.
Answer: **insufficient evidence to make this claim**

(d) By 9.24

$$F = \frac{\frac{R^2_{y.1,2,3,4} - R^2_{y.1,3}}{4-2}}{\frac{1 - R^2_{y.1,2,3,4}}{40-4-1}} = \frac{\frac{.68-.27}{2}}{\frac{1-.68}{35}} = 22.42$$

For two and 35 degrees of freedom, critical F is 3.27 so that the null hypothesis is rejected.
Answer: **yes**

D. The following answers refer to Case Study D.

9.7 Letting y represent NPZ-8 scores, x_1 PBV and x_2 CD4 values, we begin by calculating the following intermediate values (see calculations on page 329).

$$SS_y = \sum y^2 - \frac{(\sum y)^2}{n} = 374 - \frac{(54)^2}{15} = 179.6$$

$$SS_{x_1} = \sum x_1^2 - \frac{(\sum x_1)^2}{n} = 9.944 - \frac{(12.169)^2}{15} = .0717$$

$$SS_{x_2} = \sum x_2^2 - \frac{(\sum x_2)^2}{n} = 3356299 - \frac{(5817)^2}{15} = 1100466.4$$

$$SS_{yx_1} = \sum yx_1 - \frac{(\sum y)(\sum x_1)}{n} = 42.962 - \frac{(54)(12.169)}{15} = -.8464$$

$$SS_{yx_2} = \sum yx_2 - \frac{(\sum y)(\sum x_2)}{n} = 16442 - \frac{(54)(5817)}{15} = -4499.2$$

$$SS_{x_1x_2} = \sum x_1x_2 - \frac{(\sum x_1)(\sum x_2)}{n} = 4736.929 - \frac{(12.169)(5817)}{15} = 17.7908$$

By Equations 9.20 and 9.21

$$b_1 = \frac{(SS_{x_2})(SS_{yx_1}) - (SS_{x_1x_2})(SS_{yx_2})}{(SS_{x_1})(SS_{x_2}) - (SS_{x_1x_2})^2}$$
$$= \frac{(1100466.4)(-.8464) - (17.7908)(-4499.2)}{(.0717)(1100466.4) - (17.7908)^2}$$
$$= -10.8337$$

$$b_2 = \frac{(SS_{x_1})(SS_{yx_2}) - (SS_{x_1x_2})(SS_{yx_1})}{(SS_{x_1})(SS_{x_2}) - (SS_{x_1x_2})^2}$$
$$= \frac{(.0717)(-4499.2) - (17.7908)(-.8464)}{(.0717)(1100466.4) - (17.7908)^2}$$
$$= -.0039$$

Noting that $\bar{y} = \frac{54}{15} = 3.60$ and $\bar{x}_1 = \frac{12.169}{15} = .8113$ and $\bar{x}_2 = \frac{5817}{15} = 387.8$ and applying Equation 9.19 gives

$$a = \bar{y} - b_1\bar{x}_1 - b_2\bar{x}_2 = 3.6 - (-10.8337)(.8113) - (-.0039)(387.8) = 13.9018.$$

The two predictor model is then

$$\hat{y} = a + b_1x_1 + b_2x_2 = 13.9018 - 10.8337\,x_1 - .0039\,x_2.$$

(a) By Equation 9.22

$$SS_{reg} = b_1 SS_{yx_1} + b_2 SS_{yx_2} + \cdots + b_p SS_{yx_p}$$
$$= (-10.8337)(-.8464) + (-.0039)(-4499.2)$$
$$= 26.7165$$

Then using SS_y as previously calculated and Equation 9.9

$$\widehat{R}^2_{y.12} = \frac{SS_{reg}}{SS_y} = \frac{26.7165}{179.6} = 0.1488.$$

(b) Using the result from 9.6c with Equation 9.24 we obtain

$$F = \frac{\frac{\widehat{R}^2_{y.L} - \widehat{R}^2_{y.S}}{p_L - p_S}}{\frac{1 - \widehat{R}^2_{y.L}}{N - p_L - 1}} = \frac{\frac{.1488 - .0556}{2 - 1}}{\frac{1 - .1488}{15 - 2 - 1}} = 1.314.$$

Reference to Appendix C with 1 and 12 degrees of freedom gives a critical F of 4.75 for a test at level .05. The result is not significant so that we are unable to demonstrate an increase in R^2 as a consequence of adding CD4 to the model.

(c)

$$\hat{y}_1 = 13.9018 - 10.8337 \,(.791) - .0039\,(16) = 5.27$$
$$\hat{y}_2 = 13.9018 - 10.8337 \,(.782) - .0039\,(324) = 4.17$$
$$\hat{y}_3 = 13.9018 - 10.8337 \,(.646) - .0039\,(256) = 5.90$$
$$\hat{y}_4 = 13.9018 - 10.8337 \,(.740) - .0039\,(563) = 3.69$$
$$\hat{y}_5 = 13.9018 - 10.8337 \,(.804) - .0039\,(321) = 3.94$$

There appears to be little if any improvement obtained from use of the two predictor model.

(d) Using $\widehat{R}^2_{y.12} = .1488$ calculated in Exercise 9.7a with Equation 9.23 on page 332 gives

$$F = \frac{\frac{\widehat{R}^2}{p}}{\frac{1 - \widehat{R}^2}{N - p - 1}} = \frac{\frac{.1488}{2}}{\frac{1 - .1488}{15 - 2 - 1}} = 1.049$$

Reference to Appendix C with 2 and 12 degrees of freedom gives critical F as 3.89 for a test at the .05 level. Because obtained F of 1.049 is less than this value, the null hypothesis is not rejected. We were unable, therefore, to show that PBV and CD4 when used together account for any of the variation in NPZ-8 scores.

Chapter Ten

10.1 By Equation 10.1 on page 345, when five tests are administered, $P_n = n! = 5! = 120$. For six tests, $6! = 720$.

10.3 (a) From the discussion on page 350, we calculate $3! = 6$.

(b) No. The smallest possible p-value is $1/6 = .167$.

10.5 As noted on page 358, we use Equation 8.4 on page 309 with the result then being referenced to Table B with $n - 2$ degrees of freedom so that

$$t = \frac{r}{\sqrt{\frac{1 - r^2}{n - 2}}} = \frac{.88}{\sqrt{\frac{1 - .88^2}{124 - 2}}} = 20.464.$$

Critical t for a two-tailed test with 122 degrees of freedom at $\alpha = .10$ is ± 1.657 so that the null hypothesis is rejected.

10.7 The signed ranks (sr), their squares (sr^2), and the sum of each are as follows.

sr	sr^2
2	4
5	25
−4	16
−1	1
3	9
9	81
7	49
−6	36
8	64
$\sum$ 23	285

The paired samples t statistic (Equation 5.1 on page 162) with R_d substituted for d to indicate that the analysis is performed on signed ranks of difference scores rather than difference scores is as follows.

$$t = \frac{\bar{R}_d}{\frac{s_{R_d}}{\sqrt{n}}} = \frac{2.556}{\frac{5.318}{\sqrt{9}}} = 1.442$$

where $\bar{R}_d = \frac{23}{9} = 2.556$ and

$$s_{R_d} = \sqrt{\frac{\sum R_d{}^2 - \frac{(\sum R_d)^2}{n}}{n-1}} = \sqrt{\frac{285 - \frac{(23)^2}{9}}{9-1}} = 5.318.$$

Reference to Appendix G shows that the two-tailed critical value for the signed-ranks statistic for $n = 9$ and $\alpha = .05$ is ±2.704. The null hypothesis is not rejected.

10.9 The conversion to ranks results in the following.

Public	Private
18	11
14	8
5	13
16	15
19	20
10	3
9	4
2	1
17	6
12	7

The mean rank for the first group is

$$\bar{R}_1 = \frac{\sum R_1}{n_1} = \frac{122}{10} = 12.2$$

while that for the second is

$$\bar{R}_2 = \frac{\sum R_2}{n_2} = \frac{88}{10} = 8.8.$$

The pooled variance estimate (by Equation 6.2 on page 220) is

$$
s_{P_R}^2 = \frac{\left(\sum R_1^2 - \frac{(\sum R_1)^2}{n_1} \right) + \left(\sum R_2^2 - \frac{(\sum R_2)^2}{n_2} \right)}{n_1 + n_2 - 2}
$$

$$
= \frac{\left(1780 - \frac{(122)^2}{10} \right) + \left(1090 - \frac{(88)^2}{10} \right)}{10 + 10 - 2}
$$

$$
= 33.733
$$

By Equation 6.1 on page 219, obtained t, with appropriate substitutions of R for x, is then

$$
t = \frac{\bar{R}_1 - \bar{R}_2}{\sqrt{s_{P_R}^2 \left(\frac{1}{n_1} + \frac{1}{n_2} \right)}}
$$

$$
= \frac{12.2 - 8.8}{\sqrt{33.733 \left(\frac{1}{10} + \frac{1}{10} \right)}}
$$

$$
= 1.309.
$$

Reference to Appendix H shows that critical t for a two-tailed Wilcoxon test conducted at $\alpha = .05$ is ± 2.248. We were unable to reject the null hypothesis and were, therefore, unable to demonstrate a difference in waiting times in the two institutions.

10.11 Ranks for the three groups are as shown here.

Lecture	Internet	Film
8	11	15
6	5	14
2	1	10
12	3	9
7	13	4

If we let R_1 represent the ranks of the first group, R_1^2 the squared ranks of the first group and the subscripts 2, and 3 for the remaining groups then we compute

$$
\sum R_1 = 8 + 6 + 2 + 12 + 7 = 35
$$

$$
\sum R_1^2 = 8^2 + 6^2 + 2^2 + 12^2 + 7^2 = 297
$$

$$
\sum R_2 = 11 + 5 + 1 + 3 + 13 = 33
$$

$$
\sum R_2^2 = 11^2 + 5^2 + 1^2 + 3^2 + 13^2 = 325
$$

$$
\sum R_3 = 15 + 14 + 10 + 9 + 4 = 52
$$

$$
\sum R_3^2 = 15^2 + 14^2 + 10^2 + 9^2 + 4^2 = 618
$$

The sums of squares for the three groups are as follows.

$$SS_1 = \sum R_1^2 - \frac{\left(\sum R_1\right)^2}{n_1} = 297 - \frac{(35)^2}{5} = 52.0$$

$$SS_2 = \sum R_2^2 - \frac{\left(\sum R_2\right)^2}{n_2} = 325 - \frac{(33)^2}{5} = 107.2$$

$$SS_3 = \sum R_3^2 - \frac{\left(\sum R_3\right)^2}{n_2} = 618 - \frac{(52)^2}{5} = 77.2$$

By Equation 7.4 on page 265 the sum of squares within is

$$SS_w = SS_1 + SS_2 + SS_3 = 52.0 + 107.2 + 77.2 = 236.4.$$

By Equation 7.3 on page 265 the mean square within is

$$MS_w = \frac{SS_w}{N-k} = \frac{236.4}{15-3} = 19.7$$

where SS_w is the sum of squares within, N is the total number of observations, and k is the number of groups.

By Equation 7.7 on page 267 with substitution of R for x to indicate that the calculation is for ranks rather than original observations, the sum of squares between is

$$SS_b = n\left[\sum_{j=1}^{k}\bar{R}_j^2 - \frac{\left(\sum_{j=1}^{k}\bar{R}_j\right)^2}{k}\right]$$

where n is the number of observations in *each* group, and $\bar{R}_j$ are the rank group means.

From earlier calculations we obtain

$$\bar{R}_1 = \frac{\sum R_1}{n_1} = \frac{35}{5} = 7.0$$

$$\bar{R}_2 = \frac{\sum R_2}{n_2} = \frac{33}{5} = 6.6$$

$$\bar{R}_3 = \frac{\sum R_3}{n_3} = \frac{52}{5} = 10.4$$

Then

$$\sum \bar{R} = 7.0 + 6.6 + 10.4 = 24.0$$

and

$$\sum \bar{R}^2 = 7.0^2 + 6.6^2 + 10.4^2 = 200.72.$$

Making the proper substitutions in Equation 7.7 yields

$$SS_b = 5 \left[200.72 - \frac{(24.00)^2}{3} \right] = 43.6.$$

By Equation 7.6 on page 267 the mean square between is

$$MS_b = \frac{SS_b}{k - 1} = \frac{43.6}{3 - 1} = 21.8.$$

where SS_b is the sum of squares between and k is the number of groups. Finally, by Equation 7.2 on page 264 obtained F is

$$F = \frac{MS_b}{MS_w} = \frac{21.8}{19.7} = 1.107.$$

With $\alpha = .05$, three groups and five observations per group, Appendix I gives critical F as 4.072. Because obtained F of 1.107 is less than this value, we are unable to reject the null hypothesis and are, therefore, unable to demonstrate a difference in the impact of the three instructional methods.

A. The following answers refer to Case Study A.

10.13 This statement is misleading in that non-parametric tests may be less or more powerful than parametric counterparts.

D. The following answers refer to Case Study D.

10.15 (a) Wilcoxon's rank-sum test which is also known as the Mann-Whitney test.

   (b) Replacing NPZ-8 scores with ranks produces the following.

HIV +	HIV −
20.0	5.0
15.5	5.0
13.0	10.0
15.5	5.0
11.0	5.0
17.0	
19.0	
5.0	
13.0	
5.0	
5.0	
18.0	
13.0	
5.0	
5.0	

Using the subscripts 1 and 2 to represent the HIV positive and negative groups respectively, we calculate s_p^2 on the ranks via Equation 6.2 on page 220 as follows.

$$s_{P_R}^2 = \frac{\left(\sum R_1^2 - \frac{(\sum R_1)^2}{n_1}\right) + \left(\sum R_2^2 - \frac{(\sum R_2)^2}{n_2}\right)}{n_1 + n_2 - 2}$$

$$= \frac{\left(2607.5 - \frac{(180)^2}{15}\right) + \left(200 - \frac{(30)^2}{5}\right)}{15 + 5 - 2}$$

$$= 25.972$$

Then applying Equation 6.1 on page 219 to the ranks yields

$$t = \frac{\bar{R}_1 - \bar{R}_2}{\sqrt{s_{P_R}^2 \left(\frac{1}{n_1} + \frac{1}{n_2}\right)}} = \frac{12.0 - 6.0}{\sqrt{25.972 \left(\frac{1}{15} + \frac{1}{5}\right)}} = 2.280.$$

Reference to Appendix B gives the critical values for a two-tailed t test with 18 degrees of freedom conducted at $\alpha = .05$ as ± 2.101[7] which results in rejection of the null hypothesis. Ranks for the PBV data are as follows.

HIV +	HIV −
6	16
4	11
1	12
3	18
8	19
13	
2	
7	
10	
9	
5	
14	
17	
20	
15	

s_p^2 is then

$$s_P^2 = \frac{\left(1664 - \frac{(134)^2}{15}\right) + \left(1206 - \frac{(76)^2}{5}\right)}{15 + 5 - 2} = 28.763$$

---

[7]This value would be viewed with caution since it was taken from a t distribution and the two sample sizes are not both greater than 15. The exact critical value is ± 2.263 which also results in rejection of the null hypothesis.

and obtained t is

$$t = \frac{8.933 - 15.200}{\sqrt{28.763\left(\frac{1}{15} + \frac{1}{5}\right)}} = -2.263.$$

Reference to Appendix B gives the critical values for a two-tailed t test with 18 degrees of freedom conducted at $\alpha = .05$ as ± 2.101 (see footnote 7) which results in rejection of the null hypothesis.

(c) The assumption of no tied observations was violated for the test on NPZ-8 scores but not for the test on PBV values.

(d) An approximate test was necessary for the NPZ-8 analysis because of the presence of tied observations. An exact test could have been carried out for the PBV analysis but because the available table of critical values was constructed for equal samples sizes, reference was made to a t distribution so that an approximate test was conducted.

(e) An exact version of the rank test on NPZ-8 scores could be obtained by converting the original scores to ranks in the manner shown, then developing the permutation distribution of the t statistic and referencing the test statistic to that distribution. An exact version of the test on PBV scores could be obtained by referencing the test statistic to a table that includes unequal sample sizes.

10.17 Ranks for the three groups are as follows.

ADC +	ADC −	HIV −
20.0	5.0	5.0
15.5	13.0	5.0
13.0	5.0	10.0
15.5	5.0	5.0
11.0	18.0	5.0
17.0	13.0	
19.0	5.0	
	5.0	

Designating the three groups by subscripts 1, 2, and 3 respectively, we calculate sums of squares for the groups as follows.

$$SS_1 = \sum R_1^2 - \frac{\left(\sum R_1\right)^2}{n_1} = 1820.5 - \frac{(111)^2}{7} = 60.357$$

$$SS_2 = \sum R_2^2 - \frac{\left(\sum R_2\right)^2}{n_2} = 787 - \frac{(69)^2}{8} = 191.875$$

$$SS_3 = \sum R_3^2 - \frac{\left(\sum R_3\right)^2}{n_3} = 200 - \frac{(30)^2}{5} = 20.000$$

Then by 7.4

$$SS_w = SS_1 + SS_2 + \cdots + SS_k = 60.357 + 191.875 + 20.000 = 272.232$$

By 7.3

$$MS_w = \frac{SS_w}{N - k} = \frac{272.232}{20 - 3} = 16.014.$$

By Equation 7.8

$$SS_b = \frac{\left(\sum_{i=1}^{n_1} x_{i1}\right)^2}{n_1} + \frac{\left(\sum_{i=1}^{n_2} x_{i2}\right)^2}{n_2} + \cdots + \frac{\left(\sum_{i=1}^{n_k} x_{ik}\right)^2}{n_k} - \frac{\left(\sum_{All} x_{..}\right)^2}{N}$$

$$= \frac{(111)^2}{7} + \frac{(69)^2}{8} + \frac{(30)^2}{5} - \frac{(210)^2}{20}$$

$$= 330.268.$$

By Equation 7.6

$$MS_b = \frac{SS_b}{k - 1} = \frac{330.268}{3 - 1} = 165.134.$$

Obtained F is then

$$F = \frac{MS_b}{MS_w} = \frac{165.134}{16.014} = 10.312.$$

Entering Appendix C with 2 and 17 degrees of freedom, we find that critical F for a test at $\alpha = .05$ is 3.59. Because obtained F of 10.312 exceeds this value, we reject the hypothesis of equal population means.

Because this test is approximate as well as for other reasons, we should not read too much into a comparison with the ANOVA conducted on the original NPZ-8 scores (see 7.14 on page 294) but it is interesting to note that obtained F calculated on ranks is larger than that calculated on original scores.

K. The following answers refer to Case Study K.

10.19 No, for two reasons. (1) Under a true null hypothesis difference scores tend to be symmetric regardless of the shape of the pretest/posttest distributions. The paired samples t test tends to be robust to population nonnormality so long as the difference score distribution is symmetric. (2) The pre and posttest scores are means rather than individual observations and, due to the central limit theorem, means tend to normality regardless of population shape.

O. The following answers refer to Case Study O.

10.21 No, this is pure myth. There are circumstances where parametric tests are more powerful and circumstances where nonparametric tests are more powerful. Blanket statements of this sort are unjustified.

10.23 This is definitely not true. The Kruskal-Wallis test may be thought of as a generalization of the Wilcoxon rank-sum (or Mann-Whitney) test but definitely not the signed-ranks test.

10.25 No, this is done by the P-M conducted on original data but the Spearman uses ranks. As a result, the assessment by the rank based test is not for a straight line relationship but rather a monotone relationship.

Bibliography

1. R. CLIFFORD BLAIR AND JAMES J. HIGGINS, *A comparison of the power of wilcoxon's rank-sum statistic to that of student's t statistic under various non-normal distributions*, Journal of Educational Statistics **5** (1980), 309–335.

2. R. CLIFFORD BLAIR AND JAMES J. HIGGINS, *Comparison of the power of the paired samples t-test to that of wilcoxon's signed rank test under various population shapes*, Psychological Bulletin **97** (1985), 119–128.

3. WILLIAM E. BODEN, ROBERT A. O'ROURKE, MICHAEL H. CRAWFORD, ALVIN S. BLAUSTEIN, PRAKASH C. DEEDWANIA, ROBERT G. ZOBLE, LAURA F. WEXLER, ROBERT E. KLEIGER, CARL J. PEPINE, DAVID R. FERRY, BRUCE K. CHOW, AND PHILIP W. LAVORI, *Outcomes in patients with acute Non-Q-Wave myocardial infarction randomly assigned to an invasive as compared with a conservative management strategy*, The New England Journal Of Medicine **338** (1998), 1785–1792.

4. JAMES V. BRADLEY, *Distribution-free statistical tests*, 1st ed., Prentice-Hall, Inc., Englewood Cliffs, New Jersey, 1968.

5. NORMAN E. BRESLOW AND N. E. DAY, *Statistical methods in cancer research, volune i-the analysis of case-control studies*, IARC Scientific Publications No. 32, Lyon, France: International Agency for Research on Cancer, 1980.

6. CHANTAL COLES, NOEL A. BRENNAN, VICKI SHULEY, JILL WOODS, CHRIS PRIOR, JOSEPH G. VEHIGE, AND A. SIMMONS, *The influence of lens conditioning on signs and symptoms with new hydrogel contact lenses*, Clinical and Experimental Optometry **87** (2004), 367–371.

7. WILLIAM J. CONOVER, *Practical nonparametric statistics*, 2nd ed., John Wiley and Sons, New York, Chichester, Brisbane, Toronto, 1980.

8. WILLIAM J. CONOVER AND RONALD I. IMAN, *Rank transformations as a bridge between parametric and nonparametric statistics*, The American Statistician **35** (1981), 124–129.

9. Cytel Software Corporation, Cambridge, MA, *Statxact:software for exact nonparametric inference*, 2003, Version 6.

10. A. E. DUSOIR, *Statistical calculator*, Mole Software, 2004, v2.03.

11. G. C. EBERS, D. E. BULMAN, A. D. SADOVNICK, D. W. PATY, S. WARREN, W. HADER, T. J. MURRAY, T. P. SELAND, P. DUQUETTE, T. GREY, AND ET AL., *A population-based study of multiple sclerosis in twins*, The New England Journal Of Medicine **315** (1986), 1638–1642.

12. JOSIE M. M. EVANS, RAY W. NEWTON, DANNY A. RUTA, THOMAS M. MACDONALD, RICHARD J. STEVENSON, AND ANDREW D. MORRIS, *Frequency of blood glucose monitoring in relation to glycaemic control: observational study with diabetes database*, British Medical Journal **319** (1999), 83–86.

13. L. EVANS AND M. C. FRICK, *Helmet effectiveness of preventing motorcycle driver and pasenger fatalities*, Accident Analysis and Prevention **20** (1988), 447–458.

14. EXPERT PANEL ON DETECTION, EVALUATION, AND TREATMENT OF HIGH BLOOD CHOLESTEROL IN ADULTS, *Executive summary of the third report of the national cholesterol education program (NCEP) expert panel on detection, evaluation, and treatment of high blood cholesterol in adults (Adult Treatment Panel III)*, Journal of the American Medical Association **285** (2001), 2486–2497.

15. RONALD A. FISHER, *Frequency distributions of the values of the correlation coefficient in samples from an indefinitely large population*, Biometrika **10** (1915), 507–521.

16. RONALD A. FISHER, *On the "probable error" of a coefficient of correlation deduced from a small sample*, Metron **1** (1921), 3–32.

17. RONALD A. FISHER, *Applications of "Student's" distribution*, Metron **5** (1925), 90–104.

18. JOSEPH L. FLEISS, *Statistical methods for rates and proportions*, John Wiley and Sons, New York, 1981.

19. TRISHA GREENHALGH, *How to read a paper: Statistics for the non-statistician. I: Different types of data need different statistical tests*, British Medical Journal **315** (1997), 364–366.

20. SANDER GREENLAND, *Applications of stratified analysis methods*, Modern Epidemiology (Kenneth Rothmann and Sander Greenland, eds.), Lippincott - Raven, Philadelphia, 2nd ed., 1998.

21. TIM HARLOW, COLIN GREAVES, ADRIAN WHITE, LIZ BROWN, ANNA HART, AND EDZARD ERNST, *Randomised controlled ,trial of magnetic bracelets for relieving pain in osteoarthritis of the hip and knee*, British Medical Journal **329** (2004), 18–25.

22. YOSEF HOCHBERG AND AJIT C. TAMHANE, *Multiple comparison procedures*, John Wiley and Sons, New York, 1987.

23. J. M. HOENIG AND D. M. HEISEY, *The abuse of power: The pervasive fallacy of power calculations for data analysis.*, The American Statistician **55** (2001), 19–24.

24. S. HOLM, *A simple sequentially rejective multiple test procedure*, Scandinavian Journal of Statistics **6** (1979), 65–70.

25. ESKO A. KEMPPAINEN, JOHAN I. HEDSTRÖM, PAULI A. PUOLAKKAINEN, VESA S. SAINIO, REIFO K. HAAPIAINEN, VESA PERHONIEMI, SIRPA OSMAN, EERO O. KIVILAAKSO, AND ULF-HÅKAN STENMAN, *Rapid measurement of urinary trypsinogen-2 as a screening test for acute pancreatitis*, The New England Journal Of Medicine **336** (1997), 1788–1793.

26. WINFRIED V. KERN, ALAIN COMETTA, ROBRECHT DE BOCK, JOHN LANGENAEKEN, MARIANNE PAESMANS, HAROLD GAYA, GIORGIO ZANETTI, THIERRY CALANDRA, MICHEL P. GLAUSER, FRANÇOISE CROKAERT, JEAN KLASTERSKY, ATHANASIOS SKOUTELIS, HARRY BASSARIS, STEPHEN H. ZINNER, CLAUDIO VISCOLI, DAN ENGELHARD, AND ANDREWPADMOS FOR THE INTERNATIONAL ANTIMICROBIAL THERAPY COOPERATIVE GROUP OF THE EUROPEAN ORGANIZATION FOR RESEARCH AND TREATMENT OF CANCER, *Oral versus intravenous empirical antimicrobial therapy for fever in patients with granulocytopenia who are receiving cancer chemotherapy*, The New England Journal Of Medicine **341** (1999), 312–318.

27. ROGER E. KIRK, *Experimental design: Procedures for the behavioral sciences*, 3rd ed., Wadsworth Publishing Company, Belmont, California, 1994.

28. WILLIAM H. KRUSKAL AND WILLIAM A. WALLIS, *Use of ranks in one-criterion variance analysis*, Journal of the American Statistical Association **48** (1953), 907–911.

29. JAN W. KUZMA, *Basic statistics for the health sciences*, 3rd ed., Mayfield Publishing Company, Mountain View, California, 1998.

30. JAN W. KUZMA AND STEPHEN E. BOHNENBLUST, *Basic statistics for the health sciences*, 4th ed., Mayfield Publishing Company, Mountain View, California, 2001.

31. A. M. LILIENFELD AND D. E. LILIENFELD, *Foundations of epidemiology*, 2nd ed., Oxford University Press, New York, 1980.

32. KAREN D. LILLER, VIRGINIA NOLAND, PABITRA RIJAL, KAREN PESCE, AND ROBIN GONZALEZ, *Development and evaluation of the Kids Count Farm Safety Lesson* , Journal of Agricultural Safety and Health **8** (2002), 411–421.

33. FREDRICK M. LORD, *On the statistical treatment of football numbers*, American Psychologist **8** (1953), 750–751.

34. THOMAS C. MARKELLO, ISA M. BERNARDINI, AND WILLIAM A. GAHL, *Improved renal function in children with cystinosis treated with cysteamine*, The New England Journal Of Medicine **328** (1993), 1157–1162.

35. CHRISTINA MASLACH AND SUSAN E. JACKSON, *Maslach burnout inventory: Manual*, 2nd ed., Consulting Psychologists Press, Palo Alto, California, 1986.

36. WILLIAM MENDENHALL AND ROBERT J. BEAVER, *Introduction to probability and statistics*, 8th ed., PWS-Kent Publishing Company, Boston, Massachusetts, 1991.

37. SOHIL H. PATEL, DENNIS L. KOLSON, GUILA GLOSSER, ISABEL MATOZZO, YULIN GE, JAMES S. BABB, LOIS J. MANNON, AND ROBERT I. GROSSMAN, *Correlation between percentage of brain parenchymal volume and neurocognitive performance in hiv-infected patients*, American Journal of Neuroradiology **23** (2002), 543–549.

38. KAREN M. PERRIN, SOMER L. GOAD, AND CAROL WILLIAMS, *Can school nurses save money by treating school employees as well as students?*, Journal of School Health **72** (2002), 305–306.

39. PHILIP H. RAMSEY AND PATRICIA P. RAMSEY, *Evaluating the normal approximation to the binomial test*, Journal of Educational Statistics **13** (1988), 173–182.

40. GEORG RÖGGLA, BERTHOLD MOSER, AND MARTIN RÖGGLA, *Effect of temazepam on ventilatory response at moderate altitude*, British Medical Journal **320** (2000), 1–5.

41. JULIAN SIMON AND PETER BRUCE, *Resampling stats software*, Resampling Stats, Inc., 1973-2004.

42. DAVID SKUSE, ARMON BENTOVIM, JILL HODGES, JIM STEVENSON, CHRISO ANDREOU, MONICA LANYADO, MICHELLE NEW, BRYM WILLIAMS, AND DEAN MCMILLAN, *Risk factors for development of sexually abusive behaviour in sexually victimised adolescent boys: cross sectional study*, British Medical Journal **317** (1998), 175–179.

43. L. A. SMOLIN, K. F. CLARK, J. G. THOENE, W. A. GAHL, AND J. A. SCHNEIDER, *A comparison of the effectiveness of cysteamine and phosphocysteamine in elevating plasma cysteamine concentration and decreasing leukocyte free cystine in nephropathic cystinosis*, Pediatric Research **23** (1988), 616–620.

44. STANLEY S. STEVENS, *On the theory of scales of measurement*, Science **161** (1946), 677–680.

45. STUDENT, *The probable error of a mean*, Biometrika **6** (1908), 1–25.

46. THE CONTINUOUS INFUSION VERSUS DOUBLE-BOLUS ADMINISTRATION OF ALTEPLASE (COBALT) INVESTIGATORS, *A comparison of continuous infusion of alteplase with double-bolus administration for acute myocardial infarction*, The New England Journal Of Medicine **337** (1997), 1124–1130.

47. JOHN W. TUKEY, *The problem of multiple comparisons*, Mimeographed monograph, Princeton University, 1953, cited in Yosef Hochberg and Ajit C. Tamhane, *Multiple Comparison Procedures*, John Wiley and Sons, New York, 1987.

48. ANNELIES VAN RIE, ROBIN WARREN, MADELEINE RICHARDSON, THOMAS C. VICTOR, ROBERT P. GIE, DONALD A. ENARSON, NULDA BEYERS, AND PAUL D. VAN HELDEN, *Exogenous reinfection as a cause of recurrent tuberculosis after curative treatment*, The New England Journal Of Medicine **341** (1999), 1174–1179.

49. FRANK WILCOXON, *Individual comparisons by ranking methods*, Biometrics **1** (1945), 80–83.

Index